Aging, the Individual, and Society

Aging, the Individual, and Society

NINTH EDITION

Susan M. Hillier
Sonoma State University

Georgia M. Barrow

WADSWORTH
CENGAGE Learning™

Australia • Brazil • Japan • Korea • Mexico • Singapore • Spain • United Kingdom • United States

This labor of love is dedicated to my husband, Rod Ferreira, for his loving support; and to the memory of my father, Kenneth Lynn Hillier.

WADSWORTH
CENGAGE Learning™

Aging, the Individual, and Society, Ninth Edition
Susan M. Hillier
Georgia M. Barrow

Publisher/Executive Editor: Linda Schreiber

Development Editor: Melanie Cregger

Acquisitions Editor: Erin Mitchell

Assistant Editor: Rachael Krapf

Media Editor: Lauren Keyes

Marketing Manager: Andrew Keay

Marketing Associate: Jillian Myers

Marketing Communications Manager: Laura Localio

Content Project Management: Pre-PressPMG

Art Director: Caryl Gorska

Manufacturing Manager: Barbara Britton

Print Buyer: Becky Cross

Permissions Editor: Roberta Broyer

Production Service: Pre-PressPMG

Photo Manager: Leitha Etheridge-Sims

Photo Researcher: Kelly Franz

Cover Designer: Riezebos Holzbaur/ Christopher Harris

Cover Image: Corbis

Compositor: Pre-PressPMG

For product information and technology assistance, contact us at **Cengage Learning Customer & Sales Support, 1-800-354-9706**

For permission to use material from this text or product, submit all requests online at **www.cengage.com/permissions.** Further permissions questions can be emailed to **permissionrequest@cengage.com.**

Library of Congress Control Number: 2010920682

ISBN-13: 978-0-495-81166-4

ISBN-10: 0-495-81166-1

Wadsworth
10 Davis Drive
Belmont, CA 94002-3098
USA

Cengage Learning is a leading provider of customized learning solutions with office locations around the globe, including Singapore, the United Kingdom, Australia, Mexico, Brazil, and Japan. Locate your local office at: **international.cengage.com/region**

Cengage Learning products are represented in Canada by Nelson Education, Ltd.

For your course and learning solutions, visit **www.cengage.com.**

Purchase any of our products at your local college store or at our preferred online store **www.CengageBrain.com.**

Printed in the United States of America
2 3 4 5 6 7 14 13 12 11

CONTENTS

9 Finances and Lifestyles

10 Living Environments

11 The Oldest-Old and Caregiving

12 Special Problems

13 Women and Ethnic Groups

PREFACE

Now has never been a better time to be a gerontologist. Aging in America is at long last a nationally recognized reality. The boomer generation pushes at the door of "old age" and true to the cultural mores of that cohort (as well as their sheer numbers) boomers demand to be listened to by policy makers, and that shapes the country. The boomer generation certainly is not homogeneous in politics, but it is homogeneous in the expectation that they will be provided for—as the country provided for their parents—through government programs such as Social Security and Medicare.

The next great policy debate is health care, and in another 15 years, political scientists and sociologists will have a heyday analyzing the outcome of what is shaping up to be a great struggle. From the study of the well-being of older people we can see strong parallels between pro and con health care arguments and the battles waged when Social Security was coming into being, and again during the creation of Medicare.

Giant leaps forward have taken place in health care since the mechanical revolution and again with the revolution of high tech. Now the creators of new levels of medical sophistication use that technological knowledge to advance genetic and pharmaceutical research, study brain functioning, develop artificial joints and mechanisms enhancing the aging circulatory system, and artificially create other highly complicated body parts. Enormous strides have been made in perfecting, for example, an articulating ankle and flexible artificial foot—in part to address the concerns of the numbers of American soldiers with war injuries who return from the Middle East, but also to impact those older adults who have lost a foot from diabetes or an arm from an accident. Unquestionably genetic science has revolutionized our understanding of pathogens, cellular structure and function, and intricate biochemical interactions in the body.

These amazing advances give us the reality of an extended lifespan, but the question of *quantity* of life versus *quality* of life remains the domain of the psychologists. Given there are few role models, where is the older person in the American family? What is the meaning of growing very old, of outliving friends and family members who have died "prematurely" through diseases and accidents? Does personality and character continue to develop into the 80s . . . the 90s? What is the role of the human spirit in all this?

Gerontology is the study of each of these approaches and questions. It is interdisciplinary: it addresses the complete person. Gerontology includes family history; social history; genetic history; the political moment; and issues of spiritual dimension such as the meaning of life accomplishments, or relationship to things larger than oneself, or understandings of dying and death. It is a field reflecting multiple concurrent forces that interact in a complex way on the condition of the aging individual and society. No field of study more completely integrates the total mature person than does gerontology.

This edition includes greater attention to global aging. The technological advances that extend our healthy, vigorous lives also connect us globally. To an increasing extent we now recognize these connections: social, political, economic, and spiritual. The global recession of 2008 exemplifies an economic and social integration of the planet. The threat of global warming has created a multinational issue in which people from all over the world recognize our level of interconnectedness. From this starting point of shared understanding we can develop a global interest in health care, policies both formal and informal, and multigenerational families. The well-being of older people in *any* culture reflects cultural values.

The bibliography in this edition reflects scientific, professional, and popular writings—providing students an outstanding exposure to the component parts of the field of gerontology. Each chapter begins with a current news story of real-life older people whose lives or story represent an issue being focused on in the chapter. These stories can help people to make the connection between something read in a textbook and something that really does happen to people "out there." What we talk about is real.

Learning can never come simply from a textbook: words do not generate knowledge, they generate information. The text seeks to help students understand issues and concepts and to explore multiple ways of learning. Questions for discussion, experiential learning, and for exploration on the Internet end each chapter. Each set of activities is integrated topically, theoretically, and pedagogically. They require a personal interaction of student and material, thereby reinforcing the context of each chapter.

Features

- Old Is News: Jump-start student interest in each chapter. The news stories help students understand the practical reality of the key concepts within the chapter.

- Key concepts: Helps students and faculty keep focused on the main points of each chapter. They can be used as a framework for teaching and learning key topical concepts.

- Chapter summaries: An organizational tool for students to review. Each summary is linked to the chapter's key concepts, thereby reinforcing the learning framework.

- Fieldwork suggestions: Experiential learning opportunities to guide teaching and learning.

- Discussion questions: To be used either in group discussions or as assignments for written work. The questions combine need to think critically about theory as well as practice.

- Internet activities: Suggestions for explorations on the Internet to challenge students to understand the complexity and reliability of available information.

New to This Edition

- Integration of a global perspective.
- Reorganization of subject matter for a more effective sequence of learning material.
- Addition of learning aids for students and teaching aids for instructors.
- Throughout, topics are linked to policy, public will, and policy implementation.
- Greater discussion of intergenerational issues.
- Extensive use of current, professional literature as well as popular literature and Internet sources.

Supplements

Instructor's Manual with Test Bank

By Patricia G. Harvey of Horry-Georgetown Technical College. Streamline and maximize the effectiveness of your course preparation using such resources as chapter summaries, chapter outlines, key terms, video resources, and in-class activities. This time-saving resource also includes a Test Bank that offers 15–20 multiple-choice questions, 10–15 true/false questions, and 3–5 essay questions for each chapter.

ExamView Computerized Testing

Featuring automatic grading, ExamView allows you to create, deliver, and customize tests and study guides (both print and online) in minutes. See assessments onscreen exactly as they will print or display online. Build tests of up to 250 questions using up to 12 question types and enter an unlimited number of new questions or edit existing questions.

PowerPoint Presentation Slides

Created by Laura Hess Brown of SUNY Oswego, these Microsoft PowerPoint slides let you incorporate images from the book right into your lectures.

Acknowledgments

Writing and editing a textbook is never a solitary task. Thank you to the many people on the Cengage Wadsworth Publishing Company team, most especially Chris Caldeira, Melanie Cregger, and Erin Parkins, who reviewed the manuscript for this edition.

The organization and completion of much of the excellent bibliography in this edition was accomplished with help from my husband, Rod Ferreira. His meticulous and thorough style, his steady approach, and his words of encouragement (not to mention meals of delicious food) were essential. Very special thanks also to Linda Rarey, whose teaching experience as well as friendship was invaluable to me and to developing discussion questions, fieldwork suggestions, Internet exercises, and other learning aids incorporated in this edition.

Thank you for your thorough review of the previous edition and your very helpful comments and suggestions: Catherine Solomon, Quinnipiac University; Mary Ann Davis, Sam Houston State University; Patricia G. Harvey, Horry-Georgetown Technical College; Mary L. Bender, University of Nebraska at Omaha; Stephanie A. Bennett, SUNY College at Oneonta; and Laura Hess Brown, SUNY Oswego.

To my mom Alouise Hillier—my 89-year-old role model; my daughter Jennifer and grandson Ben; loving and encouraging friends Carol and Gabriele; and especially to my students, from whom I learn so much: Thank you.

Aging in America

McGriff Takes the Wheel Again at Age 81

Bob Padecky

The first thought, the most logical thought. . . . an 81-year-old racing in NASCAR is like a 13-year-old named the Dean of Students at Harvard, the task a bit beyond the reach of the participant. And then I saw Hershel McGriff Wednesday. Oh my. Maybe 81 is the new 60. Eighty-one? Come on. If Hershel is 81, then I am a Chinese acrobat. Hershel has a full head of hair, 20-20 vision, and mental acuity sharp as a steak knife. While he may not have the body of a Chippendale dancer, Hershel is not a Slim Fast ad waiting to happen either. He weighs 197 pounds, just 17 pounds heavier than when he was in his prime.

So, I want what Hershel is having. And that would be Pilates. "Gave it to him last Christmas," said Sherrie, his wife. "He looked at me and said, 'Oh, thank you.'" The look, she said, was kindly dismissive, Hershel making every attempt not to insult his wife while hiding his incredulity. "I didn't even know how to pronounce the word," Hershel said.

For two months, the gift certificate for 10 Pilates sessions sat unused. Pilates might as well have meant pancakes to McGriff. But then . . . Hershel said what the hay, this gets me off my fanny. And one thing Hershel hates is daydreaming a day away. He may be 81, but he wants more out of his life than watching tomatoes grow.

McGriff, named one of NASCAR Top 50 Drivers, manages a copper mine near his home in Tucson, Arizona. Goes to work every day. Been doing that for 23 years. He raced 27 years in NASCAR on three separate occasions, mostly out West, and retired from the sport in 2002 at 74. But the sport has never left him, or maybe it's that he could never leave the sport. Heck, he ran his first stock car race in 1945. So every once in a while, the man who raced 16,663 miles in stock cars feels the need for speed. And what McGriff does about that, well, it will more than satisfy anyone's curiosity on his ability to drive this Saturday in the Bennett Lane Winery 200.

McGriff owns a 12-cylinder, double-turbo Mercedes that goes 160 miles an hour. "I would like it if it went 200," McGriff said. But, doggone it, every time Hershel gets on that two-and-half mile road, the darn engine cuts out at 160. "I know I could reach 200 before that curve," McGriff said. "When I hit 160, there's still a quarter-mile of road left before that curve."

McGriff said that with a sigh that could only come from someone who has raced 10,643 NASCAR laps. But the way the curve, he said, needs to be taken at 80. McGriff said he has done 160 on that road just a few times and, not surprisingly, without his wife. McGriff is a loving husband and a practical man, taking into account that his wife didn't sign up to go 160 miles in an hour when she married him. So it will come as no surprise that McGriff and no one else came up with the idea of his racing three road courses this year. Managing that copper mine keeps his mind active but nothing kept his mind active like pushing it in a stock car.

Two months after his wife gave him the Christmas present, McGriff began Pilates. A month into it, he asked his wife a question that still stuns him. "I don't understand," he said, "how I can get muscles laying down."

How indeed. On the ground McGriff is going zero miles an hour. He's passing no one, and no one is passing him. It's probably as close to being 81 as he ever gets, or wants to get. No wonder it took McGriff two months before he started Pilates. He's a race car driver, and it took some time for him to comprehend he could get somewhere without going anywhere.

Source: www.pressdemocrat.com/padecky(2009).

Who is growing old? We all are! In many people's minds, however, growing old is something that happens only to others and only to individuals older than themselves. If you have not yet reached your 50s, can you imagine yourself to be 65, 75, or 90? With reasonable care and a bit of good luck, you will live to be 75 or more. With advances in medical science and technology, and an increased awareness of taking care of our bodies, we all can anticipate long lives.

But what will be the quality of our lives? As we advance through life, aging may bring either despair or enhanced vitality and meaning. Indeed, in the twenty-first century many social and medical issues of aging focus more on postponing **senescence** (age-related loss of function) and on ensuring a good quality of life than on ensuring old age itself. For example, research for the past 30 years on the health risks of environmental factors like cigarette smoking or air and water quality, on genetics and genetic engineering, and on biochemical and pharmaceutical factors in health—all combine to increase our longevity. Concern with living well in our old age is now at least as great an issue for many people as concern with living a long life.

An Interdisciplinary Topic

The study of aging is exciting and complex, and can be examined from many perspectives. It can be viewed through emotional, physiological, economic, social, cognitive, or philosophical lenses, for example. **Gerontology** is the study of the human aging process from maturity to old age, as well as the study of the older adult as a special population. Each viewpoint in gerontology adds a dimension to the broader understanding of what it means to age personally, socially, and globally. This understanding, in turn, allows us to plan for our own well-being in later life, and to consider quality of later-life issues on a social level.

Gerontology as a scholarly field has changed markedly since its fledgling beginnings in the early 1950s. Gerontologist John Rowe described a "new gerontology" in which the focus "goes beyond the prior preoccupation with age-related diseases . . . to include a focus on senescence . . . and physiological changes that occur with advancing age and that influence functional status as well as the development of disease" (Rowe, 1997, p. 367). To his concept of new gerontology, we might focus on the social issues that are inherent in any society undergoing social, interpersonal, and economic changes as rapidly as the United States and other industrialized countries are doing now because of their changing demographics.

The term *aging* is wildly nonspecific: wine ages, babies age, galaxies age, we each are aging right now regardless of our chronological ages. Clearly that does not imply a common biological process. **Aging** in the context of this text refers to progressive changes during adult years, but these changes are not necessarily negative nor do they necessarily reduce an individual's viability. For example, gray hair is a result of aging, but does not impair a person's functioning. Because of negative stereotyping, however, gray hair might have negative *social* meaning in some cultures.

Mutations that may accumulate over time in certain genes in cells in the reproductive system, on the other hand, describe age-related loss of function, which is referred to as **senescence.** Gerontologists define aging in terms of (1) **chronological aging,** or number of years since birth; (2) **biological aging,** or the changes reducing efficiency of organ systems; (3) **psychological aging,** including memory, learning, adaptive capacity, personality, and mental functioning; and (4) **social aging,** referring to social roles, relationships, and the overall social context in which we grow old (Scheibe, Freund, & Baltes, 2007).

There is no cure for the common birthday.
JOHN GLENN

Perhaps the most basic discipline in the study of aging is *biology.* Without the biological aging process, we could all theoretically live forever, but the causes of biological aging are still not clearly understood.

A biological study of aging includes all kinds of animals, as well as detailed analyses of the human body. The effects of diet and exercise (lifestyle effects) on longevity are an important focus of study, and the cutting-edge field of genetics has dramatically changed our understanding of the complexity of the human organism. We can impact our biological health through paying attention to lifestyle; however, we can do nothing—at this point—about our genetic background. Recent research indicates, however, that genes determine about one-quarter of our longevity (Singh, Kolvraa, & Rattan, 2007). That leaves three quarters of how long we will live up to factors such as lifestyle and social environment—factors that can be addressed by individuals and by society. We are more in control of what will be our well-being in later life than we even imagine.

A second component of gerontology, the *sociological perspective,* examines the *structure of society*—its norms and values and their influence on how a person perceives and reacts to the aging process. Rather than focusing on individual experience, however, sociology focuses on *groups of individuals* and the cultural context in which they age.

The impact of context is huge: a society that gives the aging person a high status can expect a more positive outcome for its aging population, whereas a society that accords the aged a low or marginal status can expect more negative outcome. Within the sociological circle are anthropologists, who, in documenting the aging process around the world, find that cultures offer elders enormously varied roles. Also in the circle are political scientists, social policy experts, and historians. Demographic and population experts provide information on the numbers and distribution of older persons in societies and countries around the world and provide projections of population trends for consideration by politicians and policymakers.

A third lens from which aging is viewed is that of *psychology.* In contrast with a sociological perspective, the psychological locus of inquiry is on the individual. Psychologists are interested in the aging mind—how perception, motor skills, memory, emotions, and other mental capacities change over time. The psychological constructs of motivation, adaptability, self-concept, self-efficacy, and morale all have an important impact on how we age. Psychologists bring a perspective to solving social problems that considers individuals in terms of their life span, or particular places in the life span rather than one point in time. They view individuals as being dynamic and interactive, existing in multiple webs of relationship, history, and culture. Psychologists focus on identifying the connections between internal (psychological) and external (social) aspects of the individual's life.

Studies of older people cannot be complete without including an understanding of *philosophy, spirituality,* and *ethics.* Virtually all theories of human development suggest that the psychological task of later life is to gain greater understanding of the life we have lived and of our own approaching death. We seem to gain greater clarity of the meaning of our lives by asking the very questions that have been asked throughout the history of humankind: what was this life all about? What is the relationship of the people I am connected with to the meaning of my life? What is my understanding of death—my own as well as the deaths of others? What is my legacy?

In a related vein, *ethical issues* are central in the care of older adults as well as in life decisions made by elders themselves. Families are the major care providers for America's frail elders and issues of competence and decision-making, or autonomy and family relationships are central to families and therefore to the larger society of which they are part (Vitaliano, Young, & Zhang, 2004). Developing an understanding of ethics and values requires that psychologists and health care practitioners be culturally competent (Wang, 2007). *Cultural competence* refers to the ability to honor and respect styles, attitudes, behaviors, and beliefs of individuals, families, and staff who receive and provide services (Yali & Revenson, 2004). Culturally competent practitioners are thus able to support and reinforce older adults in achieving their own culturally appropriate sense of self-efficacy, that is, to help elders develop

Multi-generational gatherings of friends and family are central to the transmission and validation of family and cultural values.

personal mastery in a shifting internal and external environment (Fry, 2003).

Gerontologists, then, are multidisciplinary. They examine aging from a chronological perspective (age on the basis of years from birth); they study biological and psychological processes and individual meanings of aging; they look at the social meaning of aging including changing roles and relationships brought about by moving through the course of life. Additionally they study the meaning of aging from a contextual perspective: family, community, and national/geopolitical processes and events.

Gerontologists apply their specialty in many fields—medicine, dentistry, economics, social work, mental health, religion, education, and recreation. They are practitioners in, nursing, occupational therapy, sociology, and many other fields having to do with the health and well-being of individuals and society. The field of *geriatrics*, a term sometimes confused with *gerontology*, focuses on preventing and managing later-life illness and disease. Geriatrics is less multidimensional than gerontology, looking specifically at biological and physiological health issues. Geriatrics is a medical model perspective; gerontology uses a bio-psycho-social model.

Person-Environment and Social Issues Perspectives

In this text both social issues and psychological perspectives and strategies are presented as factors in understanding aging. Social situations that are problematic or undesirable for a large proportion of older adults as well as those situations and solutions that function to promote well-being in later life will be examined. Chapter by chapter, the text addresses social issues affecting the lives of older people—issues of status, roles, income, transportation, housing, physical and mental health, work, leisure,

and sexuality and relationships. We discuss the strengths and contributions that elders bring to their families and communities and we discuss the perspectives that address targets widespread patterns of behavior affecting quality of life indicators. The causes and solutions of social problems do not remain at the individual, or microlevel, but must ultimately be found at a greater level—they require macrolevel response. Generalized problems associated with aging, and with a society that is aging, lie with large numbers of people; the causes and solutions impact everyone, not just an age-identified segment of the population.

Person-Environment

A person-environment approach views the environment as a continually changing context to which individuals adapt as they also adapt to the personal, psychological, and physical changes inherent in the aging process. From this perspective as the aging person adjusts to life's changes, this adaptation impacts the environment which, cycle-like, further changes the individual as well as the social context (Hoyer & Roodin, 2009). This reciprocity of change is the person-environment model: the context changes so the individual must change; the individual changes thereby impacting his or her context, and so the environment changes.

Environmental Press

Eventually the individual's ability to adapt or change will become exhausted. For example an 84-year-old woman has become quite frail following a hip replacement, a lingering cough, and increasingly arthritic knees. She has become more cautious when she walks (internal change), and she placed appropriate hand-rails and lighting throughout the house (external change). As time goes on her frailties increase and the internal and external adaptations available to her are no longer sufficient for her to continue to live alone with safety. To maximize her quality of life the press of the environment (cooking, navigating stairs, cleaning and doing laundry, caring for her rose garden) must be aided by external resources.

The solution to this woman's situation requires a macrolevel response. It is when the environmental press becomes too great for the individual to manage alone that family, neighbors, community, and local and state resources can be mobilized. The causes and solutions to social problems relevant to aging might be at an individual, a group, a societal, or even a global level.

No Golden Age of Aging

Nature has always been harsh with old age. Among humans, however, the ways in which the old are treated is closely tied to the culture of their society. Early in the study of aging, sociologist Leo Simmons pointed out that in all cultures it takes both values and environmental context to provide for all age segments of a society. According to his analysis, the culture must state that aging is a positive achievement and must value the aged *as individuals* in order for the aged to have status and value (Simmons, 1945, 1960).

When old age is viewed historically and cross-culturally we get a mixed picture of the position and status of elders throughout history. In ancient times most people died before the age of 35. Our general understanding is that those few survivors into their 40s were treated with respect and awe, honored ceremonially and socially as keepers of the memories.

The belief that elders were once held in high status in American society has a couple of different etiologies. First, the honoring was believed to be an inherent family value of older people and ancestors. Second, because so few people lived into later life, old age became seen as exclusive. Status was bestowed by virtue of being part of this exclusive circle. A shared religious perspective held that reaching this exclusive age was a sign of God's blessing and therefore old people were to be valued because God valued them.

A third theory of elder status suggests that because the aged were perceived to be closer to death they served as mediators between this world and the next. This valued role lent prestige to age by means of providing a respected function to the larger community, assuming the elder's mental faculties were intact (Hooyman & Kiyak, 2002).

We turn not older with years, but newer every day.

EMILY DICKINSON

Not all ancient cultures equally honored old age. The difference was huge between surviving into later years with good physical and mental health and surviving with frailty. In a subsistence culture, people outliving their usefulness were a burden, so the treatment of the frail could range anywhere from being ignored to being treated cruelly or even killed, to being honored as more godlike because they were such exceptions (Glascock, 1997).

After the fifth century BC in Greek and Roman cultures old age was generally seen as being a distasteful time of decline and decrepitude. This perspective gradually emerged from between the sixth and fifth centuries BC Previously, records indicate old age to have been associated with wisdom, but even then this positive association seems to have been tied to material wealth and social status (deRomilly, 1968). We know next to nothing about the status of old people pre-fifth-century BC who were not part of the power elite—that is to say, who were peasants rather than landed gentry.

A gradual shift to a denigration of old age seems to have emerged concurrently with the belief in social equality in which the status of no one was supposed to be elevated merely because of birthright (Hoyer & Roodin, 2009).

During the Classical period beauty and strength were idealized in art and in myth, and old age was considered to be a time defiled by physical incompetence and mental ineptitude. Old age for Classicists was ugly.

We know very little about the treatment of elders in Medieval Europe. As urbanization created population centers, life expectancy in Medieval Europe dramatically dropped: it became shorter than it was in Greek and Roman times. Nutrition, sanitation, and crowded living conditions coupled with a lack of social organization appropriate to urban living resulted in an era of tremendous social disruption. Norms and values from previous generations were no longer applicable; new norms had not yet emerged. Art from the Medieval and Renaissance eras pictured age as cruel or weak, as surely it must have been given the environmental context (Hooyman & Kiyak, 2002).

With modernization, the cultural status of elders has declined in many cultures of the world (Vanderbeck, 2007). This decline is seen as a twenty-first-century phenomenon attributed to advances in technology, communication, science and medicine, and global economic forces (AoA, 2008b).

Personal and Social Definitions of Age

Social status among Americans is related to no small extent with education, wealth, and health—and most older people are better off in all of these categories today than they were in previous generations (AoA, 2008a). With the exception of a dramatic increase in obesity among older adults, lifestyle changes including nutrition and exercise have helped them add health and vigor to the longevity experienced over the past 30 years. This trend suggests that the United States might be presently undergoing a shift in cultural values toward greater status for older people.

To understand cultural trends as quality of life indicators, it is important to note that several age cohorts exist within the elder population. When the term *older people* is used, it could refer to people from their 50s to their 90s, or it might refer to those 85 years and older. Specificity is important. For example, 78 percent of people in the over-65 age category—the bureaucratic definition of old age in the United States—rate their health to be good to excellent (AoA, 2008a). On closer examination we find that of people 75 and older, 85 percent have chronic physical limitations; nearly 41 percent have mental (learning, remembering, or concentrating) disabilities; and over 70 percent experience difficulty going outside the home without assistance (U.S. Census Bureau, 2008b). In other words, the likelihood of having a disability or physical limitation increases with age.

From a health and vigor perspective there is enormous difference between a cohort of 65-year-olds and a cohort of 85-year-olds, although members of both age groups are "older Americans." When gender and ethnicity are specified in the health data, we get yet another, more complete picture of the face and experience of aging.

The age of a person during a historical event of major proportion, such as the Civil Rights movement beginning in the late 1950s, profoundly influences the social and personal meaning of the event. This is what is meant by *historical cohort*. Normative history-graded factors including wars, economic recessions, and disease patterns such as AIDs and polio have a shaping effect on the opportunities, challenges, and developmental trajectory, shared by an entire population. The impact on personality and other developmental issues of these events is age dependent.

The personal and educational and career choices available to a 35-year-old black woman in 1960 as she grew into adulthood would greatly impact her social and personal status as an 80-year-old in 2005. If, however, a woman was 15 years old in 1960, she will come into old age from a markedly different social experience. Cultural variations in the form of changing values and norms come about through historical events ranging from epidemics and wars to scientific breakthroughs and social changes.

The magnitude of historical events to the development of a particular cohort cannot be overestimated. Consider the impact of terrorist attacks on the United States on September 11, 2001—now a part of American lexicon as 9/11. Children who were 5 at the time will grow into adulthood with the experience of their country under attack by amorphous "others"—terrorists. As they attempted to manage and decipher the event, they saw frightened and ill-equipped adults who themselves had few skills to cope with the magnitude of the event. Depending on demographic factors such as proximity to the disaster, family religiosity, parental education and stability, and so on, the event took on different meanings and a different magnitude for those children. All of them, however,

experienced this historical event through the eyes of a 5-year-old.

People in their adolescence at 9/11 had a very different perspective on the event than their 6-year-old siblings. An adolescent's cognitive abilities to draw conclusions and make moral judgments and decisions are only one aspect of their perspective. Unlike the 5-year-old, the adolescent's understanding of personal safety extends beyond the protection provided by parents and family; the world view becomes shifted. Terrorism, formerly a word with little personal meaning, suddenly became a high-focus concept.

The impact of the event on the lives of young adults—some of whose college education was being supported by their participation in the National Guard—was different again from those of younger cohorts. Apart from those whose lives were directly impacted by the bombings, the event radically modified the course of life as the national economy shifted

The bombings of 9/11 will be remembered differently by different age groups, depending on their social roles and life experiences at the time.

focus, job opportunities changed, and society absorbed the possibility of global warfare.

The age of individuals at the time of the historical event shapes the meaning people give to historical events of great magnitude. By the time they are 65, those who were 5 at the time of the disaster will essentially have lived with its outcomes for their entire lives. For those who were 65 at the time of the disaster, the event becomes included in a lifetime of other events—its meaning is modified by other life experiences. The global economic recession of 2008 will likewise impact the lives of all people—not only those in the United States but those in all countries, be they developed nations, developing nations, or nations whose economies are in transition.

Historical Perspectives on Aging

Historically the status of older people was related to property ownership that resulted in the control of political resources. Control of property by one generation created opportunity for a balanced exchange between generations: elders were cared for in their later years in exchange for the inheritance of property by their children. Clearly, this worked for property owners, but the ethic of providing care for infirm elders in exchange for some level of inheritance was the cultural norm. With the onset of industrialization, farming and the control of property became less central to a family's well-being as economic resources became more available to people independent of their age. As the economic center shifted from the family to the corporation, status of older people also declined.

This cultural shift in the status of elders is known as *modernization theory* (Cowgill and Holmes, 1972, 1986; Simmons, 1945; c) and it suggests that the Industrial Revolution was the key-pin to the decline in status of older people. The vigor and energy of young adults, whose stamina kept the industrial sector moving, became the national icon. Youthfulness was the embodiment of the nation's progress toward increased wealth and prosperity. When physical energy and strength began to ebb, older workers were less able to contribute to the industrial economy and they ultimately either died on the job or were retired from work when they could no longer do their share. No substitute social role emerged for those not in the industrial work force. The status of women not was also very low; however, a substitute social role of wife and homemaker provided women access to social status depending on the incomes of their husbands and thereby the size and beauty of their homes and the quality of the lifestyles of their families.

Equality and individualism as values coupled with crowded and costly urban living correspondingly shifted the role of older people from participant in a reciprocal system of the exchange of resources to one of dependence on a younger generation. The growing emphasis on impersonality (equality) and efficiency (through individual effort) further contributed to the status shift of older people.

In their classic articulation of modernization theory, Cowgill and Holmes (1986) identified the characteristics of modernization that contribute to lower status for elders as:

- Health technology—reduced infant mortality and prolonged adult life

- Scientific technology—creating jobs that do not depend on skills and knowledge accumulated over decades of experience

- Education—targeted toward the young

These three cultural characteristics continue to shape social values today. The aged are not perceived to be as great a "problem" today with the advent of pension reform and the introduction of Social Security. Their special needs, however, continue to make elders an identifiable group. Elders today live longer than did the previous generation. They have lived through—and sometimes been surpassed by—major technological changes. Their educations, appropriate to the twentieth century, reflect a more traditional and classical approach than one focusing on job skills and the abstractions of a society focused on information technology.

From a perspective of *social issues*, old age itself has not been seen as an issue. Until 1900 or so, only the illnesses related to old age (not the actual *being* old) were defined as problems. In the seventeenth and eighteenth centuries the issues relevant to old age were understood to be the responsibility of families. At a family level, grandparents were respected—in part because there were so few older people in society. Long life was associated with God's favor among the Puritans in early Judeo-Christian America.

People believed that God blessed with long life those few who lived truly pure lives (Achenbaum, 1996). The primary basis of the respect and power granted to older people in the seventeenth century, however, was their control of property. For a nation whose economy was based on the abundance of agriculture produced in family farm organizational structures, the control of land provided the senior family member an especially powerful position.

With the industrialization of the late 1800s, problems associated with growing old became reconceptualized, not just on a physical level but on social, economic, and psychological levels as well. Social change occurred at breathtaking speed, and the relative status of youth became elevated. Additionally in the 1800s, birth rates began to drop; this resulted in an

Family relationships are hugely important at younger ages as well as old age.

increase in the median age and the evolution of an identifiable category of "older" people as birth rates dropped and longevity increased over the next 75 years.

By the 1930s and 1940s this new conceptualization of youth and the aged by society and by individuals themselves had created an identifiable group with physical and social problems that called for collective action. For example, the right to a decent income at retirement became an issue. Applying the phrases of sociologist C. Wright Mills, a family's "private troubles" had become "public issues" (Mills, 1959). Responsibility for aging individuals became seen as belonging to society as well as to family. Older people received more public attention, but in the process they began to be viewed as helpless and dependent. These negative images were universally applied for many years in spite of improvements in the health of those relatively young-old, aged 65 to 74, and in spite of the countless people aged 75 and over who remained active and involved in society.

Ageism

A social problem is a widespread negative social condition that people both create and solve. Ageism is such a problem, defined as "the prejudiced behavior of individuals and systems within the culture against older adults, including the negative consequences of inaccurate stereotyping of the elderly" (Hoyer & Roodin, 2009).

Ageism has been called the third "ism," following racism and sexism. Whereas racism and sexism prevent racial minorities and women, respectively—and in what is called "reverse sexism," men—from developing full potential as people, ageism limits the potential development of individuals on the basis of age. Ageism can oppress any age group, young or old. If you are young, you may have been told that you are too inexperienced, too immature, too untested.

If you are an older adult, you may have been told you are out-of-date, old-fashioned, behind the times, of no value or importance. At both ends of the scale, young and old, you may be the victim of ageism. Although ageism may affect the young as well as the old, our concern here is with the senior members of society. Robert Butler, noted gerontologist, observed that "the tragedy of old age is not that each of us must grow old and die, but that the process of doing so has been made unnecessarily and at times excruciatingly painful, humiliating, debilitating, and isolating" (Butler, 1975, p. 2–3).

Ageism is a complex phenomenon affected by technology, industrialization, changing family patterns, increased mobility, demographic changes, increased life expectancy, and generational differences. A discrimination leveled by one group against another, ageism is not an inequality to be associated with biological processes alone. It is created and institutionalized by many forces—historical, social, cultural, and psychological.

Our Western cultural heritage decrees that work and financial success establish individual worth. Industrialization has reinforced the high value of productivity and added further problems for the aging worker. The speed of industrial, technological, and social changes tends to make skills and knowledge rapidly obsolete. Most people must struggle to keep abreast of new discoveries or skills in their fields. The media has used the term **Detroit syndrome** to describe older people in terms of the obsolescence that exists for cars. When younger, stronger, faster workers with newly acquired knowledge are available, employers tend to replace, rather than retrain, the older worker. Within the workforce, older persons have often been considered a surplus population. As such, they suffer the potential for being managed much like surplus commodities: devalued and discounted.

Social change can create a generation gap that contributes to ageism. Rapid social change can cause our values to be somewhat different from our parents' and significantly different from those of our grandparents. Those who grew up in a given time period may have interpretations of and orientations toward social issues that differ from those who grew up earlier or later. For example, the person who matured in the 1940s and experienced the patriotism of World War II may be unable to understand the behavior and attitudes of those who matured in the 1960s and protested the wars in Vietnam or the Persian Gulf.

People maturing in the 1990s may not understand the historical rationale for the United States intervening in small countries like Panama or El Salvador. These people, many of whom have postponed marriage and childbearing, may be unable to grasp the reasons for early marriage and large families held by the now-elderly generations. The study of intergenerational relations, which provides insights into similarities and differences in values across generations, reveals that communicating and understanding across generations are difficult when values are different.

Ageism appears in the many euphemisms for old age and in the desire to hide one's age. Older adults themselves do not want to use the term *old*, as the names of their local clubs show: Fun After 50, Golden Age, 55 Plus, and Senior Citizens Club. Some people even forgo their "senior discounts" because they do not want to make their age public.

Fear of aging is illustrated by men and women wanting to keep their age a secret, hoping that their appearance denies their age and that they project a youthful image. Indeed, a common "compliment" given to mature people is that they "do not look their age." Many people suffer a crisis of sorts upon reaching age 30 and repeat it to some extent when entering each new decade. Some even experience an identity crisis as early as their late 20s, because they are entering an age that the youth culture considers "old." Many counselors recognize the "over 39" syndrome as a time when young adults come to terms with the fact that youth does not last forever but blends gradually with the responsibilities of maturity.

Greeting-card counters are filled with birthday cards that joke about adding another year. Despite their humor, they draw attention to the fear of aging that birthdays bring. Some birthday cards express the sentiment that to be older is to be better, but then add a note that says, in effect, that no one would want to be better at the price of aging. Though birthday cards often joke about physical, sexual, or mental decline, the fear in the minds of many is no joke at all. Fear of aging can damage psychological well-being and lead us to shun older people. Ageism is a destructive force for both society and the individual.

Ageism as a concept in literature has been described in a general sense, but it has also been measured in more specific ways. Early in the study of gerontology, Alex Comfort (1976) used the term *sociogenic* to imply ageism in a broad sense. He described two kinds of aging: *physical*, which is a natural biological process; and *sociogenic*, which has no physical basis. *Sociogenic aging* is imposed on the elders by the folklore, prejudices, and stereotypes about age that prevail in our society. Thus, age prejudice, as it exists in our minds, is institutionalized in many sectors of our society.

We can find more specific evidence of ageism in our laws, particularly those dealing with employment, financial matters, and legal definitions relating to "competency" as an adult. Income differences, occupation differences, and education differences vary by age. One aspect of ageism is age inequality in education and occupation, caused by the reality that newer generations receive an education attuned to a highly technical and computerized society and are therefore better qualified for jobs. Elders are easily left behind on the "information highway" as the "high-tech" knowledge of younger age groups rises. Income inequality based on age is caused not only by younger age groups having more extensive formal or technical education, but also by age discrimination in employment. Gerontologists believe that ageism in employment dates back to the early 1800s. In the work and leisure chapter, this age prejudice will be covered in more detail.

The *critical perspective* in gerontology draws attention to inequalities in U.S. society, addressing broad and fundamental structures of the society such as the class system, economic system, race and gender issues, and age-related roles and opportunities. This approach addresses **cumulative disadvantage**—the negative effects of inequality in wealth, status, and opportunity over the life span. Gerontologists have made use of the critical perspective to understand the problems of aging in a broad political, social, and economic context (Dannefer, 2003; Sokolovsky, 1997).

Despite the social and personal issues arising from ageism, we must be careful not to view older people as more dependent and helpless than they are. That perspective is a negative stereotype and is not productive to the mission of understanding social issues and problems salient to America's aging population. We need to understand both the strengths and the vulnerabilities of the older person in our present society, and identify and give support for institutions and social structures that support strength and self-reliance.

Ageism Yesterday: The Early American Example

A look at older people in earlier times, when age relationships were different, provides us with a clearer view of ageism now and in the future. Generalizing about ageism in the past is not easy. Some historians believe the status of older people was elevated in the colonial period—the time during which early settlers, especially the Puritans, founded America and formed the 13 colonies. In contrast, other historians point to ageism and neglect of older persons in the colonial days.

Early colonial days

According to David Fischer, author of *Growing Old in America* (1977), the power and privilege of old age were deeply rooted in colonial times, when age, not youth, was exalted. To be old was to be venerated by society and to be eligible for selection to the most important positions in the community. Meeting house seats were assigned primarily by age, and elders sat in positions of highest status. According to Fischer, the national heroes were "gray champions." Community leaders and political office-holders tended to be older men, and as elders were honored during ceremonial occasions.

Older adults were believed to be in favor with God. Their long life was thought of as an outward sign they would be "called" or "elected" to heaven. Biblical interpretation suggested that good people would be rewarded with long life: "Keep my commandments, for length of days and long life shall they add unto thee." The Puritans pictured Jesus as an old man with white hair, even though, according to most theologians, Jesus died in his early 30s. Respect for age was also evident in the manner of dress. Increase Mather, the president of Harvard College from 1685 to 1701, wrote that old men whose attire was gay and youthful, or old women who dressed like young girls, exposed themselves to reproach and contempt. Male fashions during the 1600s, and even more in the 1700s, flattered age. The styles made men appear older than they were. Clothing was cut specifically to narrow the shoulders, to broaden the waist and hips, and to make the spine appear bent. Women covered their bodies in long dresses. Both sexes wore white, powdered wigs over their hair. Not until the 1800s did clothing styles begin to flatter the younger man or woman.

Fischer studied other historical data that indicate age status. American literature, for example, emphasized respect for old age from the 1600s until after the American Revolution. A careful examination of census data shows that in the 1700s individuals tended to report themselves as older than they actually were, in order to enhance their status. (In the mid-1800s this tendency reversed itself.)

The tradition of respect for elders was rooted not only in religious and political ideology but also in legal and economic reality. The elders owned and controlled their own land, which did not pass to their sons until they died. The sons, therefore, had financial reason to show respect for and deference to their fathers. In these conservative times, the young had little choice other than to honor, obey, and follow the ways of the old.

This photo of a successful middle-aged man might illustrate advantages income can bestow on the aging individual.

A word of caution must guide our consideration of the older person's status in colonial times. "Status" is a multidimensional concept, measurable in many ways, that indicates one's social ranking in society. Deference, respect, health, economic resources, material possessions, occupation, education, and political power are all possible indicators of social status. By some measures, the colonial elders had high status. They were shown deference and respect, and they had political power and financial control of their land. Not all but only a few elderly colonial citizens had financial and political power. Colonial legal records show that widows who had no means of support wandered from one town to another trying to find food and shelter. Older lower-status immigrants and minorities, and most certainly older African Americans, had an especially difficult time because of their low economic status; many were indentured servants or slaves.

Most old people suffered from health problems that medical science was unable to cure or alleviate. Benjamin Franklin, for example, was wracked with pain in his later years because of gout and "the stone" (gallstone). Yet the old, in spite of their infirmities, were expected to be models of service and virtue to their communities. The very veneration that brought older persons respect kept them from enjoying close, intimate relationships with younger people. Youth/elder relationships were distant and formal, causing the old to suffer loneliness in their elevated position.

A number of historians take exception to Fischer's rosy picture of colonial days. Haber (1983) described old age in colonial times as more dire than Fischer's work indicates. Haber believes that although select, well-to-do elderly had high status in the Puritan days, they did not live in a golden era of aging. Too many not so well-to-do fared badly; they were viewed with scorn and contempt. Haber advises that a careful sociologist or historian must try not to idealize the past, but to recapture reality by examining all of its facets: political, historical, economic, and social. Cole (1992) and Quadagno (2008) make the same point as Haber makes, emphasizing that multiple forces, some positive and some negative, shaped life in colonial times.

You are as young as your faith, as old as your doubt; as young as your self-confidence, as old as your fear; as young as your hope, as old as your despair.

DOUGLAS MACARTHUR

Shifting status of old age

According to Fischer, change throughout the 1800s altered the system of age relationships in a negative way, leading to social problems for the aged. The most fundamental change took place in political ideology. The principles formulated in the Declaration of Independence became stronger: equality for all in legal, social, and political matters. This trend affected older persons because "lovely equality," in Jefferson's words, eradicated the hierarchy of age, and hence the respect automatically accorded the old. A study of word origins shows that most of the negative terms for old men first appeared in the late 1700s and early 1800s. *Gaffer*, originally an expression of respect, changed from a word of praise to one of contempt. Before 1780, *fogy* meant a wounded soldier; by 1830 it had become a term of disrespect for an older person. *Codger, geezer, galoot, old goat*, and *fuddy-duddy* came into general use in the early 1800s.

The preeminence of religious elders began to wane as doctors and other technologists replaced preachers as the custodians of virtue and learning. The United States became more industrialized. In the 1800s, the city became a means of escape from both farming and parental control. Instead of waiting for his father to provide him with land, a young man could move to the city and find work in a factory. As long as America had remained a traditional agricultural society, in which parents controlled property until their advanced years, older adults had exercised considerable power. Urban and industrial growth led to diminished parental control over family, wealth, and possessions (Quadagno, 2008). By the late 1800s, the young pioneer and the young cowboy had become popular heroes; Teddy Roosevelt was young, rough, and ready. The youth cult began to replace the age cult.

The older population grew rapidly during the 1800s and 1900s because of advances in the medical sciences. Retirement gradually became more and more common. However, many of the older people who retired had no source of income and were often neglected. Old age became a burden to those who lived it and a social problem to those who analyzed it.

Fischer (1977) divided U.S. history into two general periods:

1. 1600 to 1800: an era of growing *gerontophilia*. Old age was exalted and venerated, sometimes hated and feared, but more often honored and obeyed.
2. 1800 to present: an era of growing *gerontophobia*. Americans increasingly glorified youth instead of age, and older people often became victims (self-victims as well as social victims) of prevailing attitudes and social arrangements.

Fischer's historical analysis suggests that we may eventually enter another period of age relations, one that will create better conditions for older adults. The goal, he states, should be to make a new model—a fraternity of age and youth, and a world in which "the deep eternal differences between age and youth are recognized and respected without being organized into a system of inequality" (Fischer, 1977, p. 199).

The example of colonial America illustrates that the position of elders in our society can be something other than what it is now. We can be aware of various age relationships and possibilities that are more positive than the situations we have created.

The Aging Revolution: Demographics of Aging

What was once referred to as the *graying of America* is now more accurately understood to be a *revolution*, and the social meaning of that revolution profoundly permeates American culture. The population of Americans 65 years and older has greatly exceeded the growth of the population as a whole, and the "oldest-old" (85 years and over) are the most rapidly growing elderly age group. Gains in life expectancy in this century are coming at the end of life. The 2000 American census found that nearly one-half of Americans age 65 or older were over 74, compared with less than one-third in 1950. One in eight were 85 or older in 2000, compared with one in 20 in 1950 (U.S. Census Bureau, 2005) While people 85 or older made up only 1.5 percent of the total U.S. population in 2000, they were about 12 percent of all of those Americans over 65. By 2007 the 65 and over population represented 12.6 percent of the U.S. population and is projected to grow to over 19 percent by 2030 (AoA, 2008a). These oldest-old are of particular interest to gerontologists because this group requires the largest number of services to remain viable in their homes and communities.

Increasing numbers of age

In the past decade, the number of men and women living to 85 years has increased dramatically; the number of those living past 85 has increased at an even greater rate. The 85 and older population in the United States is projected to more than triple, from 5.4 million to 19 million between 2000 and 2010 (AoA, 2008b).

The total number of people in the United States *under* 65 tripled in the twentieth century; however, the number of people aged 65 or *older* has increased by more than a factor of twelve (U.S. Census Bureau, 2005). Children under five were the largest age group in 1900; in 2000 the largest five-year age groups were 35–39-year-olds and 40–45-year-olds, both roughly 8 percent of the total population. By 2010 the age group 55–64 will be around 12 percent of the total population and all of those over 65 will represent over 13 percent of the population (AoA, 2009). Even as the population grows older, it grows more diverse. By mid-century, minorities (roughly one-third of the population in 2008) are projected to become a 54 percent majority, comprising 20 percent of the total population (AoA, 2008a, 2009a).

A huge jump in the over-65 population will occur from 2010 to 2030, when the leading edge of the **baby boom generation** (those born between 1946 and 1964) reaches 65. By 2030 when all the baby boomers will be 65 and over, nearly one in five United States residents will be 65 and older (U.S. Census Bureau, 2008). By the sheer force of its numbers, the boomers represent a demographic bulge that has remodeled society as it passed through it and will continue to do so as it ages.

As implied by Figure 1.1 the move of baby boomers into later life will profoundly affect regional and national policies on education, health, recreation, and economics. Older people, especially the oldest-old, are dependent on family, the government, or both, for physical, emotional, and financial help. America's national self-concept will experience a shift from being a youth-oriented nation to one of mature citizenry with the movement of this cohort into old age.

Most recent Census Bureau findings also highlight the strong growth of the older population from 1900 to 2000. Over this period of time the percentage of Americans 65 and

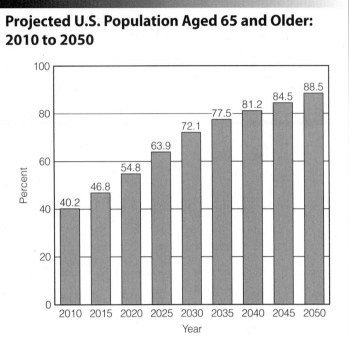

FIGURE 1.1

Projected U.S. Population Aged 65 and Older: 2010 to 2050

Source: Population Division, U.S. Census Bureau Released: August 14, 2008

over has nearly tripled (from 4 percent to 12 percent) and the number has increased 12 times (from 3 million to 35 million) (AoA, 2008a). The speed at which the oldest-old population (85 and over) is aging is staggering: in 2000 the number was just over 4 million, or 1.6 percent of the total population. By 2030 this number will have grown to nearly 8.5 million (2.4 percent of the total), and by mid-century there will be over 18 million Americans 85 and older—representing 4.6 percent of the total population (AoA, 2009). In 2007 there were nearly 81,000 people aged 100 or more, representing 0.21 percent of those 65 and older. This is an increase of 117 percent from the 1990 figures (AoA, 2000).

Figure 1.2 reflects these changes. Notice in these population pyramids representing the years 2000, 2025, and 2050 the proportional change in the 85-plus age group. Note both the shape of each pyramid and the length of each bar. All older age groups starting at age 65 show large increases. In general, the higher the age interval, the greater the proportional increase. In other words, the old as a group are becoming older, that is, not dying off as quickly, so the numbers of the very old will continue to mushroom.

Global Aging

The dramatic growth in number and proportion of is likewise increasing. Indeed, the world is at the threshold of a new revolution: global aging. From a global perspective, East and Southeast Asia have the most rapidly aging population. Whereas there were around 95 million Asian elders aged 60 and over in 1950, the number more than tripled over the

FIGURE 1.2

Population Pyramid Summary for United States

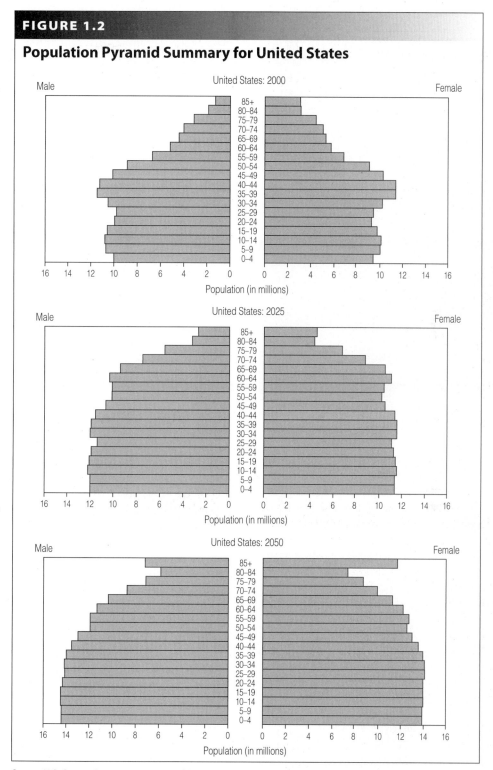

Source: U.S. Census Bureau, International Data Base. Extract data from IDB Online Aggregation

Grandmother, great-grandmother, mother, and baby: strong generational ties reinforce cultural values.

next 50 years to 322 million in 2000 (Mui, Leng, & Traphagan, 2005). The speed and timing of these demographic changes is globally unprecedented. Compare Figure 1.3, Population Pyramids of China, to Figure 1.2, United States pyramids. Japan took 26 years (1970–1996) to increase its proportion of those 65 and over from 7 to 14 percent. Countries such as France, the United States, and Sweden took nearly 115 years to do the same (Mui et al.).

Asia's older population grows merely at a rate faster than other nations. Globally, in 2009 the world population 65 and older was 518 million people. Projections indicate this number will increase to 1.6 *billion* by 2050 (U.S. Census Bureau, 2009).

Globally females are longer lived than males. Boys outnumber girls in all countries;

however, over time different mortality rates produce a changed sex balance in a population (AoA, 2008b). In other words, the female share of the older population rises with age. Differences in developed and developing countries remain notable: as infant and childhood mortality reach lower levels in developing countries as in developed countries, improvements in average life expectancy are achieved (AoA, 2008b). Of greatest concern with global aging, a twenty-first-century phenomenon, is whether living longer means living better.

Global aging can be attributed to modernization in medicine and technology thereby leading to increased life expectancy and a declining birth rate. We will consider each topic separately.

Increased life expectancy

Life expectancy in the United States has consistently increased throughout the nineteenth and twentieth centuries and continues to do so in the twenty-first century. The population 65 and over increased over tenfold from 1900 to 2000 (U.S. Census Bureau, 2009). In 2006 those reaching age 65 had an average life expectancy of an additional 19 years, and those born in 2006 could expect to live 78 years. This is about 30 years longer than a child born in 1900. From 1900 to 1960 life expectancy at age 65 increased by only 2.5 years, but it increased an additional 4.7 years from 1960 to 2006 (AoA, 2008a).

A dramatic increase in life expectancy occurred in the 1920s as a result of reduced infant mortality, health care advances, and improved nutrition, although the increase is characterized by ethnic disparities. Racial differences exist, however, in longevity patterns: white males live more than six years longer than African-American males. The life expectancy for African-American females in 1995 was 74; by 2008 that life expectancy had risen to 77 (U.S. Census Bureau, 2008).

FIGURE 1.3

Population Pyramid Summary for China

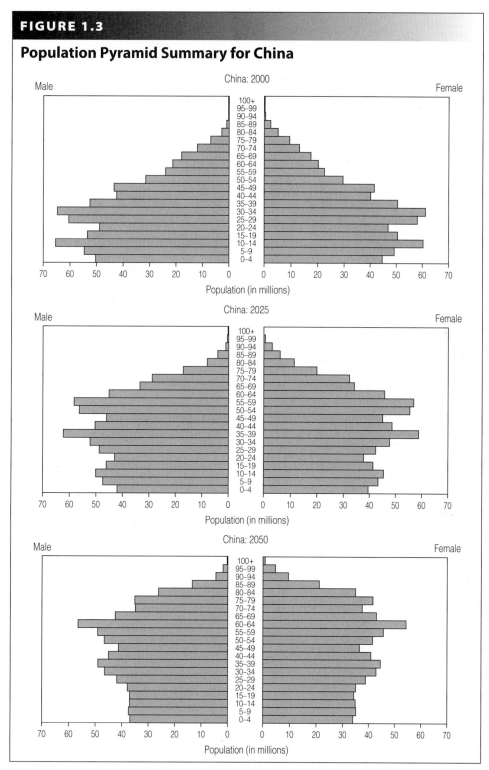

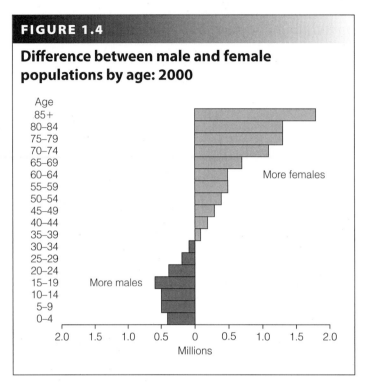

FIGURE 1.4

Difference between male and female populations by age: 2000

Age

85+
80–84
75–79
70–74
65–69
60–64
55–59
50–54
45–49
40–44
35–39
30–34
25–29
20–24
15–19
10–14
5–9
0–4

More females

More males

2.0   1.5   1.0   0.5   0   0.5   1.0   1.5   2.0

Millions

Note: The reference population for these data is the resident population.
Source: U.S. Census Bureau, 2001. Table PCT12.

American women at every age, regardless of race or ethnicity, have longer life expectancies than do men, though this discrepancy has decreased in the past 30 years. Figure 1.4 shows differences between male and female populations in the United States by age. Note the dramatic increase of females over males in the 85 and over age category.

The longer a person has lived, the greater is that person's statistical life expectancy. The reasons for this have to do with **selection for survival** (World Health Organization [WHO], 2007), meaning that members of a population are selected for survival based on their resistance to common causes of death. Those causes might be intrinsic or environmental, but studies on the genetics of long-lived people suggest that genes probably affect longevity by altering the risk of death at different ages, rather than by directly determining age at death (Moffitt, Caspi, & Rutter, 2006; Volger, 2006). In other words, our genetic programming seems more aimed at the level of "protection" we have against mortal illness than at setting age-at-birth.

Over the years there have been dramatic reductions in the death rates for diseases of the heart, cerebrovascular disease, and pneumonia. Even though life expectancy for males is not as high as for females, both genders' expectancies have increased considerably over the last several decades. This increase has been driven overwhelmingly by changes in environmental factors causing death, rather than factors intrinsic to the aging process itself. In addition to having long-lived grandparents (genetic factors), being near an ideal weight for one's stature, having low blood pressure and low cholesterol, not smoking, consuming alcohol

FIGURE 1.5

People surviving to selected ages according to life tables for the United States: 1900–1902 to 2000

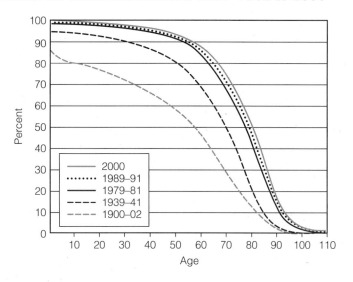

Note: The reference population for these data is the resident population. Data for 1900-02 and 1939-41 also include deaths of nonresidents of the United States.

Sources: 1900-02, U.S. Bureau of the Census, 1921, Table 1: 1939–41, U.S. Bureau of the Census, 1946, Table 1: 1979–81, Nathional Cener for Health Statistics (NCHS), 1985, Table 1: 1989-91, NCHS, 1995, Table 1: 2000, NCHS, 2001b, Table 1.

moderately, exercising vigorously three to five times a week, eating a healthy diet, and living a relaxed and unstressed life style are central predictors to a long life.

The **survival curve** graphically illustrated in Figure 1.5 shows the extension of this longer lifespan from 1900 and 2000. As a birth cohort ages a certain number of people die, of course, along the way. But incrementally fewer people died at younger ages between 1900 and 2000. This is called the **rectangularization of the survival curve.** The curve is sometimes known as the **mortality curve.**

Heart disease remains the leading cause of death for older Americans; however, the proportion of death due to coronary heart disease has fallen in the past 20 years and now death rates from cancer take a third of lives between 65 to 75, especially among African Americans (AoA, 2009; ; Wailoo, 2006). Older Americans are living longer, and they are less frail than were their parents and grandparents.

Diseases of the heart are primary causes of death for people 65 and over (He, Sengupta, Velkoff, & deBarros, 2005); however, smoking, overuse of alcohol, obesity, lack of exercise, and poor diet are some risk factors associated with mortality in later life (Barnes & Schoenborn, 2003). Many of these causes of death are linked to behaviors that our culture either encourages or finds more acceptable in males than in females: using guns, drinking alcohol, smoking, working at hazardous jobs, or appearing fearless. Such cultural expectations seem to

Because women generally outlive men, many older social groups are female-dominated.

contribute to males' elevated mortality. Men suffer three times as many homicides as women and have twice as many fatal car accidents (per mile driven) as women. Men are more likely to drive through an intersection when they should stop, are less likely to signal a turn, and are more likely to drive after drinking alcohol. But behavior doesn't entirely explain the longevity gap.

Women seem to have a genetic makeup that "programs" them to live longer. Some scientists think that the longevity gaps may be due to chromosomal or hormonal differences. Whatever the reasons, older women outnumbered their male counterparts in 1994 by a ratio of 3 to 2—20 million to 14 million—and the difference grew with advancing age. One consequence of this gender discrepancy is that older women are much more likely than men to live alone. More than eight of 10 non-institution-alized older adults living alone in 2000 were women (AoA, 2008a).

The leading causes of death for men and women over age 65 have remained stable in the past decade. Ranked in order from most to least common are (1) diseases of the heart, (2) malignant neoplasms (tumor), (3) cerebrovascular disease (stroke), (4) chronic obstructive pulmonary disease, (5) pneumonia and flu, (6) chronic liver disease, (7) accidents, (8) diabetes, and (9) suicide. Men have higher death rates in all the categories, except diabetes.

Decreased birth rate

When the birth rate declines, the number of young people decreases in proportion to the number of old people. The birth rate has gradually declined since public record keeping began in the eighteenth century. A baby boom in the

1940s and 1950s increased the birth rate temporarily but did not reverse its long-term trend. In 1972, we witnessed a near-zero population birth rate (2.1 children born for every couple): the number of live births nearly equaled the number of deaths, stabilizing the population. U.S. population pyramid in Figure 1.2 illustrates this relationship. According to the Population Reference Bureau in Washington, DC, the 1991 birth rate evened out at 2.1 children. If the United States maintains a lower birth rate, the proportion of older people will further increase. This is what is meant by an aging nation. With no increase in the total population the relative proportion of older persons will grow each year until the population pyramid becomes rectangular, as illustrated in Figure 1.2, the 2025 projection.

The post–World War II baby boomers of 1946–1964 are one of our largest age groups. Now in midlife, this age group will begin to reach age 65 in 2011, massively increasing the over-65 population. If we assume continued low birth rates and further declines in death rates, the older populations will jump tremendously by the year 2030. As we saw earlier, their numbers will double and their percentage of the population will rocket to over 20 percent.

A controversy rages as to whether medical science can do anything further to extend life expectancy at birth to more than 85 years. In the past 125 years, the life expectancy of Americans has almost doubled: from 40 to nearly 80 years. But these gains in life expectancy, most of which have come through a combination of reducing deaths of the young (particularly infants) and mothers in childbirth, may have been the "easy" ones. It is clear that in our present century longevity is a complex interaction of environmental, historical, and genetic factors (AoA, 2008b).

The population that is presently the oldest-old has had unique life experiences. They are a cohort that survived infancy when the infant mortality rate was about 15–20 times the present rate, the infectious diseases of childhood

when medical practice did not have much to offer. They survived at least one world war, and the females survived child bearing at a time when the maternal mortality rate was nearly 100 times its present level. The next generation of older Americans will have lived their adult years with many of the advantages that were only emerging for the previous older generation, so mortality rates will continue to be lower in later life. That does not, however, appear to change the maximum human life span, which seems to have fixed limits (Finch & Pike, 1996).

Some medical experts and laboratory scientists say that the period of rapid increases in life expectancy has come to an end. They argue that advances in life-extending technologies or the alteration of aging at the molecular level, the only ways to extend life expectancy, will be either improbable or long, slow processes. And, though they do agree that eliminating cancer, heart disease, and other major killers would increase life expectancy at birth by about 15 years, cures for these diseases are not in sight. Other scientists are more positive about extending life expectancy. Findings of a study by Ken Monton at Duke University, reported in 1990, predicted that Americans could very well live to age 99 if they quit smoking, drinking alcohol, and eating high-cholesterol foods. We now see clear evidence of the accuracy of that prediction. Populations with low-risk lifestyles, such as Mormons in the United States, already have achieved life spans exceeding 80 years (Hoyer & Roodin, 2009).

The longest documented human life span on record is that of Frenchwoman Jeanne Calment, who lived to be 122. There are some generally accepted records of people throughout history who have died between 110 and 120 years of age, but few that are extensively documented. The most extreme claims come from populations with the least reliable records. There is no evidence, either current or historical, that there has been much change in the rate of aging. Increases in life expectancy have been driven

Frenchwoman Jeanne Calment, who died in 1997 at the age of 122, was the world's oldest person with verifiable birth records.

overwhelmingly by reductions in environmental causes of mortality (AoA, 2009). It appears that the maximum human life span has not increased; however, the mean life expectancy in developed countries has done so tremendously.

Our Aging Nation

Identifiable regions of the country exceed the U.S. proportion of citizens over the age of 65. The Midwest has the largest number of counties with higher than national average proportions of people over 65, and the western area of the country has the lowest number of counties exceeding the national average. Although California has the largest *number* of citizens over 65 (just over 3.5 million), Florida, a retirement haven, has the highest *proportion* of over-65 elders in the nation. Several states (North and South Dakota, Pennsylvania, Rhode Island, Iowa, Arkansas, and Maine in the 2005 census), have 14 percent or more of their total population falling into the over-65 age category. Note that elder population can increase by virtue of in-migration of elders to retirement communities (Florida, Arkansas, and Arizona for example) or by the out-migration of younger citizens (represented by some Midwestern states). Many of the states are farm belt states where younger people are leaving farms for jobs in cities, whether those cities are in the same state or not.

The social implications of these demographics are broad. The increasing percentage of older people means that more and more families will be made up of four generations instead of two or three. Currently whites are more likely than African Americans to live in married-couple-only households, and African Americans are more likely than whites to live in multigenerational households (Sheeder, Lezottte, & Stevens-Simon,. 2006), although these differences balance out somewhat as people age. In the next 20 years it seems clear that more children will grow up with the support of older relatives, and more people in their 60s will be called on to care for 80- and 90-year-old parents.

Population pyramids, illustrated in Figures 1.2 and 1.3, picture the effects of a population's age and gender composition on the structure of a nation's population. The horizontal bars in the pyramid represent *birth cohorts* (people born in the same year) of 10 years. The effect of the "boomer" cohort on the U.S. population can be clearly observed in Figure 1.2 as a bulge in the middle of the pyramid in 1997, a squaring-off of the population in 2025, and a startling increase in the over-85 population in 2050. Note the relationship between lower birth rates and lower death rates, indicating fewer young people and more older people. This trend effectively reshapes the pyramid into a more boxlike image, illustrating the more balanced proportional in the population of each of the age cohorts.

Developed countries throughout the world are also experiencing an aging boom. The shape of the population pyramid for a less developed country like Nigeria, for example, would be a large base, indicating high birth rates, and a small top, indicating a high death rate with few people surviving into old age. This pyramid was typical of the United States as a developing nation in the 1800s, and it is typical of most developing nations. Countries with this pyramid form have difficulty caring for all their young, and as a result, social policy is directed toward youth.

Figure 1.3 shows population pyramids for China at 2000, projected to 2025 and 2050. Note that China's 2050 projected pyramid is roughly parallel to that of the United States (Figure 1.2) in 2000. This is a result of industrialization and all that accompanies a nation's increasing affluence.

Countries like Sweden and Japan, on the other hand, have an even higher proportion of older adults than the United States. They have achieved virtual zero population growth and thus would eventually have a "stationary," or boxlike pyramid. The social implications are that the needs of a society change with a changing age structure, as must social policy that directs national resources to various segments of the population. Housing, health care, education, and other services for elders must be balanced with services targeting more youthful age groups, as the population shifts in age.

Ageism in the Future

For many decades our society has suffered from *gerontophobia*. The term comes from the Greek *geras*, "old," and *phobos*, "fear, " and refers to fear of growing old or fear or hatred of the aged. To conceive of any status for elders other than that to which we are accustomed is difficult. We have accepted tension between youth and age. As a nation, the United States clearly identifies as youth-oriented. Respect by

the young for the old in our society is not a given. It is not deeply imbedded in the fabric of our society.

Some people view the increasing number of older adults as a burden on society, referring to the economic burden of providing care for the unemployed elders who depend on society for financial aid. The number of people 65 and older relative to the working population (those 18–64) is called the old **age–dependency ratio**. If the population age 65 and older grows faster than the working population, the cost to the taxpayer of providing for the aged population rises. The percentage of elders to the working population has increased steadily, so there are proportionally fewer employed people to support older, retired people today. In 1910, there were 10 working people per older person; in 1980, 5 or 6. If the trend continues, by the year 2010 a ratio of 0.22, or about 4.5 workers per retired person is expected (U.S. Bureau of the Census, 2008).

A larger proportion of seniors requires more Social Security and Medicare payments, which consequently means higher taxes for taxpayers. Indeed, Social Security taxes have slowly increased over the years. The reasons are multiple—one, however, is the increase of retirees in that system. This will be addressed more fully elsewhere in the text. The prospect for the future, however, rests on one simple fact: if you go to work at a young age, you will have to live a very long time to receive in benefits what you have paid into Social Security, because resources for that "enforced savings plan" will have been spent on people who are *presently* retired. If we view elders solely as an economic burden, ageism may increase as the number of retired, sick, or frail older people increases.

Some gerontologists believe that we have become an age-segregated society, with separate schools for the young and separate retirement communities for the old. Undeniably, segregation further generates misunderstandings and conflict. Other gerontologists, alternatively, maintain that ageism is declining. They point

Healthfulness and longevity together are key to a shifting image of the older person.

It is possible that the increasingly large number of older persons is eroding the youth culture. Marketing experts are skilled at analyzing the data we have reviewed in summary here. Young adults of the 1980s, many of them "yuppies," prospered from relatively inexpensive college education and the economic expansion of the times. These same baby boomers are aging in unprecedented numbers and continue to influence America's lifestyle and economy. Review again the illustrated population "bubble" in Figure 1.2. As boomers make demands on the market, our culture will modify to accommodate an older population—from changing how long it takes a traffic light to turn from green to red, to clothing styles, to increased services at airports.

Old age can be an exciting time of contributing to others and of self-fulfillment. Demographics predict that communities in the future will have a more evenly balanced age population. That means that younger people will be more accustomed to seeing older people with health ranges from outstanding to frail, just as medical advances have allowed younger people with injuries or disabilities to live longer.

to the improved health of seniors and to retirement communities composed of increasingly younger retirees who seem happy and content. In such contexts the image of older persons is improving. In addition, the increasing numbers of elders may be leading to a psychological shift away from a youth-oriented culture, toward a more life-course inclusive identification.

Chapter Summary

Gerontology is a multidisciplinary study of the human aging process from maturity to old age, as well as the study of the older adults as a special population. The key perspectives in gerontology are, *biological, sociological,* and *psychological* processes. The text uses a *social problems approach,* examining patterns of social behavior and institutional structures that negatively affect the quality of life of the aging individual.

America has always been a youth-oriented nation; however, increased longevity and lowered birth rate have transformed the population to an older one during the last half of the twentieth century. Social problems such as ageism, chang-

ing economic burdens, and the need for changes in social policy have resulted. The need to address the issues of an aging nation is upon us. Gerontologists believe that ageism can be ameliorated through education and the changing health and lifestyle of the "new" elderly, who are more healthy and vigorous than the population of elders preceding them.

The United States is aging; so too are most other developed countries, globally. As national economies become more global, so too do population pyramids begin to approximate each other. Affluence, education, health awareness, and longevity are inseparably intertwined with the global economy. America's pharmaceuticals are already often produced in another country—France or

Germany, for example. Scientists increasingly share their research and it is not unusual to see a medical research lab with American, Canadian, French, Indian, and Italian scientists working side by side.

Key Terms

age–dependency ratio	Detriot syndrome
ageism	gerontology
aging	gerontophilia
baby boom generation	gerontophobia
birth cohorts	life expectancy
chronological/biological	population pyramids
aging psychological/	selection for survival
social aging	senescence
cohort	social gerontologists
critical perspective	sociogenic aging
cumulative disadvantage	

Questions for Discussion

1. There are different ways of looking at age. How old are you in your mental outlook on life? How old are you spiritually? Chronologically? Physically? How old are your parents in each of these categories? Your grandparents? What basis did you use for assigning the ages?

2. Close your eyes and imagine yourself growing 10, 20, 30, 40 years older…Continue until you've reached "old age" as you imagine it. How old are you? Who is still in your life from your youth? Who is new in your life? Who has passed on? What do you imagine a typical day to be like?

3. What is the "Aging Revolution" discussed in the chapter? How are you affected by it?

How is our society affected by it? How is the world affected by it?

4. What are the possible positive and negative effects of increased longevity on family life?

5. What is the impact of global aging on health and human services, the world economy, cultural development in developed and developing countries?

Fieldwork Suggestions

1. Look at a selection of birthday cards for adults. Do the cards support a positive or negative attitude toward aging? How do they promote ageism? Give examples.

2. Talk to several of the oldest people you know about their aging. How do they feel about growing older? Have the same conversation with a few of the youngest people you know. How do the responses of the young people compare to the responses of the older adults?

3. Ask several young adults how many older people they have spent time with over the last week. Ask several older adults how many young people they've spent time with over the last week or month. Do you find evidence of age segregation?

Internet Activities

Using the Internet, locate the home page for the U.S. Bureau of the Census. Go to Bureau of the Census Current Population Reports, and locate data on the number of people over 65 in your home state. How does this compare with the number over 65 for your state in the 1980s? What would the population pyramid look like for your state?

2

Stereotypes and Images

Stereotypes of Aging

Explaining Stereotypes

Breaking Negative Stereotypes

Taking Age Stereotypes to Heart
People who hold negative attitudes toward the elderly face an increased risk of heart-related ailments later in life.

By Bruce Bower

Heart-felt perils await people who hold disapproving attitudes about the elderly, a new study suggests. Young and middle-aged adults who endorse negative stereotypes about older people display high rates of strokes, heart attacks, and other serious heart problems later in life, compared with aging peers who view the elderly in generally positive ways, say Yale University psychologist Becca Levy and her colleagues.

"We found that age stereotypes, which tend to be acquired in childhood or young adulthood and carried over into old age, seem to have far-reaching effects on cardiovascular health," Levy says.

Her team describes evidence for a connection between attitudes toward aging and eventual heart health in a paper published online February 13, 2009, and set to appear in *Psychological Science*. Reasons for this association remain unclear. In earlier studies, Levy found that elderly volunteers who reported negative stereotypes about old people were more likely to display heightened physiological responses to stress and to report unhealthy habits, such as cigarette smoking.

Levy's new report "is the latest in a series of well-conducted studies by various scientists that demonstrate that individual psychological differences assessed early in life predict various health and longevity outcomes many years later," remarks psychologist Howard Friedman of the University of California, Riverside.

Further work should examine whether psychological traits already linked to physical health

as people age, such as conscientiousness, influence attitudes toward the elderly, Friedman says. It's an open question whether an association also exists between negative age-stereotypes and non-heart-related illnesses, or dying at an unusually early age, he adds.

Levy's team studied 386 people, ages 18 to 49, who were participating in a larger, long-term study of aging. In 1968, these volunteers—none of whom had experienced any heart ailments—completed a questionnaire that measured the extent to which they agreed with 16 negative age-stereotypes. These stereotypes included beliefs that elderly people are "feeble" and "helpless."

Thirty years later, 25 percent of those reporting negative age-stereotypes had suffered a heart ailment or a stroke, compared with 13 percent of those who rejected the age stereotypes. Heart-related problems commonly occurred about 11 years after participants entered the study.

To get a better handle on those whose heart-related problems did not show up for decades, Levy and her coworkers winnowed their sample down to 225 individuals, ages 18 to 39 at the study's start, who experienced an initial heart ailment or stroke after their 60th birthdays. Those who had a poor regard for the elderly were much more likely to have heart-related problems in their early 60s, before those volunteers who viewed old age positively.

These findings held after the researchers accounted for other factors that can influence heart disease, including blood pressure, family health history, depression, education, sex, marital status, total cholesterol level, and cigarette use.

Absent any detailed understanding of the ways in which age stereotypes may ultimately affect heart health, it's too early to recommend any preventive programs aimed at altering negative attitudes toward the elderly, Friedman cautions. Possible mechanisms to explain why age stereotypes affect heart health include temperament, behavior, and physiological traits.

Source: www.sciencenews.org/view/generic/id/41105/title/
Taking_age_stereotypes_to_heart

Have you ever heard the following statements or made them yourself? "Old people are narrow-minded." "They are set in their ways." "Old people are terrible drivers!" These statements are negative stereotypes. Are you familiar with the phrases "a twinkle in his eye," or "old and wise?" These statements are also stereotypes. This chapter explores stereotypes based on age and provides information to explain why they exist.

Stereotypes of Aging

Stereotypes are generalized beliefs or opinions based on individual experience, often produced by irrational thinking. Stereotyping and labeling seem to fulfill our need to structure and organize situations in order to minimize ambiguity and to clarify where we stand in relation to others. Because of the complexity of our society, we need to quickly assess situations and people, based on our beliefs or previous experience: this is a person I can trust . . . this situation makes me uncomfortable . . . this person is probably not reliable, and so forth. These assessments are our "people skills" and form the basis for making judgments and shaping many of our interactions in the larger society. When observations, however, become rigidly categorized—I cannot trust this person because she looks like an untrustworthy person I once knew—then we have fallen into making assessments based on stereotypes.

Stereotyping, whether direct or subtle, is usually inaccurate. When we generalize by putting people into categories, we tend to oversimplify reality. We ignore inconsistent information and emphasize only a few characteristics. Thus, the statement "old people sit around all day" is a generalization that does not apply to the many active older individuals who work, write, paint, are physically active, or involve themselves in community affairs.

Whether positive or negative, stereotypes are emotional impressions and are not based on objective information, and they categorize people. Stereotypes can interfere in our judgment by arousing strong and sometimes negative emotions, such as hatred or resentment. Hating or resenting any person or groups of people for any reason, but especially on the basis of a trait such as age, is both ignorant and unfair.

Positive and Negative Stereotypes

Tuckman and Lorge (1953) were among the first gerontologists to study stereotypes. Using a list of statements with which subjects were asked to agree or disagree, they found that old people were perceived as being set in their ways, unproductive, a burden to their children, stubborn, grouchy, lonely, "rocking-chair types," and in their second childhood. Since the Tuckman and Lorge study, psychologists continue to find that society stereotypes older persons (Bargh, Chen, & Burrows, 1996; Hummert, 1999; Levy, 2002; 2008; 2009). Twenty years ago Palmore (1990) summarized **negative** aging stereotypes as including (1) illness, (2) impotency, (3) ugliness, (4) mental decline, (5) mental illness, (6) uselessness, (7) isolation, (8) poverty, and (9) depression (grouchy, touchy, cranky). As the population has aged over the past 20 years stereotypes are not as negative as they were in the 1950s (Palmore, 1990) or in the 1990s when Palmore's research was published. Shifting demographics, discussed in chapter 1, result in more exposure by the general population to older individuals as well as the better general health of elders in the final decades of the 1900s.

Despite this positive shift in social attitude, negative stereotyping of the elderly remains a significant social issue. Two kinds of negativism are relatively common. One is an ageism that focuses only on the least capable, less healthy, least alert aged. This focus on the sick takes attention away from the healthy aged who defy negative stereotypes. The **biomedicalization of aging** emerged over a century ago with the growth of scientific inquiry and subsequent breath-taking advances in medical sciences. It is the belief that problems associated with aging are biological rather than social and behavioral. The "problems of aging," therefore, can be addressed only by medical technology, if at all (Vertinsky, 1991). A biomedicalized perspective of age locks the aging process and the individual experiencing it into an irreversible decline of physical deterioration. Physical

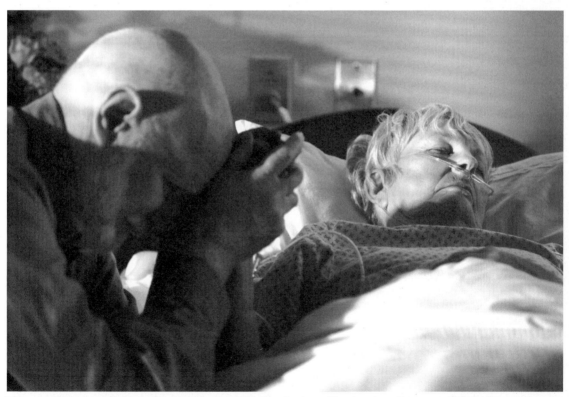

This man grieves not for his wife's declining body, but for the loss of her lovely smile, energetic behaviors, and companionship—all aspects of her spirit.

change and decline, of course, are only a piece of the process of aging yet if society and the aging adult are trapped in this stereotype of age, it limits all other possibilities for recognizing—and valuing—other physical, psychological, and spiritual growth and development in later life.

The second kind of negativism is called **compassionate stereotyping.** An early pioneer in the field of gerontology, Robert Binstock (1983) coined this term to describe images that portray all older adults as disadvantaged on some level (economic, social, psychological), in need and deserving of help by others. This may sound harmless, but consider the reaction of disability activists to posters of the Easter Seal Child—a poster designed to invoke pity for those with disabilities. All individuals, including older people, do not want pity; rather, they want the tools for being independent and self-reliant. For some the tools for independence and self-reliance are strictly inherent qualities; for others, those inherent qualities must be supplemented with external resources such as a wheelchair or hearing aid or specially designed educational opportunities. Like all people, older adults want to be seen as individuals, not as a category. Compassionate stereotypes perpetuate dependency and low self-esteem, and unnecessarily lower expectations of what older people can achieve.

A **positive stereotype** is a generalized belief that categorizes all older people in a favorable light, whereas a **negative stereotype** categorizes old people in demeaning ways. The extent to which different stereotypes elicit positive versus negative attitudes now has been studied by a number of social scientists (Crockett & Hummert, 1987; Heckhausen, Dixon, & Baltes, 1989; Hummert, 1999; Hummert, Garetka, Shaner, & Strahm, 1994; Linville, 1982). Hummert (1995) found that the two most frequently cited positive stereotypes of older people were the "Golden Agers" (lively, adventurous, active, sociable, witty, independent, well-informed, successful, well-traveled, etc.) and "Perfect Grandparents" (kind, loving, family-oriented,

generous, grateful, supportive, understanding, wise, knowledgeable, etc.). She found the most prevalent negative age stereotypes to be "severely impaired" (slow-thinking, incompetent, feeble, incoherent, inarticulate, senile) and "despondent" (depressed, sad, hopeless, afraid, neglected, lonely).

The despondent stereotype finding illustrates compassionate stereotyping, whereas the severely impaired stereotype reflects biomedicalization of aging. Like all stereotypes, positive and negative type-castings of elders emerge from a kernel of truth: older people *do* become more frail. Depression in later life *can* be a serious problem—in part because when it is categorized as a natural part of aging it goes undiagnosed and untreated as a mental health issue. The tragedy of stereotyping is that the individual becomes objectified; that objectification is internalized by the older person, and a vicious circle of loss of sense of self ensues. If a retired 78-year-old former CEO of a large corporation is treated by everyone in his environment as being incompetent or slow, that social reflection of who he or she is can profoundly impact his or her sense of well-being. We humans are interdependent creatures, deriving much of our personal sense of self from our social environment. If I am treated as being feeble, I will identify with that kernel of truth that I can see in myself, rather than identifying with my continuing strengths.

The importance of challenging negative self messages was brought to light in an epidemiological study by Levy and Banaji (2002). They found that people with more positive self-perceptions toward aging lived longer by 7.5 years than those with negative aging self-concepts! "[This] is greater than the independent contributions of lower body mass index, no history of smoking, and a tendency to exercise [and] each of these factors has been found to contribute between one and three years of added life," the authors explained (p. 273). Likewise, when older people have a role model who represents aging successfully, they show resilience in the face of negative characterizations (Horton & Deakin, 2007).

The extent to which younger people assimilate negative stereotypes of older adults is the extent to which they will have negative aging self-images (Levy, 2008). We age as we are: if I do not stereotype in my youth, I am not likely to stereotype in my adulthood, and vice versa.

The mass media is a profoundly important source of stereotypes about aging in the United States. Media influences shape the attitudes of children as well as the self-concepts of adults, and older people continue to be invisible or negatively portrayed. When they are visible, it is seldom in major roles, and we especially do not see aging women. One place where ageism is most public is in movie roles of mothers and of female love interests. Actresses often play mothers of men who are their age peers: Sally Field as Tom Hanks' mother in *Forrest Gump* is 10 years older than he. Angelina Jolie is a mom one year older than Colin Farrell in *Alexander*; Brad Pitt's screen mother Taraji Henson in *The Curious Case of Benjamin Button* is seven years Pitt's junior (http://womenandhollywood.com, 2009). *In the Line of Fire* stars Clint Eastwood, who is about 24 years older than his love interest, Rene Russo (http://www.cinematical.com, 2009).

Although the average age of Americans is increasing, most television characters are young. Older adults spend more time watching television than all other age groups (Levy, 2003; Levy, 2008). Many elders do not travel and are dependent on television for companionship and entertainment; for these older people, television is a window on the world. The more elders watch television, however, the greater exposure they have to negative images of their age group. Repeated exposure to negative images or few images with which they can identify is neither psychologically productive for the elderly nor in the long run economically productive for television networks, yet the trends continue.

Yale researchers asked older adult television viewers to maintain evaluations of how they saw themselves portrayed on television (Levy, 2003). One outcome of that study was that the older viewers were less likely to turn on the television than they had been previous to becoming consciously aware of how they were portrayed. Generally, elders found that older Americans were portrayed as forgetful and ill-tempered, or they were simply omitted. Despite numerous findings that viewers aged 50 and older were significantly more likely than younger viewers to be consumers of television news, a study of television station management described the struggle within news management regarding stories of interest to an "older" audience. In the Yale study one 71-year-old viewer said of television news: "When people were interviewed about different matters, older people were left out." Another responded "I feel like we've been ignored. I feel like we're non-existent." (Levy, 2003). It is difficult to cover many social issues of interest to an older audience with action-oriented television news. If local television were to report more stories of interest to an older audience, viewers of all ages would have a more well-developed perspective of social issues affecting a broad age range. In that way we all will be less exposed to having our stereotypes and biases reinforced through this powerful medium. The more exposure people of all age groups have to one another, the less likely are stereotypes of any sort for it is through our exposure to "others" that we are reminded of the huge range of personality, values, and behaviors of one another.

Who Is Old?

When is a person old? Do the words *old, elderly, senior, mature adult, older adult,* and *senior citizen* mean the same thing? Or do they mean different things to different people, under different circumstances? No definition of an older person has been universally agreed upon. *Old age* means different things and is assigned on the basis of chronology (e.g., age 65), of biology (how well one functions), and by social standards (the point at which, for example a woman is considered "too old" to wear a bikini on the beach). Sweeping statements about a category of people ("the elderly," "the 20-somethings," "teenagers," etc.) stereotype the individuals within that category because they bypass existing vast individual differences. Some old people are frail; some are wise; some are cranky, and some are jolly, or patient, or self-centered. We age just as we have lived—we do not become a different person because we have reached a chronological marker that makes us "old." Aging is a gradual process with many influences. The reality is that people age differently.

In addition to individual differences, the definition of "old" defies precision because over time new cohorts move into later life, and those people bring with them their unique life experiences, values, and attitudes that were shaped by shared socio-historical events (natural disasters, wars or recessions, etc.) of their times. In a classic study of life course effects of the Great Depression, Glenn Elder (1979, 1994) found notable differences in personality style between people who were young adolescents and those who were young adults beginning their families and careers at the time of the great stock market crash of 1929. Stereotypes of "cautious" and "conservative," for example, might be quite descriptive of one age cohort among the current "old," and completely inaccurate for another cohort, also currently "old."

In the twenty-first century the public face of aging has changed. Prominent women, from actresses to politicians, often appear not to grow older at all. Women are actively encouraged by their plastic surgeons to begin treatments in their 20s or 30s. Dr. Jean-Louis Sebagh, a popular plastic surgeon to prominent clients, was quoted as saying "preventing the ageing process is better, where possible, than correcting it . . . If a woman comes to me at 35 or 40 and we treat her every three to four months, I can keep her looking that way for 20 years or more" (http://womenandhollywood.com, 2009). The

public image of aging is to be forever youthful in appearance.

Botox was first used for cosmetic purposes about 20 years ago. In 2002 it won approval in the United States from the Food and Drug Administration (FDA) for the removal of frown lines (http://www.nlm.nih.gov/medlineplus, 2009). The National Institutes of Health website (2009) indicates that cosmetic surgery procedures in the United States increased by 115 percent between 1997 and 2007, while nonsurgical procedures such as Botox or Restylane Vital increased by 754 percent. The faces we compare ourselves to are unlined, the hair color is unchanging over time, and

Women's changing roles in American culture are reflected by Secretary of State Hillary Rodham Clinton, former First Lady.

the bodies older adults see of their age-peers are sculpted and bronzed. If older adults believe those images are what the aging person is "supposed" to look like, they will always come up wanting.

Longevity is behind the need for more precise descriptions of just who are the old, and what it is like socially to be "old." Those who are relatively young, about 65 to 75, are referred to as the "young-old." Older people who are vigorous, fit, and healthy have been labeled as the "able elderly." Those 80 and older are variously called the "old-old," or labeled the "frail elderly" or the "extreme aged," depending on their health and the focus of the gerontologist's work. In this century the category of "centenarian" is more frequently used by the media and the professions to address the growing numbers of people living past their hundredth year.

The Legal Definition of "Old"

In the 1890s, Germany's Otto von Bismarck established a social security system for German elders that benefited citizens 65 years or older. Life expectancy in the late 1800s was approximately 48 years for men and 51 years for women, so the political advantage of addressing the emerging Positivist social thought by establishing social security programs far outweighed any economic disadvantages.

Nearly 60 years later, in 1935, the United States passed the Social Security Act under President Franklin D. Roosevelt. In that act, according to a tradition by then established in Europe, 65 was named as the onset of old age. In line with Social Security standards, most companies as well as state and local governments developed pension programs beginning at age 65 for retiring workers.

This legal definition has become a social definition: on retirement, a person's lifestyle generally changes dramatically, creating a point of entry from one phase of life to another that has become a social event for celebration and congratulations. Retirement, in fact, is one of the few life course transitions that is celebrated throughout the United States. "Retirement age" has become somewhat standardized legally, socially, and psychologically as an initiation into "old age."

Since the 1930s, however, medical science has extended longevity and improved general health. The 65-year-old today is not the same physically or psychologically as the 65-year-old in 1935. That person today is likely to be healthier and better educated, for starters, and to be more intimately connected with the larger world through the medium of television and radio, than was his or her counterpart in 1935. Social scientists now question whether 75, 80, or 85 might more accurately mark the beginning of old age. Whatever the age, any *chronological* criterion for determining old age is too narrow and rigid, for it assumes everyone ages in the same ways and at the same time.

The Social Construction of Aging

Self-concept is the way in which a person sees himself or herself as being. It is how individuals define themselves *to* themselves, and it forms the basis for the way people maintain a sense of continuity even as their bodies age and change. It is the ongoing image we have of ourselves. Even if we see ourselves changing, that change is emerging from something that is/was also part of the self-concept. Early in the study of gerontology, Bernice Neugarten (1977) cautioned her colleagues that chronological age is an "empty variable." It is the importance of the *events* that occur with the passage of time that have relevance for the study of identity development, not time itself. Perhaps most importantly, self-concept dictates the way in which people interpret and make meaning of the events that occur in their lives. As discussed earlier, sense of self is a complex psychological function, not just one general image; our styles of coping and managing the world we live in are intimately shaped by our self-concept and that sense of the aging self has profound impact on physical and mental health (Levy, 2008, 2009; Vincent, 2008).

The **social construction of self** addresses the idea that the way we interpret events in our lives is partially a reflection of how we are treated, and partially the extent to which we have internalized the way society has defined or categorized us. So, for example, it might not be unexpected for that 70-year-old woman to believe that it is improper for her to wear a bikini on the beach. Her self-concept precludes that behavior: she sees herself as, perhaps, too dignified *at her age* to publicly expose her less-than-youthful body.

Those who see themselves as old and accept as true the negative characteristics attributed to old age may, indeed, *be* old. As sociologist W. I. Thomas (1923) stated, "If people define situations as real, they are real in their consequences" (p. 42). Current research on self-concept emphasizes the knowledge base, or structure, that helps individuals to maintain a consistent sense of who they are throughout their life experiences (Scheibe, Kunzmann, &

Baltes, 2008). Those who have a good sense of continuity of who they are appear to be better adjusted in later life. They are less likely to identify themselves as being "old," because they identify as being who they *always* have been. The dimensions of an individual's self-concept that deal with self-esteem and a sense of social worth are the very dimensions that our society is most likely to treat harshly.

A current focus in gerontology is to move away from attempting to measure "hard, scientific" data, and to listen to the voices of aging people themselves—to listen to their own narrative of their life process. This methodology is referred to as **phenomonology**, in which the meaning of an event is defined by the person experiencing that event (or phenomenon), not a researcher's hypothesis. Sharon Kaufman's classic study *The Ageless Self: Sources of Meaning in Later Life* (1987) is an excellent example of using the voices of elders to develop a description of their self-concept. "I wanted to look at the meaning of aging to elderly people themselves, as it emerges in their personal reflections of growing old" (Kaufman, 1993, p. 13). Kaufman documented the finding that many older people do not see themselves as old, but as being ageless, living *in their old age.*

The voices of older people, documented by Kaufman (1987) and in a second study by Tandemar Research (1988), best describe this ageless self:

- I don't feel 70. I feel about 30. I wear my hair the way I did then. . . . I just saw some slides of myself and was quite taken aback. That couldn't be me . . . (Kaufman, 1987, p. 8).
- The only way I know I'm getting old is to look in the mirror . . . but I've only felt old a few times—when I'm really sick (Kaufman, 1987, p. 12).
- I'm always telling my children I'm still the same inside. I just have to walk for the bus now instead of run (p. 23).

Old age is to no small extent a matter of self-concept.

Kaufman summarized the process by which people maintain a solid sense of self in later life, even as their bodies, their relationships, and their social circumstances changed over time:

> [The ageless self] draws meaning from the past, interpreting and recreating it as a resource for being in the present. It also draws meaning from the structural and ideational aspects of the cultural context: social and educational background, family, work, values, ideals, and expectations. . . . [Older adults] formulate and reformulate personal and cultural symbols of their past to create a meaningful coherent sense of self, and in the process they create a viable present. In this way, the ageless self emerges. Its definition is ongoing, continuous, and creative. (1994, p. 14)

Through the use of narrative methods, the personal experience of growing old is increasingly captured by theorists as well as the media. The cognitive act of constructing a story of one's experience for another person to hear helps the narrator to organize his or her sense of that life; it additionally helps the listener to release preconceptions (stereotypes).

Through narrative techniques the concept of **possible selves** has emerged: we have a sense of who we were, of who we are presently, and of who we are becoming (or might become if we are not careful [Bauer & McAdams, 2004; Carstensen, 2006; Markus & Herzog, 1991; McAdams & Pals, 2006; Nielsen, Knutson, & Carstensen, 2008]). This image or projection of who we might become with time can be positive, hoped-for selves, or negative, feared selves. Not surprisingly, older adults experiencing a threat to their health can project a negative possible self: I could become that bent, osteoporotic 80-year-old woman if I don't get my act together! Both positive and negative possible selves can be very motivating for making useful and appropriate behavioral change.

Studies of Children's Attitudes

Since the mid-1960s interest has grown in studying attitudes young children and adolescents have toward the aged. The body of research developed since that time has produced varied and contradictory results, but uniformly shows that children do tend to stereotype older adults, but those stereotypes have become incrementally less overtly negative in the past 30 years.

Children develop stereotypes about elders and the aging process at an early age. These attitudes toward groups of people are shaped by various outside forces: families, social interactions with peers, school influences, and media influences (Silverstein & Ruiz, 2006). Children as young as 3 exhibit ageist language (c), and by age 5 begin to have clear attitudes about aging and being old (Hoyer & Roodin, 2009). In adolescence, children appear to stereotype and adopt values in a way similar to adults (Carstensen, 2006; Carstensen, Mason, & Caldwell, 1982).

The media is a primary source of ageist messages for children. Even fairy tales instill ageist feelings in the very young. Evil, ugly, old witches and mean old stepmothers endanger children in many stories. Some books are changing this theme, emphasizing, for example, the grandparent-grandchild connection. In a study of children in four countries (Australia, England, North America and Sweden), Goldman and Goldman (1981) found that children believe that physical, psychological, socioeconomic, and sexual powers of old people decline, and that old people's skills become less useful with age. Five-year-olds in a Canadian study showed preference for seeking advice from younger, not older adults (See, 2003). These views reflect a negative stereotype based on a biological model of decline—a decline which, of course, children can observe by watching their grandparents age and die, for example. It is particularly important that children be provided greater opportunities to develop positive interactions with older people, in part because their internalized attitudes about old age will have a significant influence on their own lives and the ways in which they see themselves as they age.

Changes in Attitude

The ways in which attitudes can be changed has been extensively examined in the past 50 years (Amir, 1969; Bengtson & Achenbaum, 1993; Corbin, Metal-Corbin, & Barg, 1989; Sanders, Montgomery, Pittman, & Blackwell, 1984). Three primary ways to impact attitudes were identified in 1982 by Class and Knott as being (1) through discussions with peers, (2) through direct experience with attitude objects, and (3) through increased information or knowledge. This model has been used in several projects for school-based programs to change children's attitudes toward the elderly. In one very well designed classroom experiment (Aday, Sims, McDuffie, & Evans, 1996), older volunteers were paired with children to assist with a school-based task. The interactions also included structured and informal discussions designed to foster more intimate relationships between the pairs, and it tested attitude changes in one-year and five-year follow-ups. They found as a result of the interactions, the children and youth developed more positive attitudes about the aged and about aging, and those attitudes remained part of the children's perceptual schema five years later. Responses by the children to the question "What have you learned from the intergenerational project?" included:

I never knew some older people were so active. . . . That not all older people are mean and stingy. . . . I learned that older people were once a young person just like me. I never thought about it before. (p. 150)

In response to the question "How has this project changed you?" some responses were:

I learned that everyone on this earth is equal, no matter what their age. . . . I'm not so afraid of older people. . . . It has changed my outlook on older people to a more positive one. . . . I'm not as scared of growing old as I used to be. . . . I'm more apt to smile and speak to older people . . . (Aday et al., 1996, p. 150)

Teachers can improve children's attitudes toward old age by telling them about the physical and mental capabilities of old people. Teachers can present accurate information about old age, and they can bring active, creative elders into the classroom. Teachers can also help children explore attitudes toward older people as they explore cultural and ethnic differences among classmates' backgrounds. Children are generally encouraged to follow the traditional values of their families' cultures, whatever that cultural heritage might be (Zandi, Mirle, & Jarvis, 1990). Discussing different ways parents and grandchildren interact with grandparents is a wonderful way to honor cultural differences and to identify and explore different attitudes about aging.

Studies of Young Adults' Attitudes

Understanding the nature of how younger adults view older adults is important because those attitudes will impact intergenerational relationships, the level of concern for social programs that benefit older adults, and the self-concept of that younger adult as he or she matures into middle age and later life.

Studies of college student attitudes toward aging, like those of the population in general, show very mixed results. Some of this lack of consistency is methodological—different researchers ask different questions, and they often are not actually measuring the same thing. Attitude is a multidimensional mental schema, influenced by many factors, including exposure to older people, gender, culture, and individual differences of personality. "The elderly," too, do not comprise a single category: vast individual differences exist among elders. A second reason for mixed findings on young adults' attitudes toward their elders is that older adults may be seen to some extent as multidimensional people, with both positive and negative attributes (Slotterback & Saarnio, 1996).

The most generalized stereotype among college students is based on a biological model of decrement and excludes personality, skill, and interactional factors. This can be a particularly problematic perspective among students training for the health and healing professions. In a study conducted by AARP and the University of Southern California in Los Angeles in 2005, 66 percent of college-aged students believed that most older people didn't have enough money to live on (compassionate stereotype), and around 61 percent believed that loneliness is a serious problem for older people, when in fact only 33 percent of people over 65 agree (Gotthardt, 2005).

The attitudes of pharmacy students (Shepherd & Erwin, 1983), medical students (Intrieri, Kelly, Brown, & Castilla, 1993), nursing students (Downe-Wamboldt & Melanson, 1990; Rowland & Shoemake, 1995), and college students;) Shoemake & Rowland, 1993) have all been studied and surveyed. A conclusion drawn in one study of nursing students (Rowland & Shoemake, 1995) also summarizes the findings of other studies:

Change comes in small increments, but changes in attitudes come only after learning has occurred. Providing students opportunities to test their professional skills and meet challenges to their preconceptions of the elderly can increase the rate of change. (p. 747)

Barrow (1994) conducted a study of college students in which they developed their own descriptions of older people. In this study, stereotypes emerged directly from students' minds, phenomenologically in method. A salient finding was that students paid the most attention to the changing physical appearance and capabilities of older people. Many students responded to the word *aged* with a description of physical decline. Some responded to the word with a social position or role description, such as "bad driver," "interesting life stories," "bingo," "grandparent," or "calm life." Still others included psychological qualities such as "wise," "lonely," "fear of death," or "experienced." The students had a relatively balanced mix of both positive and negative stereotypes— for example, an equal number of responses of "wise" and "lonely."

In response to the question "What are your general thoughts or views about the aged?" students again displayed a range of attitudes, from negative to neutral to positive, and including positive, compassionate, and negative stereotypes defined earlier in the chapter.

- Their late years may be years of dependency on physicians, drugs, institutions, and ambulatory medicine. The aged sometimes do not live their last few years of life with great happiness; it is a time of stress and unhappiness, pain, sorrow, weeping, and reversals in memory.

Learning environments can provide a rich source of interaction between age groups.

- They have a lot to teach yet don't have a fair role in society. Some people, what I think, are just waiting to die because they're lonely. A lot of people put them aside. Cause—the amount of attention the elderly need seems to be a burden to a lot of families.

- They have done and seen so much. I view old people sort of as old cars, with some parts broken, but others running as smooth as ever. I feel it is always worth it to stop and talk to an elderly person. You never know when they could teach you a valuable lesson.

Explaining Stereotypes

We categorize. On meeting or merely seeing a person, we determine the person's age, race, gender, perhaps other social categories like socioeconomic status (Whitley & Kite, 2006). We can explain the existence of stereotypes on a number of levels. On one level, the historical/cultural explanation requires the gerontologist to look at the roots and cultural context of our concepts about old age. On another level, current social explanations look at elements such as social class and the influence of the media. On yet another level, psychologists ask why some individuals, either young or old, accept the negative stereotypes of old age whereas others accept the positive ones.

Ageism—age stereotyping—is developmental: it develops and is reinforced across the lifespan and is directed toward the youth as it is toward older adults. Children are told they are "too young" to participate in a particular activity or use particular language. Their task is education. Adolescents are dismissed as serious participants in social problem solving by those older than they because adolescents are seen as *evolving*. They haven't yet "arrived" and cannot fully be trusted (Gordon, 2007). Their task is to find out who they are. Adults are seen as those responsible for the productivity of the working world—society's economic structure. Rule makers, power mongers, and shapers of society. Older adults are expected to retire into a life of leisure, thus leaving space in occupational and political roles for those emerging into adulthood (Bytheway, 2005). From a positive stereotype model those older adults will evolve into wise elders, moving into a life of contemplation. Negatively projected, those older adults will become demented and a burden to family and society.

Ageism is also institutional: social policy establishes required times for school, work, and retirement. This results to no small extent in age segregation for most of the lifespan. Parents' productive work is carried out in settings that exclude both children and those now retired (Hagestad & Uhlenberg, 2005). School settings and work sites exclude older adults. Much of health care is also age differentiated. Concern for children's health will be a driving force as America grapples with the political polemics of social medicine versus healthy children who will grow into healthy adults.

Ageism is spatial. In the twenty-first century Americans increasingly occupy different spaces, based on age. Residential segregation through retirement communities, continuing care retirement communities (CCRCs), and nursing homes predominates. This reduces face-to-face interaction, thereby increasing each age group's sense of being "other."

Institutional, social, and spatial segregation result in cultural age segregation. Our language reinforces this: "youth culture" and "elderly" increase us/them distinction among age groups, thereby increasing our sense of categories (Hagestad & Uhlenberg, 2005). This becomes highly significant on individual behavioral levels. If I am "old" and you are not, we are separate—and vice versa. Ageism is therefore a two-way street. In one study young American students found themselves to be frequent targets of patronizing stereotypes by middle-aged adults (Giles & Reid, 2005). Further, those stereotypes were reported as being bothersome and unwelcome.

How do people manage ageism when they are on the receiving end? From research on racism and sexism, we find that confrontational reactions by victims can be seen in the social world as *overreaction* and therefore be easily dismissed (Czopp &Monteith, 2003). Ageism, like racism and sexism, often lies at an unconscious level: the ageist (or sexist or racist) would never consider himself or herself that. It is not uncommon for ageist behaviors to emanate from older people themselves, who—because of a life course steeped in ageism—hold the very beliefs being attributed to them. The young child who learns he or she is excluded from certain activities because he or she is "too young" grows into the middle-aged adult and the older adult who believes the prevalent stereotype of himself or herself (Hagestad & Uhlenberg, 2005; Harwood & Giles, 1993; Montepare & Zebrowitz, 1998).

Age puzzles me. I thought it was a quiet time. My seventies were interesting and fairly serene, but my eighties are passionate. I grow more intense as I age.

FLORIDA SCOTT-MAXWELL

Historical and Cultural Explanations

Understanding the relationships between generations and exploring views about growing old in a previous era are the jobs of a historian. For example, a historian in Chapter 1 described how, throughout American history, society's view of elders shifted from one of veneration and favor to one of scorn. Reasons for the change are extraordinarily complex, related to larger philosophical issues, influences of multiple cultures brought about by immigration, advancements in science and medicine, longevity, global and local economic structures, and the shift of cultural values as a result of all those factors.

Rude, insulting, or negative labels directed to anyone whom we consider "other" is not new, however, and seems to endure throughout history. The words we use to describe people provide a basis for the formation of stereotypes, and people often internalize those labels, incorporating them into their own self-concepts. Language is powerful: it shapes consciousness, and our consciousness affects our health and well-being as well as our interpersonal relationships.

Becca Levy and Ellen Langer (1994) identified a dramatic relationship between cultural beliefs and the degree of memory loss people experience in old age. They conducted memory tests with (1) old and young mainland Chinese, (2) old and young from the American Deaf culture, and (3) old and young hearing Americans They expected that the Chinese and Deaf cultural groups would be less likely than the hearing Americans to be exposed to and accept negative stereotypes about aging. They found that younger subjects, regardless of culture, perform similarly on memory tests. Older Chinese and older Deaf participants, however, outperformed the older American hearing group. They described their findings:

> A social psychological mechanism contributes to the often-reported memory decline that accompanies aging. . . . The negative stereotypes about how old people cognitively age, to which individuals starting at a young age are exposed, become self-fulfilling prophecies. (p. 996)

An unexpected finding of the research was that the old Chinese sample performed similarly to the young Chinese sample. The scores for the two groups did not differ significantly even though the memory tasks being used typically reflect memory loss with age in the United States. The results were even more surprising because the older Chinese group had completed fewer years of education than the young Chinese and the old and young American samples. Of the three groups studied, the Chinese reported the most positive, active, and internal image of aging. Because of this, the authors conclude that "the social psychological component of memory retention in old age may be even stronger than we believed" (p. 966). In other words: the internal self-concept, reflecting as it does society's judgments, directly impacts memory processes. If I believe on a very deep level that I am less competent now at age 82 than I was at age 32, that prophesy becomes self-fulfilling.

Use of ageist language is hardly unique to the twentieth century. Aristotle (reprinted, 2007) famously defined old age as a negative time of hopelessness, with little remaining but memories.

They live by memory rather than by home; for what is left to them of life is but little as compared with the long past; and hope is of the future, memory of the past. (quoted in Rhetoric, 2007)

Herbert Covey (1991) beautifully illustrated the extent to which a culture's religious and philosophical values play out in stereotypes and/or expectations in his history of the term *miser*. Covey associates the miser, a social role meaning a mean or grasping person, with "avarice," one of the seven deadly sins and thought for centuries to be the chief sin of old age. He traced the image of the miser from classic literature and religious thought through to modern times, exploring both literature and art. He concludes:

There are justifiable reasons why older people have been associated with miserly behavior, such as the need to be frugal to ensure their future survival. Other reasons accounting for this association have been proposed, such as the reluctance of older

parents to surrender family wealth to their demanding offspring. In addition, social support programs for older people were not readily available and stigma was sometimes attached to those receiving benefits. The depictions of older people dying while surrounded by their worldly possessions also fueled the image of the miser. Older people were expected to surrender their worldly concerns and possessions in order to enter heaven. Those who were reluctant were viewed as misers and avaricious. (p. 677)

Present-day researchers observe few age-specific terms that refer positively to older people. Several examples are *mature, sage, venerable,* and *veteran.* A study conducted in the 1980s of the language of aging found that even the terms *aged person* and *elderly* were considered less than positive (Barbato & Freezel, 1987). This language bias continues more than 30 years later—and it must be remembered that *language structures consciousness.* We use the words (language) we have to describe ourselves and others, and in doing so, reinforce the meaning of those words. Today as in years past, presidents of companies or people in a position of power typically do not want to be called "aged" or "elderly." Their negative connotation is cultural and based on decades—even centuries—of negative beliefs about aging. The core of the problem is that as long as there are negative attitudes about aging, even initially positive terms may develop into negative stereotypes. Currently the term *older adult* is the most positively perceived label: it reflects adulthood as well as age, in the same way that "middle-aged" does.

Historians have examined magazines, newspapers, poetry, sermons, and other written materials for information about aging in prior times. For example, sheet music of the 1800s and 1900s reflects the then-popular sentiments about age (Cohen & Kruschwitz, 1990). With few exceptions, writers of tunes popular in the late nineteenth and early twentieth centuries saw old age as a time of failing capacities, the clear preference for youth, and the growing dread of becoming old. In these songs, elders fear that their children will abandon them; they worry about spousal death, loneliness, disability, and their own deaths. "Silver Threads Among the Gold" (1873), a classic of the period, is a touching song that emphasizes the declines in old age. Another example is the song "Old Joe Has Had His Day" (1912).

A whole series of songs, such as "Will You Love Me When My Face Is Worn and Old?" (1914), echoes the fear of loss of attractiveness. Perhaps the most poignant of all is "Over the Hill to the Poor House" (1874). It ends with the following lines:

For I'm old and helpless and feeble

The days of my youth have gone by

Then over the hill to the poor house

I wander alone there to die.

In contrast, only a few songs during this time period or currently celebrate positive aging—growing old together and being "young at heart," for example.

Songs of recent decades continue the themes of the past. A well-known song, the Beatles' "When I'm 64," carries ambiguities about aging: although anticipating the joys of growing old together with his wife, the singer has doubts: "Will you still need me, will you still feed me, when I'm 64?" The Alan Parsons Project sees aging as a time to simply bid life farewell in "Old & Wise" (1982): "As far as my eyes can see/There are shadows approaching me." Bette Midler's "Hello in There" evokes the compassionate stereotypes with ". . . but old people—they just grow lonely, waiting for someone to say 'Hello in there; hello.'" One of the biggest country songs of 1990, "Where've You Been?" by Kathy Mattea, finds an old woman lying helplessly in a hospital, waiting for death and a last visit from her husband. We do not hear the full range of the status of elders in popular music, but we do see some historical roots of both

acceptance and fear. We also see various stereotypes, most of which are negative, but some of which represent positive stereotyping.

The fears of aging expressed in the songs at the turn of the twentieth century, such as going to the poor house to die, were more valid then than now. Life expectancy was lower and so was overall health status. Before Social Security and legislation requiring pension options for spouses following death of the pensioner, old age as a nonwage-earning woman was quite bleak. Resignation and sadness were more appropriate to them. Older people are now leading healthier, more active lives. Legislative protection of pension investments for widows and widowers has improved financial security for older nonwage-earning adults. The negative stereotypes still present in popular songs are an example of the **cultural lag** that makes our attitudes and cultural beliefs slower to change than the technology and social awareness that has improved our longevity.

Social Forces: The Media

Sociologists study present-day situations to find explanations for negative stereotyping. The media that both reflects and creates society's views has a strong impact on our views of life. *Redbook* magazine's bold-faced title on its cover read: "When It's Smart to Lie About Your Age." The article cited sexual attractiveness and career pressures as reasons to lie. The author said, "We all want to be young. Despite feminist assertions, 20 implies desirable, attractive, sexy and 40 doesn't" (Peters, 1994). This article reinforces a fear of aging in its readers.

The obvious message is that by the age of 40, aging has taken an insurmountable toll on women. The increased influence of tabloid news has made a profound impact on American and Western culture in general. The impact of tabloid journalism to information on the Internet cannot be overstated. Movies can have a tremendous influence of the quality of our lives. We often dismiss them as mere entertainment, not understanding their impact on the culture we live in—on our children as they grow into adolescence, young adulthood, and then as they will age as older adults.

Television. In 1972, Lydia Bragger, the articulate former public relations director for the Rhode Island State Council of Churches, met Maggie Kuhn, founder of the Gray Panthers, a group organized to fight for the rights and interests of older persons. Following their meeting, Bragger organized the **Media Watch Task Force,** supported by the Gray Panthers, to identify and protest television programs that present stereotypical and unrealistic portrayals of elderly people. In an interview in 1975, Bragger expressed her outrage at television's portrayal of older people. "Look at how rarely we see older couples on TV sharing affection or, heaven forbid, making love," (Hickey, 1990). Two decades later, discussing her role with Media Watch, Bragger discussed television's negative portrayal of elders. "People would watch old people in the commercials who did nothing but take laxatives and use denture cream. . . . On TV, they were shown as helpless, toothless, sexless, and ridiculous. They were walking in a certain way, dropping things, forgetting things" (Tanenbaum, 1997). In the late 1980s, Media Watch disbanded: television had moved from an industry in which younger actors portrayed older ones because it was assumed that older people would not be able to remember to read their lines, to one in which older people were far more well represented.

Industrial standards have improved the representation of older people, based on greater understanding of the shifting market potential. The Nielsen Television Index did not even regularly identify individuals over 55 years of age until 1977, but now that population is carefully examined (Greco & Swayne, 1999). There is a great distance to go, yet, before ageist stereotypes are eliminated from advertising and programming. A review of the literature shows that much of the information about aging that informs advertising decisions is based on research conducted in the early 1970s and 1980s, emphasizing the biological model of cognitive decline. This is another example of a cultural lag between current research in aging and the incorporation of that information culturally.

Older women are particularly slighted by television media that consistently underrepresents them. Older female news anchors are a rare species, in contrast with their male counterparts. In a widely publicized, successful lawsuit in the 1990s, Christine Craft sued a Kansas City broadcasting station for sex discrimination, claiming she was demoted for being "too old, unattractive, and not deferential enough to men." The "elderly" Ms. Craft was 38. Female broadcasters in England have spoken out more recently against what is known as "sageism": sexism and ageism combined. In 2007 well-known British news anchor Selina Scott won an age discrimination suit, giving publicity and voice to the issue of television media ageism.

Despite the increased exposure of some older performers, including older women, television still has problems with its portrayal of older Americans. Television programming commonly and unfortunately uses a comedy gimmick—a **reversed stereotype of aging.** A *reversed stereotype* refers to older characters driving race cars, break-dancing with great abandon, or referring to their amazing sex lives. The introduction of sexual performance-enhancing drugs such as Viagra made a huge impact on images of older men (and women) in advertising. Such images were initially intended to be comical or outrageous, because they were in stark contrast with the stereotype of a low energy, sedentary lifestyle. By 2009, Viagra ads depicted sensual older couples with beautifully sculpted bodies, gazing romantically at each other. "Oh, I'm so happy," declares the woman to the man as he smiles toward the camera to confirm his sexual prowess over women to the audience. The advertising image has shifted from one of giggly discomfort at the thought of older adults' sexuality to one of an image of perpetual youthful sexuality.

The public tends to believe what stereotypes, reversed or not, present. Laughing at a reversed stereotype demonstrates unconscious, uncritical acceptance of the underlying negative image, as does viewing sexual functioning and attraction among older people to be outrageous in a humorous way. Older viewers are even less critical than the young. Older people as a general segment of the viewing public do not complain about their television image. Older men, on seeing an advertisement for Viagra, likely think "Good for him!" and compare themselves to the image of the sexy, satisfied older man. Older women whose partners are interested in performance-enhancing drugs are likely to focus attention on choices they or others now have, and compare *their* experiences to the image.

The unspoken downside of organ-enhancing drugs can be seen medically. Pelvic inflammatory disease has reached epidemic proportions among women whose partners take these drugs; however, the media does not inform the public of these findings. We once again look at the biomedicalization of aging: normal age-related changes, including sexual functioning, cannot merely be addressed by taking a pill. Being an aging person is not to be "fixed." It is to be explored and recognized for what it is, which can be and is for many older people an extraordinarily sensual time of life. We explore this concept in greater detail in Chapter 7.

The television industry tends to portray of older Americans in ways no longer consistent with who elders see themselves as being. The challenge of television is to offer a true portrait of the elderly—and that portrait is a constantly changing one, as one cohort replaces the next and the oldest cohort grows larger each decade. A sensitive, realistic portrayal is the goal. On the one hand, older adults must not be demeaned; on the other hand, television must not gloss over the real issues of aging. Portrayals should attempt to balance strengths and satisfactions with the real problems of being an older person in this society.

Television viewing time increases with age. To isolated persons, television commonly acts as a companion. The widowed and lonely often prefer programs that emphasize family solidarity and a sense of belonging. To continue to be relevant in a social demography that is becoming dramatically older, television programming will need to respond to the numbers, the values, and the perspectives of an older audience. It will be an interesting ride to watch the ways in which images and stereotypes shift in the next decade, as a new cohort of active, healthy, and well-educated people move into later adulthood. Television programming in the United States that features older characters portray them in the role of the parents of a main character (*Everybody Loves Raymond; Seinfeld; The Simpsons; Frasier*, for example). British public television, however, better represents a broad age range including two outstanding programs: *Keeping Up Appearances* in which a ditsy mom is the main character; and *Judging Amy* in which Tyne Daly plays a brilliant mother. The parental role remains a theme, but the character development is age appropriate.

Advertising and Nonverbal Communication

Television advertising that urges the public to cover up the signs of aging is particularly powerful. Advertising currently tells us that aging is primarily ugly, lonely, and bothersome unless of course one looks and acts young. Advertisers create markets by instilling a fear of aging or by capitalizing on already existing fears. Commercials imply that the elderly are sluggish and preoccupied with irregularity, constipation, and sexual performance. As a group, they suffer from headaches, nagging backaches, and loose dentures. Men are offered alternatives to baldness; both sexes are urged to buy products that will "wash away the gray"; women are urged to soften facial wrinkles and smooth "old-looking" hands with creams and lotions.

Public images of older and younger people have a powerful impact on self-concept. No one wants to identify with impending frailty and incompetence.

Advertisers are beginning to catch on, however, and a more positive view of aging is emerging. Some active, happy older people are appearing in commercials on television and in magazines. As the market demographics shift, manufacturers are aware that older people are consumers; correspondingly, advertisers are devoting more commercial time and space to elders as their numbers and buying power increase.

Advertisers, however, must be careful not to alienate their target audience. One study showed that consumers in their 50s and 60s respond best to an actor around age 40. If the actors are older, advertisers may inadvertently be targeting the elderly parents of the 50-plus group (Greco & Swayne, 1992). The

internalization of those negative messages has worked to cause some of the 50-year-olds—those who might theoretically be opposed to ageism—to ignore ads if the actors are too old. They do not want to identify with the "old" faces they see on the screen.

Ageist messages are nonverbal as well as verbal. The use of **patronizing communication** (over-accommodation in communication based on stereotyped expectations of incompetence) can be as offensively ageist as directly derogatory language (Ryan, Hummert, & Boich, 1995). According to **communication accommodation theory**, people modify their speech and behavior based on their assessment of their communication partner. If that assessment is based on erroneous stereotypes, the communication can be patronizing and produce the opposite effect than that intended (Coupland, 1991).

Critics of television ads using older people observe the image of foolish, out-of-it elders. In a classic Doritos chips advertisement a gray-haired woman shuffles along munching on chips, oblivious to a steamroller behind her. Chevy Chase comes to the rescue to save the chips, and the woman is plowed into wet cement. In another spot for Denny's restaurants, an oldster stumbles over the name, repeatedly calling it Lenny's. In these examples, the communicated image is distinctively ageist. In another ad, a younger woman looks kindly and sympathetically at an older woman, pats the older woman's hand, and explains that the product she is being served is "good for you." The image is controlling, the younger woman's verbal communication evokes images of a young child, and the overall nonverbal behavior is patronizing and insulting (Kemper, 1994). Carol Morgan, president of the marketing group Strategic Directions, sees a "huge lack of sophistication" among people advertising to the mature market (Goldman, 1993). This is changing, however slowly, as the population of older adults increases.

Movies

For most makers of feature-length commercial films, a major aim is to reach young people, particularly males of ages 16 to 24. Consumer studies conclude that most moviegoers are teenagers and young adults; studios market their products accordingly. Many commercially successful movies seem to be rather mindless entertainment with a focus on high-speed chases, violence, and sexual encounters. Exceptions to the standard movie formula are rare. The typical movie reflects the larger society in which youth holds much more promise than age. It is the exceptional movie that stars older persons and promotes understanding of the challenges and joys of aging. Shifts in this pattern are becoming more evident, however.

Through her Internet blog, The New Old Age Blog, (http://newoldge.blogs.nytmes.com, 2008) Jane Gross requested and received from readers their lists of movies about old age. She was particularly interested in compiling a comprehensive list in which aging is portrayed with richness and fullness in the same way movies about younger adults often illustrate spiritual and emotional development. With the help of her readers and Dennis McCullough, a geriatrician, the movies listed in Table 2.1 represent collected favorites produced in years from 1937 to 2007. Though this is far from a scientific study, note that the proportion of movies addressing later life in a positive way increases roughly as the boomer generation has aged.

These movies describe aging, care of the ill, spousal relationships, friendships over the years, and intergenerational relationships attempting to portray older people as complex characters, not caricatures. The list contains some excellent examples of character growth and development. For example, the 1993 movie *Shadowlands* portrayed writer C. S. Lewis (played by Anthony Hopkins) as someone willing to risk a new relationship, and as having depth to his person. Paul Newman, at age 70, brought insights into the possibilities for healing family relationships in his role as Sully in the 1994 movie *Nobody's Fool*. Richard Farnsworth's portrayal of Alvin Straight in *The Straight Story* (1999) provides an outstanding example of the deepening of character over time that can develop in older adulthood.

The Psychology of Prejudice

Those who hold negative stereotypes of aging are prejudiced against older persons. The two variables go hand in hand. To explain why an individual would subscribe to negative stereotyping is to explain why a person is prejudiced. The **psychology of prejudice** draws attention to the psychological causes of prejudice, as opposed to social causes previously discussed, such as TV and magazine advertising.

TABLE 2.1

Silver Hair on the Silver Screen*

Movie Title	Year	Director	Movie Title	Year	Director
Make Way for Tomorrow	1937	Leo McCarey	Madadayo	1993	Akira Kurosawa
Umberto D.	1952	Vittorio De Sica	Nobody's Fool	1994	Robert Benton
Tokyo Story	1953	Yasujiro Ozu	Antonia's Line	1995	Marleen Gorris
Wild Strawberries	1957	Ingmar Bergman	A Thousand Acres	1997	Jocelyn Moorhouse
The Shameless Old Lady	1965	Rene Allio	Safe House	1998	Eric Steven Stahl
The Lion in Winter	1968	Anthony Harvey	Buena Vista Social Club	1998	Wim Wenders
I Never Sang for My Father	1970	Gilbert Cates	To Dance with the White Dog	1994	Glenn Jordan
Harold and Maude	1971	Hal Ashby	The Straight Story	1999	David Lynch
Kotch	1971	Jack Lemmon	Innocence	2000	Paul Cox
Harry and Tonto	1974	Paul Mazursky	Iris	2001	Richard Eyre
Queen of the Stardust Ballroom	1975	Sam O'Steen	About Schmidt	2002	Alexander Payne
The Sunshine Boys	1975	Herbert Ross	Secondhand Lions	2003	Tim McCanlies
Being There	1979	Hal Ashby	Saraband	2003	Ingmar Bergman
Tell Me a Riddle	1980	Lee Grant	Calendar Girls	2003	Nigel Cole
On Golden Pond	1981	Mark Rydell	The Memory of a Killer	2003	Erik Van Looy
The Battle of Narayama	1983	Shohei Imamura	The Gin Game	2003	Aaron Brown
The Trip to Bountiful	1985	Peter Masterson	Aurora Borealis	2004	James Burke
Cocoon	1985	Ron Howard	The Notebook	2004	Nick Cassavetes
Foxfire	1987	Jud Taylor	Mrs. Palfrey at the Claremont	2005	Dan Ireland
The Whales of August	1987	Lindsay Anderson	Elsa and Fred	2005	Marcos Carnevale
Driving Miss Daisy	1989	Bruce Beresford	Boynton Beach Club	2005	Susan Seidelman
Dad	1989	Gary David Goldberg	Venus	2006	Roger Michell
Everybody's Fine	1990	Giuseppe Tornatore	Away from Her	2006	Sarah Polley
Strangers in Good Company	1991	Cynthia Scott	Starting Out in the Evening	2007	Andrew Wagner
Tatie Danielle	1991	Etienne Chatiliez	The Savages	2007	Tamara Jenkins
Grumpy Old Men	1993	Donald Petri	Love in the Time of Cholera	2007	Mike Newell
Wrestling Ernest Hemingway	1993	Randa Haines	Evening	2007	Lajos Koltai
			The Bucket List	2007	Rob Reiner

*Adapted from Gross, Jane (2008). New Old Age. http://newoldage.blogs.nytimes.com/2008. Retrieved May 30, 2009.

One psychological explanation is self-concept. Someone having a positive self-concept may be less prone to believe the negative stereotypes of other groups. And when that person ages, he or she may well choose to accept only positive stereotypes of age. Psychologists use the term *projection* here. If we feel negative about ourselves, we project it on to others. This might explain why prejudice against elders correlates with one's personal degree of anxiety about death (Palmore, 2001).

Three well-known theories that explain racism may also be used to explain ageism (Tornstam, 2006): (1) the **authoritarian personality,** in which less-educated, rigid, untrusting, insecure persons are the ones who hold prejudices; (2) the **frustration-aggression hypothesis,** in which those who are frustrated, perhaps by poverty and low status, take it out in aggression toward others; and (3) **selective perception,** in which we see what we expect to see and selectively ignore what we do not expect to see. Our perceptions then confirm our stereotypes. For example, perhaps we "see" only old drivers driving badly. We do not "see" young drivers mishandling a vehicle. Nor do we "see" all the old drivers who do well. In fact, we may perceive as "old" only those who are stooped, feeble, or ill.

Breaking Negative Stereotypes

The negative stereotypes of age must be disproved if we are to have a true picture of older people. One way to do this is to draw attention to people who have made significant contributions in their old age. Michelangelo, Leo Tolstoy, Sigmund Freud, Georgia O'Keefe, Pablo Picasso, and Bertrand Russell, for example, continued to produce recognized classics until the end of their long lives. Other prominent men and women are still working productively at relatively advanced ages. Willie Nelson, Jane Goodall, Nelson Mandela, Henry Kissinger, Billie Jean King, Madeleine Albright, Archbishop Desmond Tutu, Buzz Aldrin, Clint Eastwood—each of these men and women continues to make rich social contributions well into their 60s and 70s. Numerous Nobel Prize winners every year are 65 or older. As a society we need to be saying, "What have you learned? What can you teach those of us who aren't there yet?"

Emphasizing the Positive

Negative stereotypes must be countered with accurate information. For example, the myth that elders as a group suffer mental impairment still persists. More specifically, commonly held beliefs are that the mental faculties of older people decline, and that most old people are or will become senile. However, longitudinal studies of the same persons over many years have found little overall decline in intelligence scores. Older people are just as capable of learning as are younger people—although the learning process may take a little more time. One longitudinal study of intelligence in subjects ranging in age from 21 to 70 showed that on two out of four measures intelligence *increases* with age. The authors conclude that "general intellectual decline in old age is largely a myth" (Baltes & Horgas, 1996).

The stereotype that most old people are senile simply is not true. Proportionately few ever show overt signs of senility. Those who do can often be helped by treatment. Although mental health is a problem for some, only a small percentage of the elderly have Alzheimer's disease or any other severe mental disorder. This will be addressed more fully in a later chapter.

Physical stereotypes are as common as mental ones and are just as false. More positive images are replacing the "rocking chair" stereotype of old age, as older Americans stay more physically active and fit. The physical fitness craze has not been lost on the over-65 generation. Aerobics classes, jogging, walking, tennis, golf, and bicycling have become very popular among this group. Of greater importance, however, is that the stereotypic model of sedentary later life is becoming flexible enough to include the possibility of *choice*: older people can choose activities for themselves just as do younger people. Those activities for *any* age group must be appropriate to physical stamina, interest, and availability.

Many sports now have competition in senior divisions. Tennis is one example. It's never too late for a shot at Wimbledon—"Senior Wimbledon West," that is, held annually in the western part of the United States. Divisions of this tournament, for both men and women, exist for those in their 50s, 60s, 70s, and 80s. Golf, swimming, cycling, bowling, softball, competitive weight-lifting, and basketball have senior events. Sports and physical fitness can extend throughout one's life.

Vigor and activity are central to a positive quality of life in later life.

The key ingredient to a long, full life, according to psychologist Lee Hurwich, a wise elder in her 80s, is not physical health but attitude (Opatrny, 1991). With an attitude of passion about life—whether this passion is found in career, friendships, or interests—a person can enjoy some of the best and most rewarding years in life. Hurwich interviewed active, socially committed women, and discovered that her subjects live in the present, squeezing from daily life all its enjoyment. They had relationships with people of all ages. Many had suffered physical afflictions that would send most people into despair, but they had optimistic attitudes and a trust in people. One woman was studying Spanish at age 87. Why? "I want to keep the cobwebs out of my head," she told Hurwich. As they reached their 80s and 90s, Hurwich's subjects still felt a richness in their lives, and they were satisfied with lives that had meaning for them.

Consequences and Implications of Stereotyping

Negative stereotyping of old people has detrimental effects on society in general and on old people in particular. First, negative stereotyping perpetuates ageism in our society. Ageism increases when society views all old people as senile, decrepit, and rigid. These and other negative stereotypes, which do not apply to the majority of elders, reinforce prejudice and lead to discrimination. Perpetuating ageism often results in polarization (a feeling of "us" against "them") and segregation. One student in an unpublished study (Barrow, 1994) had this to say:

> I can't stand old people and I don't get along with them at all. To me they seem useless and without a purpose. I try to avoid the aged.

This opinion serves as a good example of the negative stereotyping and ageist attitudes that result in the avoidance of old people. When we avoid old people, society becomes age segregated. Real communication cannot take place in a segregated society, and the cycle of stereotyping, ageism, and polarization continues.

Ageism even affects professional objectivity. In a study of psychologists, ageism was evident. When presented with clinical vignettes in which the ages of the clients varied, clinical psychologists considered older, depressed clients to be significantly less ideal than younger clients with identical symptoms and histories, and older clients were given poorer prognoses than younger ones. However, older psychologists were more favorable toward older clients than were young psychologists (Ray, 1987). An experimental program was introduced in a medical school to improve the medical students' attitudes and skills in working with elders. In another study, an experimental group participated in four 10-minute group sessions that emphasized psychological and biological knowledge as well as communication skills. The experimental group developed more positive attitudes and more socially skilled behavior in their work with older adults than did members of a control group (Intrieri et al., 1993).

Employees might relate better to older clients if they rid themselves of negative stereotypes, especially the stereotype that older people are in their second childhood, which is a very poor way to elicit the highest potential from a person. Even if the older person has cognitive disabilities or is physically dependent, the "second childhood" stereotype glosses over the ways in which he or she is not childlike.

Negative stereotyping fosters fear of aging in both old and young. Who wants to be "hunched over," "grouchy," "useless," "rejected," and "alone"? One study, which used agree-disagree statements to measure fear of aging, showed a clear and strong relationship between low fear of aging and subjective well-being (Klemmack & Roff, 1984). The study measured fear of aging with statements such as these:

I feel that people will ignore me when I'm old.

I am afraid that I will be lonely when I'm old.

I am afraid that I will be poor when I'm old.

Subjective well-being was measured by agree-disagree statements such as these:

I have made plans for things I'll be doing a month or year from now.

Compared to other people I get down in the dumps less often.

The things I do now are as interesting to me as they were when I was younger.

Those who did not fear aging felt good about themselves and their lives. On the other hand, those who feared aging did not have a good personal sense of well-being.

A question asked of the aged participants in the outstanding longitudinal Berkeley Older Generation Study was "Looking back, what period of your life brought you the most satisfaction?" This was asked when the respondents were, on average, 69 years old, and again 14 years later when the average age was 83. The findings remained consistent over time: adolescence was considered the most unsatisfactory time. The decade of the 30s was named as most satisfying time period by 16 percent of the sample. The period of the 50s was second most popular named by 15 percent of the sample. Old age was seen as more satisfying than childhood. Twelve percent said their 60s brought them the most satisfaction; 13 percent named their 70s and 5 percent described their 80s as the most satisfying period of their lives. The common stereotypes that old persons are fixated on childhood memories, that youth is best, and that old age contains few satisfactions were, thus, dispelled.

Negative stereotyping stifles the potential of older people and draws attention away from the happy, sociable, successful, active older adults. A self-fulfilling prophecy is created: older people do not do anything because they assume they are not able. Their lives, therefore, become neither as satisfying nor as fulfilling as they might be.

We have hardly begun to explore the potential of older adults in this society. Business companies try to remove older persons from the labor market to reduce labor costs and to make room for the young. Too often, we provide no alternative ways for older people to make contributions. Too often, society works against elders instead of for them. On the other hand, aging boomers are undeniably making a cultural and social impact. The summer of 2009 national concert calendar was dominated by rockers in their 60s: Crosby, Stills, and Nash (67, 64, and 67, respectively); Joe Cocker (64); the Rolling Stones' Mick Jagger (66); Doobie Brothers' cofounder Patrick Simmons (60); and Loggins and Messina, both 61 (*The Week*, 2009). The time is upon us to understand and embrace the contributions of all age cohort groups in American culture.

Chapter Summary

Many stereotypes of old age exist, many of which are negative. Sources of negative stereotyping are the language we use to describe elders, songs, speeches, television, advertising, movies, and a complex socio-historical heritage. The psychology of prejudice examines ageism to understand the roots of this prejudice.

Far greater public education and interaction with older people needs to take place if we are to develop a full understanding of the potential of elders in our culture. Action directed to diminish our cultural belief in stereotypes about aging can and must take place on a personal level, as well as the level of social policy. Emphasizing the accomplishments of older scholars, scientists, and artists is helpful. Senior sports events draw attention to the physical fitness potential of elders and their ability to enjoy competition. The entire society is benefited when we have a more holistic understanding of what the nation's older adults look and act like in the present, and who they will be in the future.

Key Terms

authoritarian personality
communication accommodation theory
compassionate stereotypes
frustration-aggression hypothesis
language of aging
negative stereotypes
patronizing communication

phenomenology
positive stereotypes
pragmatic knowledge
psychology of prejudice
reversed stereotype of aging
self-concept
social construction of self
stereotypes

Questions for Discussion

1. How might negative attitudes about aging influence public policy, health careers, and social programs for older adults?

2. Bring a variety of magazines to class. In small groups, look through the magazines for examples of aging. How many ways are elders depicted in an ageist manner? How many ways are older adults depicted in a positive way? Are there as many images of older adults as of younger adults?

3. Have you ever experienced or known anyone who experienced prejudice or discrimination? On race? On gender? On age?

4. Describe yourself at age 85: what you will look like, what you will be doing, where you

will live, who your friends will be. How will you spend your time?

5. In small groups, develop an advertisement that illustrates ageist attitudes. Then develop one that shows positive attitudes toward aging.

Fieldwork Suggestions

1. List the first 10 words that come into your mind upon seeing the words *aged, middle aged, adult, adolescent.* Analyze your words. Do they reveal your personal biases and judgments about these age groups? Ask three people—preferably representing different age cohorts—to list the first 10 words that come into their mind when they see the words *aged, middle aged, adult, adolescent.* Compare your lists in a discussion group in class. What are the common patterns and language can you identify? What differences do you see? Can you draw any conclusions from your comparisons?

2. Design a study to identify stereotypes of aging. Identify your focus of inquiry—popular music? Preschool children? Advertisements in public places? Determine a systematic way in which you will record your observations; draw your conclusions.

3. Review a variety of TV series during the week to see how many older characters are written into the scripts. What percentage of shows have an older character? If there is an older person in the story, how is he or she depicted? Do you see examples of ageist stereotypes?

Internet Activities

1. Identify the target audience for the advertisements on any search engine on the Internet. What are your criteria for determining what that audience is?

2. Use the key word *elderly* and make a list of the *range* of references—that is, the various information networks that are linked to this word. Repeat this exercise, using the words *ageism* and *ageist literature.* What are you finding? What are you *not* finding?

3. Search for blogs and other site addresses written by and for older adults: blogspot.com is a good source.

3

Social and Psychological Theories in Later Life Development

Making Music: The Key to Healthy Aging

The Norman Transcript

September is Healthy Aging Month, an annual observance designed to focus attention on the positive aspects of growing older and to inspire adults older than 50 to improve their health. But what, one might ask, is its connection to Recreational Music Making?

"Music making is linked to a number of health benefits for older adults," said Dr. Suzanne Hanser, chair of the music therapy department at the Berklee College of Music in Boston. "Research shows that making music can lower blood pressure, decrease heart rate, reduce stress, and lessen anxiety and depression. There is also increasing evidence that making music enhances the immunological response, which enables us to fight viruses."

She said with all the benefits, it's no wonder many older Americans are sitting down at the piano or picking up their guitars, woodwinds, and horns. They're discovering that making music is a perfect hobby for the empty nest and retirement years.

Hanser said it's a great way to meet new people, get exercise, and challenge the mind, all of which lead to proven social, physical, and psychological benefits. Group Recreational Music Making programs abound, such as Weekend Warriors, New Horizons, and HealthRHYTHMS, all sponsored by NAMM, the music manufacturers' association.

But what about those who've never played before, or who put down their instrument years ago, after high school or college?

Hanser said the philosophy of the Recreational Music Making movement is that anyone, regardless of age or ability, can make music and benefit from it.

For instance, Judy Murray of Winchester, Massachusetts, took up violin at age 51. "I have miles to go with my learning," she laughs, "but each step is a pleasure."

Joyce Gast of Miami dusted off her French horn 15 years ago after retirement and joined a community band. "My standmate is 90 years old and still coming to every rehearsal," she said. "I hope to be playing like her at age 90."

Where can individuals learn more about Recreational Music Making? *Making Music* magazine is a one-stop resource for the recreational music maker.

With a mission to help older Americans get the most out of music, *Making Music* and its companion website www.makingmusicmag.com features the latest news about music making and health, inspirational stories, instruction for beginner and intermediate players, product and book resources, and much more.

"People are not only living longer these days, they want to remain healthy and active for as long as possible," said *Making Music*'s editor-in-chief Antoinette Follett.

This chapter considers the human life cycle from the psychological perspective of adult development. In it, we see how developmental perspectives are useful in understanding the fear of aging and age transitions. The personality variables that affect aging and are in turn affected in the aging process are studied.

Theories help us understand and organize what we see—the empirical observations we make. Some areas of study, biology for example, have highly developed theoretical structures that allow questions, or hypotheses, to be studied with scientific precision. Mendel's theory of plant genetics integrated a series of observations and allowed for the prediction of what size or color a plant's flower would be, or how many ears a corn stalk would produce. A theory of human development would be required to have a similar capacity for outcome prediction; however, there are many competing forces affecting the growth and development of an individual. For this reason, there are many different theoretical approaches to human development.

Imagine for a moment six people watching a tennis match. The eye of a ballet dancer would probably be skilled at observing the movement of the players in relation to the space of the court and the ball. The artist might see color, shape, proportion, and intensity. The corporate sponsor's perspective, on the other hand, would include assessing the apparent popularity of the players by the crowd, perhaps calculating the value of corporate exposure to a given audience size. There are many different lenses from which to view an event, and different theoretical approaches describe that phenomenon from a slightly—or not so slightly—different lens.

Data is the information being gathered by each of these viewers, as a means for testing or developing their perspectives, whether it be economic or physical. Data might be summary scores on objective tests, used to elaborate **quantitative development**. This perspective emphasizes the changes in the number or amount of something. Data might also be the telling of a story, of the kinds of things that people do, or how people do them. This approach emphasizes **qualitative development**.

The different perspectives, or theories used to describe an event or process, shape what is observed, and they can be described by different metaphors. A **metaphor** is a figure of speech that implies a comparison: being in "the autumn of life," for example. Metaphors can describe theories, too. The primary metaphors in personality or ego psychology are outlined in Table 3.1.

Perhaps the metaphor most highly integrated into Western thinking is a view of the person as machine: development dictated by determining forces such as biology (the **mechanistic metaphor**). This model is based on eighteenth-century Newtonian physics, in which an object cannot move unless it is acted upon by outside forces (Cloninger, 2003).

The **organic metaphor** sees the individual much like an unfolding flower. All the potential of the rosebud is within itself to become a perfect rose. Our potential lies within. An associated metaphor, **information processing**, is concerned with one person's ability to function differently than another. Theoretical models emphasizing thinking, or cognitive processes, fall into this group. Although individual competency emerges from within, this metaphor

TABLE 3.1

Theories as metaphors

Metaphor	Description	Exemplary Theorist
Information processing	Personality relected by cognition of subjective experience, which differs among people; recognizes multiple potentials; is causal	Sperry (1980, 1990) Kelly Bandura Mischel
Mechanistic	Personality determined by external determinism; adopted from the physical sciences; biological in nature; empirically testable	Freud through behaviorists
Organic	Personality compared with growth of plants and animals; potential is within the person, not the environment	Erikson Levinson
Narrative	Personality as the story of a person's life; when personality changes, we rewrite our life stories; narrative has plot, characters, time progression, and episodes	Saarbin Kelly
Emergent self	Self-directed, willful personality from internal determinism; emphasis on choice and striving; free will, and individual behavior; future oriented, not past oriented	Adler Sappington Ziller Sperry Rychlak
Transcendent self	Individual personality is not separate and self-contained but connected with others on a plane of shared experience; experience beyond individual ego	Jung Maslow Rogers

Source: Cloninger, Susan. (2003). *Theories of personality: Understanding persons.* New York: Prentice Hall.

implies multiple potential, coming from both internal and external sources.

In the **narrative metaphor,** development is thought of as the story of a person's life. Our stories are rewritten by us as we see ourselves as having changed; the "new" story might be only slightly modified, or it might be an entirely new story (McAdams & Bowman, 2001). Life stories have a plot, characters, time progression, and key events, all of which describe and shape the person's development. The life story is somewhat parallel with Levinson's concept of life structure, elaborated on later in this chapter.

Identity in adulthood is an inner story of the self that integrates the reconstructed past, perceived present, and anticipated future to provide a life with unity, purpose, and meaning.

DAN MCADAMS

The **emergent self metaphor** views the individual as being highly self-directed and emphasizes choices and motivations as primary factors in shaping development. Finally, the **transcendent self metaphor** sees development as being shaped by experience beyond the individual ego, sometimes from a plane of shared experience (Cloninger, 2003). This metaphor primarily allows for the concepts of spiritual development to be addressed. The transcendent self is also the "end state" or "ultimate unfolding" that occurs in the organic perspective. It is a particularly important metaphor in the study of aging when we attempt to deal with concepts such as wisdom, or issues concerning preparation for death. In the transcendent self the focus is on making meaning of one's life as lived: it is an interior process requiring thought and often solitude. The transcendent self process has been misunderstood in the past as disengagement, thereby lending credence to disengagement theory. As Tornstam (2005) suggests, the process of gerotranscendence is indeed disengagement from some aspects of society but also intense engagement in aspects of those issues and persons having greatest meaning to the life of the older person. This will be discussed in greater detail later in the chapter.

Sigmund Freud, father of psychoanalysis.

Early Developmental Models

According to early human developmentalists, distinct stages or phases form the **life cycle** through which humans pass. We proceed from infancy to childhood, through adolescence to adulthood, and then into parenthood and grandparenthood, and the cycle begins anew for each newborn baby. The life cycle is the course of aging: individuals adapt throughout their lives to their own biological, psychological, and social role changes.

The experiences common to all people in their passage through the life cycle give life some consistency. On the other hand, individual variations supply a measure of uniqueness to each person. We are all different, but we are not different in all ways, nor are we necessarily different in the *same* ways. Yet we are like *all* others in that we are conceived and born in a given time period, we age, and we die.

Freud, 1856–1939

The influence of Sigmund Freud, father of the psychoanalytic perspective, on the field of psychology is profound and indisputable. Freud believed that it is not human reason but unconscious psychological forces that most profoundly affect our thought and behavior. These forces originate in the emotions of early childhood and continue their influence throughout our lives. The influences, said Freud, shaping all that we are or will be occur in the first 8 to 10 years of life. From that point on, we replay the fears, insecurities, and issues that were established through early interaction with our parent figures. Freud believed that human behavior and motivation are driven by instincts, the outcome of which can be either positive or negative. They are the source of our generativity, creativity, and empathy as well as war, crime, and mental illness.

Psychoanalytic theory established the impact of early life experiences on the psychology, and therefore the life choices, of the individual. The focus on the unconscious, including our understanding of sex and aggression, has transformed the way in which people in modern times understand their conscious experience.

Jung, 1875–1961

A younger contemporary of Freud's, Carl Gustav Jung initially had strong professional ties with Freud. For years they maintained an active correspondence and jointly presented a critically important seminar on psychoanalysis in the United States in 1909. Over time, however, Jung developed a strong intellectual disagreement with Freud based primarily on Freud's emphasis of the role of sexuality and his relative failure to address the potential of the unconscious to contribute positively to psychological growth. Freud, for his part, feared that Jung had abandoned scientific knowledge for mysticism (Cloninger, 2003). The two men, by this time both preeminent theorists in the field of psychology, developed a powerful personal conflict from which two separate schools of psychological orientation emerged. The Freudian school continued to focus on the shaping power of past events, whereas Jungians began to focus attention to the future direction of personality development.

We cannot live the afternoon of life according to the program of life's morning, for what was great in morning will be little at evening, and what in the morning was true will at evening have become a lie.

CARL JUNG, 1933

Jung was perhaps the first modern voice to focus on adult personality development. The first three decades of development in the individual, he said, deal mainly with the **shadow**—the repressed childhood desires and attributes that Freud first discussed. Jung preferred to describe the psyche in the language of mythology rather than of science, "because this is not only more expressive but also more exact than an abstract scientific terminology, which is wont to toy with the notion that its theoretical formulations may one fine day be resolved into algebraic equations" (Jung, 1959, p. 13).

After 40, Jung believed, individuals begin to develop their internal self-potential. Through balance of competing opposites, the personality can reach maturity; otherwise, the personality remains in struggle, incomplete. For example, all people have both a feminine and a masculine side, produced by biological and social conditioning factors. As an example of the power of opposites, Jung (1916/1965) wrote that psychologically Eros (pure love) is the opposite of the will to power. "Where love reigns, there is not will to power; and where the will to power is paramount, love is lacking. The one is the shadow of the other" (p. 63). In other words, love and the need for power are opposites—two sides of the same thing.

As individuals age, said Jung, personality archetypes change, and people adopt psychological traits more commonly associated with the opposite gender. Men become more nurturing, and women develop a sense of their masculine personality traits. This represents Jung's formulation of *animus* and *anima* (Markson, 2003). Likewise, a balance of **extroversion** and **introversion** is necessary for mature personality development. Over the life course, individuals move from self-in-society—a focus on social interactions and institutions—to a more internal focus, or **interiority**. This process of self-reflection begins to take place around midlife and becomes a central way in which the individual is able to prepare for life's final state, death. The idea articulated by Jung that individuals become more introverted with age has become one of the most studied issues in personality research on aging (Cavanaugh & Blancard-Fields, 2006).

Erikson, 1902–1994

In the late 1920s Erik Erikson, then named Erik Homberger, helped to develop a program to teach art to children of Freud's entourage. The woman who would become his wife, Joan Serson, was studying psychoanalysis; it was she who introduced him to psychoanalysis. Erikson eventually was recruited to be a "lay analyst" (because of his nonmedical training), and he became part of Freud's inner circle. In 1933 he and Serson, fearing the increased anti-Semitism, left Germany, relocating first to Holland and then to the United States. It was at this point that Erik Homberger took the last name of his biological father, Erikson. Some scholars suggest that Erikson's career-long interest in identity emerged from his own somewhat confused identity (Cloninger, 2003).

Psychoanalytic theory was the base on which Erikson extended a stage model of the full course of human development, incorporating the impact of the social environment on the maturing personality. He was concerned with the mechanism by which people develop an identity. Erikson emphasized interactions between genetics and the environment in personality development and developed the concept of the **epigenetic principle,** which refers to an innate structure of development in which people progress through stages as they become emotionally and intellectually more capable of interacting in a wider social radius (Erikson, 1963).

Erikson's model of the **stages of human development** extends beyond childhood and adolescence to include middle and old age although the adult years, from roughly 20 through 60, were described by only two ego stages (Erikson, 1963). Erikson believed the individual progresses through eight psychosocial stages to establish new orientations to self and the social world over time (see Table 3.2). Each of Erikson's stages is identified with a developmental task, or challenge, to be accomplished. There may be either a positive or a negative resolution of the challenge, and the

TABLE 3.2

Approximate Period in Life and the Corresponding Eriksonian Crisis

Period in life	Crisis
Infancy	Trust vs mistrust
Toddlerhood	Autonomy vs shame
Preschool	Initiative vs guilt
Childhood	industry vs inferiority
Adolescence	Identity vs role confusion
Young adulthood	Intimacy vs isolation
Middle adulthood	Generativity vs stagnation
Late adulthood	Integrity vs despair

From: Dunkel, C. S. & Sefcek, J. A. (2009) Eriksonian lifespan theory and life history theory: An integration using the example of identity formation. *Review of General Psychology, 13*(1), 13–23.

ego resources we gain (or do not gain) on completion of one stage are brought with us to the next stage of development (Richman, 1995). The first five stages are similar to Freud's stages of psychosexual development.

Erikson's last three stages deal with early, middle, and later adulthood. In early adulthood the main issue of growth and development of identity is intimacy. Relationships in friendship, sex, competition, and cooperation are emphasized. Mature, stable relationships tend to form in the late teens and 20s. The task of young adulthood is to first lose and then find oneself in another, so that affiliation and love behaviors may be learned and expressed.

In middle adulthood the ability to support others and in doing so to create a legacy is the primary developmental task. In this stage **generativity** involves a concern for the welfare of society rather than contentment with self-absorption. The ability to create, to care for, and to share are the positive outcomes of balance in middle adulthood. Parenting and grandparenting are manifestations of generativity, but examples also include mentoring relationships and circumstances of "husbandry" in terms of care of a garden, concern with the environment, and creative endeavors to be enjoyed by others. If we look at environmental groups, a large proportion of members are middle-aged and older people. This reflects both life course events (their children are older and thus less demanding of time; careers are established) and an interest in "the larger picture": the generational legacy.

In later adulthood, Erikson's psychosocial emphasis shifts to the considerations of being nearly finished with life and facing the reality of not being. The crisis of later adulthood is **integrity versus despair**. Erikson (1966) wrote:

> [Integrity] is acceptance of one's one and only life cycle as something that had to be and that, by necessity, permitted no substitutions. The lack or loss of this (accumulated) ego integration is signified by fear of death: the one and only life cycle is not accepted as the ultimate of life. Despair expresses the feeling that the time is now short, too short for the attempt to start another life and to try out alternate roads to integrity. Disgust hides this despair. Healthy children will not fear life if their elders have the integrity not to fear death.

We are fortunate if in old age our passage through the first seven stages has provided us with a balance of ego resources—tools for development of identity—appropriate to take on the final task of preparing ourselves for death. The primary and profound task of Erikson's final stage is to integrate all the experiences of our life in a way that provides meaning to the life as lived. It is the time during which an individual determines, in the process of reviewing his or her life, whether that life has been "successful" to that person in a social and spiritual way. In essence, the final stage is a time when an individual asks and seeks the answer to the questions, What is the meaning of my life? What difference has my life made? The pro-

cess involves the remembering and the telling of stories; sorting through and adjusting or arranging remembered events until a cohesive life "story" can be made from all the events of life. In having a witness to that story, it becomes tangible—real—and the telling of it helps to clarify it for ourselves. The process has been called **life review**, or **reminiscence**, and will be addressed more fully later.

Positive resolution of Erikson's eighth, and final, stage allows the individual to interpret his or her life as having purpose. Meaning is ascribed to those many lived experiences, and the outcome of that meaning is a sense of life satisfaction. The negative resolution of this stage is one of meaninglessness and despair, the feeling that one's life has been useless. The final stage is one of reflection on major life efforts that are nearly complete. The reason for this process is to prepare the person to leave life—to die—with a sense of peace and completion. Life has had a purpose, and I have fulfilled my small part of a larger whole in a meaningful and fulfilling way. This accomplished, I can now leave.

Erikson did not consider his stage model to be **unidirectional**; he believed that in addition to moving onward to the next level of development (moving in one direction only), we revisit various stages again and again throughout our life course by means of *remembering*. Haunting memories to which we return again and again exist because we are not at peace with them. We have not yet been able to make sense, or to understand *why*, in terms of the memory and its "fit" with the rest of our life. When we are able to make meaning of those events and experiences, we are able to gain the positive outcome that experience represents. Wisdom is the adaptive strength—the personality characteristic that allows us to face the next task at the final stage. It is "a kind of informed and detached concern with life itself in the face of death itself" (Erikson, 1982, p. 61).

Erikson's model is difficult to test. In one of the few empirical studies of Erikson's model, McAdams (2005) developed an elegant measure of generativity. His findings support that generativity peaks in middle life, consistent with Erikson's expectations. A study of women in three American cultures—Mormon, Appalachian, and Amish—found that through the activity of quilting, generativity was expressed and enhanced. The behaviors contributed to well-being and psychosocial development of the women as well as betterment of the community and rising generations, as each group mentored younger women in the craft (Cheek & Piercy, 2008).

Following Eric Erikson's death, his collaborator and wife Joan Erikson reconsidered the eight-stage model and developed a ninth stage, applicable to very late life. Based on Tornstam's (1997) theory of gerotranscendence, a ninth stage emerged through Joan Erikson's experience with herself in later life as well as being with Eric Erikson in his dying time. She observed that to accomplish full identity development the individual must go *backwards* through the eight stages, focusing on the dystonic (negative) rather than the systonic (positive) outcomes. One must learn to *distrust* rather than trust: my body is different than it once was, I cannot know when I walk across the street that my gait will be fast enough, or my "trick" knee will hold me upright. The issue of *trust* becomes challenged, and I must reidentify myself around the concept. Growth and development in personal identity comes about through a revisitation of all the developmental stages, reinterpreting each in terms of the aged person. It is the most final stage of identity.

Most compelling about this modification to the Eriksons' stage model is that it views the character and sense of self of the older adult as having a natural potential for transcending—going beyond—midlife reality. Traditional models simply project midlife developmental values into later life; therefore, as the elder's energies and interests shift, those shifts become interpreted as losses by both the elder and society.

Loevinger

Jane Loevinger (1976) addressed stages of **ego development**, each of which provides a frame of reference to organize and give meaning to experience over the individual's life course. As the adult ego develops, she said, a sense of self-awareness emerges in which one becomes aware of discrepancies between conventions and one's own behavior. For some, development reaches a plateau and does not continue. Among others, greater ego integration and differentiation continues. This differentiation process is accompanied by a growth in cognitive complexity, impulse control, and respect for autonomy and mutuality in relationships (Lodi-Smith, Geise, Roberts, & Robins, 2009). Loevinger's model has helped to establish basic ego construct definitions such as conscientiousness, individualism, autonomy, and personality integration. The model describes a continuous increase in these characteristics as higher ego development unfolds. It is worth noting that Loevinger does not necessarily associate ego level with psychological adjustment or well-being. She views ego development as emerging from previous levels of complexity, but not necessarily something that creates individual happiness or a sense of fulfillment.

Levinson, 1923–1994

Daniel Levinson's interest was development in adulthood; he published an extraordinary longitudinal study of men's lives in 1978. In it, he conceived of the life cycle as a sequence of eras, each of which has its own bio-psycho-social character. Major changes occur in our lives from one era to the next, and lesser (although equally crucial) changes occur within each era. The eras partially overlap, with one era ending as another begins. These are referred to as **cross-era transitions**, and they generally last about five years.

The eras and transitions described by Levinson form a broader **life structure**: the underlying pattern or design of a person's life at a given time. The primary components of a life structure are the person's *relationships* with others in the external world, identified as *central components* and *peripheral components*, depending on their significance for the self and the life.

It is through our life structure that we are able to address the question, What is my life like now? This question is crucially different than the question emerging from a theory of personality: What kind of person am I? The last question looks at the life from an egocentric perspective; the first one, from a life course perspective—life *in context*.

In 1996 Levinson's second major longitudinal study was published: *The Seasons of a Woman's Life*. In this study, he attempted to develop a model of the structure of women's lives, including the significance of gender. His study concluded there are wide variations in the ways in which the genders transverse each period but the basic pattern is the same for women as for men.

A valuable concept emerging from the second longitudinal study is what Levinson (1996) named **gender splitting**—a sharp division between feminine and masculine that permeates all aspects of life. Gender splitting, he said, takes many forms. It is

> *the rigid distinction between feminine and masculine in the culture and in the psyche; the division between the domestic world and the public occupational world; the traditional marriage enterprise, with its distinction between the male husband/father/provisioner and the female wife/mother/homemaker; the linkage between masculinity and authority, which makes it "natural" that the man be head of the household, executive and leader within the occupational domain, and predominant in a patriarchal social structure. (p. 6)*

It is through this cultural and psychological process of gender splitting, Levinson postulated,

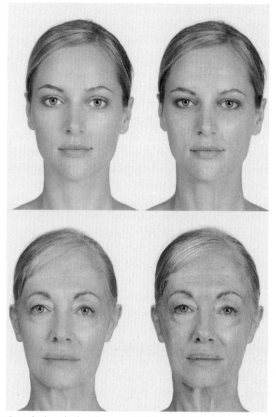

Coupled with the changes in this woman's appearance over time is a wealth of life experience and knowledge.

that the seasons of a woman's life is primarily distinguished from seasons of the life of a man. Gender differences exist cross-culturally. Male/female roles are not necessarily the same cross-culturally, but all cultures have clearly defined roles for men and roles for women.

Transitions in Adult Life: Developmental Patterns

Developmental psychology concerns the explanations for changes in personality throughout the life span. Stability or change in personality and the ways stability and change play out in the context of life shape the questions begging answers for the developmental psychologist. The perspective also incorporates human biology and the impact biology has on our progression through the human life course in explaining human psychological growth.

Up until about 25 years ago "human development" in psychology actually was in content "child development." The field of psychology at that time reflected the profound influence of Freud's model of development, which did not recognize growth and development in adulthood and later life. A new focus on early, middle, and later adulthood now characterizes the field of human development thanks, in no small part, to Jung's initial interest in midlife and beyond. Considering that we will live approximately two-thirds of our life as adults, it is reasonable that adulthood and later life are now major foci in developmental psychology.

The study of personality in middle and later life is one aspect of *adult development*. The term **transitions** describes points at which the person's development moves, or transitions, between one phase or stage and the next. The developmental perspective focuses on the unfolding *process* of the individual. As maturity develops, attitudes and behavior change in a somewhat orderly fashion, following a path unique to each individual, yet generally similar for all people. A caution must be given regarding the term *life course*, which is actually a misnomer, since there are so many courses of the life path. The very term implies one correct way for a life to pattern itself, and clearly in the study of human lives there are many ways to reach and live a full, productive adulthood. Additionally, a reliance on a central human tendency (such as a pattern of development) may reinforce restrictive age-based expectations and age stereotyping (McAdams & Pals, 2006).

Psychologists continue to disagree over the number of stages in the adult life course and the points at which they begin—indeed whether there are actual stages or not. Focusing on what happens normatively, for example, to most people, at a particular point in the life cycle can reinforce a belief that one path of development is best while another is somehow less-than normal or abnormal. With this caveat, the movement from childhood through old age is too complex *not* to have theoretical structures to help us gain a better understanding of the process. And clearly, many *stages* are points that we observe in our own lives or in the lives of people around us. They are consistent with our empirical observations, with folk wisdom and stories, and with our cultural expectations.

Some suggest that adulthood consists of four or more different life periods: young adulthood, maturity, middle age, and late adulthood (which can be divided into young-old and old-old). Others see only three categories: youth, middle age, and old age. Many young people connote middle age with being age 35 or thereabouts, whereas older people more often think of age 45 or 50 as middle age; in other words, one's own age colors one's perceptions of the stage boundaries.

We will now consider three broad stages of the life cycle: young adulthood, middle age, and late adulthood. First, the point must be made that the developmental tasks in each stage are culturally based and not clearly defined, and they are becoming even less so as regional and national culture become more global.

The Identification of Eras, Phases, or Stages

The distinctions between life periods are blurred. Some young people postpone marriage and childbearing, and some people experience a pattern of remarriage and divorce throughout their life courses. No longer is marriage necessarily reserved for young adulthood. Parenthood may happen at age 19 or age 42 or never. Consequently, one may be a grandparent for the first time before age 40, or after age 75—or not at all.

People are also more flexible in entering and exiting jobs and careers. Some retire at age 50, whereas others start new careers at age 70 or beyond. Launching a new career is no longer the prerogative of the young, nor is education. Increasingly, people of all ages are attending school. Indeed, public education has begun to address the shifting demographics to provide relevant continued educational opportunities for older adults. The number and quality of educational programs for older adults has increased dramatically in the past 30 years including the university-within-university model—Lifelong learning programs that sponsor classes held on a university campus, designed for older adults (Rogers, 2000).

Much as education is no longer only the purview of youth, events throughout the course of life are becoming less predictable. Some teenagers face adulthood early by leaving home or by becoming emancipated minors. Others live at home for long periods, getting an education or working in their first full-time jobs, saving money to get married or for a home. Teenage pregnancy forces some very young women and men into adulthood, whereas other people delay taking on family responsibilities until they are nearly age 40 or perhaps older.

In some ways, even the line between childhood and adulthood is disappearing. Young girls are encouraged to wear makeup and adult clothing. Soap operas cater to young adolescents, and children are bombarded with images and messages on once-taboo topics such as sex, drugs, suicide, and family dysfunction. Divorce, no longer taboo for the older generation, is a common occurrence for married couples of all ages. Many older persons are changing their marital status and lifestyle to "single" and engaging in social activities and arrangements once reserved for young adults. Our society is becoming more complex as many of the expectations for others' behavior are no longer valid, and the developmental tasks that once seemed to be set in a clear time frame no longer are.

Healthy middle-aged and older adults have changed many of the social norms that formerly dictated how a 65-year-old, for example, should dress and behave.

We develop not only a sense of personal identity as the life course progresses, but we also develop a sense of social identity, or who we are as a member of different groups or social categories. A 55-year-old African American woman, for example, might "belong" to the social categories of middle-aged American female; African American; African American woman; post-menopausal female; and woman, to name only some groupings. Our various social categories have different meanings to us depending on our life stage, social circumstances, and the historical time in which we reach specific age groups. Therefore identifying socially and psychologically as an African American will at times be most salient for the woman in our example; at other times, identifying and being identified as an American, or as a post-menopausal woman, will be her most salient group. It depends on the circumstances. The groups with which we identify impact our perspectives and our development, because each group has normative behavioral expectations.

Identification by social group becomes an **intergroup perspective**, which for our purposes is the shared assumption of a specific group. When we categorize others in this way, it is generally in terms of salient features such as appearance, behavior, and so forth. Once the categorization occurs, we ascribe certain attributes to that category and generalize to all members of the group. This is the process of stereotyping, and it can lead to prejudicial behavior.

Nevertheless, generational or **cohort** categories (based on historical decade, not biology) provide an easy and intuitively "valid" way to generalize for the purpose of social commentary. Baby boomers, Generation Xers, thirty-somethings, baby busters . . . each of these categories reflects a possible (and commonly used) socio-historical intergroup identification. This places the analytic focus on *generation*, not on *age*. People do not belong to an "age," they belong to a generation with its own language, music, experiences, and memories. The following section provides summary descriptions of larger categories identifying the life course: young adulthood, middle age, and old age.

Young Adulthood

Young adulthood comprises the years between 18 and 35 or so. Various challenging tasks present themselves at this time. Late adolescence in America often involves physical separation from one's family. College or military service can be the separating factor, or the young person may leave home to share an apartment or house with friends. Young people tend to have more friends than any other age group. This is perhaps to help them with the real task, which is one of psychological separation—of becoming an independent, autonomous person. Many adolescents find the passage stormy; identities are difficult to create when many options exist. "Who am I?" can bring much inner turmoil as that answer is forming.

Family and society place many expectations on young adults. After establishing an identity and an occupational goal, the young man or woman may be expected to finish his or her education, begin a job or career, get married, set up housekeeping, and have children.

Current expectations for getting married are not as clear as they once were. Parents are anxious for their children to become established in the economic structure; they are concerned about jobs and their children's future economic prospects. In response to this parental "should," young people increasingly delay marriage per se, opting instead to set up shared living with their significant other mate while they pursue education and/or concentrate on performing in their first jobs. In effect, marriage is postponed but the development of intimacy and affiliation, identified by Erikson as the task of young adulthood, is not postponed.

This pattern, it must be noted, is evident among some people in the United States but does not speak for all cultural groups in the country. This caution must be kept foremost in mind as we study human development, because in the search for similarities to describe a human process, it can become easy to overlook the differences. As we look through our particular lens, we must not forget to watch also for the individuating patterns of culture that have profound impact on our lives.

The centrality of major culture values creates a particularly intense pull for minorities. Internal and family expectations, as well as the social rules in which people are embedded, are culturally mediated. Minority-culture young adults have been asked through public education in the United States to give up their cultural roots and embrace a culture that is different than their own. This is known as **acculturation** and serves to help assimilate non-mainstream cultures into the American social and economic reward system. Acculturation, however, can place the young adult in the position of feeling the need to abandon values held by parents and grandparents and embrace the values and beliefs accepted as normative by the larger culture. This dilemma can create tremendous disruption between the generations. The task of maintaining one's sense of self over the life course is an enormous one. It is even more monumental when one's cultural identity has been systematically presented by the major culture as something to abandon and one's family bonds remain strong.

The difficulty of maturing and finding a place in the world for young adults depends on a range of sociocultural factors, including their ethnicity, family resources, place of birth, and so forth. Most people are so busy coming to terms with life they do not consciously think about growing old. But when asked, they do express attitudes and opinions about aging.

Attitudes and opinions toward aging vary, as the responses of college students show (Hillier, 2008). Students who were asked "What is your personal reaction to growing old?" offered the following responses:

- As you get older, you become wise, less caring of time, and are not in a big rush as much as when you are younger.
- I look forward to growing old, simply to have the luxury of getting to live. I work as an AIDS hospice worker and watch young people die year after year.
- I *hate* the fact that I have to grow old. I look at old people and dread getting there. My biggest fear is looking old, being ugly, and sick . . . I'm sure that sounds shallow but that's how I feel.
- Growing old scares me because I am scared to be alone. Being alone is the scariest thing that could happen to someone.
- My life is so complex right now that I look forward to sitting back in a rocking chair and watching life go by.

Some responses from these young people show a degree of acceptance and anticipation of aging, while others show fear. How and why do people come to fear aging? We can assume that fear and worry about aging does not make the coming stages of the life course easier. Apprehension makes every life stage more worrisome and less enjoyable. The two persons quoted above who were worried about getting unattractive and alone were female. Women seem to have the most fears, because their aging is judged more harshly by society. They indeed may have more to fear.

Middle Age

Is midlife a point of transition, or is it a crisis? In 1965 psychoanalyst Elliott Jaques' essay "Death and the Mid-Life Crisis" was based on analyses of artists' lives and works, and he found a dramatic change of some sort around

Across the life course, people who interact regularly and positively with multiple generations help to modify the image of later life.

the age 35 in almost every artist he studied. This change was generally precipitated by the individual's recognition of his own mortality—often generated by the death of a parent or a long-term friend.

The phrase *midlife crisis* apparently spoke to the experiences of enough people that it took hold in the language. In addition to being intuitively compelling, it is consistent with the theoretical underpinnings of adulthood developed in the early part of the century by Jung, van Gennep, Ortega y Gasset, and Erikson. Numerous psychologists took up the popular concept and developed a more popularized literature that assumed or further developed the idea of a crisis at midlife.

Gail Sheehy's book *Passages* (1976) launched the term *midlife crisis* into the popular media. She named the time between ages 35 and 45

the "deadline decade." Not much thought is given to aging during the teens and 20s, wrote Sheehy; but somewhere in our 30s we realize we have reached a halfway mark; "yet even as we are reaching our prime, we begin to see there is a place where it finishes. Time starts to squeeze." She continues:

> *The loss of youth, the faltering of physical powers we have always taken for granted, the fading purpose of stereotyped roles in which we have thus far identified ourselves, the spiritual dilemma of having no answers—any or all of these shocks can give this passage the character of crisis.* (p. 244)

Although Sheehy did not claim that everyone experiences a midlife crisis, she observed that many, if not most, find it difficult to face their own aging during these years. She further stated that somewhere between the ages of 35 and 45, if we permit ourselves to do so, most of us will have a full-out authenticity crisis, during which we may, as we did in adolescence, find ourselves desperately seeking to define our identity and purpose:

> *I have reached some sort of meridian in my life. I had better take a survey, reexamine where I have been, and reevaluate how I am going to spend my resources from now on. Why am I doing all this? What do I really believe in? Underneath this vague feeling is the fact, as yet unacknowledged, that there is a down side to life, a back of the mountain, and that I have only so much time before the dark to find my own truth. (Sheehy, 1976, p. 242)*

Sheehy clearly hit a nerve. How people felt about themselves and what they observed in other people was apparently consistent with the idea that people experience a major psychological disruption, often creating a need for change, at or around midlife. The term midlife crisis became a topic of discussion among journalists, psychologists and sociologists, television talk-show hosts, and patrons in the local café and coffee shop. By virtue of sheer repetition, a midlife crisis now existed.

Levinson's research on men's lives echoed Sheehy's sentiments:

> *For the great majority of men . . . this period evokes tumultuous struggles within the self and with the external world. Their Midlife Transition is a time of moderate or severe crisis. Every aspect of their lives comes into question . . . A profound reappraisal of this kind cannot be a cool, intellectual process. It must involve emotional turmoil, despair, the sense of not knowing where to turn or of being stagnant and unable to move at all. (1978, p. 199)*

Few corresponding scientific studies, however, could state unequivocally that age crises are to be expected within given time intervals for every individual. Soon after the publication of *Passages*, controversy ensued over whether a midlife crisis occurs or not.

Despite all the attention paid to midlife in the past 25 years, the time period is variously referred to either as "the prime of life," or as a "crisis." Most studies have found that midlife is both, depending on what life sends a person's way, and the coping skills and resources the person has to deal with life. Among those who clearly identify having had a crisis at midlife, it could have taken place not just between ages 35 and 45 but anytime between about 30 and 60. Even before 30 some individuals refer to an "early midlife crisis," and others after 60 to a "late midlife crisis"; therefore, midlife as a factor to predict an age crisis is virtually meaningless.

The midlife crisis might best be considered as a narrative form providing the person with a way of shaping and understanding the events and experiences in his or her life (Lodi-Smith et al., 2009). The midlife crisis represents a story (a plot) around which the personal narrative might be constructed. The underlying theme of the story is that there *is* a turning point—a change in the stable narrative of early adulthood. That change may be experienced as a crisis, or not. The importance in this narrative of change is in terms of the context—the particular historical time and culture. It is this context that allows individuals to understand and describe their experiences to themselves.

The words *transition* or *shift*, rather than *crisis*, more aptly describe the midlife experiences of most individuals (Roberts & Wood, 2006). Everyone agrees that we all make many transitions in life, and young adulthood to middle age is one of them. The new perspective views **midlife transitions** as normal situations likely to confront anyone. Such times, which are marked by feelings of uncertainty and instability, eventually result in some kind of adaptation. Klohnen and Vandewater. (1996) conceptualize midlife as a

> potential stressor that brings with it forces that impinge upon individuals. Individuals, in turn, posses mediating resources (e.g., ego-resiliency and coping styles) that aid in the creation of mediating conditions (e.g., social support systems and meaningful work involvement) that modify the impact of the potential stressor. (p. 432)

Ego-resiliency is the general capacity for flexible and resourceful adaptation to external and internal stressors. Klohnen et al. (1996), supposing it be an important personality resource to help individuals negotiate their lives under conditions of change, assessed ego-resiliency in women at age 43, and again at age 52. They found that ego-resiliency at age 43 indeed predicted life adjustments at age 52. Whether a midlife age transition becomes a crisis seems to depend on the ability, developed along the life path, for the ego to be resilient in face of stress and ambiguity. Let us take the hypothetical story of John (Samuels, 1997). The corporation John works for is in economic distress, but John believes that at least his job is as secure as the company. His younger sister recently died after a sudden heart attack. She lived in another state and was taking care of their mother who was in frail health. The mother of John's wife, who suffers from some form of dementia, lives with them. She frequently wakes them up in the middle of the night, telling everyone to get up to get ready for work. His wife often stays awake all night with her mother. John has tried to help them at night but is afraid that sleep deprivation will keep him from being able to perform successfully at work. The death of John's sister left him very fearful of having a heart attack, and he told his doctor:

> It's all too much. My wife isn't around, and we're not making love anymore. I have no energy, and my job performance is terrible.

> My sister just died, and I couldn't even stay with my own mother because I had to get back to help Marti with her mother. I sometimes wish that I would just die in my sleep, but I can't even get to sleep to get away from it all.

John *is* experiencing a crisis, but is it inherent to aging? Or is it a combination of his position in the life course—a working person with aging parents (probably also with children, perhaps in college)—plus the economic circumstances of this particular moment in history? Those factors, combined with his aging biology and the symbolic indicators of age (John really hates that he is becoming bald!), can feel overwhelming.

John has sought the assistance of a psychiatrist to help him with his "sleeping disorder." His physician has determined that John is quite depressed. Small wonder: John has multiple, significant stressors. It is stories like John's, of which there are many, that reinforce for us that a midlife crisis exists: we see these crises around us. Those in the helping professions who see the John's and the Jane's of the community employ an intergroup analysis, see age-related patterns, and seek therapeutic and medical treatment patterns for patient in "the midlife crisis."

There's some kind of profound something going on—a reassessment, a rethinking, a big gulp, whatever. It's not biological. It has to do with self-image and the work-place. And I find this astonishing.

A. CLURMAN, A PARTNER AT YANKELOVICH SURVEY PARTNERS, 1995

As a self-fulfilling prophecy, many people anticipate a midlife crisis because they accept its existence: it is believed to be so. Cultural norms create psychological and interpersonal pressures on individuals to conform to those behaviors and attitudes that express "human nature" as it "should." A new wrinkle—not knowing the latest music groups, having a doctor who looks too young to have graduated from medical school, or being the age of coworkers' parents—all these can precipitate feelings of being out of control, of being dated. The death of a close friend or certainly that of a parent can precipitate a profound process of self-reflection: how much time do I have? What have I really accomplished? Am I clear about what happens after death?

Those are the very questions predicted by developmental theorists including Jung and Levinson who postulate that midlife is a special time for reflection—a time when one's focus begins to become more interior as the full meaning of mortality begins to emerge and take shape in an individual's consciousness. The jury on the presence or absence of a midlife crisis is still out. It is clear, however, that midlife is a uniquely identifiable period in the life course that one moves in to, then out of.

Middle age is the *transition* between young adulthood and old age. In whatever way this time of life plays out, an individual's reactions and adjustments to changes in middle age affect his or her reactions and adjustments to old age, just as one's response to aging at any point affects the points that follow.

Those in their middle years frequently accomplish or conclude certain developmental tasks. The normative expectation is that people typically spend their 20s settling down and become further established in their 30s, 40s, and 50s. Middle age becomes the expected time to buy a house and establish an economic base. In middle age, those who are married may feel satisfaction or discontent, but those who have not married and settled may feel a stronger push to do so. Age norms constrain those who have not filled the appropriate social role at the appropriate time. Middle age can become a time to sort through which roles one might still fill and which roles to abandon, and to deal with the feelings of loss for those roles that never will

be. Tornstam (2005) would refer to this as part of the lengthy process of becoming gerotranscendent: it takes reflection and self-knowing to continue personality and spiritual growth in adulthood and later life.

Major role changes and events occur in the 40s and 50s. Our parents are likely to be in their old age, and this is our time to cope with their old age and eventual death. The older generation, in turn, must confront the reality of their children's middle age. Many books are on the popular market to provide guidance for middle-aged children who are helping and coping with older parents.

Grown children are launched during one's middle years. This event can be sad and depressing, a relief, or a mixture of both. Though the post-parental period has often been described negatively, for many, this new freedom is a cause for a celebration rather than a crisis, depending on their outlook and perhaps on the ability to redefine one's goals and purpose in life. Note, however, that the response to a normative age event depends on characteristics of the particular individual and his or her life course and context.

The concept of "empty nest syndrome" has been used to describe a midlife depression experienced by some women whose energies have been focused on childrearing (father's experience of children leaving home is unfortunately not well studied). However, the consequences of the "empty nest" phenomenon, research tells us, are as varied as the individuals experiencing it. Some people perceive their lives as having little purpose once their children have left home. Others identify a new sense of life-satisfaction—a time to renew the marriage relationship and to attend to personal interests that were pushed aside by the necessity of addressing the needs of children.

Another version of this is the return home (empty nest refilled) by either unmarried or married adult children. This is most likely to happen in an uncertain economic environment when the adult child has lost a job or accepted a very low-paying job, or if the adult child has had a divorce. The return of adult children to the parental home is not in and of itself negative, however. For many cultural groups in the United States, adult children once again living in the home might seem to be "traditional," as opposed to being aberrant. The event itself has no good/bad value. It is the cultural load, or *meaning to the individual* and the culture, that establishes value.

In some families, the needs of aging parents for care giving are synchronous with the return of adult children—indeed, those needs become determinants in the decision to "move home." In other families, the older parent moves in with the middle-aged "child" because of the parent's frail health or for economic reasons. The impact of increased longevity substantially increases the likelihood of responsibilities for elderly parents on the part of people in middle age as well as grand-parenting responsibilities for rearing grand children. Midlife is a time to come to terms with what is not possible. We may acknowledge, at this point, the impossibility of becoming president of the company or of fulfilling any of a host of goals.

Midlife is likewise a time of peak competence, professional respect, and earnings. Those who are unemployed or in low-paying jobs, or have been victims of an economic downturn and are unable to find a job, may have more trouble squaring their dreams—expectations— with reality. Aligning dreams with the facts can be a daunting task or none at all, depending on the dreams and the degree of success we expect. Some individuals have both the inner strengths (self-confidence and motivation) and outer resources (monetary savings and cultural support) to find new jobs or careers in their 40s and 50s. Changing jobs may require risk taking and sacrifice and, for many, can literally be a time for starting over—or the challenge can become crushing.

Middle age brings biological changes, as well as changes in career and family. For women, menopause—the ending of menstruation—is a major physiological change: the childbearing years have ended. The decrease in estrogen in menopausal and post-menopausal women can create numerous physical changes and symptoms, such as hot flashes, genital atrophy, urinary tract changes, and loss of bone density.

Deciding whether to take hormones (HRT) or not to alleviate menopausal symptoms and reduce risk of osteoporosis is not an easy one, for there are health risk trade-offs. Done well, the decision takes a thoughtful, self-reflective approach. This creates an opportunity that many women take to reflect on who they see themselves as being as they come to terms with the ending of one stage of life and the beginning of another. In this way, a biologically determined experience can become a time of self-reflection and examination. Women in their 40s and 50s may mourn the loss of their fertility, but at the same time be relieved. It is a profound change in the rhythm of a woman's life, and sometimes women and their partners need reassurance that menopause does not, in itself, diminish sexuality. Rather, it can free partners from concern about pregnancy; many couples report improved pleasure and joy in their sexual relationship after menopause.

Tornstam (2005) would see midlife as a decade of reminders that bring age consciousness to new heights. These reminders come from the reflections of self that come from the appraisals of other people, a "looking-glass self" that gives rise to self-appraisals about aging. Those in their 50s are often surprised by their own aging. In their minds, they are young; but either their bodies or people around them send different messages. For example, man who plays on a basketball team is referred to as "sir" by a younger team member. It is startling! Is he talking to *me*?

Each to each a looking glass, reflects the other that doth pass.

COOLEY, 1902

Self-concept, recall, is socially constructed at least in part. Who we are and how we respond is mediated by the way we are treated by others, and how we believe we *ought* to be in the world. We have a tendency to become the person others say we are.

One category of reminders is generational—50th-birthday bashes focus on the fact that the birthday "boy" or "girl" has made it to the "big 5-0." Children grow up and become independent; they have children of their own. Parents die, long-term friends die, and the reality of being at or near the alpha generation is ever-present. The person in midlife is surrounded by social reminders of age: being the oldest member of a group or club; self-consciousness about going to places such as singles dances or bars where most people are young; or being old enough to be the parent or even grandparent of students, clients, or patients.

Research on the perception of self over time suggests that many people expect a peak in integrity in late life and a peak in generativity in midlife. They anticipate high levels of well-being in midlife and later (Fleeson & Heckhausen, 1997). The way in which people anticipate their future self can profoundly affect the choices they make in the process of becoming old. A projected self-concept of well-being in later life is, perhaps, half the battle.

Late Life

In the mid-twentieth century, when psychological theories of human development began to expand inquiry from child development to the adult experience, late life remained the least examined part of life. Late life has been characterized by decline and loss: loss of physical health, life-long partners and friends, mental capacity, creativity, social roles—in short, depressing, discouraging, and barely worth spending much time on.

Late adulthood is still the least studied portion of the life course; however, the interest in social gerontology brought about by longevity and demographics in the past 30 years has created a parallel interest in the psychological development of later life. Today, many people will spend about one-third of their entire lifespan in their "old age"—a fact that the French address by calling old age *le troisième âge*, "the third age" (Hoyer & Roodin, 2009). Although the student of gerontology is still likely to be met with the incredulous response of "Why are you studying *that?*" the exploration of growth and development in later life is now acknowledged as an exciting field of inquiry attracting increased numbers of scholars.

Loss does occur in late life, but it is important to distinguish between normal, pathological, and optimal late-life experience. Although physical changes do occur with advancing age, physical and mental decline are not necessarily part of normal aging. Just as wearing prescription lenses to maintain 20/20 vision is considered to be an adaptive compensation, so too are the compensations an individual develops for minor dysfunctions that accompany the normal aging process (Hoyer & Roodin, 2009). Walking more slowly on a hike to compensate for a reduced energy level might be an example. Being unable to take a walk at all, however, is not a consequence of normal aging, but an indication of pathological aging—a physical state based on disease or injury rather than as an outcome of the aging process. This serves as a reminder that "the aged" are a very diverse population: well-being in old age is largely a function of the physical and psychological processes that preceded it. Additionally, losses tend to be counterbalanced with gains: the death of spouse, for example, can lead to remarriage and a regeneration of a loving relationship. Retirement from a full-time career may lead to part-time work at the same place or another job elsewhere, or to the expansion of a creative hobby.

Developmentalists generally divide *le troisième âge*, into categories of the young-old (around 65 to 75 or 80), the old-old (75 or so to about 90), and the very-old (around 90 and older). This distinction is very important when we talk about "old age," because if we are talking about a 95-year-old and a 68-year-old, we are speaking of people in profoundly different developmental places. Clearly, 40 years ago this was almost a moot point, because so few people lived past what we now call the young-old group. Today, the young-old are more like the middle-aged than they are like the old-old.

As we learned, Erikson (1979, 1982) defined old age as a time when one is seeking balance between the search for ego integrity and feelings of despair. Other scholars have elaborated on Erikson's theme. Havighurst (1972) identified six tasks of late life:

1. Adjusting to decreasing physical strength and health
2. Adjusting to retirement and reduced income
3. Adjusting the death of a spouse
4. Establishing an explicit association with one's age group
5. Adopting and adapting societal roles in a flexible way
6. Establishing satisfactory physical arrangements

The function of these tasks is to promote well-being in later life: optimal aging. Note the extent of adaptation required of older people in Havighurst's model.

Extreme old age is a very special time. In his 90s, Erikson wrote from personal experience about the final stage of the life cycle model that he presented with the help of his wife, Joan, 40 years earlier. He expanded on his late-life stage, "integrity versus despair," by describing the wisdom that comes with age if one completes the developmental tasks that began earlier in the life cycle. For example, the basic trust learned in infancy evolves into the knowledge in old age of how interdependent we are—of how much we need each other. In early childhood, the life cycle's second phase, learning physical autonomy and control of one's bodily functions, versus shame and doubt in not learning them, paves the way for coping with deterioration of the body in old age. Old age is basically the second phase in reverse. One must "grow" in order to avoid the shame and doubt that can accompany decline, just as one "learned" bodily development without having shame and doubt. Every stage offers lessons that can apply in old age.

In old age, we can develop humility by comparing our early hopes and dreams with the life we actually lived. **Humility** is a realistic appreciation of our limits and competencies. On the individual level, in old age, one might achieve a sense of integrity, a sense of completeness, a sense of personal wholeness strong enough to offset the downward psychological pull of inevitable increased physical frailty (Fleeson & Heckhausen, 1997).

Transitions in Late Life

As a result of the longer life expectancy in the late twentieth century, most Americans can expect to pass through these age-linked events: long-term survival, empty nest, retirement, and for women an extended term of widowhood characterized by solitary living. Actually, a transition can be movement either from one age to another or from one age-linked event to another (e.g., children leave home or a spouse dies).

Every decade has its share of tasks and challenges. In later life, illness, death of a spouse, or increased frailty may take its toll; but personal growth and joy in living is still possible and probable. While Socrates awaited death in prison, at the age of 71, he learned to play the lyre. Transitions do not end at age 65. We continue making them until the final transition. We can begin early in life to teach ourselves and our children that growing older is a natural event, one not to be feared, but rather one to be anticipated for new roles, new avenues of expressions, and new opportunities.

Activity Theory

Activity theory is a dominant theoretical perspective. It implies that social activity is the essence of life for all people of all ages. Early studies found that positive personal adjustment correlates highly with activity: the more active people are—mentally, physically, socially—the better adjusted they are. Early proponents of this theory believed that normal aging involves maintaining the activities and attitudes of middle age as long as possible. Any activities and roles that the individual has been forced to give up should be replaced with new activities. The theory predicts that those who are able to remain socially active will be more likely to achieve a positive self-image, social integration, and satisfaction with life, and that, therefore, they will probably age successfully.

Having no clearly defined role in society is similar to Durkheim's concept of "anomie," a condition in which some individuals in a society are in a normless state. These individuals lack a consensus on rules to guide their behavior and, therefore, receive no support or guidance from society. The result is that they are excluded from participation in social activities. Research seems to indicate that examining the integration of an individual network into a larger community might be an important way to understand how people maintain a sense of belonging and well-being as their social networks change.

Generally speaking, the last 50 years of research have found a positive correlation between being active and aging successfully. In many of these studies **successful aging** was defined in relationship to life satisfaction: people with strong reports or measures of life satisfaction were considered to be aging "successfully." As the complexity of human aging has become more clear, however, questions have arisen about those life satisfaction studies: what is the relationship of other important factors to life satisfaction, such as health, gender, culture, socioeconomic status, desire to maintain active? Might not one person's internal experience of "activity" be differ-ent than another's, based on their histories and interests? Indeed, Jacob and Guarnaccia (1977), in a study of life satisfaction and social engagement, suggested that *life satisfaction* is probably a misnomer that might better be termed "momentary contentment" (p. 816). They concluded that the culture reinforces disengagement among elders; therefore, commitment to new goals and relationships is less necessary to psychological health in old age than it is in early life.

Activities might enable people to confirm their identities and participate in roles they highly value. In those cases, activity is likely to bolster self-esteem and life satisfaction. Many older people, however, seek a more relaxed lifestyle and are quite happy when they achieve it. For example, a 65-year-old woman may long for the time when she can retire or work half time, and devote more time to her aerobics classes and to reading the newspaper at a local coffeehouse.

Disengagement Theory

Disengagement theory is an explicit theory developed through research and explained in the book *Growing Old* (1961) by Elaine Cumming and William E. Henry. This book, one of the best known in the history of social gerontology, contends that it is both normal and inevitable for people to reduce their activity and seek more passive roles as they age. Disengagement is a mutual withdrawal of the elderly from society and society from the elderly to insure the optimal functioning of both the individual and society. Aging individuals, wishing to escape the stress of recognizing their own diminishing capacity, collaborate in the withdrawal.

Disengagement theory has generated a great deal of criticism. Some say the theory is ethnocentric, in that it reflects the bias of a male-dominant industrial society. Others have suggested that it discourages interventions to help old people. Still others have questioned why some elderly choose to disengage and others do not,

This woman, a music lover, continues to grow and develop in her later life. Her interests remain much the same as in her childhood, but have blossomed with experience.

contending that society pressures people into disengagement against their will. However, it must be remembered that the theory emerged from a particular context of social thought—one in which biology-as-destiny prevailed, and the extent of interacting variables (such as gender, socioeconomic and cultural factors) with the process of aging was not yet clearly understood.

Gerotranscendence

Do relationships and our need to be connected in order to maintain psychological well-being alter with time? Particularly in most recent research, support is emerging for a more complex form of disengagement. In 1994, Lars Tornstam used the term **gerotranscencence** to refer to the older adults as selectively investing in some relationships over others, rather than comprehensively withdrawing. In his model, older people do seem to disengage, but do so more *at will*, choosing where their priorities lie and divesting themselves of superfluous relationship to focus on a more transcendent view of experience. In Quinnan's (1997) study of older religious men:

> *Thus the elderly demonstrate a higher degree of autonomy by dispensing with forms of social intercourse which have little value for them. This exercise of autonomy, rather than breaking connectedness, selectively enhances those relationships which the gerotranscendent find filled with meaning. (p. 118)*

Gerotranscendence is rooted in stage theories such as Erikson's that postulate a movement from dependence to greater autonomy with maturity. From this perspective, growth in autonomy takes place through a shift in connections—for example, reducing connection (the process of individuating) from the family occurs among adolescents in conjunction with a growing connection with peers and people outside the family (Hyse & Tornstam, 2009).

In a related line of inquiry, comfort in being alone was found to be related to lower depression, fewer physical symptoms, and greater life satisfaction in a survey of U.S. adults in 1995 and again among older adults participating in recreation programs (Dunkel & Sefcek, 2009). These findings are consistent with our intuitive observations: some people deal with stress by secluding themselves from social contact—spending time reflecting and engaging in self-care activities like reading, gardening, listening to music. It appears that solitary activities for some are a balance for their social engagement activities.

Clearly solitude does not imply that it is healthy to be involuntarily isolated from others, but consistent reports continue that people do spend less time with others as they age, and those who are able to enjoy this segment of their lives are better adjusted and have a greater sense of well-being. An anthropological study of patterns of interaction in a nursing home, where elders have little choice to pull away from social contact, showed residents to engage in what the author referred to as *sitting* time and *giving* time (Gamliel, 2001). Sitting time was characterized by silence in which "[residents] transcended the borders of past and future time to live in a 'sacred present' or a 'limbo' time" (p. 107). Giving time was characterized by "limitless concern" for the health and well-being of one another. The author concluded that sitting time and giving time combined to help residents transcend the circumstances of their own health and environment. In this way, the nursing home residents were able to maintain a higher degree of self-controlled social activity and thus a higher degree of life satisfaction than they would otherwise experience, consistent with Tornstam's theory of gerotranscendence (2005).

Continuity Theory: "You Haven't Changed One Bit"

Continuity theory is broad enough to be considered a sociological theory as well as a psychological one. Continuity theorists propose that a person's adaptations to young adulthood and middle age predict that person's general pattern of adaptation to old age. According to continuity theory, the personality formed early in life continues throughout the life span with no basic changes. This theory implies that neither activity nor disengagement theory explains adjustment to aging: adjustment depends on personality patterns of one's former years. This approach is consistent with the core of personality theory. Some therapists believe that significant personality change after about the age of 30 is unlikely. And, although some researchers continue to debate the degree to which the personality remains stable throughout the life course, continuity theory maintains that the individual achieves a core personality by adulthood. By adulthood, people have adopted coping mechanisms, established stress and frustration tolerance levels, and defined ego defenses.

Surely enduring personality traits . . . form the core of our identity. For better or worse, we are what we are, and the recognition of that fact is a crucial step in successful aging. Costa,

TERRACCIANO, & MCCRAE, 2001

Trait theorists look for consistency in the personality: **traits** are the enduring response patterns exhibited by a person in many different contexts. A **state**, on the other hand, more accurately describes something that is transient in the personality. Not unexpectedly, trait theorists find consistency in personality development over a lifetime. They define personality in terms of the basic tendencies that are the core potential of the person.

Two primary trait theorists, Paul Costa and Robert McCrae, state that personality stability is crucial for successful aging. Personality traits are central to an individual's self-concept; they form the core of how people see themselves, and the continued sense that we are the "same" person provides the basis by which we can move on to make meaning of that life as lived. Enduring dispositions provide a dependable and necessary basis for adaptation to a changing world (Costa, Terracciano, & McCrae, 2001).

Trait theory holds that lives change but fundamental personality characteristics do not. We are born with certain characteristics—shyness or extraversion, for example. If we are shy, in the course of our lives we might learn to modify that shyness socially in ways that work well for us, but we remain fundamentally shy. Those more enduring traits provide solidity for the ongoing sense of self, even in the midst of change.

Those who argue with the continuity theorists point to individuals who believe they have made dramatic personality changes later in life. Consider the 78-year-old tycoon, describing himself as a youth: "I didn't like myself. I thought I wasn't as good as others." Now he sees himself as being not fearful, not hostile—as having changed a great deal from his youth.

It would appear that he indeed has become more open and self-accepting. Continuity theorists would respond to this by saying that we can alter our ways of behaving, but the fundamental personality remains the same. Personality characteristics that make us—or others uncomfortable—can be modified. Tempers can be controlled, fears lessened, and skills learned. Thus, personalities can improve with age but that temper will always need to be controlled; it is always there from a trait perspective.

Most people believe they remain "themselves" over time, and that belief seems to be reinforced in research studies (Adler, Kissel, & McAdams, 2006). By feeling a sense of continuity people view change as being connected to their past and linked to their present and future. This sense of the ongoing self is consistent with Kaufman's findings (previously reported) of the ageless self.

Research supporting people's skills at adaptation with age can help to shatter the stereotypes that older people become less flexible, more cranky, more conservative, and less satisfied with their lives. Continuity theory suggests, as do other developmental psychologists, that growing old is a process of *becoming*; who we are in late life is a culmination of who we have been throughout our lives (Blagov & Singer, 2004; McAdams, 2005).

Exchange Theory

Exchange theory is based on the premises that individuals and groups act to maximize rewards and minimize costs; interaction will be maintained if it continues to be more rewarding than costly; and when one person is dependent on another, that person loses power. This model explained decreased interaction between the old and the young in terms of the older generation having fewer resources to offer in the social exchanges, and thus less to bring to the encounter. Power is thus derived from imbalances in social exchange. Social exchanges are more than economic transactions. They involve psychological satisfaction and need gratification. Though this perspective sounds rather cold and calculating, social life according to exchange theory is a series of exchanges that add to or subtract from one's store of power and prestige. The concept has been particularly useful in current studies of leadership (Camplin, 2009; Carney, 2009; Garcia, 2009; Slaven, 2008).

Exchange theory is based on the idea of reciprocity. The **norm of reciprocity** means to maintain balance in relationships; goods or deeds are "paid for" with equivalent goods or deeds. This social rule requires us to return favors to those who do something nice for us. It does not imply an open-ended obligation to return a favor. Rather, it requires only that acts of kindness be returned within a reasonable period of time (Burger, Horita, Kinoshita, Roberts, & Vera, 1997). It can be applied in business or in relationships with family and friends. For example, I'll trade you this CD and five blank ones for that CD: the goods' values are equal; we're "balanced." The "goods" might be less tangible: I will come to stay with your frail mother while you go shopping, because you were there for me when my husband was ill. A kindness is exchanged for a kindness. A classic study used exchange theory to interpret differences over time in a group of older women's close friendships (Roberto, 1997). They found that the essential elements of the women's friendships (understanding, affection, trust, and acceptance) endured over time, although the balance of the exchanges of those elements shifted: I can be trusted to keep your confidence because, in the past, I have been able to trust you.

Some groups in society are unable to repay what they receive—children and mentally disabled people are two examples. In these cases, beneficence becomes the norm. The person giving then does not expect a material reward but does expect love or gratitude. The **norm of beneficence** calls into play such nonrational sentiments as loyalty, gratitude, and faithfulness. The norm is particularly relevant for care providers of frail and vulnerable people. In a classic analysis of interactions made nearly five decades ago, Gouldner (1960) distinguished between reciprocity as a pattern of social exchange, and reciprocity as a general moral belief. The **reciprocity norm** dictates that one does not gain at the expense of another's beneficial acts (a moral belief). **Equity theory** suggests that people react equally negatively to under- and to overbenefiting: balanced benefit is the moral standard. The concept has implications both for service providers and for those who receive senior services. Exchange theory and the norms of reciprocity and beneficence remain valuable concepts with which to view the position of elders. We can apply exchange theory at a small-group level, between one older individual and another person, or at a societal level.

Theoretical Roles of Individuals and Society

Roles, Gender, and Ethnicity

Role is one of the most basic concepts in all of sociology and one that you will find used in almost every sociological framework. A role is a status or position, which carries known attributes, accorded to an individual in a given social system. "Doctor," "mommy," "sports fan," and "church-goer" are all roles.

Roles are modified, redefined, and transformed as people age. They change over the life course, in marriage, families, careers, and the community. The interplay of race, gender, and ethnicity serves to shape life opportunities and lifestyles. People have informal roles in friendship and neighboring. They have formal roles in institutional settings like schools, hospitals and professional settings, and more informal roles in less structured settings. Transition from one role to another is a major focus in gerontology. **Gender roles** have to do with the cultural aspects of being male or female. Because aging men and women are looked upon differently in our society and in other cultures around the world, the study of gender is very important. Being old is a different experience for men than for women.

Ethnicity refers to one's identification with a subgroup in society having a unique set of values, traditions, or language. The role of ethnicity as it affects aging and the aged is often overlooked in the studies of gerontology. Particularly among gerontologists, generalizations were previously made on the basis of white, middle-class respondents. Clearly these generalizations did not apply to all racial and ethnic groups. Numerous international studies now add richness—and reality—to our understanding of aging (Kvale 2003; Pitts & Nussbaum, 2006; Plath, 2009), and newer studies of cultural groups within the United States further develop an understanding of the experience of all older adults (Robine & Michel, 2004; Takamura, 2001).

Gender Development

The developmental model emphasizes the psychology of the individual. Developmental psychologists who look at how individuals change over the life course focus on the individual by studying variables such as personality, motivation, cognition, and morale. Social and cultural factors are examined for their role in personality changes. Maturation, or some inherent biological mechanism, may also be a strong force in creating change in persons as they progress through life. As theoretical models of aging are set forth, empirical studies test and refine them. The question of which attitudes, behaviors, and individual characteristics are **intrinsic** (biologically mandated) and which are **extrinsic** (formed by changes in social structure and roles) is still an issue for study and discussion.

One example of the question of intrinsic or extrinsic causality is that of gender differences. Carl Jung postulated the existence of opposites in our personalities, with young adults generally expressing only one sexual aspect as they developed into adulthood. The expression of that sexual self is usually defined by sex-role stereotypes.

As a person ages, he or she may become more self-accepting and more comfortable exploring all sides to one's personality. Both sexes, according to Jung, move psychologically closer to middle ground.

Psychologist David Gutmann (1987) found that this path toward the opposite characteristic is intrinsic. Regardless of the cause of the initial gender behavior, he says, at around midlife, generally all males begin to grow more nurturing, and all women become gradually more "executive." This finding is cross-cultural, according to Gutmann's research (1987). He found a "cross-over effect" from one gender-defined set of behaviors closer to the other, which he named the **post-parental transition** (1992). Older men become more free in expressing behaviors that would be considered feminine, and women show more "masculine" behaviors.

Gutmann proposed the initial reason for gender role differentiation (and the subsequent cultural norms for male/female behavior) is the "parental emergency" in which the arrival of children requires an extensive period of "parental service" where clearly differentiated gender roles are necessary. Therefore, for the good of society, women become the nurturers and fathers, the providers. When this period of time is over, men and women can release previously unused potential. It is hard to know necessarily that this is a developmental outcome or change in personality. Older men may have more time to give the love and tenderness to their grandchildren than they were not able to do with their own children. Older women may become more assertive and confident not because of the aging process, but because they have ended parenting roles and entered the job market, or are now in a role that does not require them to simultaneously balance the different needs of many individuals. One provocative implication of Gutmann's hypothesis is that the mother role and role of worker/provider are inherently incompatible. On the other hand, research based on interviews with older adults often identify a "shift in the politics of self," in Gutmann's words (1987, p. 203).

We have all seen the old person who is bitter, depressed, and very anxious about life and living. We have also all seen the old person who has a clarity and grace, with a special understanding of the meaning of their existence, and the willingness to share the story of their understandings. Clearly, we want to be like the latter person when *we* are 70 or 80 or 90. Psychologists are only beginning to ask some of the questions that are central to a better understanding of the development of well-being in later life. They are not in agreement with which theoretical perspectives most accurately describe the growth and development taking place in later life, possibly because there is no single unifying theory to be found. The human life and human psyche is dynamic and infinitely resilient. We return, after crushing life events, after multiple experiences and cultural ways of interpreting those experiences, to a central human core. The task of gerontological psychology is to better understand and describe that core, as well as the process for getting there.

Age Grading

We live in an age-graded society. **Age grading** means that age is a prime criterion in determining the opportunities people may enjoy. Our age partially establishes the roles we may play. Both children and old people are welcomed to or barred from various opportunities because of the beliefs society has of the young and the old. Such beliefs can prevent the young and the old from expressing individual differences and, thereby, lead to injustices. In other words, older people may be less active not because of their biology or the aging process, but because they are expected to present an image of their age they do not feel and that feeling is not comfortable. It may feel easier not to attend a public gathering place than to do so and not be able to "be oneself." Role expectations at various age levels are called **age norms**. Society expects individuals to engage in activities such as schooling, marriage, and child rearing at a socially approved age.

Age norms are a form of social control. If one follows the age norm, one receives approval; if not, disapproval and possibly negative sanctions result.

We observe a continuing shift, however, toward a loosening of age grading and age norms in the United States. Life is becoming more fluid. There is no longer a definite age at which one marries, enters the labor market, goes to school, or has children. It no longer surprises us to hear of a 23-year-old software company owner, a 34-year-old governor, a 36-year-old grandmother, or a retiree of 52. No one is shocked at a 60-year-old college student, a 50-year-old man who becomes a father for the first time, or an 80-year-old who launches a new business. Our ever-advancing technologies continue to test and stretch the limits of what people find acceptable.

Age Cohort

For our purposes, an **age cohort** is a group of individuals exposed to, or who experience, a similar set of life experiences and historical events at a similar state of biological and physical development. Demographers perform cohort analysis to study the effects of events or demographics on a broad group of individuals born during specific time periods, usually separated by 5- or 10-year intervals. We expect that age cohorts will show similarities to one another. Cohort analysis permits the sociologist to study the effects that events or demographics may have on a broad group of individuals, all of whom have experienced the same events at a similar state of biological and physical development.

With the tendency of the media to homogenize social groups, a cultural image has evolved of baby boomers as the free-loving hippie generation who dodged the draft, protested against the Vietnam War, attended Woodstock, and enjoyed economic prosperity. Likewise, young adult cohorts in their 20s have been referred to by the media as Generation X (*Times of our Lives,* retrieved May 2, 2009). The "30 somethings" who represent Generation X are stereotyped as whiners and slackers, complaining about the national debt they have inherited and totally unappreciative of "how good they have it." This cohort has spent more time watching television and, as a result, has probably witnessed more violence and murder than any generation in history; it has had more time alone as young children and is the first generation to spend considerable time in day care. (Many Gen Xers also grew up with stepparents and stepsiblings, and with both parents in full-time employment.)

Generation X is the smallest cohort since the early 1950s and is thus labeled a "baby bust" generation as opposed to their boomer parents. They are the first generation in the United States to be smaller than the generation that precedes them (Generation Birth Years, retrieved June 6, 2009). For this reason their opportunities in the job market were increased from previous generations, and they experienced less competition for admission into careers as well as colleges, universities, and professional programs. They were primed to be economically *the* most successful generation. It was also the first generation to be profoundly influenced by the media. They grew up with televised images of perfect families, perfect ways to look, and the best product to use. The recession of 2008, however, might have shifted that expectation to a new reality. The generation born ten years following Generation X: Generation Y. They are more technologically saavy than those who precede them. Generation X grew up with television and babysitters. For Generation Y, computer games and friendship networks are a reality they have always known.

Despite the many similarities in people as members of cohorts as a result of growing into maturity within a specific cultural and economic moment in history, the differences among people remain greater than their similarities. The reminder to avoid overgeneralization must be restated. Subgroups within a cohort experience the world in different ways based on such variables as class, gender, ethnic background, or region of residence.

Generations and Events

The concept of **generation** is more complex than the concept of a cohort. A generation has common beliefs and behaviors, a common location in history, and perceived common membership (Pitts & Nussbaum, 2006). Some use the term to identify the values of those who are older or younger than a particular designated age: the "over-30 generation" or the "under-30 generation," for instance. Historians and sociologists have applied the term in additional ways. For them, *generation* may mean distinctive life patterns and values as they emerge by age; or it may connote not only distinctive life patterns but also a collective mentality that sets one age group apart from another.

An age group that has lived through a major social event, such as the Great Depression or 9/11, may exhibit characteristics that are not due to internal or biological aging and that are not found in other age groups. Thus, historical events in our younger years play an important role in shaping feelings and attitudes that persist throughout our lives. The impact of widespread social events on their survivors has not received enough systematic investigation; indeed, events that your parents or you have experienced may well continue to influence your lives. What, for example, are the effects of having lived through the Vietnam or Watergate era? The war in Iraq? If you were a young adult during the Iran-Contra affair or the Savings and Loan crisis, will you be more skeptical of politicians than your children, assuming no comparable incidents occur in their early adulthood? Would a young adult of the Vietnam era be less patriotic than you if you were a young adult during the 1990 war in Iraq? If so, one might conclude that events, rather than the aging process, make one more or less skeptical or patriotic.

Longitudinal Studies

How do we use theory and concepts to check whether they are accurate? To facilitate changes in behaviors, social policy, program design, and implementation, we need to understand from real people whether an idea really describes the human condition. Studying a group of people over several years is a longitudinal study. This research model provides tremendous amounts of information because individual changes can be observed over time and within context. Longitudinal studies are excellent examples of research tapping both psychological and sociological theoretical perspectives.

Longitudinal studies with a focus on later-life development are few and far between. They are costly to support, requiring groups of researchers to remain dedicated to the ongoing process over a period of several years. The data gleaned from such studies, however, cannot be replicated in any other way and are a tremendous source of information about growth and development over time.

The Maas-Kuypers Study

The Maas-Kuypers study tracked personality change in women and men over a 40-year period. They found personality to be stable throughout adulthood. Evidence in the research suggested those who had negative attributes in young adulthood had various negative personality characteristics in old age; fearful oldsters were rigid, apathetic, and melancholy as young mothers; anxious mothers were restless and dissatisfied in old age; the defensive elderly were withdrawn in early adulthood. The personality types that seemed most firmly connected to early adult-life behaviors were those with the most "negative" features. On the positive side, cheerfulness, lack of worries, and self-assurance in young adulthood seemed to match high self-esteem and self-satisfaction in old age.

The Elder-Liker Study

Another 40-year longitudinal study assessed the **coping** mechanisms and consequences for women who lived through the Great Depression of the 1930s (Elder & Liker, 1982). During the Depression, financial losses were more severe for working-class women than for middle-class women; and many working-class women, lacking the educational, financial, or emotional resources to master their circumstances, were too hard hit to make a comeback. The researchers suggest that the hardship of the Depression offered these women a trial run through the inevitable losses of old age. Economic hardships meant new challenges for women: coping with unemployed husbands, taking in boarders for pay, looking for work, borrowing money from relatives, and getting along on less by reducing purchases to a bare minimum. In short, women had to become more self-reliant.

The women under study experienced the Depression as young married adults in their early childbearing years. On the whole, the Depression diminished the emotional health of the lower-status women but increased that of the middle-class women. The middle-class women who struggled through hard times turned out to be more self-assured and cheerful than a "control" group of women who had not been deprived. They were also less fretful, less worrisome, and less bothered by the limitations and demands of living. Hard times left them more resourceful, with more vitality and self-confidence. In contrast, Depression losses added to the psychological disadvantage of working-class women, lowering their self-esteem and increasing their feelings of insecurity and dissatisfaction with life.

The conclusion was that a life history of mastery enables women to manage traumatic experiences throughout their lives. Women like the middle-class women in the study are lucky if they have the resources for such mastery—economic resources do matter significantly. Coping skills are acquired in hard times, not in tranquil ones. One who experiences no hard times until old age may not have developed coping skills. "Neither a privileged life nor one of unrelenting deprivation assures the inner resources for successful aging" the study concluded (Elder & Liker, 1982).

The Baltimore Study

The Baltimore Longitudinal Study of Aging, begun in 1958, is an ongoing study of how biology and behavior change as people age. Every two years participants go to the National Institute of Aging's Gerontology Research Center in Baltimore, Maryland, to undergo two-and-one-half days of tests. In addition to measuring physical functioning, the tests assess attention, problem solving, memory, personality, and other behaviors. It is the longest running study of human aging and hundreds of professional journal articles have been generated from its assorted measures and findings. In the personality realm, the study supports the idea that stress appears to be a key modifier in the extent to which personality continuity takes place. Adults seldom respond passively to stress: they change the circumstances if possible (problem-focused coping) or they work to change the meaning of the stressful situation (emotion-focused coping) (Cacioppo, Hawkley, Rickett, & Masi, 2005; Hoyer & Roodin 2009).

Cross-Sectional Studies: Coping and Adaptation

Stress, we can see, accompanies the need for change—and the need to adapt or change is integral with progressing on the life course. One classic study monitoring the stress patterns experienced when older adults were relocated to different nursing homes found the very old to be unique. Continuity theory did not apply. Their roles had changed, their bodies had declined, important people in their lives had died. Issues of finitude and death were close to their hearts. Thus, the psychology of the very old assumes new dimensions. Discontinuity with earlier periods of life is the rule as the very old adapt to stresses such as relocation. Their coping patterns have changed, and their psychological survival seems to hinge on preserving a sense of self. Survival strategies involved adopting myths of control and turning to their past to maintain self-identity. In other words, imagining themselves to be in control of their health and surroundings, and "living in the past" were successful coping strategies for the very old (Lieberman & Tobin, 1983; Parker, 1995).

Emotional problems are not physical in origin, but they may be related to physical losses. The older one is the more likely one is to face physical disability and imminent death. Loss of hearing, loss of vision, and loss of the use of limbs have a strong psychological impact at any age. So does being told you have only five years, six months, or two weeks to live. Other age-related losses are the death of a spouse, of siblings, and of friends. Retirement, the loss of one's driver's license, the sale of the home, or a move to an institution are the types of losses that scare people about aging. They do not happen to every older person, but the longer one lives, the more likely are these events to happen. Cross-sectional research indicates considerable difference between the young-old (65 to 74) and the old-old (75 and over) as to the losses they can expect.

Even in the face of loss we continue to find that older persons, on the whole, adapt well to loss. Most older adults adapt to the loss of their spouse without severe psychiatric repercussions, though the grief must be lived through. Adaptation does happen: the widowed older adult has learned through past experience that life does go on, even when one feels sad, and that sadness dissipates over time. The *caring* about the person does not necessarily dissipate, but the deep sadness does.

The personality characteristic **locus of control** came under careful study in the 1980s and continues to be examined in the literature. Locus, or center, of control is considered internal if the person sees that his or her own actions bring about a reward or positive change. If, however, the person sees rewards due to fate, luck, change, or powerful others, the locus of control is external. Research has found locus of control to be a long-standing personality component developed over years of positive and negative reinforcement. A person with an internal locus of control feels more control over the environment and is more likely to attempt to improve his or her condition.

Older people experience higher life satisfaction if they possess an internal locus of control (Diener, Lucas, & Scollon, 2006; Lucas, 2007; Lucas, Clark, Georgellis, & Diener, 2003). Thus, both psychologists and sociologists have reason to study locus of control. Changes in the social environment can facilitate the development of internal control. By allowing older individuals more self-determination and administration on policies that affect them and by increasing their involvement, control, and power in all aspects of their social and political lives, we can help them to develop greater satisfaction with their lives.

An older person who has received positive support from a cohesive family tends to have a strong internal locus of control. Social support, such as tangible assistance and emotional help, also tends to increase feelings of control, but only to a point. Beyond this point, additional support can decrease feelings of personal control. We need a word of caution concerning beliefs about both extreme internal and external loci of control: both extremes may hinder older adults' abilities to cope with stressful life events.

Integration into a social group, be it family or friend networks, is central to high life satisfaction.

Chapter Summary

This chapter considers the progression to late life from social and psychological perspectives. From this view, individuals move through stages, or identifiable eras, in their development. A complex interaction of internal (biological) and external (social) factors combine to shape the person as we age. Expected behavior patterns exist at every level of adulthood, whether young adulthood or advanced middle age. Ageism in a society can be internalized to fear of aging within the individual. Some individuals experience a traumatic transitional crisis at midlife, but it is unclear whether or *if* this crisis affects

all people, whether it affects every person in the same way, or to what extent awareness of "midlife" is cultural. Psychologists coming from a biological perspective say it is universal to our human condition; psychologists coming from a social perspective and sociologists say it is cultural. Increasingly more people studying older adulthood say it is both.

Personality theory draws attention to how personality changes as a function of age and how individual variations in personality may affect one's own aging process. Continuity theory suggests that personality remains stable throughout life and is based on trait theory. Whether stable or changing, the struggle to cope with the world interacts with personality, the historical moment, and the context of the individual to shape the aging process. Studies of locus of control address the degree to which older people feel in control of their lives.

Perhaps most importantly, all theorists studying aging believe there is profoundly important human task to be dealt with in later life and that is the hard work of making coherent meaning of all of life's experiences. It is the task of pulling together a meaningful life story, ending with a sense of *integrity*.

Key Terms

emergent self
epigenetic principle
Erikson's stages of
    human development
extroversion
gerotranscendence
interiority

introversion
mechanistic metaphor
metaphor
narrative
organic metaphor
theories
transcendent self

Questions for Discussion

1. In small groups, make up one list of everything you fear about aging and a second list of everything you look forward to in life as you age. Discuss the basis of some of your fears about aging and the reasons you look forward to certain milestones. What are some of the changes you feared or looked forward to when you were younger?

2. Can you think of an example of reciprocity in your life? How might that change as you get older?

3. In small groups pretend to develop a character for a movie script. Your character will go through all of the stages of human development as described by Erikson from birth to old age. Decide what sort of personality your character has and how he or she responds to each stage of life. What occurs in your character's life that leads to "successful" or "unsuccessful" aging?

4. Imagine you have unlimited resources to develop a longitudinal study that examines some aspect of aging. What would you like to study about aging? How would you set up your study?

5. Imagine you have unlimited resources to develop a cross-sectional study that examines some aspect of the differences in behavior between two or more generations. What would you like to study about generational differences? How would you set up your study? Consider, for example, differences in social habits, family involvement, spiritual practice, or exercise habits.

Fieldwork Suggestions

1. Interview an older relative or friend about their aging experiences. Ask in what ways they feel they have remained the same with age and in what ways they have changed. How have they been affected by their physical changes? Their social changes? Has their outlook on life changed over the years? Do the responses of your interviewee seem to support the Continuity theory described in your text?

2. Watch a movie about midlife crisis and discuss how the characters respond to their own aging and the aging of those around them. Some good suggestions are *Shirley Valentine*, *Bridges of Madison County*, *City Slickers*, *Buena Vista Social Club*, *American Beauty*, and *Lord of War*.

3. Ask six middle-aged or older women their ages. Did you violate a norm? Did they answer you? What was their reaction? Ask six middle-aged men their ages. Are responses the same? What did you expect?

4. Interview an older friend or relative and record his or her life story, with a recording device or notepad. . Encourage the story-teller to relate things to you that gave meaning to his or her life.

5. Visit a senior center and interview several members there. Choose one or two questions to ask each person and see how their answers compare. For example, ask several

people what is the most meaningful thing in their lives and who is the closest person in their lives.

Internet Activities

1. Beginning with the key search words *narrative therapy*, search the Internet for information about this emerging therapeutic field. Repeat this, using key words *Jung* or *Jungian psychology* and see what comes up.

2. Locate the Internet site of a university that is not in your state. Find a psychology department that has a faculty member doing research in human aging. What are the research interests of this professor? What classes does the department have that would provide more specific information about some of the topics touched on in this chapter?

4 Physical Health and Well-Being

No Slowing Down

By Dianna Smith

Frances Woofenden is an 84-year-old competitive trick water skier who also bikes 10 miles a day.

But it's her functioning knees that are her greatest source of pride.

She brags about them like some would their grandkids.

And why shouldn't she? While others with far younger joints dread a flight of stairs, Frances gracefully twists, turns and twirls like a dancer, while balanced on a slat of carbon graphite and tethered to a boat doing 28 miles an hour.

This unusual talent landed her a gig as the face and body and lifestyle of V8, the vegetable drink. The grandmother from suburban West Palm Beach zips across the water in television commercials and smiles from full-page ads in Women's Day and Martha Stewart Living.

People who know her wonder how long she'll last. Those who've never met her have trouble believing she's as old as she is.

Frances doesn't care.

"I never even think about my age," she says, "until my birthday rolls around and I see all those candles."

But when you're a water-skiing octogenarian and a "hot grandma," as Frances has been called, it's your age that gets attention, especially from the national media.

The questions those reporters come up with, though! They baffle Frances.

A CBS News correspondent, for example, said he was surprised she'd wear a bathing suit on national television.

"What was I going to wear?" she asks, still incredulous three years later. "Bloomers?"

Not Frances, who refuses to sacrifice style for her sport or sex appeal because of her age. She favors cotton-candy colored lipstick, pearly pink nail polish and a backless violet bathing suit. (She likes to match her suits to her beige and purple Nautiques ski boat.) And always, in water or on land, her gold hoop earrings, which make her feel "complete."

When she's not on water skis, she's cruising around in her Mercedes SL 500, which still smells of fresh leather.

The widow invites you to peek inside. "Isn't it cool?" she giggles.

Frances has two laughs. One so high pitched and bubbly you'd think you were talking to a high school cheerleader and another so loud and deep you wonder how it came out of her tiny 105-pound frame.

The bubbly laugh is the one you get when you ask for her secret to looking so good.

"My diet and exercise. I never overeat, no sweets, a lot of fruit," she says.

And, ever the pitchwoman for the company now sending her a handsome check, "I drink V8 juice."

Frances talks about her life like it's a novel nowhere close to being finished. And why should it be? She didn't even start skiing until age 50, when her late husband bought a boat after she voiced an interest in learning the sport. (Stewart, a doctor who died of cancer in 2005, was like that, she says fondly, surprising her with lavish presents.)

Since then, she's racked up over 100 medals.

Age simply doesn't matter, she says. It's really, truly about how you feel.

And Frances feels great. So great she's telling America.

She curls up on her cream leather couch and fumbles with the remote control.

She has the V8 commercial on a DVD and airs it often for visitors. But before she pushes play, she gives an enthusiastic preview of what's about to come.

She skied on a lake in California! The crew provided a professional driver! Even a chef for the day! And, best of all, she is the oldest person in the commercial!

There's a 66-year-old sky diver and a 67-year-old stunt driver. And then there's 84-year-old Frances.

Her bubbly laugh slips out again.

"Watch me, I come out of the mountains," she says, studying the commercial like she hasn't seen it before.

"I love it!" she yells, slapping her knee as she sees herself on the screen.

Source: Dianna Smith, *The Palm Beach Post*, Published: Sunday, June 28, 2009 at 10:33 P.M., Last Modified: Sunday, June 28, 2009 at 10:33 P.M.

Underscoring a fundamental truth about biological functioning, the phrase "Use it or lose it" is often used to refer to physical abilities. What causes the decrease in physical activity that we see in people as they age? Is it inevitable?

Statistically, good health declines with age. We do not yet fully understand, however, the role of the aging process in contrast to other factors that affect this decline. Poor diet, overeating, smoking, excessive drinking, misuse of drugs, accidents, and stress all affect our health. Some of the health problems that the elderly have may be an inevitable consequence of the aging process. Others clearly are not. Physical fitness and good nutrition are two critical factors that can affect the aging process.

To best fulfill their individual potentials, people of all ages, especially elders, need factual information about physical health, and their body's changing dietary and exercise needs.

The Aging Body: A Description

Aging is a gradual process beginning at birth. As guaranteed as biological aging is to all life, we do not know for certain what causes aging. For starters, we know that our life span is finite: we will not survive indefinitely. That period between birth and death, then, is our "life," no matter how long or short it might be. The **absolute human life span** is the maximum possible chronological age that the human can live. Although currently under debate, humans are considered capable of living about 120 years (Robine, Crimmins, Horiuchi, & Zeng, 2007). Because aging is species-specific, some geneticists say, we are not programmed by our genes to live indefinitely: the human maximum is around 120 years; the tortoise, about 150; the domestic cat, about 30 years; and the mouse, a little over 3 years.

There may be a biological limit to the number of times human cells can divide and replace themselves, but if so, we do not know what that limit may be or whether it is possible to extend it (Rattan & Singh, 2009; Wade, 2007).

Longevity is a complex trait influenced by genes, environment, and chance. A more useful measure of how long humans live than absolute possibility is the **mean human life span,** which is the chronological age by which 50 percent of humans will have died, according to statistical projection. Another way to think of longevity in this sense is how many years people are projected to live after they have reached the age of 65. From 1980 to 2003 there has been a steady increase in life expectancy for women and men in the United States, Canada, England and Wales, Japan, and France. The Japanese and French have longer life expectancies after the age of 65 than do Americans, the English, and the Welsh (Centers for Disease Control and Prevention [CDC], 2008a). At age 65, Japanese are projected to live an approximate additional 23 years; France is not far behind with a projected 21 additional years to live. Americans live approximately 17 years past their 65th birthdays—very close to that of England and Wales, which fall just below 17 years (CDC, 2008a). In all these countries, women outlive men by approximately three years.

As a person ages, his or her body systems reach peak levels of operation or performance and their functioning then remains constant or begins a slow decline. For some, peaks are reached in youth; for others in young adulthood or middle age; and for some others in old age if the potential of the younger years was never developed. Physical declines may result not just from the aging process but also from various pathologies (diseases), lack of proper diet and exercise, smoking, overweight and obesity, stress, and other factors. Some declines that once were attributed to aging itself take place at very different ages or not at all once disease can be controlled

for. An example of this might be the aging brain, most especially specific aspects of the brain such as memory.

This chapter will examine only that physical change that is *correlated* with age. These physical problems generally are considered to be pathological or disease related and may or may not be *caused* by the aging process.

Normal Aging

Many changes that take place in the body are observable: the skin loses its elasticity and becomes more wrinkled, the hair grays and thins out, and the body becomes less erect. Individuals get tired more easily and quickly. As early as the 30s and 40s, most people develop **presbyopia,** a condition in which near vision is impaired and the fine print of a book or newspaper becomes difficult to see at close range. Hearing loss may occur, especially among men. As the aging process continues, teeth may be lost or gums may develop disease. In addition, an older person tends to gain fat and to lose muscle strength, especially if he or she does not reduce caloric intake and is not physically fit. In most cases, the older people become the fewer calories their bodies need.

Further, the ability of the body cells to absorb calcium declines. This loss of calcium results in more brittle bones that are more easily broken. Aging brings more wear and tear on joints; thus, the likelihood of rheumatism and arthritis increases with age. Loss of calcium in bone mass resulting from lack of exercise and a poor diet, in addition to normative hormonal shifts, can result in loss of bone mass in the jaw, which, in turn, results in the loss of teeth. Dental issues or loss of teeth are not uncommon among older adults although tremendous advances have been made in educating the public about oral hygiene. Additionally, advances in technology have improved techniques for saving damaged teeth. In this century more older people are retaining their natural teeth than ever before in history to our knowledge.

Some health declines are not as apparent as others. The capacity of the body to achieve homeostasis (physiological equilibrium) declines with age. This means that older adults have greater difficulty "getting back to normal," biologically speaking, after a stressful event. Blood pressure and heart rates, for example, take longer to return to prestress levels. Various organs operate at reduced efficiency: the lungs decrease their maximum breathing capacity; the kidneys decrease the speed at which they can filter waste out of the blood; bladder capacity declines; and the level of sex hormones decreases. The nervous system also changes: reflex action remains constant with age, but reaction time declines. The digestive juices decrease in volume; consequently, the body takes longer to digest its food. The body's **immune system** decreases in its ability to protect a person from disease; hence, the older person is less immune to contagious diseases, such as the flu, and is likely to become more ill and take longer to heal than younger adults. Thus, body systems take longer to perform a task. In many instances this additional time is not apparent in the real world but can be measured clinically. In some cases, the slowed reaction time is apparent, as in the speed of emergency braking in a car for example.

Although we know that these changes take place, we do not yet fully understand why they take place. In other words, we do not know what causes aging. Some scientists believe that aging is not an inherent process at all but merely describes a "medley of unhappy outcomes" (Posner, 1995, p. 17). From this perspective, age is strictly the result of illness, and if we were able to keep ourselves completely disease-free, we would not "age."

A more useful understanding of aging, however, is to view it as a process in which body changes naturally occur, resulting in a general physiological decline in body functioning. This model of physical decline is somewhat offset by an intellectual and psychological increase in competence. In other words, body systems slow down and life experience accumulates with age. It is our wise use of previous experience that allows us, regardless of age, to continue to learn and develop mentally and spiritually. Remember: longevity is influenced by genes but also by our environment and sheer chance. So, for example, at the very biological time one's metabolism is becoming less efficient, the person may be more emotionally competent to take seriously the idea of self-care—more attuned to messages from the body and knowing the importance of taking responsibility for his or her health.

Biological changes such as slower reaction time or shifts in visual acuity can call into question the functional ability of older adults. One major social and family concern is how long or whether older adults should be permitted to drive. For example, of the changes the aging body undergoes, visual changes are among the most pronounced. We must have visual acuity as well as depth perception, in addition to sufficient reaction times, to be competent drivers. These systems change with age. At the same time, however, we know that **chronology** (age by year) is a poor indicator of **functionality** (ability to function). Declaring that people at age 70, for example, could no longer drive would be grossly unfair to those 70-year-olds whose functioning is unimpaired. This would be somewhat like declaring that 17-year-olds do not have the experience or judgment to make quick, critical driving decisions, and therefore the driving age should be increased to 20. Although drivers aged 55 to 65 are the safest drivers on the road by almost every measure, after age 75 collision rates increase dramatically. What are the social implications of age changes that might endanger the safety of others?

Although physical changes may significantly affect the ability of some older adults to drive competently, improved structural changes—such as larger road signs with more legible lettering and raised pavement reflectors—in driving conditions would create a safer driving environment for older adults. As more of our citizens reach healthy later life, we will need to adjust traditional transportation practices, policies, and options to accommodate the mobility needs of this group.

Physiological changes with age occur so gradually that many go unnoticed much of the time. The deterioration of body organs and systems with age may be fairly insignificant as it relates to an individual's ability to function independently, to get around, and to carry out normal activities. Nevertheless, the process of physical decline proceeds.

Health Status

The majority of people over the age 65 are in good health. The **health status of the aged** has improved markedly in the past 50 years. Although the health status of the average 45-year-old and the average 65-year-old probably differ, this difference is not great and is largely dependent on their lifestyle options and choices. People do not reach age 65 and suddenly become decrepit. However, physical decline does become more apparent with advancing age: those aged 75 and over usually have noticeable physical declines compared to the middle aged, and those over 85 can be seen to have even more noticeable declines. With advanced age come more numerous and longer hospital stays, more doctor visits, and more days of disability. But declines are typically very gradual.

Ninety-five percent of Americans aged 65 and over live successfully in the community; a smaller percentage are confined to institutions. Among noninstitutionalized elders, 86 percent have relatively minor chronic conditions, but the vast majority of older people with chronic conditions suffer no interference with their mobility. Nevertheless, functional limitations do increase with age. In 2005, 32 percent of women and 19 per-

Health status of the aged is far better in 2010 than it was in 1950.

cent of men aged 65 and older reported inability to perform one or more of the following indicator tasks defined by the Centers for Medicare and Medicaid Services: stoop/knee, reach over head, write/walk 2–3 blocks; or lift 10 pounds (CDC, 2008b).

Of noninstitutionalized elders functional limitations range from mild arthritis conditions to totally disabling ailments that interfere in their ability to perform **activities of daily living (ADLs)** or **instrumental activities of daily living (IADLs)**. ADLs include washing our hair, feeding ourselves, dressing, and taking a bath. IADLs are the abilities that allow a person to live independently: examples include doing light housework, preparing food and doing cleaning, shopping for groceries, and managing money. It is important to understand here that health statistics indicating the number of elders suffering from chronic conditions hide the fact that most older people manage quite well. The very old, 80 or older, are those who suffer most from disabilities and because they have adapted to their aging bodies over a long period of time, most continue to express life satisfaction: I'm doing pretty well for my age.

Disability among older adults has social consequences. As the population ages, the number of older people requiring medical and social assistance with disability management will increase proportionally, and if we add to this the fact that women generally outlive men by several years, the need for health care services becomes clear. In short time, especially as boomers reach older adulthood, health care and social programs will be impacted and there will be tremendous strain on established resources for well-being care of adults.

Chronic and Acute Conditions of Aging

The key health problems facing middle-aged and older adults today are those that are chronic, exacting a particularly heavy health and economic burden on individuals and on society. Young people tend to have **acute conditions,** that is, short-term illnesses in which the cause is known and the condition is curable. Chicken pox, colds, and influenza are examples. In contrast, the number and severity of chronic conditions increase with age, whereas acute diseases decline with age. **Chronic conditions** are long term. Their causes are typically unknown but the ultimate cost is profound on health, quality of life, and increased health care expenditures. Hypertension, arthritis, heart disease, diabetes, and cancers top the frequency list of chronic conditions affecting older adults.

Alarmingly, much of the disability and death currently associated with chronic disease is avoidable through prevention measures and lifestyle choices. Some chronic diseases emerge after years of poor eating, over eating or drinking, smoking, and lack of exercise. By practicing healthy lifestyles including regular physical activity, proper diet, and so forth coupled with the use of early detection practices such as mammography, and cervical and colorectal examinations, much of the chronic diseases of later adulthood could be avoided or the impact reduced (CDC, 2008a). Addressing health issues, however, requires a certain health literacy, and older adults are proportionally more likely to have poor basic health literacy than any other age group. Almost two-fifths of people age 75 and over have a health literacy level below basic, compared to 13 percent of people age 50–64 (CDC, 2008b). This fact can serve as a reminder to those not yet in middle or later years that the life we live *now* gives shape to our later old age life. Lifestyle, including educating ourselves, matters—and it is happening. When those health literate 60-year-olds reach 75, they will have entirely different middle- to later-life health experiences because of that literacy.

Table 4.1 lists nine suggestions for living a longer and more healthy life. This "Power 9 Pyramid" was developed by Dan Buettner following his global study of centenarians (Buettner, 2008). His quest was to determine what factors were common among very long-lived people. His question was "why have you lived this long?" and the answers he and his research team gleaned were distilled to nine points, repeated again and again in various forms appropriate to the elder's culture.

- Exercise, do not eat too much and eat mostly fruit and vegetables,
- drink red wine but do not drink excessively, do something you love,

TABLE 4.1

How to live longer: The Power 9 Pyramid*

1. Add simple activities throughout your day like walking farther than you must, doing gardening or home repairs, or running around with your pets or grandchildren.
2. Try eating from a smaller plate to decrease your portion sizes and reduce calories.
3. Limit the number of servings of meat you eat in a week.
4. Drink a glass or two of red wine most evenings.
5. Know your passions in life and take time to enjoy them most days.
6. Take quiet time to relieve stress.
7. Belong to a spiritual community and gather with them regularly.
8. Make your family and loved ones a priority; express that through your actions.
9. Surround yourself with friends who have healthy habits and support you in your goals.

Source: *Adapted from Dan Buettner's Power 9 Pyramid in *The Blue Zone*, 2008.

- have a meditative practice,
- belong also to a spiritual community and meet regularly,
- surround yourself with family and friends you love who support your goals,
- make these people your priority.

This is advice our grandmothers could have—and perhaps have—given us. Thoughtful, sensible, and something people can accomplish if they choose to. But not something one can suddenly "get." It reflects a lifestyle.

A chronic condition may or may not be disabling, depending on the type and severity of the condition. Loss of teeth can be considered a chronic condition but is rarely disabling. A chronic condition might be progressively debilitating. Parkinson's disease is one example. An older person may have a number of chronic conditions but not be severely limited by any of them. Imagine the 70-year-old with mild arthritis, mild diabetes, minor visual impairment, loss of teeth, and a mild heart condition. Although this person has five chronic conditions, he or she may remain active, vigorous, and unaffected in a major way by any of them. Conversely, another older person might become completely bedridden by just one severe chronic condition.

Sometimes the discovery of a medical cure for a disease can transform it from a chronic to an acute one. For example some forms of cancer, now considered acute, were once thought to be chronic. Some chronic conditions do not seem to be the result of pathology (disease). Instead, they are the result of the normal aging process. Several forms of arthritis fit into this category as does visual acuity.

Major Health Problems

Though people tend to assume that old age brings sickness, this is not necessarily true. Most major health problems of old age result from pathology—the presence of disease, most of the causes of which lie outside the aging process. Poor living habits established early in life, as discussed previously, cause many of the "diseases of old age." With preventive measures, these diseases can be avoided. About 75 percent of all deaths are caused by heart disease, cancer, and stroke. Although still the primary cause of death in older people, death rates from heart disease have somewhat decreased due to medical advances, modifications to diet, reduced smoking rates, greater knowledge of self-health, and better exercise habits (American Cancer Society, 2009).

Differences in mortality based on race and ethnicity are evident. In 2006 the death rate for the black population was 1.3 times that of the white population. That means the average risk of death for the black population was about 30 percent higher than that for the white population. This disparity reached its widest point in 1989 but since then death rates for both populations have tended toward convergence, declining by 23.0 percent for the black population and by 16.9 percent for the white population (CDC, 2009).

Heart disease is a term incorporating conditions of ischemic heart disease, heart attack, arrhythmias, heart failure, and hypertension and stroke. The most widespread form of heart disease, **coronary artery disease** or **ischemic** (deprived of blood) **heart disease**, is now the major killing disease in the United States and in other industrialized nations as well. Worldwide, coronary heart disease killed more than 7.6 million people in 2005 (CDC, 2007). It is the leading cause of death for older adults in the United States. Ethnic and racial differences exist, however, probably as a result of lifestyle differences including diet and exercise. Heart disease remains the leading cause of death for American Indians and Alaska Natives, African Americans, Hispanics, and whites (Kung, Hoyert, Xu, & Murphy, 2008), but among Asians and Pacific islanders, cancer is the leading cause of death with heart disease a close second (CDC, 2007). The incidence of heart disease increases with age.

If coronary heart disease results in deficient blood supply to the heart, heart tissue will die, producing a dead area called an **infarct** and the disease can lead to myocardial infarction, or **heart attack**. Heart attack can be acute, sudden, and painful—clearly identifiable; or it can be more subtle, creating a more generalized dizziness, weakness, confusion, or numbness. The symptoms have a broad range, and they are different in men than in women. Men and women have similar rates of heart attack after 65, but women are less likely to suffer this attack

TABLE 4.2

Risk factors for heart disease among adults (age 20 and older): 2003–2004

Risk Factor	Percentage of Persons
Hypertension or taking hypertension medications	32.1
High blood cholesterol	16.9
Physician-diagnosed diabetes	10.0
Obesity	32.0

Source: CDC (2008). Heart Disease Facts and Statistics.

before that age than are men. Even at younger ages, women are more likely to die from a myocardial infarction than are men—a fact that is still not clearly understood because of the limited research conducted on women's health issues (Hoyer & Roodin, 2009).

Risk factors for heart disease among adults are shown in Table 4.2. Note the roles of blood cholesterol, obesity, and diabetes—all risks that can be addressed if not eliminated, by lifestyle.

Two major disorders of the circulatory system are **atherosclerosis** and **hypertension**. Atherosclerosis is one of a group of cardiovascular disorders called arteriosclerosis (hardening or loss of elasticity of the arteries) (CDC, 2007). It occurs when fat and cholesterol crystals, along with other substances, accumulate on the interior walls of the arteries, thereby reducing the size of these passageways. The accumulation of some deposits of fat in the arteries seems to be part of normal aging; however, the increased incidence of heart attack in Western countries suggests that excessive deposits are linked to factors such as smoking, obesity, and serum cholesterol levels that might be controlled by lifestyle changes.

A major health problem associated with atherosclerosis is *thrombosis*, or blood clotting. Blood clots occur when undissolved fatty deposits in the arteries cut off the blood supply to the heart. Some of the factors that lead to atherosclerosis are thought to be a high cholesterol count, a diet high in refined sugars and saturated fats, high blood pressure, obesity, stress, hereditary factors, lack of exercise, and smoking.

Hypertension refers to excessive arterial blood pressure. It is present in about 31 percent of people age 65 and older, and 22 percent of those in the 45- to 64-year age range (CDC, 2007). Factors associated with risk of hypertension include obesity, smoking, and excessive alcohol consumption. Some pain medications (nonsteroidal anti-inflammatory drugs) may also increase blood pressure when taken over a long period of time. Hypertension is responsive to medications that rid the body of excess fluid and sodium, as well as to diet and exercise. Once again the issue of lifestyle choices emerges in relation to positive aging.

Cancer may affect the breast, the skin, the stomach, bones, blood, or other parts of the body. Its common characteristic is an uncontrolled, invasive cell growth at the expense of normal body systems. There may be numerous causes of cancer. Some are thought to be the inhalation or ingestion of chemicals, smoking, improper diet, and radiation. Although the basic cause of cancer is not fully understood, a living cell somehow becomes a cancer cell. This cancer cell then transmits its abnormality to succeeding cell generations. When the wild growth of cancer cells is not eliminated from the body, tumors develop.

Anyone can develop cancer: it remains one of the major killing diseases of older people in the United States. Seventy-eight percent of all newly diagnosed cancer cases occur at 55 and older in the United States (CDC, 2008a). That makes cancer the second leading cause of death among older adults and also a leading chronic condition among older persons, following heart disease. Many cancers are treated for some time before the reserve capacity of the elder is gone, or the elder has passed the five-year point of "cure." This has implications for treatment: a 70-year-old woman, for example, has a remaining life expectancy of 15 years. If she has physiologic declines in addition to cancer, her ability to function in daily life may become impaired (Parkin, 2006).

A survey of more than 1,500 older adults with cancer sought to determine the impact of the disease on their abilities to function in addition to their utilization of health care (Stafford & Cyr, 1997). ADLs such as getting out of a chair and IADLs such as going shopping were determined. It was found that cancer increased the use of health care resources but only modestly reduced physical function. Elders with cancer might require not only the care of an oncologist but also the care of a geriatrician to help determine appropriate services to maximize their functioning with the disease.

Some cancers can be completely eliminated—those caused by smoking and heavy use of alcohol for example. About one-third of cancer deaths in 2008 were related to obesity, physical inactivity, and poor nutritional patterns (American Cancer Society, 2008). Lifestyle and personal choices make a difference in terms of living into a *healthy* old age.

Arthritis, which results from the inflammation of a joint or a degenerative change in a joint, is one of the oldest diseases known and is widespread today, affecting all age groups, but mostly older persons. There are numerous types of arthritis, with different causes, different symptoms, and differing degrees of severity. Rheumatoid arthritis and gout, though rare, are painful and troubling forms. The type of arthritis that most often poses problems for those 65 and over is osteoarthritis.

Osteoarthritis, the most common form of arthritis, is fairly widespread in middle age and almost universal in old age. According to annual estimates, about 46 million adults in the United States—that's about one in five—report doctor-diagnosed arthritis (CDC, 2006). The joints most commonly affected are the weight-bearing ones: the hips, knee, and spine. The fingers and big toe are also commonly affected. With osteoarthritis, elastic tissue (cartilage)

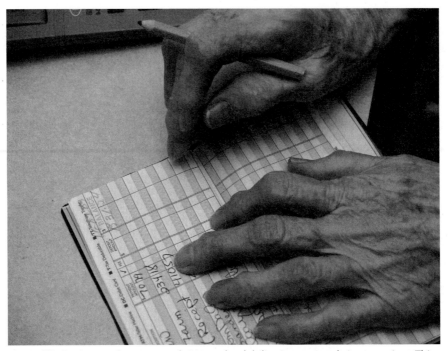

Physical limitations such as osteoarthritis can be debilitating or merely inconvenient. This woman continues with ordinary life tasks despite her arthritic body.

becomes soft and wears away, and the underlying bones are exposed, which causes pain, stiffness, and tenderness. For some, osteoarthritis starts early in life and affects mostly the small joints. For others, osteoarthritis results from injury or vigorous wear and tear (the athletic knee is an example). It occurs later in life in the large or overused joints. Therapies for osteoarthritis include anti-inflammatory drugs, steroids, exercise, reduction of strain on critical joints, and surgical procedures such as hip and knee replacement (Hoyer & Roodin, 2009).

Osteoporosis itself, another form of joint and bone degeneration, has no specific symptoms. The main consequence of this silent disease is risk of bone fractures, especially of the vertebral column, rib, hip, and wrist (Tang, Eslick, Nowson, Smith, & Bensoussan, 2007). It is characterized by a gradual loss of bone mass (density) that generally begins around the age of 50, and the number of people affected

and the severity of the disease increase with advancing age. Estrogen deficiency following menopause is correlated with rapid reduction in bone density. In men, a decrease in testosterone has a less pronounced but comparable effect. The condition is more common among people with European or Asian ancestry, especially Northern European women (Bolland et al., 2008).

Older women with smaller body size are at increased risk of osteoporosis because of lower hipbone mineral density (Shapses & Riedt, 2006), but some risk factors are potentially modifiable. Some of these include moderate alcohol consumption, appropriate vitamin D blood levels, no use of tobacco products, appropriate body mass index (BMI), good nutrition, physical activity—neither too little nor excessive, and limited consumption of soft drinks (Cauley et al., 2007; Mayo Clinic, 2007; Wong, Christie, & Wark, 2007).

Physiological change happens with age, but health and vigor are maintained through exercise. Many geriatricians now believe that physical flexibility is one of the most important keys to a healthy body later in life.

For most people, osteoporosis can be slowed or prevented by a high-calcium diet. The vitamin D metabolite calcitrol has been shown to help correct problems with calcium absorption and has been particularly useful with steroid-induced bone loss. There is growing evidence of the effectiveness of estrogen treatment for women following menopause; however, uncertainty and controversy remains about whether estrogen should be recommended to women in the first decade following menopause (Bolland, 2008; Lyles et al., 2007; Mayo Clinic, 2007).

Scientists continue to search for biomedical treatments of osteoporosis and arthritis. In 2004 markers were identified that are early indicators of progressive disease in rheumatoid arthritis (RA), a form of arthritis. From there early and aggressive baseline treatment was developed to stop or slow the development of the arthritis (Goronzy, 2004). As our biomedical understanding of the process of arthritis and musculoskeletal disease continues to develop, health and well-being in later life will be dramatically altered in a positive way.

National campaigns are mounted each year to teach about prevention of osteoporosis. Weight-bearing exercise should start in younger years, as well as a good diet, and both be continued in old age to slow this disease. Exercise helps maintain bone strength: new bone cell formation is stimulated by exercises like jogging, walking, and weight lifting.

Accidents

A health problem of major proportion is disability or death due to accidents, resulting in hospital admission (CDC, 2008b). In the

Good health, activity, and meaningful activity are key to successful aging.

because bones are often weakened by osteoporosis, a hip fracture is a common ailment to be feared. One-third of women (those most likely to have osteoporosis) and one-sixth of men who live to age 90 will suffer a hip fracture (CDC, 2008b).

Approximately 250,000 hip fractures—the most serious fracture—occur yearly at and above 65 years of age. Such needless tragedy among advanced age groups illustrates a need for more preventive measures. Organizations working with older adults increasingly incorporate "fall clinics" or training programs designed to reduce falls. These programs include exercise training for strength, balance, and flexibility. A review of an older adult's medications to determine if any side effects inhibit balance and home modifications like installation of grab bars and better lighting are just some ways to help older people to reduce or avoid home falls (CDC, 2009).

Other factors of age that contribute to accidental death or injury are failing eyesight and hearing; reduced muscular strength, balance, and coordination; and increased reaction time. If these limitations are combined with impaired judgment, the very old are especially vulnerable to accidents. They may try to lift loads that are too heavy or poorly balanced, or they may climb or reach overhead without sufficient strength to manage the task. Changes in automobile traffic conditions can happen too swiftly for the older driver to react. The swallowing reflex diminishes with age, making choking on food or objects in the mouth more likely. Because of diminished olfactory senses danger signals of fire or leaking gas may not be perceived as readily, and the few minutes or seconds of delayed reaction may prove to be fatal.

Obesity and Diabetes

Since 1991 rates of obesity among older adults have increased dramatically, as illustrated in Figure 4.1 based on data from 1991 to 2001. The substantial obesity increase in adults of all

United States one of every three people 65 and older experiences falls each year resulting in hospital admission (CDC, 2009). Falls are the most common cause of accidental death among those 65 and over, followed closely by motor vehicle accidents. Suffocation from an object that has been ingested, surgical and medical mishaps, fires, and burns are also major causes of accidental death among elders. The 90-and-over group has an alarmingly disproportionate share of deaths due to accidents in two of these categories: (1) falls and (2) suffocation by ingestion of food (CDC, 2008b).

A real danger exists when an older person falls that a hip will crack. Because we instinctively use the hip to absorb most falls and

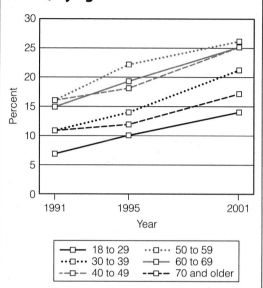

FIGURE 4.1

Proportion of adults who are obese, by age

Percent (y-axis: 0 to 30)

Year (x-axis: 1991, 1995, 2001)

Legend:
- 18 to 29
- 30 to 39
- 40 to 49
- 50 to 59
- 60 to 69
- 70 and older

Source: Centers for Disease Control and Prevention (2003). Overweight and Obesity Trends. Available at: http://www.cdc.gov/nccdphp/obesity/trend/prev_char.htm

ages in 2009 suggests that obesity among older Americans will be an even greater problem in the future. Some groups of people, such as those with lower incomes and less education, have higher rates than do others. Non-Hispanic whites age 51 and older have the lowest rates. Longevity and quality of life factors are profoundly impacted by obesity: obese adults have numerous chronic conditions, medical utilization rates are higher, symptoms of depression often accompany the condition, and general symptoms of illness are more common than for those who are not obese (Center on Aging Society, 2003).

Excess body fat has long been considered to be a risk factor in coronary heart disease, but the exact relationship is still unclear. A study of over 3,700 Japanese-American men age 71–93 from the Honolulu Heart Program suggests that the relationship of body weight to heart disease might be more complex than previously believed. The BMI, or weight divided by height in square meters, is commonly used in research as an estimate of overall weight. The Honolulu Heart study found that among men the waist-to-hip ratio (WHR) with elevated cholesterol (HDL-C) were more strongly related to coronary heart disease than was peripheral fat accumulation, or excess weight accumulated more uniformly on the body (Kung, 2008). This appears to be a particularly indicative factor for men but not necessarily among women. The data are yet incomplete regarding these gender differences.

The solutions to difficult problems are oftentimes simple, logical, and wrong.

H. L. MENCKEN, 1939

Although it may have a more pronounced impact on **morbidity** (quality of life) than on **mortality** (death rates) in later life (Kaplan et al., 2003), obesity is linked to diabetes and hypertension and all are risk factors for heart disease (Table 4.2). Although obesity is a major risk factor for chronic disease, economic, and quality of life issues, there is uncertainty in obesity treatment in older adults because of possible harmful effects of weight loss on muscle and bone mass (Villareal et al., 2005). We can see that the way in which one lives one's life before reaching older adulthood can have profound impact on the shape of that final third of life. Health management needs are cumulative and are complex in older adulthood.

Some people survive an emergency cardiac episode and some do not. Timely access to emergency cardiac care and survival is partly dependent on early recognition of heart attack symptoms and immediate action (CDC, 2008a). In a 2005 survey most persons (92 percent) recognized chest pain as a heart attack symptom but only 27 percent classified all symptoms and knew to call 911 when they thought someone was experiencing these symptoms (CDC, 2007).

Why We Age: Theories of Aging

The cover of the American Association of Retired Persons (AARP) bimonthly magazine declared in August 2004 "Sixty Is the New Thirty!" Does that mean then that 110 is the new 80? Must we grow old despite the number of years lived? Is there any way to stop aging? Human beings have asked these questions for centuries.

The longest-lived person on record was Jeanne Calment of France who died in August 1997 at the age of 122. This age currently represents the **maximum life span** of the human species—the greatest age reached by a member of a species. Biological aging is *senescence,* or the onset of the degenerative process, one that usually becomes apparent between the ages of 40 and 45. Graying at the temples, crow's feet around the eyes, and the need for reading glasses to correct nearsightedness (**presbyopia**) are among the early indicators that the process is underway. This is part of normal aging described previously.

Medical technology is slowing the aging process by increasing our fitness and vigor along with our **average life span,** or average age reached by the members of a species. No physician or anyone else has ever saved another's life—they have only prolonged it. A child born in the United States today has a life expectancy of approximately 80 years. Someday, life expectancy may be 130 or more years—an increase that will require profound social, political, and economic changes. The extension of life, although desirable, also challenges society to ensure that the quality of that life will justify the efforts to extend it. As Thomas Perls, a Boston University geriatrician, stated: "We are not try-ing to find the fountain of youth. If anything, we're trying to find the fountain of aging well" (Smith, 2005).

Research on the causes of aging has been undertaken from two primary directions: one is to search for the "aging gene," or understanding what causes cells to die. The second is to search for the "longevity gene," or a determination of what stimulates cell repair and regeneration. These two approaches represent a "damage" theory or a "programmed" theory point of view.

Damage theories are based on the idea that aging is the result of accumulated errors from sources such as free radicals. Through constant bombardment of metabolically generated free radicals, structural damage is accumulated in our cells. Oxygen free radicals are believed to increase the severity of, if not cause, diseases such as diabetes, stroke, and heart attack (Mayo Clinic, 2007).

Almost 40 years ago, Leonard Hayflick discovered that our cells duplicate up to 50 times in vitro (in the test tube) before they become senescent and stop reproduction. That finite number of replications, his theory holds, is based on damage as a by-product of normal metabolism. On the basis of that breakthrough in cell longevity geneticists have been able to isolate genes that cause certain cells to either age faster (go through their 50 duplications sooner) or extend the number of divisions to 100 and more (Sinclair & Guarente, 2006). The interpretation of research findings has proved difficult because discriminating between *causes* of aging and *effects* of aging is often impossible (de Magalhaes & Sandberg, 2005).

Cell damage may result from changes in tissues due to intrinsic cellular mechanisms, or changes in one tissue may predominate. Mattson and colleagues (2002) argue that aging is located in one tissue such as the brain; others argue that the aging process takes place in all cells (Mokni, Elkahoui, Limam, Amri, & Aouani, 2007; Trifunovic et al., 2005; Yang et al., 2007).

Programmed theories of aging state that aging has a strong genetic component and is not a product of a random process of damage, but is driven by genetically regulated processes. A family of genes have been discovered that are related to an organism's ability to withstand a stressful environment. They keep the organism's repair activities going strong regardless of age and, in so doing, maximize the organism's chance of getting through a crisis, such as reduced food or water or excessive cold or heat (Sinclair & Guarente, 2006). These genes, longevity genes, represent the opposite of aging genes.

One major study on the causes of deaths of centenarians found that 100 percent of those over 100 years of age succumbed not to "old age" but to organic failure cause by cerebrovascular disease, respiratory illnesses, cardiovascular diseases, and gastrointestinal disorders (Berzlanovich et al., 2005). By eliminating disease in the body as it occurs, medical intervention theoretically might keep the body in good health indefinitely. Still, most scientists try to separate the aging process from disease, because it is apparent that people age even when no disease is afflicting the body.

Let us consider these issues of the role of genes in longevity. We know that human longevity runs in families (Warner, Miller, & Carrington, 2002). Investigations of twins confirm that human life span is inherited. Identical twins die within a relatively short time span of one another, whereas siblings (sharing less genetic patterning) have a greater variation in life spans. A number of genetic theories have arisen to explain these differential rates of aging. One group is called programmed theories. They emphasize internal programs, or coding inherent to the cell. Some genetic theorists presuppose a **biological clock** within us that begins ticking at conception. This clock may be in the nucleus of each cell of our body, an idea that advances the proposition that the body is programmed by specific genes to live a certain length of time.

Geneticists once speculated that the biological clock was governed by a single gene. They now believe that thousands of genes are involved. The study of genetics became dramatically more sophisticated with the advent of genetic cloning and gene splicing. The identification of genes, such as SIR2, that are related to stress resistance and life span suggests they are part of a mechanism inherent in surviving adversity (Sinclair & Guarente, 2006).

Do we really know that the human life span cannot be extended well beyond 120 years? What is the nature of the biological clock? How does it affect growth, development, and decline? Scientists in the fields of molecular biology and genetics continue to search for answers to these questions.

Related to the search for longevity genes is the search for genes that are responsible for hereditary diseases. More than 150 mutant genes are now identified and by midcentury, DNA tests are almost certain to be a part of medical exams. From a sample of the patient's blood, doctors will be able to spot genetic mutations that signal the approach of hereditary diseases and also breast cancer, heart disease, and diabetes (Mokni et al., 2007).

Genes for disease are being identified at a rapid rate, including mutations responsible for Alzheimer's and colon cancer. Genes of neurodegeneration have been identified in the mouse in the form of a defect in a copper ion transporter (de Magalhaes & Sandberg, 2005). Most likely multiple genes are involved in Alzheimer's, and all these have yet to be found. Once a gene or gene cluster is found, scientists will work to develop a test to determine its presence or absence in the body. Discovering the genes for longevity in mammals cannot be far behind.

It is clear that there are multiple mechanisms to aging: it is a complex biological process that is characterized by disorder and decline and requires the approach of integrative biology rather than the single-focus approach of a distinct biological discipline (Yang et al., 2007). With support for research on the human genome, however, biogerontology is set on an invaluable path of gathering basic information on wellness, disease, and longevity.

Programmed immune system

A most promising "programmed" theory about aging is the immune system theory. Many aspects of immune function decline with advancing age as the body progressively loses its ability to fight off disease, and this decline is related to many kinds of disease, such as cancer, that attack the body. If the body's immune system becomes decreasingly effective with advancing age, harmful cells are more likely to survive and do damage. Some scientists are trying to revitalize the ailing immune systems of elders with hormone therapy, specifically the hormone DHEA (Trifunovic et al., 2005). Studies of mice have shown that this hormone restored their immune systems to youthful levels in warding off certain diseases such as Hepatitis B. Experiments with testosterone show it to increase muscle strength and counter anemia. In addition AIDS research is bringing us more information about strengthening the immune system and hopefully will help prolong lives.

A related immunological theory of aging suggests that as the body ages, the immune system seems to increase its capacity for autoimmune reactions and it develops more and more autoimmune antibodies that destroy cells, even normal ones. Several diseases, such as midlife diabetes, are related to autoimmune reactions, thus leading to the theory that such reactions cause aging. This idea is criticized because most autoimmune diseases, in fact, begin to develop at younger ages, but the impact of their *consequences* affect the quality of life of elders. Once again we see the complexity of untangling cause and effect: does disease cause cell damage and therefore aging, does the aging cell cause damage, or might it be a complex interaction of both "longevity genes" and "aging genes?"

One of the oldest and most enduring of theories is the cellular theory of **wear and tear,** which is the idea that irreplaceable body parts simply wear out. The cell is viewed as a highly complex piece of machinery, like an automobile. Some organisms live longer because they maintain themselves more carefully. Those who live more recklessly will wear out sooner. This idea ignores the fact that cells can repair damage caused by wear and tear.

Free Radical Damage

People seem to vary in their ability to fend off assaults to the body such as smoking, eating too much fat, and abusing alcohol. Some smokers, for example, manage to live long lives with no impact on their longevity. Winston Churchill and the longest-lived person, Jeanne Calment, are examples. Yet others who are health conscious and diligent in self-care succumb to cancer or other diseases before reaching old age.

One explanation for variability in longevity is that some people are more susceptible to free radicals than others. **Free radicals** are highly reactive molecules in the body that can damage cells, proteins, and DNA by altering their structures. They are the by-product of normal metabolism, produced as cells turn food into energy. A way to combat aging is to trap the damaging free radical molecules before they can do harm.

Recent research has been investigating the relationship of free radical theory to Alzheimer's disease. Harman (1995) hypothesized that Alzheimer's disease is caused by increased free radical reaction levels in brain neurons that over time advance patterns of cell loss. His conclusions suggest that the incidence of Alzheimer's may be decreased by efforts to minimize free radical reactions. Harman further speculates that antioxidant supplements taken by the general population may decrease the incidence of Alzheimer's.

Scientists have discovered chemical agents that absorb free radicals and thus prevent

Health in later life is no secret. Our grandmothers told us: Eat right, get plenty of sleep, speak kindly, and exercise.

cell and tissue degeneration. For example, a compound called PBN administered daily to aged gerbils restored the function of oxidized proteins in their brains (Spindler, 2005). The gerbils' ability to run through mazes improved, and they had fewer strokes from brain damage.

The older we become, the more free radicals we produce. Fortunately, naturally occurring antioxidants can be found in colorful fruits and vegetables: spinach, broccoli, red apples, cranberries, blueberries, cherries, grapes, as well as in chocolate and red wine. That is the reason why people eat five to six servings of fruits and vegetables every day. The role of antioxidants has been documented as critical to brain health because the brain has one of the highest percentages of fats in the body and

it is in our fats that free radicals inflict much of their damage (Nakazawa, 2006).

Longevity: The Role of Diet, Exercise, and Mental Health

There is a saying that everyone wants to live forever, but no one wants to grow old. And there is another: If I'd known I was going to live this long, I'd have taken better care of myself! Research on longevity holds a fascination for all of us—wondering whether science can find the key to knowledge that would keep each of us on the planet awhile longer. The previous section covered some biological theories that hold promise for understanding the mysteries of aging. Scientists are also looking at some other factors: diet and exercise, and social, emotional, and environmental factors. There is no one answer to living a good long life.

Diet

Research has come up with some astonishing findings regarding diet and longevity. In 1935 scientists at Cornell University discovered that reducing an animal's usual diet by 50 to 70 percent could extend its life span by 30 percent or more. These studies have been replicated numerous times in the years since then. Consistent findings demonstrate the animals not only live longer, but they are also more healthy, exhibit fewer incidents of cancer and heart disease, have better immune systems, and have a much lower incidence of diabetes and cataracts (Smith, 2005).

Calorie restriction clearly increases longevity in a number of species, though it is still unclear how this works (Beckman, 2004). Reducing a human's food intake by 30 to 50 percent might surely cause some serious interpersonal problems, if not physiological and psychological ones. Caloric restriction to the extent of the rat studies is not recommended for humans, but some intriguing findings continue to emerge. It appears that disease and other pathology become delayed to the end of the absolute maximum life span of the animal: the animals remain remarkably healthy up to the point that they become ill and die. Morbidity is delayed, and mortality has become compressed in these research animals (Spindler, 2005).

"The outcome of caloric restriction is spectacular," stated Richard Weindruch, a gerontologist at the University of Wisconsin and a pioneer in this particular field. He tested caloric restriction in animals from protozoa to rats, from dogs to monkeys. The studies may be telling us we are tampering with fundamental aging processes. The animals on caloric restriction acted friskier and suffered fewer diseases than their control groups. Tumor growth was reduced by at least 30 percent and some cancers were virtually eliminated. Thus, many usual causes of death were stripped away. A 22-year study of 19,297 men revealed that those at their ideal weight (determined by height) live longer than those who are only slightly above (2 to 6 percent) their ideal weight. Those men 20 or more pounds above their ideal weight suffered a major loss in years lived (Manson & Gutfeld, 1994). And from yet another culture: those who live on the island of Okinawa eat 60 percent of the normal Japanese diet. People live longer on Okinawa and have half the percentage of heart disease, diabetes, and cancer of those on the main island of Japan. It would seem that by restricting caloric intake, scientists can cause age-sensitive biological parameters— such as DNA repair, glucose regulation, and immune functions—to work better and longer. It appears that the decline in the immune function is at the root of many of the health problems faced by elders. Flu, for instance, tends to be more severe with age because immune responses are less vigorous. Low caloric intake helps protect the immune response system.

One outspoken voice among geneticists is that of Aubrey de Grey, a Cambridge University biologist who believes that venture capitalists and pharmaceutical companies have not invested enough in anti-aging research (de Grey, 2005). He states that because there is no short-term profit in it, the funding has not been made available for this research. He and his colleagues have created the Methuselah Mouse Prize to encourage competitive development of the longest-lived laboratory mouse. The prize, worth tens of thousands of dollars, will be granted to the genetics team that first engineers a strain of mouse that lives at least five years rather than the normal two years (Smith, 2005).

De Grey advocates what he calls an engineer's approach to aging: rather than slowing down the process of senescence, we become better at fixing damage as it happens (de Grey, 2005).

Nutrition

Research on nutrition is now uncovering how poor diet contributes to pathology (disease). Establishing the relationship between diet and disease is difficult, partly because the time that elapses before an inadequate diet results in disease can be substantial. Individuals may not be able to remember accurately their eating habits over a period of years. Nutritional cause and physical effect is difficult to determine.

Nevertheless, diet is increasingly being implicated as a factor in numerous conditions and diseases. Saturated fat contributes to atherosclerosis. A lack of fiber in the diet is thought to be one cause of cancer of the intestine or colon. With a low-fiber diet, the cancer agent remains in the intestine for a longer period of time. High fiber protects against constipation, intestinal disease, gallstones, and cancer. Diverticulitis, an infection or inflammation of the colon, may be caused by a deficiency of vegetable fiber in the diet. Research has shown that various nutritional anemias are almost certainly the result of poor diet. Similarly, studies have shown that proper dietary programs can control 80 percent of the cases of diabetes mellitus. As we grow older, our metabolic rate slows down. We require less energy intake, or fewer calories. Because of reduced kidney function, elders should eat somewhat less protein to help avoid kidney strain.

Only in recent years have we begun to understand the level of nutritional adequacy of food eaten by the average American. Good nutrition in early life is directly related to health and well-being in later life; therefore, the issue of healthy eating is also a social issue of health care for older adults.

Presently, in the twenty-first century, our awareness of the importance of fresh fruits, vegetables, and whole grains in our diet has increased. Americans are decreasing their intake of salt, red meat, and saturated fat. The statistics presented in chapter 1, showing reduced rates of heart disease, bear this out. Nutritional supplements, though controversial, are often recommended to older people whose diets might not provide adequate food-source nutrition. Additionally, specific nutrients have been associated with the enhancement of physical capability. For example, studies of older sportsmen find that vitamin C intake was associated with maximal oxygen uptake, as a measure of physical fitness, as is calcium citrate (Quinn, 2006).

Food and nutrition awareness has a new place in the American psyche. Books critical of eating patterns and government dietary standards have become nationwide best sellers, including Michael Pollan's *In Defense of Food* (2008), Wayne Roberts' *The No-Nonsense Guide to World Food* (2008), and the T. Colin Campbell and associates book *The China Study* (2006). In picking up a newspaper we are bombarded with headlines: Eat blueberries! Avoid fats! Drink green tea! One day soy is good, the next day we find the benefits have been exaggerated or taken out of the research context. Pollan's (2006) prescription for health and vitality is "Eat food. Not too much. Mostly plants." A healthy, sensible, doable concept.

Good nutrition, however, is not necessarily just an issue of preference. Some 38 million people in America are considered "food insecure"—they have trouble finding the money to keep food on the table (National Public Radio [NPR], 2009). Between 8 and 16 percent of old people in the United States experienced food insecurity in a six-month period, and this number does not reflect changes in life circumstance following the economic crash of October 2008. This means that somewhere at least roughly five million older people experience being hungry and not having access to adequate nutrition appropriate for the aging body.

Factors Affecting Nutrition and Diet

Physiological and sociopsychological factors can compound nutritional difficulties for the very old. Digestive processes slow down as part of the aging process. Dental problems can limit one to foods that are easily chewed. Reduced keenness of taste, sight, and smell can diminish enjoyment of food and dampen the appetite. Physical handicaps, such as arthritis, can complicate the preparation and consumption of meals. Lack of transportation to markets or loss of mobility poses further problems.

Less obvious, but also of great importance, are social and psychological factors. For example, a widow who has spent many years cooking for and eating with her family may find little incentive to shop and cook for herself when she is living alone. Older men living alone are even less inclined to cook for themselves than are women. Older adults on limited budgets who seldom leave their homes because of fear or disdain for shopping may settle for a diet of crackers, bread, or milk. Many have lost olfactory (taste and smell) acuity as well.

Food is a huge part of our culture. Even those older adults who live with families or in institutions may find it difficult to achieve adequate nutrition because of being offered foods that are not familiar or not consistent with their preferences. In some institutions the hurried, impersonal atmosphere of meals served at precisely 5:00 P.M. can discourage residents from eating as they had in their pasts. Fortunately, public exposure to the importance of fresh, local fruit and vegetables is evident. Cooking with these ingredients is not practiced by all groups of people or areas of the country, but it is making inroads in the American diet. Time will tell whether the current emphasis on healthy local foods is a short-termed fad or will become truly integrated in the American culture.

Exercise and Physical Fitness

A classic study of physical activity and health more than 40 years ago compared the incidence of coronary heart disease between London bus drivers and music conductors (Morris, Heady, & Raffle, 1953). Coronary heart disease was less by half among the conductors, compared to the bus drivers. The questions arising from the study were: do specific occupations produce a differential effect on heart disease morbidity, or are healthier and sturdier men selected for one job over another? From this study and others like it in the 1950s, research grew to assess the relationship of on-job and leisure-time physical activities and heart disease. The Framingham Heart Study in Massachusetts is one of the better known studies of this type. More recent studies, armed with more sophisticated capabilities to measure biological functioning, replicate the findings that physical activity is positively associated with lower heart disease. Additional research establishes that physical activity is a component in rehabilitation following cardiac illness, reduces the risk of total mortality by 20 percent; of cardiovascular mortality by 22 percent; and fatal re-infarction (second heart attack) by 25 percent (O'Connor et al., 1989). Paffenbarger and Lee (1996) summarized the state of this research:

> Since time immemorial, a physically active and fit way of life has been conceptualized as promoting health and longevity. But not until the 19th century did investigators use numerical quantification to show health benefits of physical activity by occupational categories demanding different degrees of energy output, and to demonstrate longevity differentials . . . (p. S12)

Patterns of exercise throughout the life and even exercise at any point *during* the life produce positive physical and mental outcomes: *some* exercise is better than *none*. Exercising for the first time in later life is better than believing it to be "too late" and not exercising at all. For most older adults, including those who are frail or ill, a program of strength training and flexibility exercises helps maintain mobility, improve quality of life, and prolong independence (CDC, 2009). Loss of mobility is a significant cause of loss of independence among the elderly.

It is a myth that older individuals are unable to exercise or to profit from it. Actually, exercise helps maintain good health, improves circulation and respiration, diminishes stress, preserves a sense of balance, promotes body flexibility, and induces better sleeping patterns at any age (Prohaska et al., 2006). It is now clear that most old people benefit substantially from exercise. The person who exercises reduces the risk of heart attack and, should one occur, increases the chances of survival. Swimming, walking, running, bicycling, and playing tennis are all valuable and inexpensive forms of leisure activity.

Much of the deterioration and many of the health problems and physical disabilities associated with age have been thought to be inevitable. However, many of the problems found in older people result directly from disuse of body systems, which results in decline. Disuse affects muscle mass, for example. Between the ages of 30 and 80, mean strength of back, arm, and leg muscles drops as much as 60 percent. Age-related changes in joints can lead to stiffness, which leads to limited range of motion. These trends can be slowed and, in some instances, actually reversed with proper exercise (Prohaska et al., 2006).

We commonly associate youth with supple, strong, erect bodies and old age with weak muscles, drooped posture, and low energy. Older adults, just as younger adults, can experience dramatic benefits from strength training exercises. Older individuals who participate in physical activities that constantly work their muscles will have a larger muscle mass than younger individuals who follow no physical fitness program. All unused tissues and functions atrophy. This can happen very quickly; it can occur in a matter of days while one is bedridden. With disuse, muscle tissue turns to fat tissue. Exercise prevents this from happening, but the exercise should be appropriate for the body's condition. For those who feel that high-impact aerobics are harsh, jarring, and harmful, low-impact aerobics may be ideal. Geriatric

medicine should address these issues; unfortunately, there are not enough geriatric physicians in the United States at this time.

An older person does not have to run marathons or enter competitions to get exercise, feel better, and stay fit. Many programs offer more moderate degrees of exercise. Light forms of yoga, stretching and relaxing exercises, and all kinds of dance and aerobics have been standard fare for elder fitness enthusiasts. Body fat and blood pressure may decrease with regular exercise. Nervous tension can also be reduced with vigorous physical exercise. People who become sedentary and who overeat lay the groundwork for the development of disease. Complaints of aches and pains in joints and muscles, low-back strain, high blood pressure, and other symptoms could be eased or eliminated with a physical fitness program.

Changing Activity Patterns

What are the constraints for elders to participating in exercise programs? The mantra "diet and exercise" has been repeated sufficiently that people of all ages surely understand the primacy of these two factors to good health and healthy aging. What, then, holds people back from participating in health-enhancing activities?

Constraints to exercise appear to be both universal and individual. One particularly useful study took a phenomenological approach by asking older people what they liked or did not like about their participation in an exercise program and the option offered them to participate (Whaley & Ebbeck, 1997). Some of the constraints that were voiced by elders were not those assumed by the researchers: "I don't like this activity—it's a women's class" and "I want exercise with a *purpose*." Clear gender differences emerged. Men felt that classes offered were "for women" and were not appropriate for them. Women were more likely to cite health-related reasons for inactivity. Women are also more likely to live in poverty and to have more chronic illnesses than men; this affects women's ability to participate. In addition to these, more universal themes developed related to the *lived experience* of the participants, which includes their history and the social context of their life course. Participating in an activity that has some product, like gardening, painting a house, or doing volunteer activity seemed important to some study participants, whose life experiences included a highly developed work ethic. To others, activities such as hiking, dancing, or sailing are most important.

It is never too late to start a strength training program (Prohaska et al., 2006). An early study of participants 87 to 96 years old showed dramatic improvements after eight weeks, several no longer needed canes to walk and all experienced "three-to-fourfold increases in strength" (Evans & Rosenberg, 1991). Similarly, other studies indicate that lifting weights greatly improves muscle strength, walking speeds, balance, and mobility (Mayo Clinic, 2007). Indeed, the American College of Sports Medicine now recommends the following as minimum exercise routine: some type of aerobic exercise for 20 minutes or more three times a week and some form of resistance training at least twice a week that exercises all the major muscle groups in sets of 8 to 12 repetitions each.

A study begun in 1987 of almost 200 master athletes aged 40 and over who compete in one sanctioned event each year (e.g., runners, swimmers, field athletes) asked the question, "Just how old and how fit can one become?" The study was extended to 2008 but early findings found support for the predictions that (1) speed and muscle strength will endure longer than assumed, (2) athletic performance will not decline significantly until age 60, (3) death rate will be reduced, (4) falls and injuries will be reduced, (5) the heart and lungs will not have to lose function as quickly as previously thought, and (6) the incidence of osteoporosis will be reduced. In general, we have underestimated the ability and potential of older people. The study concluded that to date nothing can retard the aging process as much as exercise (CDC, 2008b).

Mental Health

The connection between physical well-being and mental health is strong. Most mental health professionals and geropsychologists recognize, in Niederehe's words (1997), that "biological factors become more saliently intertwined with psychosocial ones in the mental disorders of late life, relative to typical problems of younger adults." (p. 102). Biology and psychology have the paradoxical relationship of being the same and being different, simultaneously. Older people who are involved in aerobic sports—sports that increase the heartbeat and respiration—for several hours a week felt healthier and happier than people who were not involved in such exercise (Payne, Mowen, & Montoro-Rodriguez, 2006). The subjective sense of being physically fit—feeling good about one's health and body—predicts better mental health. Being a happy, optimistic person in turn contributes to longevity. The 72-year-old longitudinal study beginning in 1937 of Harvard sophomore men (Paffenbarger & Lee, 1996; Shenk, 2009) found the following:

- Optimists had better health in middle and old age than pessimists.

- Men with a healthy outlet for stress (healthy use of humor, or physical activity like sports) reported being happier and lived longer.

- Those who did not take themselves too seriously, but expressed humility, were healthier and lived longer.

- Happiness must be shared: those with meaningful, sustained, healthy relationships with friends and family were happiest and healthiest.

A study by Costa at the National Institute on Aging found that a personality trait—"antagonistic hostility"—predicts premature death. This person is easily provoked to anger and is vindictive (reported in Segell, 1993).

Other related personality characteristics have been determined by social scientists to be life shortening: repressed anger, depression, egocentricity, and various other negative attitudes (Morris, Robinson, & Samuels, 1993; Shephard, Rhind, & Shek,, 1995). One review of the study begun by Lewis Terman in the 1920s reported that degree of psychological maladjustment was related to higher risk of all causes of mortality over a 40-year follow-up period. The review indicated that mental health problems were significantly more strongly related to deaths from injury and cardiovascular disease among men, although not among women. The finding was not mediated by alcohol consumption, obesity, or smoking (Martin et al., 1995).

In addition to longevity-limiting psychological characteristics, within any given age population will be those who are chronically mentally ill. Mental illness of any proportion is alarming and requires appropriate intervention, and it is anticipated that with the baby boom generation moving into late adulthood in the twenty-first century, the sheer numbers of people with chronic mental illness (as opposed to late-onset mental illness) will be large. To add to the possible load for mental health professionals, increased depression and suicidal ideation in older people with serious medical diseases has been widely observed (Niederehe, 1997). In some instances, the difficulty of letting go of one's self and of others—of recognizing that death is nearby—can explain sadness and depression. In other cases, however, it cannot be understood simply as the older person's psychological reactions to the experience of illness, and must be looked at from a medical, as opposed to a mental health, perspective (Zeiss & Breckenridge, 1997). Medications often are a major issue to deal with among older adults, particularly in anxious older patients: shifts in medication tolerance and negative cognitive effects are part of a psychological symptoms complex associated with older adults (Beck & Stanley, 1997). If we expect larger numbers of people requiring mental health services in the future, we must train and prepare for that in the present.

A whole set of findings concerns stress. In a nutshell, some stress may be a positive factor in life but too much of the wrong kind is bad. And those who are good at coping with stress will live longer. In the same vein, a good sense of humor helps, as well as a strong sense of self and purpose and a zest for life. The feeling that one is in control of one's life also adds years. Even in a nursing home, those who have choices and assert their will live longer.

In the presence of disease, guided imagery has been shown to lengthen life for some. They imagine their body parts becoming well again—perhaps their immune cells are warriors fighting off the enemy. Or a totally healed lung is pictured in their minds. Here, positive mental attitude is used to get the immune system activated and fighting. Hypnotism, meditation, and other relaxation techniques can be very useful, particularly among cohorts of older adults with previous exposure to such practices.

Social and Environmental Factors in Longevity

The environment we live in plays a role in how long we live. Noise and air pollution; pesticides; radiation; secondary smoke from cigarettes; and other adverse chemicals in our air, water, and food bring disease and shorten life. The ultraviolet rays of the sun age the skin and can cause skin cancer. Living in a high-crime area can be life threatening. For rats, overcrowding in cages alters behavior and shortens life. Likewise, living in overcrowded cities may be harmful for humans.

A positive, hopeful, stimulating social environment adds years to life. Rats in cages with lots of wheels and mazes live longer than those with no outlet for activity. Likewise an active physical and mental environment is important for humans. Social class is correlated with longevity. Those with more money for health care live longer.

A shortened life is statistically correlated with the following: divorce (for the man only); accidents (car, especially); a lifestyle that includes smoking, heavy use or abuse of alcohol or drugs; too little or too much sleep; an imbalance of work and leisure; frequent risk-taking and/or self-destructive behavior; and being a loner.

Centenarians

Rather than study animals in a laboratory, some scientists have focused their attention on centenarians as a way of learning about longevity. Twenty-five or so centenarians, the "oldest-old," with a balance of white and black Americans, were in the Georgia Centenarians Study (Poon, 1992). With regard to personality and coping, the oldest-old scored high on dominance, suspiciousness, and imagination, and low on conformity and personality traits that served as protective functions. The centenarians were described as assertive and forceful. There were a number of extraordinary persons who wrote, published, performed musically, gave guided tours, invested in the stock market, earned a living, and coped well regardless of the adequacy of support systems. In terms of cognitive skills, they rated high on practical problem-solving tests, but lower on intelligence and memory tests. Religion was important and a common coping device: "I don't worry about the future; it's in God's hands," said Charles C., 101 years old. Regarding nutrition, most were moderate, healthy eaters who did not go on diets. None were vegetarian; they did not smoke and drank very little; most ate big breakfasts. Surprisingly, they had high intake of saturated fat, especially the African American men who continued lifetime patterns of consuming pastries and whole milk. This speaks to the power of genetic predisposition: the frail and sick have died off, leaving behind a cohort with remarkable survival power.

The American demographer James Vaupel claimed that genetic factors account for no more than 30 percent of variance in life spans, with the remaining 70 percent related to life style and environmental factors (quoted in Kirkland, 1994). Beard (1991) found, in studying individual differences among centenarians, that long life was correlated with good health habits, stimulating physical and mental activity, spirituality, moderation, tolerance, integrity, and interacting with others. International studies of centenarians support the suggestion that heredity is important, but it explains far less about that which constitutes the "sturdy disposition" of these survivors into a second century of life (Italian Multicentric Study on Centenarians [IMSC], 1997). Further research of centenarians will provide us with more answers, and in the decade to come, we will have centenarians from a wider range of cultures and life styles from which to gather information.

As science continues to document habits that increase stress and decrease nutritional and physical health, the wellness movement will gain strength, encouraging prevention and the adoption of good health-related habits. The long-term effects may well result in larger populations of older adults who will experience fewer debilitating illnesses and, as a result, higher levels of life satisfaction.

Chapter Summary

The health status of the population aged 65 and over is far better than most people would predict. Most elders are able-bodied and are not limited in a major way by physical impairments—indeed, only 5 percent of people over 65 live in nursing homes or other long-term care institutions. Poor health in old age is not caused by the aging process, but lack of exercise, inadequate diet, stress, and disease are huge contributors. Heart disease, hypertension, cancer, strokes, and accidents are the leading causes of death. With the elimination of these and other factors longevity would increase, and in many instances the lifestyle one has chosen to live impacts the occurrence of stress, cancer, and so forth. Exercise and nutrition are vital in maintaining health and longevity; however, caution is given to avoid the pitfalls of "blaming the victim" for lifestyles not conducive to well-being in later life. Culture, life experience, historical time, and poverty levels each have an impact on choices made and available to each age cohort.

Wellness is the key emerging concept in the study of aging, and in the planning for one's own aging. As baby boomers enter late life in the twenty-first century, they will be the first generation to come into old age with a life time of information about the relationship of nutrition, exercise, and life style choice on the aging process.

The wellness movement is based on anticipating and taking measures to prevent health-related problems as we age. Although modifying a person's lifelong behavior patterns is not always easy, it can be done. More and more of our young, middle-aged, and older citizens are losing weight, exercising, and monitoring their diet than ever before. Learning to avoid foods with higher levels of calories or fat, reducing hypertension, and discontinuing the use of tobacco, while simultaneously including a diet of nutritional foods and maintaining higher levels of physical activity, reduce health-related risks in later life. Extending the number of healthy years of life, what has been called the **"health span,"** has become a viable goal for all Americans.

Key Terms

absolute life span	biological clock
acute condition	cancer
aging as disease	cerebrovascular disease
arteriosclerosis	chronic condition
arthritis	Duke longitudinal
atherosclerosis	study
average life span	health span

health status of the aged
heart disease
hypertension
immune system
ischemic heart disease
lived experience

maximum life span
osteoporosis
primary immunological change
wear and tear
wellness

Questions for Discussion

1. What changes could you make in your life right now that would promote healthy aging? How would these changes affect your life now?

2. List some of the normal biological changes that occur with age. What lifestyle changes can counteract these changes? What can't be changed?

3. With a partner, take turns simulating some of the handicaps that can occur with age by wearing ear plugs, blurred glasses, gloved or taped fingers, and walking with stiffened neck and limbs as if there were arthritis in the joints. Attempt various tasks such as walking around, getting change out of a wallet and opening a jar. What is it like to carry on a conversation with the "old" person sitting in a chair and the other standing? How did you feel as the old person? How did you feel as the "assistant?"

Fieldwork Suggestions

1. Interview elderly athletes at a golf course or tennis club. How is their general health? When did they first take up their sport? Did they have to make any adaptations to their sports performance as they aged?

2. Contact a number of health clubs in your area and find out what programs they have for seniors.

Internet Activities

1. Locate the home page for AARP, formerly known as the American Association of Retired People. What sort of practical health-related information does the site contain? How recently was it updated?

2. Log into Senior Com: The Source for Seniors (http://www.senior.com). This is a site designed to be used by seniors to promote their health and well-being. Compare what you find, and the way it is presented, with the AARP site, and with other sites for and about seniors.

3. Locate the best source of information you can find on the Internet related to nutrition for older people. How persistent did you need to be to find this source? Is it a site that older people might use? Is it one that someone who plans menus for an assisted living facility might use?

Mental Health

Secrets of a Good, Long Life

Sarah Mahoney

It's official: stress makes you old.

While researchers have long been piecing together all the ways chronic stress undermines our health, a new study from the University of California at San Francisco (UCSF) confirmed what we suspected all along: stress really does age you.

What happens, researchers learned, is that constant stress causes the telomeres—tiny caps on cells' chromosomes that govern cell regeneration—to get smaller. When a cell's telomeres get too short, the cell stops dividing and eventually dies.

Researchers discovered that the telomeres (pronounced teal-o-meers) of women with chronically ill children were much shorter than those of women the same age who weren't caregivers. Moreover, the greater the women perceived their stress levels, the shorter their telomeres—and the "older" their cells. "These telomeres are one of the few biological markers of aging we have," says Judy Moskowitz, Ph.D., a psychologist at UCSF who worked on the research.

But wait, you're probably saying: what happened to the women who didn't perceive their lives as stressful? Stress didn't age them nearly as much. "For them, stress is like water off a duck's back," says Thomas Perls, M.D., an associate professor of medicine at Boston University and the director of the New England Centenarian Project, a nationwide study of 1,500 people over the age of 100 and their children. "It isn't the amount of stress that matters but how you manage it."

In fact, a number of the centenarians Perls has studied have endured plenty of stress. After all, they lived through the Great Depression and World Wars I and II, not to mention the usual array of divorces, deaths of loved ones, and even job losses. "Yet they don't seem to internalize it," Perls says. "They just let it go."

AARP decided to ask a few stress veterans—chosen from the 4 million lucky Americans who have sailed past their 85th birthday—for their secrets to staying young, both mentally and, as it turns out, physically as well.

"Will this take long?" says Pauline "Dully" Kirn when asked about stress. "I'm in the middle of a game." While the Lancaster, Ohio, woman has plenty of interests, such as collecting antiques, what keeps her going is the steady swirl of her bridge calendar. She has been playing two regular bridge games for decades. "One is a little more cutthroat, and one tends to be a little more chatty," she says. While she thinks missing out on those outings—both the games and the gabfests—might make her cranky, she can't say for sure. "We don't skip games, so we don't have a chance to get grouchy."

Games zap stress Ever since psychologist Mihaly Csikszentmihalyi pioneered the study of mental "flow" back in the 1970s, researchers have studied the best ways to achieve this elusive state, in which people are so fully focused on what they are doing that time seems to stand still. In fact, when Harvard University researchers followed people over the age of 65 for 13 years, those, like Kirn, who enjoyed games found almost as much stress relief and prolonged life expectancy as did those who exercised regularly.

Ask Brownie Brown if he has stress and he deadpans, "Not yet." But the Chicago native still has plenty of hoofing left in him: he teaches occasionally at the Chicago Rhythm Project and is part of a two-man tap-dancing act that performs around the country. Humor comes naturally to Brown, and he decided to rely on it professionally early in his vaudeville career. (He was best known as half of the duo Cook & Brown and later danced with the Copasetics, a famous tap-dancing troupe.) "Telling jokes on stage gave us a break, so we could save our legs," he says. "I was the funny one, and Cookie was the straight man." Recently, recalls Reggio McLaughlin, Brown's current partner, the pair were performing, and "Brownie starts a bit. 'Reggio, did you know I used to be a boxer named Horizontal Brown?' So I say, 'No, I didn't know that,' and he says, 'Yes—when I come in the ring, everyone is screaming. The men are screaming, the women are screaming.' So I say, 'Why are they screaming, Brownie?' and he says, 'Because I forgot to put my trunks on!' He's just a natural."

Humor zaps stress Studies have shown that people who can appreciate humor are less stressed and anxious. But those, like Brown, who have the ability to make jokes too have an added advantage. According to research from Western Illinois University, they tend to be more secure and confident in their interactions, less lonely, and more likely to see the stress in their lives as lower than that of people who aren't able to joke.

Agnes Dill (right) often feels she was born 50 years too soon. She was the first American Indian woman to go to New Mexico Highlands University "at a time when most Indians were doing menial jobs. Today we're doctors and lawyers." Although Dill became a schoolteacher in Okla-

homa after graduation, she never stopped trying to expand educational opportunities for other American Indians, especially women. Her stress-fighting secret is simple: even though macular degeneration prevents her from doing many of the things she used to enjoy, "I'm very optimistic about life," she says. "I accept things as they happen and make them better if I can. Most days, I don't feel my age. I just feel pretty happy."

Optimism zaps stress Researchers from the National Institutes of Health have learned that optimism is a protective trait, but there's still a lot of work to be done in the burgeoning field of positive psychology. "While we know optimists live longer," says UCSF's Moskowitz, "it's not as simple as saying, 'Be happy, damn it.' People may be optimistic because they're healthy, not the other way around." Still, the numbers are impressive: Dutch investigators followed 1,000 people between the ages of 65 and 85 for nine years, and the optimists had a 55 percent lower risk of death.

Master leather carver Bob Brown (left) knows more about Hollywood cowboys than almost anyone alive: he crafted the holsters worn by Hopalong Cassidy, Montgomery Clift, and John Wayne. He also befriended many of the actors he worked with. His work is on display in the National Cowboy and Western Heritage Museum, and over the years, he's taught leather carving to hundreds of artisans. And though he thinks some of his longevity might come from clean living ("I don't smoke, I don't drink, I don't carouse"), he believes it's his work that keeps him happy and strong. Long retired, he still keeps regular hours on whatever project he's pursuing. "Right now, it's watercolors. I get up, and I start to work. And if it interests me, I stay up working until midnight."

Work zaps stress In some ways, work and stress have gotten a bum rap—everyone has read studies linking type A workaholism with an increased risk for heart disease, depression, and other health problems. But researchers have found that work people are passionate about—so meaningful that it becomes a calling rather than a nine-to-five—helps reduce stress as well as the risk of depression, according to the *Journal of Humanistic Psychology*. What's more, work that is flexible, such as that done by Brown, can negate the stressful effects of long hours, according to research from the University of Arkansas.

Marie and Lilly Clifford have always been extremely close. The twins, raised in northern North Dakota, both earned teaching certificates, taught for many years (while they lived apart from each other), and then traveled together in retirement. Today they live in an assisted living facility and still rely on each other: Marie doesn't see very well, so Lilly helps her out; Lilly's ears aren't the greatest, so Marie is often the spokesperson, say the nurses who help care for them. Among the oldest surviving twins in the world, neither ever married ("Well, not yet, anyway!"). And they still enjoy each other's company, whether they are attending weekly Mass or watching *Bonanza* reruns.

Close friendships zap stress Though loneliness has been linked to making all people more susceptible to stress, depression, loss of cognitive ability, and other ills, friendships seem especially protective for women. Shelley Taylor, a researcher at the University of California at Los Angeles (UCLA), has found that most women deal with stress exactly as Marie and Lilly do—they rely on long chats with one another. While men are more likely to go into "fight or flight" mode, women are more likely to "tend and befriend." It works like this: though males and females produce the soothing hormone oxytocin under stress, estrogen tends to enhance the hormone, while testosterone inhibits it. When oxytocin levels are high, people are calmer, more social, and less anxious. These friendships not only help fight stress but may partially explain why women tend to outlive men.

Electrical engineer George Gless retired from teaching at the University of Colorado back in 1982, but he didn't give up the passion he developed during his tenure there—electric cars. For 25 years while at the university, he edited a newsletter on electric vehicles, and after retirement, he continued to be involved in the experimental-vehicle community. "I'm a scientist, basically, and the universe intrigues me," says the Boulder, Colorado, man. "I feel that we ought to honor the environment. The Lord gives it to us, so we should take good care of it." (And yes, besides advocating for electric vehicles, he drives a Prius, the Toyota hybrid. "We're actually on our second," he says.)

Altruism zaps stress Adaptive social behaviors, such as Gless's concern for the environment, contribute to stress resiliency, according to research from the National Institute of Mental Health. The research focused on measuring hormones and neurochemicals among the U.S. Army Special Forces and U.S. Navy SEALs and found that people with "a set of core beliefs that are not easily shattered, who exhibit strong faith or spiritual beliefs," are more resilient. What's more, when people look for meaning in their lives, they seem to get a boost in immune function that may keep them healthier, according to research from UCLA.

Thais Crowell has always appreciated music: she studied piano and voice in college and loved the beautiful feeling of "knowing your tone was improving, of letting the music come from not just your hand on the keys but your whole body." But after college, Crowell's life got tough. Her husband was killed by a land mine in 1945, leaving her with four small children to raise. "It was a difficult time," she says, "but the children got me through it." Crowell soon remarried, and rhythm came back into her life. She and her second husband loved to dance—and even got their kids involved. After moving into a retirement community 10 years ago, she signed up for drumming lessons and was immediately hooked. "I feel myself relaxing right away when I start drumming," she says. She currently belongs to two drumming groups and performs in concerts in the area; the groups have gained enough notoriety that filmmaker David N. Brown included them in a recent documentary. "Drumming gives me the same inspiration I used to get from the piano," Crowell says.

Music zaps stress While learning anything new has been proven to beat stress, music has a special ability to calm people. Decades of research have shown that listening to music can lower blood pressure and heart rate. A new study at the Mind-Body Wellness Center in Pennsylvania also found that playing music can significantly reduce stress.

Although Grace Nunery has been a minister's wife since graduating from college, it wasn't until she was 50 that she found her true calling. A deaf person came to see her husband about a family funeral, and Nunery recalls how awkward she felt having to write a note for something as simple as "Would you like a piece of cake?" The woman taught her to sign the word "milk" (which looks like milking a cow), and Nunery was hooked. Before she knew it, she had embarked on a ministry to the deaf that would eventually expand to a summer-camp program for deaf kids, drawing participants from all over the world. There were doubts and mishaps—like the time she accidentally translated "man's decay" as "someone here has B.O." in front of the entire congregation—but she got over it. "God put the deaf ministry in my path, I'm quite sure of it. And when it got tough, I prayed," she says. "My favorite Bible verse is 'Don't worry about anything, pray about everything, and always be thankful.'"

Prayer zaps stress Despite decades of research, there is still a great deal of conflicting evidence about the health benefits of prayer. But researchers do know that among older people, spirituality—which covers not only faith and prayer but also the close-knit support of religious communities—significantly lowers stress and improves the chances of recovering from serious illness, according to the *Journal of Health Psychology*.

Every morning, Clarence Custer hits the courts for a game of tennis. On good days, he plays for an hour and a half. On better days, he plays for three hours. And three times a week, he hikes over to the local health club and lifts weights. In fact, he's so active that he was asked to carry the torch for the Winter 2002 Olympics. The energy is nothing new. Custer began playing tennis and squash as a young man back in Youngstown, Ohio, and the familiar routine sustained him through the ups and downs of starting and selling several businesses, as well as the death of his first wife. "I love tennis, and I really look forward to playing with the same group of guys each morning—they're great fun, all in their 80s," says Custer. Best of all, there's the satisfying knowledge he can still finesse anything that comes over the net. "I can't beat him, and I'm 20 years younger," admits his wife, Joanne.

Exercise zaps stress While everyone knows exercise promotes a healthy heart, researchers are constantly learning more about how it minimizes stress. Multiple studies from the University of Colorado at Boulder have shown that physiological responses to stress from the brain, hormonal system, and immune system are all moderated by regular exercise. And while all exercise is healthy, moderately intense exercise, like tennis, significantly reduces anxiety too, according to research from the University of Missouri.

Sarah Mahoney last wrote for *AARP: The Magazine* about becoming fit later in life ("Real Fitness," May–June 2005).

Social scientists continue to unravel the effects of aging on the mind. In the past, the ultimate fate of an elderly person was considered to be senility—whatever that meant. No one was surprised if Grandpa forgot where he put his glasses, that cousin Herman was coming to visit, or even that he was married. No one became too alarmed if old Mrs. Jones down the street had lengthy and loud arguments with her deceased husband or was afraid to come out of the house because she thought the world was going to end in six months. The reasoning went thus: this is what happens when people get old—their minds go. At the very least, they become set in their ways, stubborn, and cranky. Nonconformist, bizarre behavior was tolerated and rationalized: What can you expect at her age? Old age must take its course. What can you do but accept?

These ideas about old age still persist. Today, scientific evidence suggests that declining mental health is not a natural consequence of the aging process. The vast majority of people aged 65 and over are in good mental health. If they are not, specific causes other than the aging process itself usually can be pinpointed. This chapter will examine the psychology of normal aging and the most common mental health problems of older adults, including functional (behavioral) and organic (physiological).

The Psychology of Aging

In this chapter, we will discuss areas traditionally covered in the behavioral aspect of the psychology of aging. These are changes with age in perception, motor performance, intelligence, learning, memory, and personality. Gerontological psychologists also study changes in the brain and central nervous system. Many have backgrounds in physiology, biology, physics, math, or some combination thereof. Others are more interested in personality characteristics and social behavior and, thus, may have social psychology backgrounds.

A distinction must be made between **age-related changes** and **biologically caused changes.** A gerontologist or **geriatrician** (physician specializing in older adult health) may find a positive correlation between age and depression—the older the age group, the higher the incidence of depression. In this case, depression is an age-related phenomenon. Once this statistic is staring the scientist in the face, it must be interpreted. Is it that the biological aging process causes depression? That conclusion certainly cannot be made without further investigation. What would be some possible causes of the correlation? Are society's values the cause? Are there cumulative stresses that some persons resist in their younger years but to which they finally succumb in old age? How do factors such as retirement, poor health, or the death of a spouse affect aging? There may be other possible causes for the correlation. The gerontologist must look further for the cause of the depressed state.

A major review of diagnoses in Medicare, published in 2008, claims to identify depression in later life (Crystal, Sambamoorthi, Walkup, & Akincigil, 2008). It reported that though depression rates have been thought to be under-recognized in older adults, rates of diagnosis increased dramatically in the 1900s along with increases in treatment. Recognition and treatment are critical to ongoing health: unrecognized depression is associated with increased health care costs as well as worse outcomes following acute medical events such as hip fracture and stroke (Robinson, 2008). However, has the actual rate of depression among older adults increased or has the reporting of depression become more common? The National Institutes of Mental Health [NIMH] (2009) report four major groupings under which causes of depression in older adults can be placed:

(1) *Physical factors* such as predisposition to depression, constant pain, or a major health issue such as cancer;

(2) *Psychological factors*, including increasing fear of death (especially following the death

of a loved one), lingering feelings of guilt or sadness for unaccomplished goals, or problems adjusting to changes inherent in frailty of old age;

(3) *Personality factors*, such as difficulty with becoming dependent, or loss of self-esteem or self-confidence; and

(4) *Medications*, some of which can trigger depression in the older adult.

We can see that all people who age into later adulthood will face at one time or another the loss of a loved one or recognition of a dream unfulfilled. All older adults will experience changes in physical competence compared with their younger years; and the need to adjust to a changing sense of self is an ongoing process of life that accelerates when social responses to a person are startlingly different than how that person sees himself or herself. Why then are not all older people depressed?

While some disability results from normal, more general losses of physiological functions, *extreme* disability is not an inevitable part of aging (NIMH, 2002). Many older people have an internal capacity to develop new adaptive strategies, and they do so. This strength is not something that suddenly appears, but is rather one that develops as the person ages. Those experiencing loss often move in a positive direction through their own internal strengths or through the support of their family and friendships. Both informal support and formal support from mental health professionals are huge resources to older adults (NIMH, 2002) experiencing difficulty coping with physical and social changes related to aging.

Normal aging includes stable intellectual functioning, capacity for change, and productive engagement with life. Memory complaints are nevertheless common among older people, but upon examination, complaints do not correspond with actual performance. One major study found that some who complain about memory actually display performance superior to those who do not complain (Willis, 2006).

Metamemory is the self-appraisal or self-monitoring of memory; it is our own evaluation of how well we can remember. Memory complaints are thought to be more a product of self-appraisal than of actual performance. The person who *believes* himself or herself to have a less effective memory reports memory loss. Some older adults have been found to exaggerate their memory failures (Hertzog, Lindenberger, Ghisletta, Oertzen et al., 2006; Hoyer & Roodin, 2009), while other studies show people's assessments of their memory to correspond fairly accurately with memory measures (Turvey, Schultz, Arndt, Wallace, & Herzog, 2000). What explains these differences?

Older people may hold beliefs—attitudes and stereotypes—about themselves that distort memory assessments. Older adults are just as likely to have incorporated ageist social beliefs about memory as are younger adults. Having a "senior moment" is a common light-hearted comment that reflects a deeper belief in the inevitability of memory loss with age.

So what supports positive mental health in aging, given inherent losses and changes for the individual? One ongoing longitudinal study found high cognitive performance to be associated with education, strenuous activity in the home, peak pulmonary flow rate, and most importantly a strong sense of self-efficacy, which is confidence in the ability to organize and execute actions to deal with situations likely to happen in the future (Albert et al., 2007a; NIMH, 2002). It appears the importance of education is less one of how much information a person possesses, and more one of its role in a lifelong motivation for learning—for exercising mental functioning throughout life (NIMH, 2002; Rowe & Kahn, 1997; Salthouse, 2006; Schooler, 2006). Once again, successful aging is largely the outcome of lifestyle rather than genetic privilege.

The new model for successful aging includes physical and mental well-being and contains three elements: avoiding disease and disability, sustaining high cognitive and physical function, and engaging with life (NIMH, 2002; Rowe & Kahn, 1997). Note that these three elements require maintaining personal relationships and productive activities and that successful aging requires all three elements to work together—to act in concert. None by itself is sufficient for healthy aging. These variables are key to the tremendous variation in individual experiences of the aging process and of memory and mental health.

Cognitive Processes

Cognition is *thinking* about a situation. It is our awareness of the world around us—how we absorb stimuli and information, and how we make sense of it. Our perception (what we think of something) is related to our behavior in various situations. Did you see the traffic accident? Did you hear the crash? Do you know who was at fault based on what you saw? What was your response? How quickly did you react? Cognitive processes involve our senses, our arousal, attention, information processing, reaction time, and motor performance. Those studying aging want to know whether people see and hear situations differently because of their age: does attention span or speed change? Is information processed in an altered or different way because of aging?

Some psychologists compare our brains to computers. Information enters our brain, is coded, and then is stored at various levels. If our brains are working efficiently, we can retrieve information when we need it. This section

deals with cognitive functioning. First, the basic senses, sensory memory, and steps of information processing are described; second, studies of intelligence are reported; and third, learning and longer-term memory are reviewed.

Basic Cognitive Functions

Information processing describes the kinds of cognitive processes involved in memory. The loss of memory that many people assume will occur in later life seems to be more highly related to *individual differences* as well as the *context* in which human development occurs than with age itself (Albret & Killiany, 2001; Allaire & Willis, 2006; Carstensen, Mickels, Mather et al., 2006). But certain types of memory loss do seem to occur. Some of the complex processes having to do with thinking, remembering, and making judgments (perceiving) are discussed in the next sections.

Figure 5.1 illustrates a three-stage model of memory. This model describes memory as beginning with sensory storage: we see, feel, smell, or hear something. If we pay attention to that input the memory is moved to short-term memory. If we actively think about that item in short-term memory, it is moved into long-term memory.

The Senses

Sensation is the process of taking information in through the senses. **The five senses** (vision, touch, hearing, smell, and taste) that relay environmen-

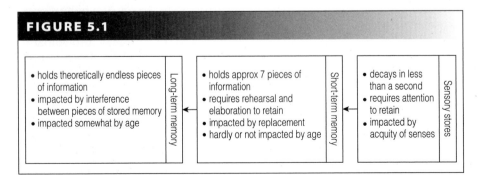

FIGURE 5.1

Long-term memory	Short-term memory	Sensory stores
• holds theoretically endless pieces of information • impacted by interference between pieces of stored memory • impacted somewhat by age	• holds approx 7 pieces of information • requires rehearsal and elaboration to retain • impacted by replacement • hardly or not impacted by age	• decays in less than a second • requires attention to retain • impacted by acquity of senses

tal information to the brain generally lose acuity with age; therefore one set of pathways through which information is gathered for processing by the brain is modified as we age. Vast individual differences exist, however. Some very old people experience no sensory declines. The social and psychological consequences of sensory declines can be enormous, and it is the oldest-old—those over age 75—who are more likely to experience noticeable declines than are the young-old, that is, those between 65 and 75 years of age.

Older persons vary in how they deal with sensory loss. Some people can compensate for losses in one sensory area by elevating enjoyment using another. For example, a person losing vision might choose to focus on an existing appreciation of symphony music. Compensation by habit and routine is probably even more frequent. Though adaptability may vary, most older adults can compensate very well for minor sensory losses. For example, loss of visual acuity can be compensated for with corrective lenses, more light, less glare, and clearly printed reading material.

Those who care for older adults could easily respond inappropriately to such sensory decline: why fix a nice meal for one who cannot taste it, or why take a vision-impaired elder outside? Rather than encouraging withdrawal, sensory decline might challenge caretakers to find ways of enhancing life for those who have suffered losses. Perhaps the person might like spicy Thai food, or stronger hugs, the scent of a garden, or a warm fire in the fireplace. The highest quality of food and meal preparation, excellent not just adequate lighting, taking the time to accompany an older person on a walk or shopping trip are all appropriate—most especially in the presence of sensory decline.

Sensory Memory

Each memory starts as a sensory stimulus. You can remember certain aspects of things you see and hear for only a fraction of a second. This initial short-term sensory experience is called a **sensory memory**. Sensory memory takes in large amounts of information so rapidly that most of it gets lost. Do you recall what is on a one-dollar bill or a dime? Could you draw the details? Much information that passes through sensory memory is never processed to a storage place in the brain because we cannot pay attention to everything hitting our senses. We make choices through giving our attention to some sensations and ignoring others. Although not as well studied as other memory phases, research indicates that the amount of visual information one can handle at a time seems to decline with age; however this decline is very subtle and possibly not noticed by the individual.

Attention

Paying attention is an important part of information processing. **Selective attention** means focusing attention on the relevant information while inhibiting response to irrelevant information. If we are distracted—unable to focus a specific stimuli among many—our ability to code the stimulus into memory is impaired. If we pay attention to a stimulus, the experience moves from our sensory memory to the short-term memory (see Figure 5.1). We vary in the capacity of information that can get to our short-term memory at any one point in time. Very old people have a lesser capacity to absorb stimuli into working memory and thus to maintain concentration over a long period of time.

Studies of **divided attention** (doing two things at once, such as watching television and listening to a conversation) show that when divided-attention tasks are easy, age-related differences are typically absent. But when the tasks become more complicated, performance differences owing to age emerge. Attention studies show that older people are more distractible and are less able to disregard the clutter of irrelevant information than are younger people. Age-related differences in vigilance also exist (Wolters & Prinsen, 1997). Older adults tend to have lower physiological arousal (Lustig, May, & Hasher, 2001; Mutter & Goedert,

1997), which lowers their alertness. Vigilance does not appear to be age related unless the task is sustained over an extended period of time, when, beyond age 70 or so, fatigue can result in appreciable performance loss.

Perception

The process of evaluating sensory information carried to the brain is called **perception**. Individuals may perceive the same stimulus differently; further, the same person may react differently at different times. **Sensory threshold** is the minimum intensity of a stimulus that is required for a person to perceive it. Mood, activities, and personality may all influence perception.

Sensory decline affects perception significantly. If a person is unable to hear parts of a conversation, for example, the content of the conversation is more likely to be misunderstood. Perceptual differences among age groups are frequently reported, but research does not explain the reason. The differences may be caused by biological changes, or they may be age related. Social isolation also tends to affect perception; therefore, isolation, not aging per se, may cause perceptual changes.

Psychomotor Speed

A person first experiences the environment through the five senses; second, the person perceives what is happening; and third comes reaction. A physical reaction to stimuli is called motor performance. Motor performance may be simple, that is, a reaction requiring very little decision making or skill. Pressing a button to turn on the television and turning off a light switch are simple motor skills. Some types of motor performance, such as dancing, riding a bike, playing tennis, or driving a car are more complex.

Very old drivers may be affected by changes in vision and hearing. Their vision does not adapt as rapidly to dark, and they are more affected by glare. Older people also take longer to read road signs but additionally, the processing of information is slower from the point of sensation

(seeing the sign) to the perception (its meaning registers, and action is taken). Researchers suggest designing cars with older adults in mind: headlights with less glare, well-lit instrument panels, easy entry, and so on. The sheer numbers of aging baby boomers with discretionary dollars to support their later life will surely drive creative approaches to signage, lighting, and training programs to address slower psychomotor speed.

Though it is common belief that older drivers pose a hazard on the road, they have far fewer accidents than do younger drivers. This is partly because older people seem to exercise greater caution (reduced speed and greater conscious awareness given to the task), drive fewer miles, and drive at less hazardous times than do younger people. Most younger adults have experienced the frustration of driving behind an older driver, but the common belief that the older driver is less safe on the road is incorrectly based on the observation of the older driver's slower pacing and use of caution.

One aspect of motor performance is reaction time—the length of time between the stimulus and the response, directly related to psychomotor speed. Reaction time begins to increase with age, at about age 26 for some tasks and not until the 70s for other tasks. The general slowing of response with age is one of the most reliable and well-documented age-related changes. This includes slowed psychomotor speed and the resultant lengthened reaction time. The more complex the task, the greater the difference in reaction time by age, generally speaking. However, individuals differ substantially: a given 70-year-old might respond more quickly than a 30-year-old.

A noted researcher in the field of cognition has introduced two terms to categorize the variables affecting cognition:

1. **Cognitive mechanics**—the hardware of the mind—reflects the neurophysiological architecture of the brain. It involves the speed and accuracy of elementary processing of sensory information.

Self-esteem and self-confidence are maintained through a stimulating environment, the continuation of enjoyable activities, and caring social support.

2. **Cognitive pragmatics**—the software of the mind—reflects the knowledge and information of one's culture. It involves reading, writing, education, professional skills, and life experiences that help us master or cope with life (Baltes, 1993).

The first term represents the genetic–biological influence and the second represents the cultural–social influence. Table 5.1 itemizes separately the biological factors, biological mechanics, and the cultural–social, pragmatic, factors associated with the slowing of information processing in old age. The cultural–social factors tend to be more reversible than the genetic–biological ones.

Many of the variables in Table 5.1 do not universally accompany aging. Physical pathologies (e.g., high blood pressure) do not go hand in hand with aging. Neither does reduced activity.

Aerobic exercise and other forms of physical activity have been shown to increase blood flow and increase speed of reaction. Activities such as tennis and dancing are helpful. Physically fit older people have shorter reaction times than less fit young adults (Colcombe & Kramer, 2003). Practice at a task can compensate for slowness. Many a grandchild has been amazed, for example, at the speed with which a grandparent can peel a potato or whittle a wooden object.

Psychologists have observed that that the slower reactions of older people have more to do with changes in the mental processes of interpretation and decision rather than at the initial level of sensory input (Marsiske & Margrett, 2006; Mather, 2006). In other words, the course of slowing reactions has a lot to do with the brain and central nervous system (CNS). It is important to remember, however, that not every older individual experiences a slowdown in

TABLE 5.1

Variables thought to be related to slowing of information processing in old age

Genetic–biological factors	Changes in sensation acuity
	Changes in physiological arousal to stimuli
	Changes in attention
	Changes in motor capacity (stiffness; reduced strength)
	Lower levels of physical activity
	Changes in blood flow to the brain leading to neural malnutrition
	Changes in the central nervous system
	Changes in cortical levels of the brain
	Gradual loss in brain mass
	Neural/metabolic changes
	Decline in physical health
Cultural–social factors	Changes in self-esteem and self-confidence
	Lower levels of mental activity
	Lessened familiarity and experience with the task
	Lifestyle characteristics such as lack of travel, lack of stimulating environment including social exposure

reaction. It is a pattern of aging that has many exceptions.

The implications of CNS slowing for the everyday life of the older person are various. Experience and familiarity are great levelers—giving the older person an edge on the job or at various tasks at home. Very old people may need more time to get a job done, and they may need to be more cautious to avoid accidents. A speedy reaction is needed to catch falling objects, to drive a car safely, to escape a fire, to get across a busy street, or to play a good game of ball. Many older adults would benefit from education or retraining to minimize the consequences of reduced cognitive efficiency. Environmental modifications may be in order—for example, stronger lighting, brighter colors, grab bars, hand railings, door handles as opposed to knobs, a louder telephone ring, larger sturdier step stools. And people should not be so ready to assume that if an individual cannot think quickly, he or she cannot think well. That is a form of ageism.

Intelligence

As we compare the learning abilities of babies, elementary students, college students, or people in general, we find that within each group, some people are better able to learn than others. **Intelligence** was conceptualized by Guilford (1967) in three dimensions: *content* (figures, symbols, words); *operations* (memorizing, evaluating, coming up with solutions); and *products from the operations* (relationships, systems, implications). The dimensions are measured by various tests to determine an individual's "intelligence." Although tests have been devised to measure an intelligence quotient (IQ), the validity of such tests has been debated for years. They have been variously criticized for being ethno-centric (biased toward the major culture), sexist (biased toward boys/men), and ageist (inadequate measures of applied knowledge) (Lee, Blando, Mizelle, & Orozco, 2006). The jury is still out, even as tests become more clearly defined for diverse populations. Two major questions about IQ tests arise: (1) is there

One of the most powerful ways to maintain a sharp mind is to learn completely new information. This man's joy is obvious: he's mentally engaged with something pleasurable.

such thing as intelligence; and (2) do IQ tests really measure intelligence?

The tests actually measure a number of specific intellectual abilities. Twenty-five **primary mental abilities** have been established, composed of five primary and independent abilities: number, word fluency, verbal meaning, reasoning, and space. Tests of six or so secondary mental abilities—skills composed of several primary abilities—have also been identified. Most research focuses on two secondary mutual abilities: fluid and crystallized intelligence (Schaie, 2005).

Intelligence is most often measured by a standardized test with many multiple-choice items on vocabulary, reasoning, and the ordering of numbers and spaces. Using such a test represents the psychometric approach in contrast to other tests that examine thought processes—the

quality and depth of thinking and the ability to solve complex problems (the cognitive process approach). The most widely used psychometric test of intelligence for older adults is the Wechsler Adult Intelligence Scale (WAIS), developed in the 1950s. This test compensates for increasing age. The test norms assume declining speed with age; an older adult can perform worse than a younger adult and still have the same IQ! Critics point to the fact that this test emphasizes skills learned in school and not skills of everyday life. For this reason, all things being equal, the test favors the younger adult. This test has been revised (WAIS-R) and is used widely in clinical settings. Another IQ test, the Primary Mental Abilities (PMA) test has often been used to examine adult intellectual development. The WAIS measures a "verbal intelligence" aspect that involves comprehension, arithmetic, similarities, vocabulary, and the ability to recall digits and a "performance intelligence" aspect that involves completion, block design, and assembly of objects. On the performance scale, speed is important. Bonus points are awarded for quick solutions.

Both the WAIS-R and the PMA test include tasks that evaluate fluid and crystallized intelligence. **Crystallized intelligence** is a measure of knowledge you have acquired through experience and education. Vocabulary tests are a clear example. **Fluid intelligence** refers to innate ability—the information-processing skills described in the previous section. Each of these types of intelligence taps a cluster of primary abilities. No standardization test measures one alone, although a given test can emphasize either crystallized or fluid intelligence (Cerella, Onyper, & Hoyer, 2006; Kaufman & Horn, 1996).

We once assumed that a decline in intelligence with age is to be expected. Scientists are now questioning this assumption. Longitudinal studies have found that some IQ components such as verbal skills remain stable with time and some can even increase with age. Schaie (2005) reported that virtually no individuals show deterioration on all mental abilities, even in their eighties. At age 60, 75 percent of subjects maintained their level of functioning on at least 80 percent of the mental abilities tested (as compared to seven years before). More than 50 percent of those aged 80 had maintained 80 percent of their mental abilities over the seven-year period.

Fluid intelligence does appear to decline with age, whereas crystallized intelligence does not. Similarly, the verbal scores of the WAIS show no declines with age, whereas the performance scores do decline with age. The high verbal score and declining performance score of older adults is so consistently found that it is called the "**classic aging pattern**" (Schaie, 2005; Tulvig, 1993). The drop in fluid intelligence indicates a decline in the ability to process information and complete tasks in an efficient manner, and might reflect the complex-

ity of age-related changes in functions such as sensory and perceptual skills, or something as simple as lack of practice or fear of tests. More than likely, changes reflect a combination of both: neuropsychological shifts with age, plus the context of the individual's life.

To understand fully what happens to intelligence with advancing age, we need to know exactly what primary mental abilities decline with age and why. We need to know at what age the onset of decline occurs and the rate of decline. We need to know whether such changes are inevitable and whether they can be reversed. And we need to account for individual differences. Schaie (2005) reported that it is not until people reach their seventies that declines in intelligence take on significance. Decline in speed of response is the major effect. There is mounting evidence that perceptual speed may be the most age-sensitive mental ability. Older people are as proficient as ever when it comes to situations that demand past experience or knowledge—given that they have enough time.

Though cross-sectional, "one-shot" studies of different age groups often find that older people have lower IQs than younger people, such studies may not acknowledge that young adults may have benefited from upper-level schooling. Many of the elderly studied in the 1950s and 1960s had not attended either high school or college. Some had no formal schooling at all. Thus, cross-sectional studies that infer a decline in IQ with age are suspect. Test makers should distinguish between actual IQ (the quotient the test actually measures) and potential IQ for the young as well as for the old.

We emphasize that intelligence tests must be used with caution. IQ tests measure specific skills to generalize about an all-inclusive concept of intelligence. Their results may be biased, and scientists must continue to investigate why, how, and in what direction such instruments are biased. The evidence today supports the finding of little decline with age in skills that one acquires through education, enculturation, and personal experience. We do not really know what happens to IQ as one approaches extreme old age, say after age 85. Although researchers believe that IQ begins to decline at some point in very old age, they are yet to determine the typical course it follows.

A problem with cross-sectional studies for the future is that the new generation of young students today are not doing as well on standardized tests in their general education. If this translates into lower scores on IQ tests, then it appears that IQ increases with age because older generations today do better at tests, and it is not necessarily true that IQ increases with age. To avoid the problems in cross-sectional research, longitudinal studies appear desirable.

A few words of caution about longitudinal studies: sample size tends to be small, and repeat presentation of material, which brings about improved performance, can produce a learning effect. Additionally, the people from whom the final measures are obtained are survivors: they are more robust and healthy than were those who had died over the duration of the study. In one major study on intelligence and aging, only one-sixth of the sample were alive after seven years (Savage, Britton, Bolton, & Hall, 1973). One must be aware of the research methodology employed and recognize that any given study is bound to have some inadequacies in addressing issues of age-related changes.

Another area of intelligence study is in practical problem solving. Researchers have done studies using tests such as the Everyday Problem Solving Inventory (Denney & Palmer, 1987) to measure the ability to solve problems such as a grease fire breaking out on top of the stove. When the problems are familiar to older people, they do extremely well on these tests, but not necessarily better than middle-aged adults. Once again, let it be said that intelligence comes in various forms. Also, retraining is becoming a major focus of study. The decline in primary abilities in old age can be slowed or even reversed in many cases. Future studies will clarify which abilities can be maintained or improved and the circumstances under which it is possible.

Further research is needed to clarify the relationship between age and intelligence. A theory of decline or "**terminal drop**" refers to the measured tendency of the efficiency of a person's biological and psychological processes to decrease precipitously in the last few years of life. According to this theory, a decrease in cognitive functioning is related to the individual's chronological distance from death rather than to chronological age. "Critical loss" in cognitive functioning is experienced starting from two years before death. Chronic diseases reduce the ability to focus attention and to provide clear thinking (Deary, 2006; Hoyer & Roodin, 2009; Wilson, Beck, Bienias, & Bennett, 2007).

Learning and Memory

Learning and memory, important components of mental functioning, are separate yet interrelated processes. When an individual can retrieve information from his or her memory storage, learning is assumed to have occurred; alternatively, if someone cannot remember, it is assumed that learning has not adequately taken place.

Learning is the process of acquiring knowledge or understanding. For purposes of psychological research, scientists speak of learning in terms of cognitive processes, which are intellectual or mental. As in studies of intelligence quotients, learning is measured by tests of performance, particularly verbal and psychomotor performance. Again, the same sort of question surfaces: do such tests really reflect learning?

Let us consider two general questions related to age: do learning skills change with age, and how do older people's skills compare with younger people's? The answers to these questions depend on the skills being learned and the conditions under which they are being learned. Up until 1960, and even later, members of the scientific community generally assumed that learning declined with age.

In normal aging intellectual competence is retained through intellectual exercise.

Many factors affect learning abilities. **Pacing** (the rate and speed required for learning) is an important factor. Older adults learn better with a slower pace, and they perform best with self-pacing. They also do better with a lengthened time to respond. Another factor is **anxiety**. Some studies suggest that older adults are more uncomfortable with the testing situation and therefore experience increased anxiety. The **meaningfulness** of the material makes a difference, too. Older people do better when nonsensical or abstract syllables and words are replaced with actual, concrete words. They tend to be more interested if the material makes sense to them. Further, they are more susceptible to distraction. **Motivation** and **physical health** are contributing factors as well. Here again the factors in Table 5.1 may inhibit learning. The finding that younger adults are better learners than older

adults, however, should be viewed cautiously. Under certain conditions, older people can learn as well as or better than younger people.

Memory varies enormously among individuals of all ages (inter-individual differences), and types of memory vary greatly in the same person (intra-individual differences). There appears to be a very slight and progressive deterioration in memory efficiency as people grow older; however, the extent of this deterioration remains controversial.

Three types of memory (see Figure 5.1) have been described.

1. *Sensory memory*. Sensory memory is the initial level at which all sensory information is registered but not stored. It is fleeting, lasting less than a second unless deliberate attention is paid to the informa-

tion and it is transferred by means of working memory to our short-term memory. Countless numbers of stimuli bombard us every fraction of a second in our waking hours and if we gave attention to it all, our senses would be quickly overloaded.

2. *Secondary or short-term memory*. Secondary memory holds pieces of information such as the phone number you just looked up; the name of a person just introduced; and so on. If that phone number is repeated (rehearsed) several times, or the new person's name associated with another person with the same name, the memory moves to long-term, or *tertiary* memory.

3. *Long-term memory*. Tertiary memory is long-term memory, sometimes called remote memory. It is the stored facts and words, learned years ago—the past life experiences such as weddings, births, deaths, and episodes from childhood.

The findings on sensory memory have been described earlier in this chapter. The findings on the other levels of memory follow.

Primary Memory and Working Memory

The content of primary memory depends on the acuity of the senses to pick up the initial signal. If I have a hearing deficit, I am less likely to hear a noise—or perhaps more likely to misinterpret a particular noise source—than if my hearing is not impaired. **Primary memory** itself shows little change with age. Once information is transferred to primary memory, people of all ages seem equal in being able to recall it. But age differences are found in **working memory**, or the processing of sensory stimuli to give it meaning and transfer to longer-term storage. If information does not get to storage, it cannot be recalled. Two examples of using working memory are (1) remembering a phone number given to you orally long enough to dial the number a few seconds later, or (2) unscrambling let-

ters of a word in your head. Some researchers believe that working memory is vital to understanding declines in the cognitive functioning of older adults (Bechara, Damasio, Tranel, & Anderson, 1998). Neuropsychologists generally believe that a major portion of age-related decline is due to decreased speed of working memory to process information. Research on working memory is relatively new but holds great promise in solving some of the mysteries of the mind.

Secondary Memory

Research indicates that secondary memory is another major area of age-related decline. Younger adults are superior on tests of secondary memory, which is the major memory function in everyday life. The processing involved is deeper than primary memory, enabling recall to take place over a longer period of time. If the processing is too shallow, a person will not be able to recall the information a year, month, or even a day later.

Numerous kinds of memory tasks have been developed to study secondary memory. For example, in "free recall," people are given 20 words and asked to recall as many as they can in any order they wish. The results from hundreds of studies are that younger people do better on this task. There is great variation with the type of task and pace involved. In tests of "recognition," people are given a list of 20 words to look at, and then given a list with those 20 words plus 40 new words. They are then asked which are old and which are new. Older people do as well as but not better than young people on this test. On some memory tests, such as those that require the subject to recall facts out of context, older persons do worse than middle-aged persons.

Starting somewhere in middle age, many persons suffer from "tip of the tongue" (TOT) syndrome, in which one is on the verge of recalling a name, date, or event and unable to do it. Later, the correct memory is recalled.

For example, imagine that you are watching Clark Gable on late night television. You *know* who he is but cannot for the world recall his name. This is an example of tip of the tongue: at the moment you choose to recall the name, you draw a blank instead. Older adults experience more TOTs, and come up with more blockers (related words) during the TOT than do younger adults (Sethi, 2008). In old age, interference hinders secondary memory. For example, if a neighbor rings the doorbell while an older person is reading the paper, he or she may forget what was just read. Studies suggest that eliminating distractions can optimize short-term memory for the older person. Still, some older people have excellent memories.

Older adults do not usually have to engage in free recall because they have external cues, such as date books or notes of reminders on the refrigerator. A psychometric test situation may not have "ecological validity" in a real-life setting. A good deal of recent research focuses on natural memory, which includes studies of autobiographical memory. Most people believe they can remember where they were on the day of a particular, historic event, such as the start of the Gulf War, or the destruction of the space shuttle *Challenger*. These searing historical moments are called "flashbulb memories" but they seem to be no more accurate three months later than any other memory (Sethi, 2008). Instead, flashbulb memories serve as "benchmarks in our lives that connect personal histories to cultural history" (McAdams, 2006).

Tertiary Memory

Information is stored for years in tertiary, or long-term, memory. The study of this memory has some problems: if the person constantly recalls the old days, say, going to college, is it still tertiary memory compared with someone who has not thought about his or her college days for years? Some people may retrieve certain events of the past often, whereas others never do. Another problem is to know the accuracy of long-term memories. Researchers often have no way of validating the memories as recalled by the subjects.

One method of study is to ask people of all ages about a historic event—such as a major earthquake or a popular television show. Typically, no age-related differences exist in the results of these studies. Some autobiographical information can be verified, such as naming high-school classmates, for example. Again, no age-related differences appear in tertiary memory. What does appear to change with age is perceptual speed, or the speed at which long-term memories are processed and retrieved. When perceptual speed was controlled in one major memory study, all age-related differences in working memory span were eliminated between young adults and older adults (Hess, 2005). The speed of retrieval, not recall of memory itself, is the difference.

Exercising the mind is important. Learning and recalling what is learned (in other words, keeping the mind active and stimulated) is a good way to preserve memory. Learning and memory are related; memory is the ability to retain what is learned. The more one learns, as a general rule, the more one can remember. Reading books, talking to friends, and seeking new activities are fun ways to exercise the mind at any age. People adapt differently to recent memory loss. Some might compensate very well by carrying a notepad to jog the memory—keeping a written record of phone numbers, plans, and appointments. Another person might give in and permit what he or she perceives as "senility" to become a self-fulfilling prophecy. As is so often the case, positive attitudes and self-evaluations can lead to constructive, corrective actions.

Memory retrieval skills may decline with age but only if they are not used. If used, memory will be maintained or even improved throughout one's lifetime. Memory expert Tony Busan (1991) advocated the use of mnemonics, that is, memory techniques to improve memory. One technique is to use a vivid imagination, calling forth wild, colorful, and exaggerated images, even sexual ones, and associating them with important dates or items on a shopping list. A second technique is to use a linking system with a special list of key words to which all other items are linked.

Many studies of memory view the person as a machine that senses, acquires, processes, stores, and retrieves information. Using this image, older adults generally show a pattern of decline in the speed and efficiency with which they can process, store, and recall information. Movement from one "task center" to the next requires a complex set of psychoneurological communications. Error at any one of the points will create disruption in the remaining processing points. So, for example, if a person misinterprets the initial sensory information, it is moved to short-term memory (acquired), processed, stored in long-term memory, and ultimately retrieved inaccurately. Likewise, error might be introduced at any point along the way in the process.

If, however, the image of a cleanly oiled machine is substituted with an image that excludes speed and includes philosophy of life and the ability to integrate a lifetime of experiences, the older person makes gains. There may be a trade-off: losing some details but gaining the broader outlook. This has been referred to as the race between the bit and the byte, with older adults giving up the bit for the byte. We have just begun to explore the potential of the aging mind. Using our increased cultural/social knowledge, we will possibly be able to "outwit" the biological limitations and deficiencies of old age.

Psychopathology and the Myth of Senility

Psychopathology is the study of psychological disease or, in other words, of mental disorders. Some mental disorders have a physical cause; others seem to be entirely emotional in nature.

In still others, the physical and emotional aspects seem intertwined. A specific diagnosis, though difficult, is critical in the treatment of patients at any age.

Unfortunately, the term *senility* has been used as a catchall term for any mental disorder of old age. Any symptoms of confusion, anxiety, memory loss, or disorientation readily receive this label, and the labeling is done by people of all ages, including old people. **Senility**, however, is not an inevitable consequence of growing old. The word masks both the possibility of multiple causes of certain mental disorders and the fact that treatment for them may be available. In some cases, for example, persons diagnosed as senile actually suffer from thyroid deficiency. Thyroid hormone treatment eliminates the "senility." In other cases, the "senile" patient may be suffering from anorexia or emotional problems, which antidepressant medication can alleviate.

It is particularly problematic when doctors and other health-care professionals accept the myth that the elderly become senile as a result of the aging process. This erroneous belief forms the basis for ignoring an older person's complaints, rather than attempting to diagnose the problem thoroughly. This tendency is not from disregard for the older person's well-being: medical professionals are generally inadequately trained in geriatrics. In reality, the symptoms of "senility" have many causes. Older patients can show confusion and disorientation as a result of infection, pneumonia, heart failure, heart attack, electrolyte imbalance, anemia, malnutrition, or dehydration—to name only a few causes. Older people more often have cognitive contraindications to drugs than do their younger counterparts, and they are more likely to be taking a multiplicity of drugs, including such "benign" over-the-counter drugs as decongestants. Depression may also be a reason for **pseudo-senility**. Most mental illnesses can be more precisely described—and thereby treated—than the label "senile" suggests.

Functional Disorders

A reference book published by the American Psychiatric Association was designed to improve the diagnosis of psychiatric disorders. The fifth edition of the *Diagnostic and Statistical Manual of Mental Disorders* is referred to as *DSM-V* for short. Psychologists use the term *functional disorder* to denote emotional problems of psychological rather than physical origin that interfere with daily functioning. Such disorders are more serious than the emotional problems (such as those based on widowhood or other such losses) previously described. To differentiate among functional disorders, psychologists use the following terms: anxiety disorders, depressive disorders, personality disorders, affective disorders, and schizophrenia.

Anxiety Disorders

Anxiety is a cluster of feelings of uneasiness, nervousness, tension, and dreading the future. Trait anxiety is related to the individual's personality, and state anxiety is more related to a transitory situation. Persons with **anxiety disorders** are often anxious, rigid, or insecure personality types. There seem to be no age differences in either trait or state anxiety, as measured by psychological tests. Anxiety disorder, which is a clinically significant form of neurotic anxiety, is quite low for elderly people (NIMH, 2009). Among those elders with this disorder, however, it can be debilitating—especially when it is undiagnosed and assumed to be the result of aging.

Generalized Anxiety Disorder

When a person becomes so anxious that fear and dread of things or events begin to impair his or her ability to function, a generalized anxiety disorder is present, exaggerating any real danger. An older person may fear being robbed or mugged. Consequently, he or she may not leave home; the person may get dozens of locks and constantly check them. Psychotherapeutic intervention would help this person face the threat—if actual—rationally and realistically.

Obsessive-Compulsive Disorder

A person obsessed with a single act, such as washing hands, walking back and forth across a room, looking for something, or touching something, exhibits obsessive-compulsive disorder. Persons who continually wash their hands may think they are dirty when they are not. The "dirt" may be internal—for example, repressed feelings of guilt that need to be resolved.

Phobia

A phobia is a fear that displaces fears that a person cannot face. Claustrophobia, the fear of being closed in, may, for example, mask a more potent fear, such as fear of death. Numerous kinds of phobias can manifest numerous fears, and left untreated, phobias can be especially limiting for an individual whose life circumstances include other limits, such as physical frailty or impaired vision. Social phobias, characterized by the fear and avoidance of situations where an individual is subject to the scrutiny of others, have been shown to be more than 10-fold higher among first-degree relatives, implying its biological basis (Stein et al., 1998). A clear understanding of the relationship of biology with all phobias still eludes psychologists.

Depressive Disorders

Many people of all ages are occasionally depressed. A depressed person feels sad, has low self-esteem, is lethargic, and finds life confusing, hopeless, or bereft of meaning. Physical symptoms may be insomnia or difficulty in sleeping, loss of appetite, and/or the inability to concentrate. Depression in later adulthood is often triggered by the death of family members or

Depression is no more common for older people, but the social consequences might be more debilitating for them than for younger adults.

other disappointments accumulated over time. The person may or may not have had bouts of depression throughout his or her life. Most people over 65 who suffer from depression are not receiving any formal psychiatric treatment.

It is a common belief that older adults suffer from depression more than do younger adults. This belief is a myth, much like the belief that memory loss is a function of aging; however, it is well established and held by older as well as younger adults. A recent study regarding older peoples' attitudes to mental health concluded that these attitudes are linked and mediated by personal experience (Quinn et al., 2008). Attitudes develop within a context: an older adult who in his or her youth knew a depressed or despondent elder will associate old age with depression.

Early studies indicated that as much as 65 percent of the older population suffered from depression. It is now recognized that the tests used to obtain those scores were not valid for an older population. For example, emotional symptoms ("I feel sad") were included with physical symptoms ("I have difficulty sleeping"). Among younger adults a strong relationship exists between feelings of sadness and lack of sleep. Among older adults this relationship does not exist *necessarily*. More recent studies reveal that roughly 15 to 20 percent of older people suffer from mild (but noticeable) depression or despondency, and only 1 to 2 percent are clinically (severely) depressed. The incidence of clinical depression among younger adults, at 4 percent, is actually higher than among older people (1 to 2 percent) (Goldner, Hsu, Waraich, Somers,

Social interaction is a crucial component of good mental health.

2002; Kant, d'Zurilla, & Maydeu-Olivares, 1997; Richardson & Hammond, 1996).

Depression can be treated, but first it must be recognized. More mild forms often spontaneously disappear, but more severe (clinical) depression requires treatment. Treatment for older adults includes psychotherapy, especially cognitive therapy, chemical antidepressants—which can be problematic if the individual is taking any other medications, the administration of electroconvulsive therapy (ECT), or a combination of these treatments (Williams, Ghose, & Swindle, 2004). The diagnosis and treatment of personality disorders including depression have been broadly addressed for younger adults but remains critically under-examined for older adults (Segal, Coolidge, & Rosowsky, 2006; Zweig, 2008).

Hypochondria

A hypochondriac is someone who is overly concerned about his or her health. The person generally has bodily complaints for which no physical cause exists. He or she may be depressed, fear physical deterioration, need attention, or be otherwise expressing emotional issues through a series of somatic (body-oriented) complaints. Although the complaints may be real, the appropriate treatment involves dealing with the underlying emotional problem.

Personality Disorders

We age into the people we have been over the course of our lives. Personality disorders are not characteristics of older people, but are characteristic of a group of people, *some of whom have aged into late adulthood*. This distinction becomes important when treatments are considered: a psychotherapeutic treatment plan for a young adult is not any more appropriate for an older adult than it is for a child. The way a person characteristically thinks and acts and how this relates to his or her adaptation to stressful situations in life form the basis of coping strategies. . The longer lived an individual is, the more strongly reinforced are his or her coping strategies. Thoughts and strategies for coping are distinctively large pieces of personality.

Personality disorders are believed to occur in approximately 10 percent of the adult population; yet they are seldom diagnosed in primary care settings (Widiger & Seidlitz, 2002; Zweig, 2008). People who have developed very rigid styles of coping that make adaptive behavior difficult or impossible fall into the category of having personality disorders. Those with personality disorders typically have held long-standing, maladaptive, and inflexible ways of relating to stress and the environment throughout adulthood. We can describe a number of personality types that lend themselves to disorders.

The *paranoid* personality is extremely suspicious and mistrustful, preoccupied with being alert to danger. A person with this personality tends to be stubborn, hostile, and defensive.

The *introverted* personality tends to be a solitary person who lacks the capacity for warm, close social relationships. Situations that call for high levels of social contact are especially stressful for the introverted individual.

The *antisocial* personality is characterized by a basically unsocialized behavior pattern that may conflict with society. Such people have difficulty with social situations that require cooperation and self-sacrifice.

These personality types and others involve behavior from childhood or adolescence that has become fixed and inflexible; for each, certain situations cause stress and unhappiness.

Affective Disorders

Affective disorders are sometimes called "mood disorders" because depression and mood swings are typical. *DSM-V* uses the term *bipolar disorder* to describe behavior that includes both a depressed phase, of sadness and slowed activity, and a manic phase, characterized by high levels of excitement and activity. An individual generally first manifests this type of disorder in his or her twenties and thirties. *Depression*, without the manic phase, is more common. Depression is most severe as an affective disorder; as an anxiety disorder, it is moderate. Some event—a great disappointment, for example—sets it off. For the older person, depression might follow the loss of a spouse or the onset of a terminal illness. In nonclinical depression, feelings of melancholy are normally appropriate; normally, the feelings will eventually wane. An unusually long duration of depression and intense, continued sadness mark clinical depression.

Schizophrenia

Schizophrenia, another category of **functional disorder**, is more complicated, severe, and incapacitating than any of the disorders previously described. It may affect up to 1 percent of the general population (Goldner et al., 2002). Typically suffering serious disturbances in thinking and behavior, schizophrenics are often unable to communicate coherently with others. Their language seems to be a means of self-expression rather than communication, and their talk is filled with irrelevancies. Feelings have no relation to verbal expression: fearful topics may be discussed with smiles; a bland topic may incite rage. Schizophrenia is characterized by an impaired contact with reality, at least during the disorder's active phases, and it often takes the form of hallucinations or delusions.

Late-life onset of schizophrenia is fairly rare; typically, the sufferer has evidenced the disorder in earlier years. Surprisingly little is known about the treatment needs of patients who remain symptomatic and functionally compromised in late life, despite the debilitating effects of the illness (Zweig, 2008). Older people with schizophrenia often develop cognitive impairment that seems similar to, but less severe than, Alzheimer's disease (Goldner et al., 2002). Postmortem studies of tissue from patients who were chronically hospitalized with schizophrenia, however, show notably little neuro-degeneration or other pathology to explain the impairment (Arnold & Trojanowski, 1996). We apparently are not yet close to understanding the path of this devastating form of mental illness in late life.

Organic Mental Disorders

Organic disorders arise from a physical origin that impairs mental functioning. About 11 percent of older adults are believed to have mild disorders of this type (NIMH, 2009; Vasavada, Masand, & Nasra, 1997), and geriatricians estimate that 6 percent of Americans over 65 have severe intellectual impairment based on physical causes. This rises to 20 percent for those over age 80 (Zweig, 2008), with some estimates as high as 50 percent.

The difficulty in assessing dementia prevalence is partially due to the many different paths of brain disorder, including reversible and irreversible dementias with a multiplicity of causes and symptom clusters (Bowen et al., 1997). The diagnosis of dementia overlaps with that of a mild cognitive impairment known as CIND (cognitive impairment, no dementia), which previously was believed to be a mild, initial form of dementia (Graham et al., 1997).

An acute organic brain disorder is short-term and reversible. An infection, heart condition, or drug reaction, liver condition, or malnutrition may cause an acute disorder. Anything that interferes with the nourishment of the brain—the supply of oxygen or food by the bloodstream—can produce an acute disorder. If not treated promptly, it may become chronic.

Chronic organic disorders are brain disorders with a physical cause for which no cure is known. Thus, such disorders characterize an irreversible, chronic, and progressive deterioration of the brain. One should not assume, however, that these disorders go hand in hand with old age; they do not. Organic brain disease is so debilitating for the minority who suffer it that we should direct our efforts to finding cures, rather than merely fostering acceptance. Of those with chronic brain disorders, 50 to 60 percent are living at home rather than in institutions and are being cared for by relatives and neighbors. Additional in-home geriatric services are essential to help maintain families and older adults with chronic brain disorders.

The two manifestations of organic brain disorders are delirium and dementia, which are general terms for two syndromes or symptoms. Delirium is characterized by a lack of awareness about oneself and the surroundings, hallucinations, delusions, and disorientation. Caused by the atrophy and degeneration of brain cells, dementia was once labeled as senility and was thought to accompany normal aging. Now no longer recognized as "normal," symptoms of dementia can result from many disorders, though causes for many types of the disorder still remain a mystery. Dementia can result from numerous conditions. For example, years of alcohol abuse can lead to dementia. However, the largest contributor by far is Alzheimer's disease.

As defined by the online resource Family-Doctor.org, dementia is

> . . . [A] word for a group of symptoms cause[d] by disorders that affect the brain. It is not a specific disease. People with dementia may not be able to think well enough to do normal activities such as get-

ting dressed or eating. They may lose their ability to solve problems or control their emotions. Their personalities may change. They may become agitated or see things that are not there. Memory loss is a common symptom. . . . However memory loss by itself does not mean you have dementia. People with dementia have serious problems with two or more brain functions, such as memory and language.

In the early stages of dementia, emotional responses to ordinary daily affairs, previously handled without difficulty, may be extreme to inappropriate. Memory, judgment, social functioning, and emotional control are impaired. Problems become more difficult to solve, and decisions become harder to make. One may lose interest in life and become apathetic or irritable. Further declines come as one has trouble receiving, retaining, and recalling new information. A newly acquired fact may be forgotten in minutes: a person may forget, for example, what he or she saw on a television program minutes after the program ends. As time passes, progressive disorganization of personality follows, accompanied by disorientation with respect to time, situation, and place. Some patients can no longer recognize even family members, friends, and neighbors.

Symptoms of dementia include the following—not everyone who has dementia will have all of these signs, and one or even two indicators can be experienced by people who do not have an organic brain dysfunction (FamilyDoctor.org).

- **Recent memory loss:** people with dementia often forget things but they never remember them in the first place
- **Difficulty performing familiar tasks:** might cook a meal but forget to serve it or even forget they cooked it
- **Problems with language:** misuse of simple words or use of the wrong words making it hard to understand what the person wants

- **Time and place disorientation:** getting lost on their own street or forgetting how to get to a certain place and back home
- **Poor judgment:** forgetting simple things like putting on a coat before going out in cold weather
- **Problems with abstract thinking:** balancing a checkbook and even knowing what the numbers are and what needs to be done with them.
- **Misplacing things:** placing things in the wrong places; might put an iron in the freezer or watch in the sugar bowl.
- **Changes in mood:** fast mood swings, from calm to tears to anger in a few minutes
- **Personality changes:** possible major changes in personality, exhibiting irritability, suspicion or fear
- **Loss of initiative:** becoming passive; avoiding going out or seeing people

Alzheimer's Disease

The most common form of chronic organic brain disease, accounting for 70 percent of all such diseases, is **Alzheimer's disease**. It has been estimated to affect more than 4 million middle-aged and older persons in the United States: double the number in 1980 and expected to reach 16 million by 2050, due to increased life expectancy (Larson, 2008). "Statistics show if we could delay the onset of Alzheimer's by five years, the number of people with the disease would be cut in half," says Yaakov Stern, cognitive neuroscientist at Columbia University (quoted in Larson, 2008). That is a huge difference, affecting social programs and policy planning for the future, not to mention the reduced impact of this condition on thousands of families.

The standardization of diagnostic criteria has improved estimates of the prevalence of this disease; however, estimates still vary widely, from 3 to 11 percent of those age 65

The ethical issues of the potential for a sitting president to become a victim of Alzheimer's disease engaged a nation when former president Ronald Reagan's Alzheimer's disease was made public.

people in a population), has increased because it is more commonly developed in later life and people around the globe are living into later life. It has been established to be more common among women than among men.

The Framingham Study estimated the life-time risk of developing Alzheimer's disease and dementia based on their longitudinal population studies. For a 65-year-old woman, the life-time risk was calculated to be 12 percent; for a man the same age, the risk was 6.3 percent. The cumulative risk was much higher from age 65 to 100, at 25.5 percent in men and 28.1 percent in women (Seshadri et al., 1997). In contrast to risk, overall prevalence has been reported to be 26 percent among women and 17 percent among men (Preidt, 2009).

The symptoms of Alzheimer's disease were first described in the early 1900s by Emil Kraepelin, a German psychiatrist. It was named in 1906 after Alois Alzheimer, another German psychiatrist who described the neuropathological features of a 51-year-old woman (Johns Hopkins, 1999). The disease was thought to be rare and was relatively unknown as late as the 1970s. In the 1980s, it was identified as the fourth leading cause of death among adults.

As a result of Alzheimer's disease, the brain gradually atrophies, shrinking in both size and weight; neurons are lost; fibers become twisted in the neuron cell bodies (**neurofibrillarly tangles**) and abnormal masses (**senile plaques**) develop. Affected individuals gradually lose their memory. Thought processes slow, judgment becomes impaired, speech disturbances emerge, and disorientation results. In the disease's more advanced stages, the individual can suffer emotional disturbances, delusions, deterioration of personal and toilet habits, failing speech, and finally total loss of memory. This disease, which heightens anyone's fear of brain disorders, is tragic for all concerned.

The major symptoms of Alzheimer's disease are gradual declines in cognitive functioning

to 75 years. It is apparent the risk rises with age—Alzheimer's is estimated to afflict between 30 and 50 percent of those over 80 (Griffin, 2005). One recent study found the prevalence to increase from 30 percent at 85 to 95 years, to 39.5 percent at 90 to 94, and to 52.8 percent among those over 94 (Preidt, 2009). As can be seen, accuracy in estimation is difficult and varies by methodology.

The overall risk of developing Alzheimer's disease during one's lifetime depends on disease incidence and on life expectancy. The *incidence* of Alzheimer's disease (number of people who contract the disease) does not appear to be increasing. Its *prevalence*, however (proportion of (memory, learning, reaction time, language), disorientation, declines in self-care, and inappropriate social behavior such as violent outbursts. In the beginning the symptoms are mild and may mimic depression or mild paranoia, but they often develop into profoundly disruptive behaviors. Extremely disruptive behaviors have been termed **catastrophic reaction**, a reaction occurring when the organism is unable to cope with a serious defect in physical and cognitive functions (NIMH, 2009). The catastrophic reaction is defined as "short-lasting emotional outburst characterized by anxiety, tears, aggressive behavior, swearing, displacement, refusal, renouncement, and/or compensatory boasting" embedded in the inability of the individual to cope (Tiberti et al., 1998). Catastrophic reactions are often manifest by people with Alzheimer's disease. We might only imagine the terror of incrementally losing our senses.

Outstanding progress has been made since the 1970s, when scientists had just begun to understand that Alzheimer's disease affected the brain in specific regions. The regions least affected are the primary neocortex, which is on the surface of the brain and receives auditory, visual, and sensory information or originates movement. Regions most affected are those integrating various functions of primary and "associational" cortical areas such as the hippocampus, to regulate memory, attention, and other higher thought processes. Disturbed nerve-to-nerve communication appears to be the major cause of disturbed thinking and behavior. The most significant neurotransmitter deficit (the chemical necessary for inter-nerve communication) is acetylcholine. With these findings, researchers began to seek and test drugs to enhance acetylcholine transmission (Internatonal Longevity Center-USA [ILC], 2009).

In 1993 another protein, apolipoprotein E (ApoE), was identified in senile plaques and neurofribrillary tangles of the Alzheimer's disease brain. The ApoE gene was found to have three variants, one of which (ApoE4) seemed to make people more susceptible to late-onset Alzheimer's disease. Inheritance of the ApoE2 variant appears to provide protection (ILC, 2009). The tangles described by Alzheimer in 1906 have now been identified as chemically altered tau protein aggregates, which form the characteristic neurofribrillary tangles of the disease. These tangles are a major factor in nerve cell death.

Genetic tracing is only part of a broad range of approaches targeting Alzheimer's disease. Research on all fronts continues. The discovery that people treated with nonsteroidal anti-inflammatory drugs for arthritis had a reduced incidence of Alzheimer's disease led to the suggestion that brain cell inflammation contributes to the disease. Exploring the possibility of environmental toxins, searching for a virus, looking at causes for the penetration of the blood–brain barrier—all of these factors continue to be explored in the search for a cure to this debilitating illness.

In 2007 final test results were released for a new generation of drugs to address underlying causes of Alzheimer's disease (AARP, 2007). The Chief of the Dementias of Aging Branch at the National Institute on Aging said, "We've gone from drugs that help for a time with the symptoms of Alzheimer's to trying to develop drugs that will actually slow down or reverse the disease itself" (quoted in AARP, 2007). Today's drug trials are the fruit of 30 years of scientific research. Government and private support for this research remains crucial.

Because a prevention or cure for Alzheimer's disease is yet to be found, attention must be directed toward improving the functioning of the ailing person and helping family members to cope. The disease is gradual and progressive; the length of survival ranges from 2 to 20 years. Several stages are involved— some say seven stages: (1) normal; (2) forgetfulness; (3) early confusional; (4) late confusional; (5) early dementia; (6) middle dementia; and (7) late dementia.

In the first stages, cognitive declines are not readily apparent. A midway stage is characterized by recent memory loss and personality changes. For example, a person may become hopelessly lost while walking to a close and familiar location. Abstract thinking can also become impaired; the difference between an apple and an orange can become confusing, for example. Ailing individuals are typically aware of their intellectual decline, becoming anxious, depressed, or angry.

The next stages of Alzheimer's disease advance the deterioration of thought processes. Further memory loss and drastic mood swings are common. Speaking may become difficult, and paranoid symptoms may appear. During this stage, complications often force the patient to relocate to housing where care is provided. He or she may have trouble remembering close family members.

The final stage is terminal and usually very brief, lasting one year or less. Though many patients with Alzheimer's disease at this stage stop eating and communicating and are unaware of surroundings, years, or seasons, they are still sensitive to love, affection, and tenderness. This is an enormously stressful time for friends and relatives; at this stage, they are most in need of support groups and counseling.

Multi-Infarct Dementia (MID)

Arising from problems with blood flow to the brain, this vascular dementia is caused by a series of small strokes (infarcts) that damage brain tissue over time. The disease is typically chronic, and a person may live with it for many years. For most, impairment is intermittent, occurring when a stroke occurs. The strokes are so small that one is unaware of them. A patient may have sudden attacks of confusion but then recover. Another may remember something one minute, then forget it the next. Gait difficulty, urinary incontinence, and palsy may accompany dementia symptoms. The brain area in which the stroke occurs corresponds with the impaired ability.

Alcoholism

Lifelong alcoholism or the onset of alcoholism in late life may yield changes that indicate a dementia syndrome. For example, chronic alcoholism may cause Wernicke-Karsakoff's syndrome, a type of dementia resulting from the lack of vitamin B_{12}, which results in memory loss and disorientation.

Creutzfeld-Jacob Disease

Creutzfeldt-Jakob disease (CJD) is a rare, degenerative, invariably fatal brain disorder. It affects about one person in every one million people per year worldwide; in the United States there are about 200 cases per year (National Institute of Neurological Disorders [NINDS], 2009). It usually first appears between ages 20 and 70, with average age at onset of symptoms in the late 50s. Creutzfeld-Jacob disease, though far less common and much more rapid than Alzheimer's, follows a similar course. Within a year, the cerebral cortex degenerates to a fatal point in the sufferer.

Parkinson's Disease

About 50,000 Americans are diagnosed with Parkinson's disease each year, but getting an accurate count of the number of cases may be impossible because many people in the early stages of the disease assume their symptoms are the result of normal aging and do not seek help from a physician. Rarely diagnosed before age 40, it increases in prevalence in those between the ages of 50 and 79. The disease is about 50 percent more common among men than women, but the reasons for this discrepancy are unclear. It occurs throughout the world; however there appears to be a higher incidence in developed countries, possibly because of increased exposure to pesticides or other toxins in those countries. Studies to date are not conclusive and the reasons for the apparent risks are not clear (NINDS, 2008).

Tremors and rigidity of movement characterize the disease. Parkinson's progresses through stages, and some sufferers are eventually confined to bed or a wheelchair. Between 20 and 30 percent of patients develop dementia. Parkinson's has been treated with some success using the drug L-dopa, which relieves symptoms but is not thought to slow the progression of the disease (Turjanski, Lees, & Brooks, 1997).

Huntington's Disease

A rare disease, Huntington's disease is inherited as a defective gene. Its most famous victim was Woody Guthrie (1912–1967), a popular folksinger and composer. The disease starts unnoticed in one's thirties or forties and proceeds over a 12- to 15-year period, ending much like Alzheimer's disease, with total deterioration of memory and bodily functions.

Caring for Older Adults with Mental Illness

Care for elders with mental illnesses, whether functional or organic, can be provided in clinical settings of mental health professionals, hospitals, institutions for the mentally ill, nursing homes, or the homes of relatives. Some environments are supportive of elders with mental impairment, whereas others are hostile or indifferent. Those who are mentally ill, whether living in group homes or boarding homes, have been shown to benefit by

(1) activities to remain busy, such as music, dance, cards, and handiwork;
(2) activities to engage in the community— even if only to have a cup of coffee; and

(3) programs and goals for reducing dependency and passivity.

Comparisons of U.S. and Canadian funding and organization of psychiatric services for elders show services to be more accessible in Canada because of universal health insurance. For acute mental health problems, hospital benefits are free of charge in Canada. The many limitations posed by Medicare in the United States do not exist there. However, in both countries, the small number of professionals trained and interested in the mental health of the elderly limits such services. In both countries, long-term care is generally inadequate, especially for those with serious behavioral disturbances.

For families caring for the mentally-ill elderly at home, the burden can be great: caretakers can expect no expanded government programs in the immediate future; if anything, programs are being curbed. The social impact of inadequate public funding is enormous given the demographic projections for the immediate future. Families and gerontologists look to volunteer programs and support groups to help infirm older adults and their families. Fortunately, support groups for caretakers of patients with Alzheimer's and other dementias are often available and highly beneficial.

Outreach programs can serve the mental health needs of elders very effectively. A program in Iowa, for example, sent workers on request to the homes of rural mentally-ill aged, most of whom were single women between the ages of 65 and 85, living alone and experiencing depression, some form of dementia, and/ or adjustment problems. The program, which assessed clients, then treated them or referred them to more professional help, was deemed a success in helping older persons and in keeping them out of institutions (Buckwalter, Smith, & Martin, 1993). Just about any form of therapeutic assistance is more helpful than many older people with histories of chronic mental illness have received in the past 15 years in the United States.

Community networks—cadres of "good people"—have been tremendously helpful by forming volunteer programs designed to consider the unique needs of mentally-ill elders, and in organizing the assistance of friends, other residents, and family members. Through network building, programs can expand to involve churches, schools, and senior-citizen groups in caring for the mentally ill in institutions or at home.

Peer counseling is a somewhat similar concept. Older persons are trained by a professional to reach out to other elders in need of mental health services. These volunteer peer counselors are taught to deal with a variety of problems, to offer advice, and to serve as a bridge, through referrals, to more formal mental health services. In some communities, peer counselors have organized hotlines for elders in crisis.

Good Mental Health

Most older people are in good mental health. One safeguard against emotional debilitation in later years is good mental health in youth and middle age. Many mental disorders in old age represent continuing problems that have gone untreated. A safeguard against emotional debilitation in old age is to maintain an active interest in life and to keep one's mind stimulated. A third safeguard is to seek professional health-care services when they are needed.

Getting mental health care is less likely in the older years. Ageism perpetuates the myth that mental illness in old age is untreatable. The problems of younger persons have often received more priority in mental hospitals, community organizing, and other care organizations. Older adults are also more reluctant to seek help. Fortunately the mental health of our elders is beginning to receive more attention, driven largely by demographics. More geriatric specialists are offering encouragement and assisting those older individuals with mild emotional problems or more severe disorders. Depression and anxiety are conditions that can be treated and cured.

At every point in our lives, we can have the goal of maximizing our potential. Even a person in a very debilitated mental state can generally respond to help, even if that help is simply a gentle touch or soothing word. Human contact, a touch or hug, adds meaning to everyone's life.

Caretakers in the field of mental health can work toward maximizing a patient's potential. In the On Loc Nursing Home in San Francisco, even the most impaired people attend programs, form friendships, and are encouraged to attend music groups, old movies, and arts and crafts workshops. They are fully dressed in street clothes every day, even though most are incontinent. In this setting, the worker's goal is to help patients reach and maintain maximum functioning potential. With a patient who deteriorates continuously, day-by-day, specific goals may have to be adjusted downward. If the goal, however, is to see Mrs. Smith smile, or to hear Mr. Juarez sing, the task remains rewarding.

Community Mental Health Clinics (CMHCs) are mandated to serve the old as well as the young; yet a disproportionately small number of older persons use these services. Medicare pays only a minimal amount for outpatient mental care. A tragic irony exists in that although nursing homes represent a major setting for mentally-ill elders, they receive, for the most part, inadequate mental health care in these homes. Geriatric mental health is an evolving field desperately in need of growth and improvement.

Chapter Summary

The psychology of aging is a broad field covering cognition and its many aspects ranging from perception, information processing, and learning and memory on the one hand, and psychological and spiritual development on the other. Older persons generally suffer a decline in reaction time but not necessarily in intelligence, learning, and other areas. This decline is accompanied by an expanded understanding of self and others. Mental-health problems may be functional (behavioral) or organic (a physical basis). Functional disorders may be moderately debilitating as with a temporary emotional problem, or they may be quite severe, as with a psychotic breakdown. The most common form of organic brain disease is Alzheimer's disease. The bulk of the aged are in good mental health. A substantial minority of elders need mental health care by a professional.

Key Terms

affective disorders
age-related changes
Alzheimer's disease
anxiety disorders
biologically caused
   changes
catastrophic reaction
classic aging pattern
cognition
cognitive mechanics
cognitive pragmatics
crystallized intelligence
dementia
depressive disorder
divided attention
fluid intelligence
functional disorders
geriatrician
hypochondria
information processing
intelligence

learning
memory
metamemory
organic disorders
perception
personality disorders
primary memory
primary mental abilities
pseudo-senility
psychomotor speed
schizophrenia
secondary memory
selective attention
senility
sensation
sensory memory
sensory threshold
terminal drop
tertiary memory
the five senses
working memory

Questions for Discussion

1. Get a sheet of paper, a pencil or pen, and a small mirror. Looking into the mirror only, draw a house. Do this with the hand you don't usually use (left if you're right-handed, etc.) Using the same method, draw a clock with the correct time. This exercise gives you some idea what the world might seem like to a person with dementia. What did it feel like to draw through the mirror?

2. What can you do to reduce the risks of developing dementia in late life?

3. List some personality traits that would be helpful in old age. List some personality traits that might not be helpful in old age. Which traits so you recognize in yourself?

4. What emotional problems might you experience in your old age based on your present personality?

Fieldwork Suggestions

1. Visit an Alzheimer's facility in your area and request a tour. In what ways has the facility adapted to the needs of the Alzheimer's patients? Do you see anything they could do better?

2. Imagine you are designing a facility for patients with Alzheimer's disease. How would you design it to best meet the needs of people in all stages of the disease?

3. Put in earplugs and a thick pair of gloves as you go through your daily routine for four hours. Note the loss of sensory perceptions and how these losses influence your ability to function and to do things for yourself? How long are you able to withstand the sensory loss? Did you find any ways to compensate for it?

4. Call your county mental health agency to find out what services are available for elders.

Internet Activities

1. Locate the Internet site for the Alzheimer's Disease and Associated Disorders Association (ADRDA), and research two therapeutic treatments for the disease. Are you able to tell whose perspective represents the information you have found? (Was it developed by academicians or by a pharmaceutical company, for example.) How useful is the level of material to you?

2. Find all that you can about *learning*, from as many different sources as possible. Keep a journal of your search: Which locations are the most useful? Was it easy, or more difficult for you to gather a range of material? Did your search provide you with cross-cultural resources? Cross-species resources?

3. Develop a list of Internet address links that you would use if you were designing a home page on learning and memory in later life.

Friends, Family, and Community

6

Teens Give Their Elders Some Video Gaming Tips

Tania Karas

Richton Park resident Gloria Cox, 63, has not been able to get around without the help of a walker in years.

But with a video game console and a few helpful teenage volunteers to share their knowledge, Cox was able to bowl again.

True, the bowling was being done on a Wii video game, but it nonetheless gave her a long overdue chance to play the sport she loves.

"It's easy for me to do because I'm not as mobile as I used to be," Cox said. "It got me going and got me active."

Cox, who had never played a video game before, joined a group of adults at Matteson Public Library recently for an hour-long tutorial on video gaming led by the library's teen advisory board.

Teen teachers taught their adult students how to play popular video games such as "Rock Band," "Dance Dance Revolution" and "Wii Sports."

The teens started off teaching the adults how to play "Rock Band," a game in which players pick up instruments—actually video game controllers designed like actual instruments—and play along with songs by hitting notes as they appear on a screen.

"We'll set you up on 'beginner' because on that level you can't fail," Asheton Mayfield, 14, said to the adults as they struggled through Michael Jackson's "Beat It" in honor of the late performer.

When the adults grew frustrated with the coordination required to play "Rock Band," the teens moved them to "Dance Dance Revolution."

To play "Dance Dance Revolution"—or "DDR"—players stand on a floor mat and move their feet to dance steps presented to them on a screen.

Tyler Mayfield, 17, said DDR is probably the most interactive of all video games she has played.

"I like more interactive games because you can get more into it and you get exercise while you do it," she said.

For the older generation, certain video games allow them to perform tasks they have not been able to do since they were younger and more agile, such as bowling, baseball and golf.

Cox, who has arthritis and used to bowl, enjoyed using the Wii, a game console by Nintendo that features a wireless controller that detects motion.

Using the Wii, she swung a controller in the same motion one used in real bowling to knock down pins on the screen, all while sitting comfortably in a chair. She even rolled a spare.

"It's not the same as the actual sport but it's the same principles," Cox said.

"I'm moving my arms in the same direction, and I could feel my arm going a little to one side the way it used to when I used to bowl."

Though it was only Cox's first experience with a video game, she said she would recommend the Wii to anyone, adults of her generation especially. She doesn't think video games are necessarily only for her grandkids.

"I recommend it for people who are just sitting at home watching the boob tube," she said. "It gives me something to do and keeps me active."

Source: AARP: The Magazine. September 13, 2009. www.aarpmagazine.org. Downloaded September 15, 2009.

Family and friendship connections take on a special significance when it becomes clear that they might not be in our life throughout *all* our life. Those significant connections include romantic and sexual relationships. Maintaining relationships can be a challenge when people become ill or frail, when friends die or move, or when family pressures for traditional grandparenthood take precedence over developing an independent lifestyle in later life.

Family Development in Later Life

Later life is not a static, stagnant time for the older family member. Transition events such as widowhood, retirement, remarriage, or a child's departure punctuate the life course. Transitions lead to changed perceptions of one's identity, to new ways of behaving, and to shifts in interdependence with kin and community. The older person's life is also influenced by family development events in the past, such as whether he or she was childless or a parent, the person's culture and ethnicity, and certainly the gender and socioeconomic status of the older person.

All of these events add to the ever-changing character of the older person's role as a family member. Older persons may face adjustments equal to or more difficult than those younger family members face. For the newly married couple, the birth of a first child may require a difficult adjustment; but for the middle-aged or elderly couple, learning to relate to an independent adult offspring who was once "my baby"

might be equally traumatic. Adjusting to the death of a spouse can be the most challenging of all changes in the life cycle. Each event in the life cycle calls for relating to others in new ways and facing the problems inherent in every transition.

Because of increased survival through childhood, adulthood, and into old age, Americans have experienced predictably longer lives in this century. Marriages are more likely to last 50 or 60 years, and parents are more likely to survive to see their children become adults, and to see their grandchildren grow up. Whole stages of life that were brief and rare in the past, such as the post-parental stage, are now long lasting. Many kin relationships last much longer now than they did in past generations.

Elders in the Kin Structure

The family is a vital part of the older person's life. Elders give a great deal to their families, and they receive a great deal in return. Family members tend to exchange emotional and financial support throughout their lives. A person's confidant in life is typically a close family member—a spouse or a child. Only when a closely related family member is unavailable does a friend rather than a relative act as confidant (Connidis, 2001). In this section, we will discuss siblings and grandparenthood. Parent-child ties are discussed in chapter 7.

Blood is thicker than water.

SHAKESPEARE, 1581

Grandparents can have a huge impact on the developing child's understanding of the world and his or her place in it.

Siblings in Older Adulthood

Brothers and sisters are brothers and sisters for life, no matter how intimate or how estranged they might be. The relationship with siblings, solid or shaky, is likely to be the longest relationship in an individual's life. If we have a brother or sister, it is likely that we will be a sibling and have a sibling from the time we are very young through to the time either we or they die. In later life, because of the duration and the shared experiences of childhood, the sibling relationship can be emotionally very intense, and that intensity can be positive and life-confirming, or bitter. There are as many different types of sibling relationships as there are types of families, but the fact remains that a sibling is a close kin connection over which we have no choice, and from which we can gain pleasure, irritation, or rage.

Although **siblings** may cause problems for one another at any time in the life course, they can also extend support to one another in a social environment that does not always foster the development of social bonds (Connidis, 2001). Bonds between siblings typically extend throughout life and are reported to be second only to mother-child ties in intensity and complexity.

Family systems in particular, because of their culturally ascribed and socially recognized status, implicate or entangle their members in a form of involuntary membership.

SPREY, 1991

Most older adults have a living sibling. The parents of today's older persons produced more children than did current older adults themselves: any given married adult probably has more siblings than offspring. If the birthrate is now at the replacement level of two children per couple, we can expect there will be fewer children, and thus fewer siblings, in the future. In fact, the baby boomers will be the first cohort in history to have more siblings than children (Butrica, Iams, & Smith, 2003, 2004). The family support system will be smaller in the years ahead, and for this reason alone, the responsibility one family member feels for another is likely to increase with time. If there are no other resources for you, and you are my sister [or brother], can I *not* be there? Because adult siblings fall outside the nuclear family structure, their impact on the life course has been woefully overlooked by family researchers.

Exploring the sibling tie can highlight the importance of continuing family ties over time. Family scholars point out a conceptual distinction between feelings that siblings may harbor for one another and the obligations that are shared by virtue of family ties (Connidis, 2001). When sibling ties are ranked by level of obligation within the family, the sibling tie is typically less binding than that of marriage or of parent/child, but is present nonetheless. The tie is important, and it is unique: siblings share biological and/or familial characteristics, values, and experiences, under comparatively egalitarian status (Bauer & McAdams, 2004a). Given their shared experiences, siblings can be a major resource for life review among older adults (Bauer, McAdams, & Sakaeda, 2005).

Studies on sibling relationships have had somewhat inconsistent conclusions regarding sister/sister/brother/brother interactions. The variety of sibling unit types is nearly endless if we consider variables such as the marital status, birth order, proximity, number of siblings in the family, living (or nonliving) siblings and their birth order, gender, blended families, and so forth. Theory helps to identify some defining characteristics of being a sibling, however, to guide investigations.

Connidis and Campbell (1995) established the hypotheses for a study on sibling relationships after conducting a major review of the literature on sibling relationships in middle and later life. Using interviews with 678 people over 55 years of age, the researchers wanted to understand more about the relationship of sibling gender, marital status, emotional closeness, and geographic proximity. Their conclusions, generally consistent with most of the less recent studies in the literature, are itemized in

The sibling relationship is often the longest relationship in life. The oldest of these sisters knows only 25 months of life without her "little sister," who has never known the world without her "big sister."

Table 6.1. Their findings imply that to understand the support and emotional closeness of family in later life we must also understand sibling relationships. Because siblings are not considered to be extended family, their impact can be underestimated—both as providers of instrumental assistance and as assistants in maintaining a continuing sense of self when a sibling is in physical or emotional trauma, such as widowhood bereavement.

The gender role of older adult siblings in families varies by ethnicity and across cultures. In Western societies women are kin keepers. The cultural norms governing family and kinship systems in more traditional countries reflect those cultures accordingly. For example, in Taiwan a patrilineal system (descent through male line) gives priority to sons over daughters in inheritance (Lu, 2007). Accordingly, sons,

TABLE 6.1

Closeness, confiding, and contact among middle-aged and older siblings

Variable	Finding
Gender differences	1. Women's ties with siblings are more involved than those of men
	2. The greater emotional attachment of women to their siblings was confirmed
	3. Respondents with sisters only are closer on average to their siblings than those whose networks include brothers and sisters
	4. Women are closer to their brother(s) and sister(s) than are men
	5. Telephone contact is more frequent if the highest contact sibling is a sister
	6. Women seem to have greater emotional investment in their ties to siblings and are more engaged in sibling ties—possibly due to an assumed level of obligation
Marital status	1. Contact is more frequent between single siblings
	2. Emotionally closer ties among those whose closest sib is widowed than among those whose closest sibling is single
	3. Being single affects overall level of involvement with both the sibling network and the sibling seen most often but does not alter feelings about siblings
Parent status	1. Childless respondents confide more in their primary sibling confidant and in siblings overall than do parents
	2. Respondents with networks including parents and childless siblings confide in their siblings more than those whose siblings are all childless
	3. No greater emotional closeness to siblings among the childless
	4. Childless seem to have greater emotional investment in their siblings (higher levels of confiding)
Emotional closeness	1. A powerful relationship to confiding, telephone, contact, and personal contact; emotional closeness appears to be a primary love/friendship binding tie
Relationship over time; education; proximity; sibling number	1. Growing attachment to sibling network as a whole, over time
	2. Higher education associated with greater closeness to closest siblings but not sibling network overall
	3. Educational level inversely related to contact with sibling network overall
	4. Proximity enhances emotional closeness but not to emotionally closest sibling
	5. Greater opportunity for selectivity within larger families
	6. Network size not related to overall closeness and confiding

Source: Connidis, I. A., & Campbell L. D. (1995). Closeness, confiding and contact among siblings in middle and late adulthood. *Journal of Family Issues*, 16(6), 722–745.

not daughters, are looked to for support of their elderly parents. Cross culturally the economic structure of the culture appears to have significant weight in sibling relationships over the life course.

In addition to having a sibling, many older people have a sibling living nearby. A review of the literature on siblings (Bank, 1995; Bengtson, Giarrusso, Silverstein, & Wang, 2000; Cicirelli, 1997) showed that sibs generally maintain contact with one another and that sibling contact increases in old age. In all of the studies it was quite rare for siblings to lose contact and in no study did more than few respondents report they had completely lost contact with their siblings. The majority of older adults view their relationships with brothers and sisters positively (Bryson & Casper, 1999; Waldinger, Vaillant, & Orav, 2007). Indeed, one of the greatest life regrets reported by older adults is a failed sibling relationship (Waldinger et al., 2007).

Older people are more likely to confront the death of a sibling than of any other kin (Connidis, 2001). The two greatest times of interest and involvement in the life course in a sibling relationship generally are in youth and in old age. When an older person experiences the death of a sibling, the loss can be far more consequential than it might have been earlier in life because siblings hold shared memories, a common cultural background, and early experiences of family belonging, and are therefore sources of a sense of continuity for the older person (Cicirelli, 1982). These qualities of relationship can never adequately be shared to another intimate of the older adult, such as a child or a spouse. Sibling relationships are part of the lived experience of the older adult, helping to define the adult that person has become and to give shape to the sense of self in the future.

As parents die, children grow and leave home, friends die or move, and health becomes frail, the sibling bond endures and perhaps deepens. The bonds siblings share tend to be more forgiving, mutually warmer, and more interested in one another in the last years of life (Bank, 1995).

Very little exploration has been done on the impact of sexual orientation on sibling relationships. The focus has been primarily on partnerships and parenting in discussions of gay family ties, in no small part because of political struggles for gay marriage and rights for gay and lesbian couples to adopt children (Connidis, 2001). As the population ages, however, understanding gay and lesbian family patterns will become more important in theories of adult and family development. For example, accounts of rejection due to sexual orientation are common; however, Joan Laird (1996) argues:

> Lesbian and gays come from families and are connected to these original families. . . Most of us are not cut off from our families—not forever rejected, isolated, disinherited. We are daughters and sons, siblings, aunts and uncles, parents and grandparents. Like everyone else, most of us have continuing, complicated relationships with our families. We participate in negotiating the changing meanings, rituals, values, and connections that define kinship. (p. 60)

Over time, most gay and lesbian people with families seem to work out the rough edges of relationship with parents and siblings, as do most heterosexual people. Indeed, it is reported that dominant family forms among older lesbian and gays include the family or origin (Fulmer, 1995, reported in Connidis, 2001). Clearly, however, a complete understanding of family patterns and relationships of older gay and lesbian people is yet to be developed if we want to understand diverse family types and development of aging families.

Variations in Kin Relations

Types of kinship systems vary widely. Some older people have large and extensive kin networks with many relatives nearby, whereas other older people have managed to outlive siblings, children, spouse, and other kin. The relationships among kin members vary by sex, culture, social class, community structure (urban, rural, small town, etc.).

Gender differences are clearly apparent in family relationships. Females maintain closer relationships with other family members than do males. There appear to be stronger norms of obligation and feelings of attachment to extended family members among women than among men, as well as what might be called an *ethos of support* among women in a family (Connidis, 2001). Families with more sisters have an increased likelihood that parents—both fathers and mothers—will have assistance from their children. While having more brothers in a family decreases the likelihood among the men of being the primary care provider for a frail parent or parents, having more sisters in a family increases the likelihood of being primary care provider (Connidis, 2001). This finding supports the conclusion that family contact remains consistent and intact through the parents for many years after children reach adulthood.

Blue-collar families tend to have close extended family ties, which members maintain by living near one another. Visits from kin often constitute the major, if not the only, form of social activity for such families. White-collar family ties are also fairly strong; however, such families are more likely to be geographically scattered by career opportunities. Contact is often maintained, however, in spite of distance.

The literature is inconsistent in reported differences between rural and urban families. Rural areas are shown to be bastions of traditional values, including family responsibility and respect for elders. We can picture rural family reunions at which the aged are celebrities looking with pride at the family line of children, grandchildren, and great-grandchildren. In this picture, kinship ties are strong and meaningful. Small-town newspapers often reflect the apparent importance of family and neighborly ties by featuring articles on who has visited whom, who is in the hospital, or who went out of town to visit relatives.

Another picture of rural areas is one of poverty, isolation, and despair. In spite of the assumed traditional values, rural elders interact less with their children than do their urban counterparts. Indeed, compared to urban older people, rural elders have substantially smaller incomes, are more restricted in terms of mobility, experience poorer physical health, and have a more negative outlook on life—probably because of less robust health and the vicissitudes of poverty and rural life.

Likewise, the quality of life for older adults living in urban settings depends on health issues, availability of social support, perceived sense of community safety, and income. Indeed, income is probably the variable of greatest distinction among people everywhere, despite their culture and their age.

Grandparenting

The three-generation family is becoming common in the current population, and four- and five-generation families are on the increase. About 80 percent of those over 65 have living children, and 80 percent of those, or 60 percent of all older people, have at least one grandchild. If one becomes a parent at age 25, one may well be a grandparent at age 50. If parenthood at age 25 continues for the next generation, the individual would be a great-grandparent at age 75, thus creating a four-generation family.

Note that middle age, not old age, is the typical time for becoming a grandparent—indeed, half of Americans become grandparents by age 50 (Szinovacz, 1999). Most grandparents

Family dynamics in the 21st century are vastly different than in previous times. This family represents some of the changes of shape and form of "family."

do not fit the stereotypic image of jolly, white-haired, bespectacled old people with shawls and canes: many are in their 40s and 50s.

> *My grandmother was the radiant angel of my childhood, and, you know, now my grandchildren are just the same. They are the part of my life that is most joyous, that gives me most pleasure.*
>
> GRANDFATHER (IN KIVNICK, 1982)

Rapid cultural changes characteristic of the second half of the twentieth century have profoundly affected the North American fam-

ily structure in the twenty-first century. Some analysts say these changes are negative for the family; some say the changes are for the better; some analysts say they are *change*, without being necessarily positive or negative but clearly being something to which people and society must adapt. In the tapestry of life, the color and design are variations brought about by change, but it is the continuity across generations that forms the fundamental shape and form. Through that continuity people learn how to be a grandchild, a parent, and a grandparent. The degree to which grandparents are involved in playing their role and just what type of role they will play with their grandchildren is significantly influenced by having known their own grandparents (King & Elder, 1997).

Living with intergenerational relationships is critical to the passing on of that cultural and family continuity, for these intergenerational exchanges serve as socializing influences as well as emotional influences (Wiscott, 2000). Grandparents can be playmates, storytellers, friends, advocates, providers of unconditional support, or mentors (Lussier, Deater-Deckard, Dunn, Davies, 2002).

There are as many different ways of being a grandparent as there are family types and styles. Grandparents can be a point of reference and identity for the grandchild; they expand the age range and number of adult role models available to children. They can be that connection that provides a sense of historical and cultural rootedness for the child, and they can provide a secure and loving adult/child relationship for their grandchild (Mueller, Wilhelm, & Elder, 2002).

Grandparenting styles differ in frequency of contact; the activities and types engaged in with the grandchild; the level of shared intimacy (confidant or advisor; friendship?); the extent a grandparent is the voice of wisdom or experience; whether financial assistance is provided; and if the grandparent holds the role of authority and discipline (Mueller et al., 2002). For some, grandparenting is a time to have fun or to indulge grandchildren, or perhaps even a chance to re-experience one's youth—the **fun-seeker** grandparent. At the opposite end of the spectrum would be the more formal, distant grandparent—perhaps one who dispenses discipline and authority.

For some, grandparenthood is personally meaningful because it represents the continuation of the family line—it provides a sense of purpose and identity, and is a source of self-esteem as the keeper of family history. The grandparent role can be central to helping children maintain identity in single-parent and step-families, and has been shown to be especially important in adolescents' well-being in the face of parental conflict or unhappiness (Attar-Schwartz, 2009). Greater grandparent involvement is associated with fewer emotional problems and more prosocial behavior in adolescents under circumstances of family disruption (Dunn, 2002; Lussier et al., 2002).

In short, one person might find great joy and meaning in the role of grandparent, whereas for another person the role has little relevant meaning. For some grandparents, their role is essentially that of substitute parent. Because their adult child is unable or unavailable to parents—because of drug abuse, death, divorce, AIDS, a mental health problem, career choice, or a host of other modern-day complications—grandparents sometimes end up with full-time care of their grandchildren (Szinovacz, DeViney, & Atkinson, 1999). A study of grandmothers' family role and the HIV/AIDS pandemic in Africa illustrated, cross-culturally within sub-Saharan Africa, that contributions of older and younger women to the household are both qualitatively and quantitatively different (Bock & Johnson, 2008). In the face of the disruption of parents' health, grandmothers' supervision and transmission of cultural skills and knowledge is crucial to developing children.

The emotional responsibility of raising grandchildren can be complicated by the complex relationships with the adult child of that grandparent. Two-thirds of children living with grandparents live in homes where at least one of the child's parents also live (Pollet, Nettle, & Nelissen, 2006). This means that grandparents who have taken responsibility for raising grandchildren also have responsibilities for (or the residential impact of) that child's parent. The stress of this complex set of relationships can be enormous. One-third of grandparent caregivers report their health to have deteriorated since becoming primary caregivers, and a vast majority report feeling depressed (Wood & Liossis, 2007).

Research has variously described grandparents who parent their grandchildren to be extraordinarily committed to childrearing, angry, in despair about their own child, guilt ridden about their role as parent to their "failed" child, and frustrated regarding the imposition of un-expected childrearing responsibilities in later life (Lussier et al., 2002; Ruiz & Silverstein, 2007). The full range of these feelings is likely to be experienced.

Women are much more likely than men to look forward to the role of grandparent. Women often visualize themselves as grandparents well ahead of the birth of the first grandchild, and the grandmother role is positive and desirable to most women, even though it can make young grandmothers feel old. Men typically become grandfathers when their primary identity is still with the work role (Pratt, Norris, Cressman, Lawford, & Hebblethwaits, 2008). Consequently, they may postpone involvement with their grandchildren until retirement.

Anthropologist Dorothy Dorian Apple (1956) explored grandparent/grandchildren relationships in seventy-five 20th-century societies and found that in societies where grandparents retain considerable household authority, the relationship between grandparents and grandchildren tended to be stiff and formal. In societies such as the United States, where grandparents retain little control or authority over grandchildren, the relationship is friendly and informal. Others have likewise observed that North American grandparents, more than grandparents in other societies, engage in companionable and indulgent relationships with their grandchildren and usually do not assume any direct responsibility for their behavior.

One of the most representative samples of grandparents of teenagers affirmed many of these results. Grandparents often played a background, supportive role, helping most during times of crisis (Attar-Schwartz, Tan, Buchanan, Flouri, & Griggs, 2009; Lussier et al., 2002). A recent study of psychological adjustment among youth from divorced families found that attachment to maternal grandmothers provided a protective factor to the children during and just following the divorce of their parents (Henderson, 2003).

Among many families the role of grandparent is somewhat distant and does not include disciplining the grandchildren, for example, but in a family crisis that grandparent is there with financial and/or instrumental support for the grandchildren. Additionally, even if they do not live with their grandparents, millions of American children receive childcare from their grandparents each day (U.S. Bureau of the Census, 2002), and many grandparents express their gratitude for the companionship and love of their grandchildren (Dunn, 2002; Pollet, Nettle, & Nelissen, 2006).

Grandparents are the child's roots, a sense of belonging to a larger family and a larger community. Grandparents are there when things go wrong for the child—sometimes the child needs that protection.

GRANDMOTHER, QUOTED IN KRUK (1994)

One can conclude that grandparenthood carries meaning—biological, social, or personal meaning—for some elders, but not for others. The degree of involvement with grandchildren varies. Current evidence seems to indicate that in the United States, grandparenthood, for many, is not a primary role, although it is enjoyed by most. It may be a primary role, however, for grandparents who are major caregivers for their families' children. Becoming a grandparent is not going to fill voids left by shortcomings in marriage, work, or friendships.

The outcomes to grandparents of their adult child divorcing are of centrality to understanding life course family development. If a couple divorces and the in-law has custody of the offspring, grandparents often cannot enjoy the relationship they desire with their grandchildren. Existing patterns of contact with grandchildren as well as gender of the custodial parent and composition of the family all influence the role grandparents will take upon the divorce of their child (Dench, Ogg, & Thompson, 1999; Drew & Smith, 1999).

In some families grandparents are emotionally distant enough from their adult children's lives to maintain cordial relations with former in-law adult children. One study found that half of the grandparents interviewed were "friendly" with their child's former spouse. These grandparents tended to be friendly especially if the former spouse was a female, but they tended to lose contact if the spouse was male, even if he had custody of the grandchildren. Recent studies of grandparents and grandchildren of divorce indicate that grandchildren and grandparents see their relationships in highly similar ways. They desire increased time together spent in specific activities. The two groups associate emotional bonding with the grandparents' listening, keeping them safe, and gift-giving. Grandparents generally believe that a good

relationship with the custodial parent is essential (Wood & Liossis, 2007; Lussier et al., 2002). Not surprisingly grandparents who are related by blood to the custodial parent and have a good relationship with him or her are also more likely to have a positive, supportive relationship with their grandchildren.

The North American norms of individualism and social ties based on mutual interest can make negotiation of kin relationships a necessary task. Good relationships with former daughters-in-law, for example, may determine whether grandparents can remain close to their grandchildren. If informal negotiation fails, grandparents have an option to argue in court for the right to spend time with a given child. Whether this process is beneficial for the grandparent/grandchild relationship remains unclear.

If for some reason grandparents are denied access to their grandchildren, they have tra-

Positive kinship ties benefit all ages within a family and in a community.

ditionally had no standing in cases involving visitation and adoption of grandchildren; the courts clearly uphold the supremacy of the parent-child relationship. A legal relationship between grandparent and grandchild was seen to exist only when both parents were incapable or dead, meaning that the judicial system had to judge the child's parents unfit before the grandparents had any rights.

Today, at least one legal precedent has been set: if provisions for visiting grandparents had been made prior to divorce, visitation is generally permitted after divorce. Thus, grandparents can, under limited circumstances, get visitation rights. (They have no rights, however, if parents choose to give their children up for adoption.) Grandparents are increasingly pressing for rights with regard to their grandchildren. The recent growth of the "grandparent rights" self-help movement in North America suggests that a significant proportion of grandparents have concern about access difficulties and contact loss to their grandchildren following divorce, or in other situations. There is currently insufficient social policy to address the phenomenon of grandparent-grandchild contact loss, although it has become an issue in the legal system for family mediators. At this time, legal mediation is seen as the best possibility for resolution of grandparent-grandchild access problems.

We should also note that elders add new relatives to their kinship system when their adult children remarry. Remarriage, which may take place more than once, brings not only new sons- and daughters-in-law but may bring step-grandchildren. When older parents can maintain ties with their childrens' former spouse(s), grandchildren lead more stable lives. Grandparents who can remain flexible and friendly have the benefit of enjoying an expanded kinship system. The ability to incorporate nonblood grandchildren as "real" grandchildren by being a step-grandparent may lead to strongly reinforced kinship ties. Some grandparents report that there is no difference between step-grandchildren and biological grandchildren.

Social Networks

One stereotypical picture of old age is of a solitary person, in a tiny hotel room, staring silently into space. Another is of oldsters in a rest home, propped up in adjacent armchairs but worlds apart. Either way, the picture is one of isolation and loneliness. Do older people have friends—really good friends—they can count on? Simply stated, like other age groups most do but some do not. Numerous factors affect the likelihood of close friendships in old age.

Positive networks of friends are equally important in the development of the person, from infancy through late old age.

Older Adult Friends and Neighbors

Friendship is extremely important in the lives of elderly people. Let us contradict one stereotype about friendships among elders by saying that most older people maintain active social lives. In one of the first major studies of friendship formation in widowhood, respondents were asked how many times in the past year friends had helped them with transportation, household repairs, housekeeping, shopping, yard work, during illness, car care, important decisions, legal assistance, and financial aid. They also were asked how many times they had given help. Only 8 percent of the respondents indicated that they did not have a close friend. For the first close friend, 84 percent of the respondents named a woman. The average length of acquaintance with the friend was over 20 years; 81 percent keep in touch at least weekly (Roberto & Scott, 1984, 1985).

Although nationally in the past two decades geographic mobility has increased, older people are still most likely to live in the communities in which they raised their children (Lussier et al., 2002). Note that the date of the Roberto and Scott study was the mid-1980s. It has only been in the past 20 years that the role of friendship in later life has been seen as an important enough factor in well-being in later life to be a topic of research. This is not necessarily because social science research is insensitive to later-life development. It is because, once again, we can see a consequence of longevity in the patterning of our lives. Friends have clearly always been important; however (as we were reminded earlier), the adage that "blood is thicker than water" is culturally ingrained, and it becomes easy to overlook the true impact that friends—particularly friendships of very long duration—have on our psychological and social well-being.

Roberto and Scott (1984, 1985) used equity theory to formulate hypotheses and analyze results. According to **equity theory**, an equitable relationship exists if all participants are receiving gains and participants will be distressed if they contribute too much to, or receive too little from, a friendship. Equity is related to high morale, and, as expected, the equitably benefited women had the highest morale. An unexpected finding was that the over-benefited (those who received more help than they offered) had the lowest morale: receiving goods and services that one cannot repay may leave one feeling uncomfortable or inferior, resulting lower morale. Also unexpected was the high morale that existed among the under-benefited, the ones giving more than they received. Perhaps the under-benefited woman feels good that she does not need help and enjoys giving without receiving. The single women who lived alone compensated for lack of a marriage companion through extensive and meaningful friendship networks with other women.

A distinction can be made between a confidant and a companion. A **confidant** is someone to confide in and share personal problems with, whereas a **companion** is one who regularly shares in activities and pastimes. A companion may be a confidant, but not necessarily. The objective of one study examining older women's friendships was to identify changes and stabilities in the qualities older women attributed to their close friendships (Roberto, 1997). They found that, over time, women's interactions with friends did change, and the change was based on the woman's work, marital, and health status. Although those interactions changed, they also found that the essential elements of friendship—understanding, affection, trust, and acceptance—endured over time. It appears that the expressive domain of friendships in later life predominates.

Social Organizations

Are elders "joiners"? To what rates do they participate socially in voluntary organizations? Many, in fact the majority, join and are actively involved in voluntary organizations (nonprofit groups that elders join only if they choose to do so).

Lodges and fraternal organizations such as the Moose Lodge, Elks, Eastern Star, and the like are the most commonly joined volunteer associations. Membership in them tends to increase with age, and leadership positions are concentrated among older persons, perhaps because these organizations have a respect for elders that other organizations lack. The second most frequently reported form of participation is in the church. Church-related activities involve more than attending services; they include participation in church-sponsored groups such as missionary societies and Bible-study classes. Except for lodges and church, other voluntary associations generally decline in participation with age. When social class is controlled, however, no decline is evident. This means that being a "joiner" is related to a middle- or upper-class lifestyle, and aging by itself does not necessarily change this lifestyle.

Senior centers, as opposed to clubs, offer services and activities in addition to recreation; they may offer libraries, music rooms, health services, counseling, physical exercise, and education. Many senior centers operate under the auspices of churches, unions, or fraternal organizations. Often, individuals who began senior centers in their own communities now find them successful and popular years later.

The demand for senior clubs and centers continues to rise. Someday every community may have a senior center, and it will be as natural for older people to use their senior center as it is for children to go to school.

Network Analysis

We can conceive of the combination of social ties—organization memberships, friends, neighbors, and family—as a **social network.** Each person has some kind of social network. The process of analyzing the strengths and weaknesses and the sources of various functions of an older person's social network, called **network analysis,** enables caretakers to tell whether social or health intervention might be necessary. A distinction might be made between social support and social network. **Social network** describes the structural characteristics of support relationships (e.g., size, composition). Social support, on the other hand, is assessed as a more qualitative aspect of the relationship, including how satisfied individuals are with the support they receive and whether supportive others understand them (Antonucci & Fuhrer, 1997).

In recent years, a great deal of research has focused on social networks. A classic study on social support networks uses a **convoy model to** provide detailed information showing that older individuals are in frequent contact with both family and friends. The term *convoy* (Antonucci & Akiyama, 1987) is used to evoke the image of a protective layer of family and friends who surround a person and offer support. Convoys made up of those family and friends who travel through time with the individual, with some members of the convoy traveling the entire way, some dropping out along the way, and some entering the convoy after it has begun. In the late 1980s, the average older convoy was reported in one major study of older persons to consist of about nine members who offer themselves as confidants, give reassurance and respect, provide care when ill, talk when upset, and discuss health.

The role of women as keepers of the network emerges in many ways when the networks of older people are examined. This finding is to some extent an effect of the difference in longevity between men and women: more older people are women, so women are more available as friends and family members. It is also due to social expectations and learning. Women tend to be more engaged in network support activities across the life span than are men. The care giving role of daughters to aging parents has been well documented, as has the finding that married older people tend to maintain closest relationships with their same-sex children and siblings—a finding consistent with the **sex commonality principle**. Women play more central roles in the network of unmarried people (including widowed parents), a pattern referred to as the **femaleness principle** in network analysis (Akiyama et al., 1995).

A recent study on older people's networks (Akiyama et al, 1995) concluded that although older people tend to have more women in their network and receive more support from those women, elders are not necessarily closer—either psychologically or geographically—to the women than to the men in their networks. Their data also showed a noticeable shift from same-sex relationships to female-dominated networks when older people became widowed and required more support. When normative expectations for family member interaction were considered, the study concluded that relationship norms seem to be more powerful than sex norms—that is, the filial responsibility of a son to his parents is demonstrated by the son's interactions with aging parents.

Religion and Spirituality

The religious dimension of aging encompasses the spiritual, social, and developmental aspects of a person's life, and is an important dimension of well-being for the elderly. Efforts to study the relationship of the spiritual dimension to aging are hindered by an inability of the research community to use consistent terms and definitions. Inconsistent conclusions abound in the literature. As more scholars become engaged in the study of spiritual development, however, a concerted effort has been made to share definitions and methodology.

Thomas (1994) conceptualized an intrinsic-extrinsic pole of religiousness for the individual in which the intrinsically religious find within themselves their ultimate meaning in life from religion. For these types, religion is the fundamental motive for living. The extrinsically religious use religion for more superficial social purposes or to justify their politics and prejudices. Their religion is extrinsic to their fundamental reason for being. Others have developed the concept of intrinsic-extrinsic religiosity by considering higher or lower levels of intrinsic religiosity (importance of structured religion to the individual).

Fehring and colleagues (1997) differentiated the religious dimension by spirituality and religiosity:

> *Spirituality. . . connotes harmonious relationships or connections with self, neighbor, nature, God, or a higher being that draws one beyond oneself. It provides a sense of meaning and purpose, enables transcendence, and empowers individuals to be whole and to live life fully. Religion, on the other hand, is a term used for an organized system of beliefs, practices, and forms of worship. (p. 663)*

In untangling these two concepts the study was able to determine that religiosity (interaction with organized religion) and spiritual well-being (psychological sense of purpose and meaning) are independently associated with hope and positive mood states in elderly people coping with cancer. Subjects with high-intrinsic religiosity and spiritual well-being had significantly higher levels of hope and lower levels of negative mood states than subjects with low spiritual well-being and low intrinsic religiosity. The researchers felt that subjects with high religiosity could use their religious patterns for coping (prayer, religious objects) and subjects with high spirituality were able to maintain a greater sense of wholeness than those with low spirituality (Fehring, Miller, & Shaw, 1997).

Church and synagogue attendance for older people exceeds that of other age groups (a measure of religiosity). People aged 65 and over are the most likely of any age group to belong to clubs; fraternal associations; and other church-affiliated organizations, such as widows

Spiritual development plays a key part of growth and potential in later life.

and widowers groups, Bible-study groups, and volunteer groups serving the sick and needy. It is uncertain whether this is a cohort effect (i.e., people over 65 have been more religious throughout their lives than are other cohorts) or an effect of aging. Evidence for shifts in religiosity in aging is inconsistent: some studies find that religiosity changes significantly as one ages; others find that it does not (Krause, 2008). Studies including centenarians, however, indicate that among the oldest-old, "nonorganizational" aspects of religion (less active aspects of religion such as beliefs, faith, and influences of religion on daily living) are more common than are the organizational aspects (Krause, 2008). This is likely due to more limited ability of the oldest-old to attend organized religious functions. These findings highlight the importance of the multidimensionality of the religious dimension.

Findings from Courtenay, Poon, Martin, Clayton, and Johnson (1992) might differ significantly for ethnic groups. More than 75 percent of all older African American adults are

church members, for example, and at least half attend religious services at least once a week (Stolley & Koenig, 1997). Traditionally, religion and church have been powerful sources of social support for older African Americans—in fact, church attendance is the most significant predictor of both frequency and quantity of support received.

It is important to view elders in a multicultural sense. In a study of health-related decisions and the role of religion, Laurence O'Connell (1994) noted, "A healthy baby, a healthy marriage, a healthy sex life, a healthy set of social skills, a healthy appreciation of our mortality and so forth, are both humanly meaningful and religiously significant" (p. 30). To this we would add that they are also *culturally* significant, and it might be difficult to disentangle culture, human meaning, and religion in this context. Individuals are multicultural, *and* cultural and ethnic groups are heterogeneous. Knowledge of the impact of religion for ethnic groups can be critical to, let us say, designing intervention programs that are culture specific. However, categorizing by ethnicity comes dangerously close to stereotyping by race. It is always necessary for the practitioner working with older people to remain sensitive to variations in cultural values, including religious and spiritual states and needs, as well as to individual differences within cultural groups.

Elders benefit from the services of religious groups that minister to the needs of the sick, frail, disabled, and homebound. Carl Jung said:

Among all my patients in the second half of life—that is to say, over 35—there has not been one whose problem in the last resort was not that of finding a religious outlook on life. (Jung, 1955)

This quote brings to mind questions about why and how and who enters a "spiritual journey" in the second half of life. The awesome challenges of facing loss, suffering, pain, and death; of finding ultimate meaning and

Cultural and personal bonds are often strengthened through participation in meaningful social activities.

purpose; of setting priorities and integrating the threads of one's life are possible reasons for such a quest. These are developmental tasks of aging, and religion may be useful in handling them. Jung's quote also brings to mind Erik Erickson's developmental concept of "integrity versus despair" discussed in Chapter 3: if a person can come to terms with life they have integrity; if not, they experience despair.

In contrast to increased religiosity with age are the observations of a minister to older people in Cambridge, Massachusetts. He reported being distressed that so many people in old age lack "self-knowledge and spiritual discernment." He found a lack of "the peace of soul which would make old age much more satisfying." One of the most neglected aspects of old age in America is the need for spiritual develop-

ment (Krause, 2006a). With spiritual development one can experience a growing benevolence and a deeper empathy with fellow human beings. It is not a given, then, that individuals become more religious with age. More studies are needed to document the circumstances under which this occurs. Such studies are now being conducted by means of participant observation (Eisenhandler, 1994; Schaie, Krause, & Booth, 2004) and interview and survey research (Krause, 2006b; Rubenstein, 1994).

Strengthening Social Bonds

Although elders are not as lonely and isolated as stereotypes would have us believe, many live out their last years without the close emotional or social bonds that they need and desire. For some, such isolation may result from their inability to establish and maintain intimate relationships with others. For many, however, isolation results from the new social situation that old age brings. Changing family patterns means less need for the services of older members within the family. The trend toward smaller families means fewer siblings and children with whom to interact. Very old age may bring the loss of a driver's license and car, physically curtailing social opportunities. Physical disability and illness can also hinder one's social life. For other aged, social organizations may not be available to provide friends and companionship.

We need more commitment to strengthening the social bonds of elders and providing them with the resources to develop the intimate ties and friendships they need to enjoy a meaningful life. The formation of supportive groups among people with common experiences could produce age-integrated as well as age-segregated groups. Fostering connections across age groups is vital, along with encouraging alliances among older people.

We need to explore other solutions—from nightclubs and centers to communes that cater to seniors. The suggestion of nightclubs is valid, not because older people need to drink, but because they have few places to interact socially. Perhaps expanding senior clubs and centers would be more to the point. The motivating factor is the same: to increase the older person's opportunities to find sociable companions. The provision of transportation is critical to those who are homebound.

The communal concept is being emphasized in some retirement housing, designed with rooms clustered around a communal kitchen and dining facility. The elderly share meals together but find privacy in their own rooms. With some imagination, communal concepts could find further approval in the aged community and, in turn, enhance social relationships of the elderly who want to participate.

Social, legal, and financial pressures on older people discourage remarriage. Many retirement programs pay the surviving spouse a monthly income that remarriage voids or reduces. A second factor is pressure from children, who may discourage an aging mother or father from remarrying for fear that, upon the parent's death, his or her assets will be in the hands of the second marital partner. Reluctance to marry out of respect for the deceased spouse poses another barrier to remarriage. Finally, old people may fear ridicule or condemnation if they choose to marry in old age. The barriers to marriage or remarriage are also barriers to the formation of intimate relationships. But, as living together without legal marriage becomes a more acceptable lifestyle for young people, it also becomes more acceptable for the old.

Widow-to-widow programs exist in many communities. These programs, which locate widows and help them to get together to share experiences, also provide legal assistance, social activities, and employment counseling. Churches are another source of counseling and other services for widows. We can strengthen social bonds and friendships through a genuine effort to provide these and other services and interaction that elders need.

Chapter Summary

Old or young, we all need intimacy and social bonds with others. Our social structure and values are such that old age can reduce opportunities for social interaction. Elders are at the latter stages of the family life cycle and may have experienced children leaving home, retirement, job changes, death or divorce of a spouse, death of dear friends, remarriage, the birth and marriage of grandchildren. Despite these changes—or perhaps *because* of these changes—older people have the coping skills for getting through life's transitions. Connections with others—maintaining the ability to express intimacy, sexuality, and emotionality—are central to these coping skills.

In the area of social bonds, gender differences clearly emerge. Women seem to be keepers of the support network—they are more involved with friends and family members when help is needed, and they themselves have more friends than do men. The bonds that men have, however, are as emotionally strong as the bonds that women make.

The importance of sibling relationships re-emerges in later adulthood; aged siblings often provide support and companionship. Grandparenthood is a role that most elders experience; however, there are many different types of grandparenting and the role of grandparent for most people is not a central one. Older adults report themselves to be happier if they form intimate friendships with other people: kin, friend, and community. Neighbors and social organizations such as lodges, church, and senior centers are important sources of social interaction. Structured religion is a source of companionship, social stimulation, and spiritual support.

Key Terms

confidant
convoy model
elderly sibling
equity theory
network analysis

religiosity
senior centers
social networks
spirituality

Questions for Discussion

Did your parents have more siblings than children? How does that differ from your grandparents? How many children do you have or do you expect to have? What effect does family structure have on late-life relationships and support systems?

The model of a social "convoy" suggests that we have a group of people in our lives who make up our social and support network. This continues to evolve throughout a lifetime. Make a list of everyone who is in your convoy right now and how they support you. How might this change over the next decade? How did it look a decade ago?

Fieldwork Suggestions

1. Locate the nearest senior center to your home. How many people does it serve? What services does it offer? Are there transportation services available to those who can't drive to the center?

2. Do you know someone who was raised by a grandparent? Interview that person and find how being parented by a grandparent has affected his or her life.

3. Do you know any grandparents raising his or her grandchild(ren)? Interview him or her to learn how parenting the second generation has affected his or her life.

4. Interview several seniors to learn how many siblings, friends, and family they have. To whom do they have the closest bonds? To whom do they go to for affection and intimacy? Who could they rely on for support if there was a crisis?

Internet Activities 🌐

1. Search for resources for grandparents on the Internet. Look for AARP Grandparent Information Center, Grandparents as Parents, Grandparents who Care, national Coalition of Grandparents, and any other sources you can find. If you had friends with sudden responsibilities for parenting their grandchildren, would you be able to provide information to them about these organizations?

2. See what kinds of Internet resources you can find that are designed to promote social

contact specifically for older people. Keep a log of your search path. How difficult was it to find information? How accessible do you think this information is?

3. Develop an idea for an Internet home page designed to promote social interaction among seniors. Remembering that older people as a group are very heterogeneous, incorporate in your home page various links to sources that might be interesting or useful to different types of interests.

7

Intimacy and Sexuality

The Taboo of Senior Sexuality

ThirdAge News Service

John DeLamater teaches a University of Madison Wisconson sociology course called "Intimate Relationships," during which he shows a movie called "Tonight's the Night." The film depicts three couples, all in their 60s, having either "a very nice romantic relationship," or sex.

"Here are these six regular people hugging and kissing and crying and talking about each other, and at the end, this one couple climbs into bed, and it just blows the students away," DeLamater says.

Many have no image of their own parents as sexual, much less their grandparents.

"Ours is still a youth-oriented society, so there's a certain amount of ageism and stereotypes of older people, including the belief that older people are not sexual."

The Viagra Generation

The co-author of a major study of "Sexual Desire in Later Life," published in 2005 in *The Journal of Sex Research*, DeLamater became interested in sexuality and aging about six years ago, as the male performance-sustaining medication Viagra was taking off.

"Viagra was obviously tremendously popular, which suggested there were a lot of men out there, presumably in their 50s, 60s, and 70s, who were interested in improving their sex lives," he says.

At the same time, at conferences with other researchers and clinicians, DeLamater heard about couples who sought counseling over Viagra-related issues: "The woman essentially said, 'This guy got hold of a pill and now he thinks I should have sex with him again. We haven't had sex in 10 years. He hasn't looked at me, hasn't talked to me, this is unacceptable.' It brought home the fact that couples make some kind of arrangement, and what Viagra did for many couples was upset their existing arrangement. So it called attention to the character of that relationship."

"When I asked myself, 'What do we really know about the sexual behavior of people over mid-50s?,' I came to the conclusion, not much."

The gist of his recent study's findings is that much of the existing research—not a great amount in the first place—tends to "medicalize" sexual issues and look at the negative impact of age, hormone levels, illness and medications.

Mitigating Factors

What DeLamater and co-author Morgan Still of the University of Michigan discovered, in their study of 1,384 persons over age 45, was that psychological and social issues were overlooked. They found that the principal influences on strength of sexual desire for women are age, importance of sex, and the presence of a sexual partner. For men, strength of desire depended on age, the importance of sex, and education.

Within his own marriage, at age 65, DeLamater finds sexual relationship to be "part of the broader relationship. The intimacy you experience physically feeds into the intimacy of the relationship as a whole. So it's important as a kind of connecting with your partner."

DeLamater said he is also convinced that "a certain amount of sexual activity is just good for your physical and mental health. Masters and Johnson said, 'Use it or lose it.' The sexual apparatus is part of the body and if you use it, it's going to keep that part of the body healthy." If sexual activity occurs "in a context in which you feel valued and you feel loved, that certainly increases your mental health."

Source: http://www.thirdage.com/sex/the-taboo-of-senior-sexuality.

The Need for Intimacy

Intimacy is the need to be close to, to be part of, and to feel familiar with another person. Old or young, we all need intimacy and social bonds with others. We may believe our ability to maintain close relationships is strictly a personal problem. However, from a sociological viewpoint, the social environment affects the maintenance of close or primary relationships, as well as the larger network of friends and of kin. The norms, values, and social structure of a society may either foster or retard the development of social bonds.

Social scientists have long discussed the positive relationship of connection with others and psychological well-being. Those with whom we are emotionally connected might be either friends—**achieved relationships**, people whom we have chosen to be in our networks; or kin—**ascribed relationships**, people in our networks over whom we have little choice or whom we may have not chosen. The number of friends and active family relationships a person has traditionally has been viewed as an indication of how well an individual is aging. One must remember, however, that sheer number of interactions or connections is not an accurate reflection of the *quality* of those interactions. Psychological well-being is enhanced when the connections being maintained are positive and support the elders to maximize their potentialities, and by interactions in which the elder's self-concept is positively reflected and maintained.

Adams and Blieszner (1995) identify **friendship relationships** in terms, first, of their process—the important attributes of the relationship, the level of enjoyment gained through the connection, and the activities conducted. Second, friendships are identified by structure, referring to the network size, how similar the network members are, and network density or the proportion of friends who know one another. **Family or kin relationships** are identified in terms of their *process*, similar to that of friendship networks; and of their *structure*, referring to family size and generational composition, marital and parenthood status, household size and living arrangements, and the functions members serve for one another.

Although most older people have strong social bonds of some sort, certain events are more likely to put constraints on their relationships. Disability and illness limits visiting, as does their loss of mobility or lack of transportation. Death takes friends and neighbors and eventually a spouse. At the very point in life when retirement brings free time for social interaction, the opportunities for it may be reduced. Malcka Stern experienced various losses, including her husband. She lives in a nursing home and she is hearing impaired, but she continues to reach out to others and respond to a warm social environment. Stern (1987) described her intimate group of friends in an article she wrote for the *Washington Post*:

> *When I count the many blessings accrued to me in my long life of 93 years, high up on the list is the fact that I am a resident*

in the Attic Angle Tower, a senior citizens' apartment complex in Madison, Wisconsin. There are about 70 of us, average age 85, mostly widows.

We have a beautiful dining area, and when we all sit together at dinner—the one meal we take together—four of us to a table, we really present a picture of a group of elegantly coiffed and attired older women.

True, at the tables lucky enough to have among them one of our few men, there always seems to be much more animated conversation and much more gaiety. We do indeed miss our men.

Our group is impressive. We have among us professional women, all retired, of course, from all walks of life. Teachers, social workers, scientists. Many are widows of renowned professors, doctors, judges, lawyers and businessmen. In our midst we have talented artists, knitting and weaving experts, even a poet in residence.

But lest we become too smug and too satisfied with our way of life, we all remember that attached to our apartment complex is the nursing home to which sooner or later we will all have to enter at the last stop. We don't talk about it very much. . . .

Our friendships are warm and close. We have all experienced the same troubles, lived through losses of loved ones. Our own health fails. You complain to your neighbor about your arthritis and she doesn't say a word, but holds out her own gnarled and twisted hands. And we both smile and pat each other on the shoulder and go on about our business.

I am very hard of hearing and often fear that I must seem pixilated when I respond inappropriately to someone's question. One does get tired of saying "What? What?" all the time. . . .

Some of us go to a discussion group every week, and last week it was about grieving.

We read James Agee's A Death in the Family, *and then our group leader asked each of us to recall our first experience with grief.*

When it came my turn I talked about the death of my first child, my Barbara, a baby of 2 who died of diphtheria. I began to tell them and I couldn't finish the story. To my embarrassment, I burst into tears. It was 60 years ago. It was yesterday.

Marital Status

The marital status of older people shapes a great deal of their roles, their patterns of interaction, and the social bonds they form. This section considers older couples and three categories of older singles: the widowed, divorced, and never married. An unknown percentage of older people live together without being married. Remarriage as a result of divorce or death of a spouse is becoming more common and is also discussed in this chapter.

Later-Life Couples

Being married is a reality for many older Americans: just under half of the population over 65 is married and lives with a spouse (U.S. Census Bureau, 2007a). The place of marriage in the life course has changed in recent decades, however, and this shift can be seen among people who are at the cusp of older adulthood: the baby boom generation. Family life patterns of boomers are different than those of their children, parents, and grandparents. Marriage in the life course then and currently is different, and this changes ways in which attitudes toward marriage are shaped. Changes in marital behavior is an outcome of those shifted attitudes and values and will be seen as cohorts from different decades move into older adulthood (Sassler & Schoen, 1999).

American men and women today are not marrying in their early 20s as their parents did, nor are they necessarily remaining married

FIGURE 7.1

Marital status of persons 65+: 2007

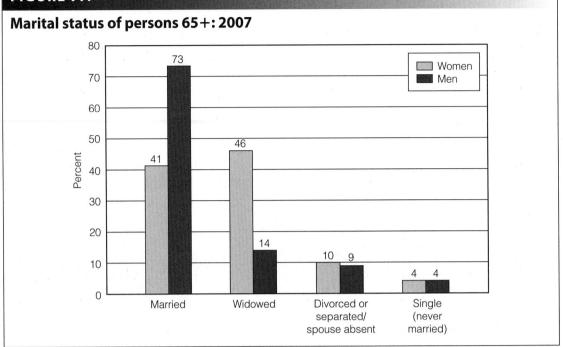

Source: Based on Internet releases of data from the 2007 Current Population Survey of the U.S. Bureau of the Census.

never to again enter the marriage market. They delay marriage, and more than half of marrying people today will divorce at least once. Not only are people delaying marriage but fewer divorced people are choosing to remarry, especially single men and women age 55 to 69 (Fitch & Ruggles, 2000; Mahay & Lewin, 2007).

The experience of divorce varies by age. Divorce rates increased significantly in the 1970s; therefore, younger people are more likely to have experienced the divorce of their parents, and older people are more likely to have experienced their own divorce (Cherlin, 1992). In 2005 roughly 67 percent of adults between the ages 35 and 45 were married compared to 87 percent in 1970 (Whitehead & Popenoe, 2006). Life expectancy is longer for women who are therefore more likely to be widowed than are men; men are more likely to be married to women younger than themselves.

The total proportion of married people over 65 does not provide an accurate social picture, however, because over three-quarters of men aged 65 and older are married, whereas about half of women in the same age group are married. In other words, older women are roughly half as likely to be married as are older men (U.S. Census Bureau, 2007a). In many respects, the older couple can be considered lucky. Most couples hope they will grow old together, but divorce or death can intervene.

Marriage maintains health: married people tend to have higher levels of well-being and better health than unmarried people. Among men, marriage is associated with a lower risk of coronary heart disease, and socially isolated men and widowed women seem at particular risk of a fatal cardiovascular event. The longevity advantage of marriage is nearly always greater among men than among women, indicating that men gain greater health benefits from marriage than do women (Lillard & Panis, 1996; Liu & Umberson, 2008; Walker & Luszcz, 2009). These health benefits might be due to the older married person being happier, less lonely, and financially more stable than older single persons (Goldman, Korenmaan, & Weinstein, 1995; Lillard & Panis, 1996). Together, a couple can usually live out their lives in a satisfying way and be a source of comfort and support to each other.

Although across the board married people are healthy, the gap between married people and those who never married appears to be closing. Therefore current evidence suggests that encouraging marriage to promote health may be misguided. Stronger indicators of the marriage benefit might be improved nutrition, care in times of illness, the encouragement of developing healthy behaviors and avoiding unhealthy ones such as smoking and excessive drinking, and a stress-free or stress-reducing atmosphere at home (Lilliard & Panis, 1996). In general, the health status of the never married has improved for all race and gender groups, and the health of married women has improved while the health of married men has remained stable (Liu & Umberson, 2008).

Social relationships have unquestionable effects on health. Interactions between spouses and partners, children, or friends can have mixed effects on health (Kurdek, 2005; Pinquart & Sorensen, 2005; Ryff et al., 2004), but it is clear that the absence of relationships contributes to poor health (Cacioppo et al., 2002). A sense of social connectedness inherent in those of friendship or positive family ties contributes to high socioemotional quality.

When social ties become disrupted because of retirement or disability, the role of the spouse takes on greater importance. The relationship can become the focal point of the couple's everyday life and continue to develop in commitment, affection, cognitive intimacy (thinking about and awareness of another), and mutuality (interdependence) (Blieszner, 2006; Huyck, 2001).

Physical and emotional intimacy and romance continue to the end of our lives.

The probability of an older spouse providing caregiving for a more frail spouse is very high for late-life married couples. As the caregiving can demonstrate a gift of love, it can be deeply satisfying for both parties. However, if the caregiving partner is himself or herself afflicted with any chronic illnesses, it can be an extremely exhausting and stress-producing situation for both partners. One particularly interesting study analyzed the quality of the marital relationship and the effectiveness of caregiving for a spouse with cognitive impairment (Townsend & Franks, 1997). The study found that people providing care to a spouse with cognitive impairment had a greater sense of effectiveness if the relationship was characterized by emotional closeness. On the other hand, emotional closeness did not mediate the sense of effectiveness when the frail spouse had a functional impairment—let us say, the spouse was incontinent, or unable to walk without assistance. In case of either functional or cognitive impairment, conflict had a direct, negative influence on a spouse's perceived care-giving effectiveness. In other words, marital emotional closeness is a critical component to understanding spousal caregiving effectiveness. This highlights once again the necessity to examine the relationship of interpersonal ties as well as their presence or absence to gain a full understanding of the meaning of issues we explore.

Either retirement or disability might suddenly precipitate a shift in the amount of time couples spend together. Disability is clearly not a planned-for event, and it can be expected that shifts in personal and couple identity and in other patterns would complicate a marital relationship. Research evidence consistently indicates, however, that retirement can be a major transition, requiring adaptation to the loss of work and reduced income, change in social status, and changes in identity. In fact, early literature in gerontology emphasized retirement as a life crisis; however, more recent studies seem to indicate that retirement is, for most people, less a crisis than an *adjustment* (Kalet, Fletcher, Ferdman & Bickell, 2006). More recent literature describes retirement not as a single transition so much as a process including having friends who are retired, thinking about retirement, and making specific plans for being retired.

Today workers must be prepared for job changes several times in a career because of job outsourcing, changed technology, or economic shifts requiring mergers or downsizing. Some workers change jobs to climb the corporate ladder or reduce the high stress levels in their current jobs. Not all older people have had or choose that option, however. Retired workers who wanted to change jobs but did not, are said to experience **occupational regret**—an outcome of the process of self-evaluation (Schieman et al., 2007). Black men often express occupational regret because of having felt trapped into keeping the same job over their working life. Many older black workers have experienced job discrimination that restricted their career advancements and choices. Rather than risk loss of income and a repeat of discriminating behaviors in a new job, they held on to one position their entire working careers thereby limiting their career potentials (Hoyer & Roodin, 2009). Both older black men and women report greater levels of occupational regret than do white men and women workers. A sense of occupational regret might make adaptation to retirement more difficult for both members of a couple.

From that relationship perspective both retirement and disability alter the interaction pattern of the couple. Prior to retirement or disability, one or both partners may have been very active outside the home; consequently, the number of hours each week they actually spent together may have been limited. It is particularly important, given the current pattern of early retirement and increased numbers of years of healthy life following retirement, to understand about spousal relationships following this life transition. People who retire involuntarily, not surprisingly, have been found to adjust more poorly to retirement than people who have planned for their retirement and have chosen the timing. This is in large part because the bulk of involuntary retirements occur when an individual has health-related disabilities and can no longer reasonably work full time. More Type-A personalities (people who attempt to control their environment and appear to be aggressive and ambitious on the job) are among those involuntary retirees than Type-B personalities (people who take events more calmly and are not very assertive and ambitious in job-related situations) (Kausler & Kausler, 1996). Type-A personalities are particularly susceptible to coronary diseases; the link between personality, work patterns, and adjustment at home in later life is evident in this regard. A survey in California indicated that voluntary retirees were more likely to exercise regularly, have lower stress levels, be in better physical condition, and report a higher-quality life than did involuntary retirees. The circumstance of the retirement, then, will have a direct effect on family relationships.

Most current research indicates no trend of increased marital strife when a husband retires and his wife is not employed. In many such cases, marital satisfaction increases following the husband's retirement—a factor probably due to increased recreational shared time. A slight trend toward dissatisfaction and/or conflict was evident in a recent study on marital satisfaction among wives who were still working after their husband's retirement (Kalet et al., 2006). Women report that their housework increases following their husband's retirement: changing the roles of home management responsibility appears to be as large a transition as changing from the role of worker to retiree.

Among husbands and wives with high levels of marital satisfaction, retirement neither enhances nor reduces that satisfaction. Primary issues found to affect marital relations and satisfaction in later adulthood include the equality of roles, adequate communication, and the transition of more time spent together (Walker & Luszcz, 2009). Although older women do more household work than do older men, dissatisfaction seems to emerge more when a perception of unfairness exists than over the actual division of labor. It would seem that traditional sex roles die hard. If these roles become more equal for younger couples, we will undoubtedly see a more equitable division of household labor between couples in their retirement years.

One common stereotype of marriage is that, after the early stages of romantic love, the relationship begins to deteriorate. Recent studies have found the opposite: namely, an increasing enchantment with each other in the later years. Marital satisfaction appears to follow an inverted bell-shaped trend: declining satisfaction after the initial years, leveling off in the middle years, and increasing again during the post-retirement years. Satisfaction may, in fact, return to its initial high level of the early marriage years, once couples have adjusted to their *new* relationship.

An inverted bell-shaped pattern also charts a husband and wife's opportunity to share time and common interests and to develop greater mutual respect and understanding. The time before the birth of their first child offers husband and wife maximum opportunity for mutual involvement and marital cohesion. The increasing time demands of careers combined with the new demands of parenting substantially reduce the amount of time husband and wife can spend together in their middle years. The last child's leaving home and the retirement of the husband and wife once again allow time for greater involvement, shared activity, and marital cohesion.

Not all studies show that marital relationships become rosier with advanced old age. The deteriorating health of both husband and wife can bring problems to a relationship. Outside intervention can help troubled couples in advanced old age. Too much stress, whether financial or health related, is hard on a marriage at any age. Marriages based on years of negative patterns of behavior and interaction probably continue to be negative into late life and through the death of one or the other spouse.

Older Adult Singles: Widowhood

Older singles may be divorced, never married, or widowed. Three-quarters of older men are married, and more than half of older women are widowed. Of women over the age 75, over two-thirds are widows, whereas two-thirds of men over 75 live with their spouses (U.S. Bureau of the Census, 2007b).

As shown in Table 7.1, widowhood is the predominant lifestyle for women who comprise the "old-old" (age 75 and above) and the "oldest-old" (age 85 and over). We can expect that, in the future, three out of four women will ultimately become widows. Of men who are 85 and over, 60 percent are married whereas among 85 and older women, 15 percent are married. Fundamentally, women can plan to be widowed in their later years and given the odds, older couples must include that possibility in their planning.

The transition from married to widowed status can bring with it both personal and familial problems. The transition is not always successful. Having an intimate relationship over many years impacts an individual's sense of self: the simple knowing that one matters in someone else's life can have a huge impact on well-being and personal happiness (Liu & Umberson, 2008; Pearlin & LeBlanc, 1997; Walker & Luszcz, 2009). Childless widows especially lack a kin-based system of support. The feelings of loss after the death of a spouse are enormous, especially if the couple had been married 50 or even 60 or more years. The empty chair and the empty space in bed reinforce memories of sharing family rituals. What does one do with the memories? How does one manage the grief? Many turn to other family members for emotional support; others simply suffer alone.

Loss of a spouse may cause the most difficult role change that a person must cope within a lifetime. The widow has lost the support and services of an intimate person in her life. If she doesn't find substitute supports, she is on her own. Her social life will be altered. She, as a single person, may be either uninvited to, or feel uncomfortable in, social settings where she had once been welcome. She may lose contact with her husband's friends and relatives. She then must form new relationships and make new friends. With this process in mind, Lopata (1988) was among the first to identify the **stages of widowhood**:

- Official recognition of the event
- Temporary disengagement or withdrawal from established lines of communication
- Limbo
- Reengagement

Official recognition of the event of widowhood typically begins with the funeral and the initial mourning period. The term **grief work** describes healthy confrontation and acknowledgment of the emotions brought about by death. The widow must accept the finality of her loss to get on with living. Grief work for the widow takes time and may bring a temporary withdrawal from past social activities and responsibilities as she reassesses her life. Once she answers the question "Where do I go from here?" she can re-engage in society.

Studies of widowhood typically characterize the period beginning about six months after the spouse's death as a reorganization stage different from the phase of intense grief that comprises the first few months after the event. Research on the later phases of coping with widowhood has zeroed in on the importance of friends and family as sources of social support. This research has emphasized the early stresses associated with grieving and the support required to help the recent widow maintain a sense of identity during profound loss. The tight-knit structure of a family network seems indicated for this kind of support. The less highly structured network of friends has been associated with help on decisions for building a new life as a single woman. A good deal of research, in fact, has taken place to describe the beneficial aspects of both of these support system types for widows. Recently, however, studies have begun to explore the finding that not all support network interactions are satisfactory.

Family dynamics, whether or not they have been tight-knit, have both negative and positive aspects, as do friendship dynamics (Morgan, Neal, & Carder, 1997; Umberson, Williams, Powers, Chen, & Campbell, 2005). One major study (Morgan et al., 1997) found there seemed to be no shift from family toward friends as a time-related pattern of reliance on support networks following widowhood. Widows were given a diagram of concentric circles and asked to place people who "have either a positive or negative effect on your life" into one of three circles: an inner circle for those who had the most effect; a middle circle for those who had effect but not quite so much, and an outer circle for anyone else who still had an effect. Although this methodology did not specify what the negatives and positives were, it allowed people to identify the existence of network members who did not have a positive impact at this time in the widow's life.

Family members, it was discovered, were the primary source of both positives and negatives.

Response to grief depends on relationships. This man's wife has died; the woman's mother has died. Both grieve, differently.

Differences in the number of negatives were notable because they increased over time in both family and nonfamily segments of the network. A shift was evident between those in the first year of widowhood where only family negatives had a significant effect on depression; and those in the third year, where only nonfamily negatives had a significant effect on depression.

Clearly, the relationships of family and nonfamily have a differential impact on the process of grief work among widows. As Morgan and colleagues point out, there is no theory to adequately address the finding of timing and the interaction of positive and negative support. Their research serves as an important reminder that the assumption that personal networks are positively supportive is not accurate and can be myth-generating, as is the belief that stressful events mobilize "positive support." Even so, research continues to support the finding that higher levels of perceived social support are related to lower depression scores for both widows and widowers (deSpelder & Strickland, 2009).

The perception of having a supportive network, regardless of the contact and the amount of exchanges made, seems to serve a critically important function for widows. "Positive illusions" (Taylor & Brown, 1988) assist in adapting to difficult situations, and we might suggest that the *belief* that one is fully supported is by far more than half the battle. Changes, be they positive or negative, will naturally occur for the widow following the death of her spouse—reduced contact with friends of the husband or with in-laws, for example, or changes in the relationship of couple-companionate friendships and new and different relationships with children (Hoyer & Roodin, 2009).

The effect of death of a mate has been well studied but the long-term effects of bereavement are not as well understood. Further, almost all studies concern themselves with heterosexually paired relationships while ignoring same-sex relationships. The grief following death of a same-sex partner may be complicated by conflict in either or both partner's families who were not comfortable with their family member's sexual orientation. Laws in most states provide same-sex partners no legal authority over the body and the property, thereby disenfranchising the bereaved partner's grief. Grief over the loss of a loved partner must be recognized independent of legal or social sanctions about the nature of the relationship (deSpelder & Spalding, 2009).

Relationships between the older widow and her adult children may develop a negative side. The older widow might feel unappreciated, that she is making too many sacrifices for her children when their children should be offering comfort to her (Hoyer & Roodin, 2009). This is partially an outcome of the identity loss of being a primary focus in the life of someone else. On the other hand, a classic finding emerging in the early 1980s is that older women are more likely to be distressed and dissatisfied if they were over-benefitted by their children, that is, when their children provided them with considerably more support than they reciprocated (Beckman, 1981). That finding appears to be consistent through recent decades also. Even in bereavement—or it might be suggested *especially* in bereavement—a consistency in the role of parent is important to the maintenance of self-identity.

Widowhood creates changes in the widow's support system beyond that of the absence of a spouse. In one longitudinal analysis of children's responses to their mother's widowhood, Roan and Raley (1996) found that contact increased between mothers and their adult children. This is a finding we would expect because adult children and their mothers have the intimacy of shared grief over the death of a father and death of a husband. Another study of parent-child communication on the death of a parent-spouse found that both parents and children seemed concerned with protecting each other from the pain and sadness associated with the loss (Silverman et al., 1995). The study identified two types of families.

In the open family, consoling and informing language was used. Less open families used language to influence and to avoid feelings and confrontation with death. In these families the surviving parent often saw the deceased partner as the competent family caregiver. Two key points are relevant to all these studies, however: (1) network contact is not necessarily always positive, and (2) study findings are generalized and can mask the vast individual differences that exist.

Friends can offer more useful support by not trying to speed recovery, emphasizing strength, and not forcing a new identity. They can just listen, not argue, and provide unconditional love.

Interviews with 300 widowed women in an urban community showed them to be self-sufficient in managing their daily lives. They did not lean heavily on anyone for help with the basic tasks of daily living unless they were in advanced age and/or experiencing poor health. (Mroczek et al., 2006). Research suggests that too much independence and self-sufficiency might lead to social isolation. Asking for help and giving help to others results in *interdependence*, which is the most valued lifestyle when balanced with self-sufficiency (Terracciano, McCrae, Brant, & Costa, 2006).

Different women react differently to the loss of a spouse. Women in traditional marriages, who see their role of wife as central and who invested their identity, time, and energy primarily in this role, suffer immensely. If the marriage was close, the loss cuts especially deep into all aspects of the survivor's life. Reactions of widows may also vary by social class. The generally lower-income levels of widows can promote isolation and loneliness. Indeed, many widows have lower morale and fewer social ties because they are poorer than their married counterparts and are unable to interact in as many social events as more financially women can. Generally speaking, the higher the widow's personal resources, such as income and education, and the higher her social and community participation, the better she can cope with her status.

Changes in self identity are necessary when a life partner dies. The structure of a day, the most familiar patterns of life are different.

The process of adjustment to widowhood is a winding path of redefinition of roles and self-concept. People may move back and forth between feeling secure and comfortable with their self-concept and feeling quite the opposite or experience several feelings simultaneously. Those with previously happy marriages are more likely to consider remarriage or to make a successful life for themselves alone; in contrast, women whose marriages were not fulfilling are less likely to consider remarriage.

The loss of identity experienced by a woman when she becomes a widow can be tremendous, most particularly among women who have maintained traditional marriages throughout their lives. One study used autobiographical accounts of widowhood to determine the cost of widowhood through the voices of 10 women who were widowed (van den Hoonaard, 1997).

The women described their experience as one of transformation, not one of recovery. Van den Hoonaard named the process described by the widows "identity foreclosure" to refer to a shifted sense of personal identity. Some examples of this foreclosure in the voices of the women:

After Martin died, I had learned that my identity had been derived from him. I did not know who I was (Caine).

Who am I without Judd? Who will define my existence for me? (Seskin).

My sense of self, to a great extent, became linked to being Leonard's wife . . . We reassured, validated, reinforced, and encouraged. We mirrored the best part of each other. I lost that mirror when Leonard died. It was a double death. When he had said, "You're wonderful," I believed him . . . But with Leonard gone, I felt paralysed. It was as if he had taken a major part of me with him (Rose).

"Sign here," the girl in the office of vital statistics said when I went to pick up a copy of your death certificate . . . "Right here. In the block that says widow of the deceased." The word pierced me like a lance and my sharp intake of breath was audible . . . Later as I walked home, I tried to give voice to my new label. Widow! Widow! I mouthed the word over and over and although I could hear it thundering in my head, no sound would leave my lips . . . Until two weeks ago, widow was only a word in the English language. Now it is me . . . (Dohaney, quoted in van den Hoonaard, 1997)

A number of organizations offer outreach services to widows and widowers. They may be religious, social service, or mental health groups. The American Association of Retired Persons (AARP) has a program called Widowed Persons Service (WPS) in which volunteers who have been widowed 18 months or more are trained to reach out and offer support to the newly widowed. This program is offered in communities all over the nation. WPS conducts outreach to minorities and other groups who are typically less active in such programs. Other widowhood bereavement groups are available through most hospice organizations, many senior centers, and many church organizations. Most programs are based on the healing interactions of people with shared experiences, and they can have a profoundly healing effect. Participation in outreach programs can be a literal lifesaver for older adults willing to seek help.

Gender Differences in Widowhood

Both men and women experience difficulty in coping with the death of a spouse. On the one hand, a husband as surviving partner of a traditional relationship may have greater difficulty

because he may need to play the new role of housekeeper—cook, cleaning chores, and hosting if he wanted to entertain friends or family. Further, the widower is less likely than the widow to move in with his children, to have a high degree of interaction with relatives, and to have close friends. He is more likely to have somewhat increased interaction with sons, however, than is a widow with her sons. If the wife was the initiator of family contacts, her death typically lessens the widow's interaction with extended family members. Statistics paint a grim picture for the husband who survives his wife. Widowers die four times as often from suicide, three times as often from accidents, ten times as often from strokes, and six times as often from heart disease as do married men of the same age.

On the other hand, the widow's situation is also difficult, both socially and psychologically (Stevens, 1995). Many widows spend a significant period of time providing care to ill husbands before becoming widowed because caregiving often precedes widowhood in later life. Physical and economic tasks of caregiving often deplete social and emotional resources.

The emotional costs of losing a life partner are high for men as well as women. A longitudinal study of widowhood found that long-term (five or more years) widowed older men and women reported more loneliness than married individuals, and women reported a lower sense life satisfaction (Mroczek et al., 2006). An anticipation effect of increased depression just prior to the spouse's death is often apparent. Although longitudinal analyses show that it is more stressful to be bereaved when young-old than when old-old, there are no age differences in how long it takes to feel well-adjusted again following a spouse's death. Adjustment to bereavement includes personality and social, economic, and family history factors. It varies from one person to another.

Men and women seem to change in different directions at the loss of a spouse. Men are often more "in search of others." Over time, they become more aware and appreciative of friends and relationships. Women tend to be more "in search of themselves." Over time, they develop a new sense of themselves: more confident and assertive, and more comfortable in their independence. In general, widows are more identified with the role of wife than are widowers in the role of husband. The new relationships that might be developed by both older men and older women seem to allow for more flexible roles and less dependency. Both men and women can emerge from bereavement into a new excitement about the changes in their lives (Blieszner, 2006).

Older Adult Singles: Divorce

Divorce is a dominant social reality in the United States. By 2010 the proportion of people 65 or older who have been divorced is projected to be 50 percent (Quadagno, 2005). In over one-third of all married-couple households at least one spouse had a previous marriage that ended in divorce or widowhood. Couples in a first marriage constitute one-quarter of black households, and just under half of all white and Hispanic households (Holden & Kuo, 1996). Implicit in these trends is the shifting role of extended family members for the care of a single, elder parent—most especially the role of adult children and their mother.

Strains in marriage are complex and will not be dealt with completely in this review; however, one should keep in mind that marital strains are probably heightened by longevity. At the end of the nineteenth century, the average length of marriage when one spouse died was 28 years; now it is over 45 years. There have never before in history been so many couples attempting to live a lifetime with one another for such a long period of that lifetime. Rare and widely celebrated at one time, the golden wedding anniversary (50 years) is now reached by many couples, and although its occurrence is still celebrated and honored, it is no longer a rarity. To the extent that it is a rarity is increasingly due to the greater proportion of divorce and remar-

riage but not because of the death of a spouse. One couple was overheard to be talking about celebrating 60 years of marriage between them: 22 years with his first wife, 20 years with her first husband, and 18 years together. This couple joked that the years were additive because they represented *experience* being brought to the present marriage. A new perspective on an old tradition is apparent in this century.

Each year at least 50,000 people over 60 years of age dissolve marriages of 30 to 40 years' duration. Not only that, but divorce will continue to increase. The rate has climbed in the last 30 years from 5 percent to 50 percent, and a high proportion of those divorces are taking place among the baby boomers—the cohort for whom divorce has become commonplace, and the next cohort to enter the population of older adults (Mulroy, 1996). For the older person who is remarried, the risks of that marriage ending in divorce are 10 times those of the older person in his or her first marriage (Lanza, 1996). The consequences of these numbers are enormous: when one considers both widowhood and divorce, only half of all women entering old age in 2025 will be in a marriage. This figure could be less than half if divorce rates continue to increase in middle and old age and if remarriage rates continue to decline (Uhlenberg, 1996).

Though divorce has gained acceptance as a solution to an unpleasant or difficult marriage, the impact of the increasing divorce rate in old age still has profound impact on the financial status of elders and the conformation of kinship networks. Many of the changes in social interactions and patterns of relationship that are only now becoming worked out—relationships with ex-in-laws, for example—will become "traditional" in the next couple of decades, and new norms will emerge. Until that time, divorce is traumatic, economically, emotionally, and socially.

Divorce is a particularly difficult event for older people. It is rarely a sudden development and might result from many years of deepening emotional disruption. Preexisting attitudes toward marriage and divorce, probably as much as the circumstance of the dissolution, shape the adjustment to divorce. Among people for whom marriage and the family have been both vocation and avocation, divorce is not only a failed promise but also a broken covenant (Lanza, 1996). The social shame felt by some men and women at being divorced can be very powerful—particularly so among older women, for whom remarriage rates are very low.

In an extensive study of divorced older women in 1988, Cain found that the length of marriage significantly contributed to postdivorce despair and slow recovery. Some women described feeling envious of widowed women they knew: society knows how to respond to the needs of a widow but is uncertain by those of the divorcee. Widows are more likely to have acceptance by peers, support of family, economic stability, and the anticipation of living alone than are divorced women (Lanza, 1996). It is easier for the widowed person to re-live and reshape positive partnership memories than for the divorced person. This helps the widow to develop a sense of contentment and happiness of the life she has lived and gives encouragement to continue that sense of happiness.

Not as much support exists for the divorced single person, male or female, as for widowed elders. Men have a higher probability of remarriage, but their friends are usually in longer-term marriages, and most of those friends were friends as couples. Adult children sometimes have a difficult time dealing with the divorce of their parents after many years. Uncomfortable and complicated family interactions commonly erupt, and the isolation of one or both of the elder parents by children and grandchildren can result. Thanksgiving might seem less uncomfortable if *neither* grandparent attends; birthday celebrations for the grandchildren are more comfortably celebrated within the nuclear family only. There are few norms for dealing with the divorce of an older couple, for the divorcing couple or their friends, or for the kin network. If we assume rates of no change in current marriage statistics, these norms will become better evolved in the next few years because of the large number of older people who have experienced marriage dissolution.

Older Adult Singles: The Never Married

Never-married persons 65 and over constituted about 4 percent of the population in 2007 (see Figure 7.1). Few people who were never married do so in old age; first marriages make up less than 10 percent of all marriages of elders. One might expect older singles who have never married to be unhappy and lonely, but the never married have typically adjusted to being single in their younger years and are well practiced in those skills of self-reliance and independence that make living alone a desirable and workable lifestyle. They report parallel levels of life-satisfaction with married persons (Bennett, 2005), and because the never married will never experience the death of a spouse, they are spared the grief and loneliness that follows such an event.

Those who have never married do suffer from losses, however. To measure the connections of people who are married and those who were never married, one psychologist asked older people to write "the one person or persons who are the very closest" in the innermost of three circles on a piece of paper. People not quite as close but who are still important were to be listed in the middle circle; and those who are less close but still important were listed in the outer circle. The results were that married people had more connections in the inner and middle circles (their spouse being the most intimate, with children and grandchildren close), whereas single adults had more outer-circle relationships (Rubenstein, 1987).

Older people who have never married often have highly valued friends and relatives, including nephews and nieces. When these important people in their lives die or drop contact, single older adults suffer greatly. Like all other subgroups the never married are a diverse group—some *are* isolated (sometimes by choice); others have many friends. Some wish they were married and feel they have lost out on something; others are happy and fulfilled with their lives and have no regrets.

Remarriage

Today we hear a great deal more about the occurrence of late-life marriages. Remarriage in old age is becoming more common for two reasons: (1) more older people are divorcing, which places them in the remarriage market; and (2) remarriage has become more acceptable for the widowed. Although remarriage of widowed and divorced women over age 65 currently constitutes only 4 percent of all marriages (U.S. Bureau of the Census, 2007b), remarriage seems to work well for most older people.

The changing family network as someone ages often focuses on loss, but remarriage can be a strong way of building one's kin network. Over time close family ties can and do develop in blended families. Adult children are often very happy to see their parent in a loving relationship, which can also ease a sense of filial responsibility for that parent.

Many older people have had long, reasonably happy second (and third) marriages and are glad to have found another compatible partner. They do not take relationships or marriage for granted. Older single people in search of relationships meet partners at church functions, local clubs and organizations, parties, and dances. The extroverted might try video dating, online dating, or advertise in the personal columns of newspapers.

The Internet has become an especially important resource for meeting people among adults 50 and older. *Online Dating Magazine* has a special section for the "50-plus demographic" that includes reviews of dating sites appropriate for older adults. In addition to site reviews, they released tips for older adults:

It cannot be assumed that a group of older women are all widows. Increasing numbers of divorced or never married women reach later life following many years of independent living.

- Share special interests: focus on what you've learned over the years
- Don't take rejection personally: don't allow a bad experience ruin the opportunity of meeting others
- Don't share too much too soon: before meeting the person don't give out too much personal information which could be dangerous or embarrassing
- Be cautiously optimistic: understand not everyone will be honest; know that the online person may not be the one you meet in person
- Be realistic: it doesn't happen overnight

These dating tips might seem unnecessary for the younger adults who are familiar with Internet communication (and perhaps more current familiarity with dating). The tips might, however, be useful for older adults for whom communication through the Internet is a foreign concept.

The numerous dating sites for older adults highlights the point that desire for courtship and meaningful relationships lasts a lifetime, yet we know surprisingly little about the process of courtship in later life. Although intimate relationships continue throughout life to be important and older people seek intimacy and meaningful relationships, the fields of psychology, sociology, and gerontology have not fully examined the process leading to remarriage.

One positive aspect of remarriage in old age is that children are grown and out of the house; being stepparents, the couple does not typically experience great daily strain. An increase in the use of prenuptial agreements has helped to re-

Sexual identity is an integral part of a person. This woman feels lovely, looks lovely, and is helping to change the image of later-life sexuality.

duce potential for family conflict over inheritance. A larger number of attorneys are now familiar with issues of particular relevance to an older client: pension rights, whether both parties do or should have nursing home care insurance, clarity on issues around durable medical powers of attorney, tax issues, Social Security benefits, and insurance.

Some older people have by-passed remarriage because of concern for children's inheritance rights and competing filial obligations among adult stepchildren. Mediation before remarriage to establish clear understandings among all family members can help eliminate misunderstanding before it surfaces. Families are reassured, and the older people can feel secure in being able to develop their own late-life plans as a couple. Remarriage can enhance older partners' abilities to adjust to changes in physical competence by providing steady and caring companionship. Caring for someone, feeling cared for, and being touched are important factors that enhance quality of life at any age. Perhaps they are most precious in later life when the possibility of being without them is likely.

Sexual Relationships and Sexuality

Sexuality in later life is gaining public interest given the size and culture of the emerging population of older people (baby boomers). Findings from earlier studies demonstrate the importance of sex to older adults (e.g., Bretschneider & McCoy, 1988; Butler & Lewis, 1983; Schlesinger, 1995). Despite this interest, literature in this area has been slow to expand. Although several studies on human sexuality across the life span have been published in medical texts, health care providers continue to report their personal discomfort regarding older adults' sexuality (Bouman & Arcelus, 2001; Sadovsky et al., 2006; Skultety, 2007; Walker & Harrington, 2002).

Sexuality is not merely a product of biology. Sexual identity is a mix of emotional, developmental, and cultural aspects of life. It is part of an individual's identity and an important characteristic carried with us throughout our lives, over the life course. Sex and sexuality are essential parts of the relationship with oneself and with others and the continuity of sexual identity remains a dynamic part of our self-concept throughout the life course.

In childhood we learn in what ways our own boundaries are separate from those of other people. Understanding those boundaries is an important part of developing a solid sense of the *self*. In adolescence, the developing sense of self learns to enter into relationships of love, in which the experience of being a separate individual is simultaneous with the experience of being emotionally merged. Sexuality is an integral part of loving and bonding. Over time the identity shifts and changes, but it remains a central aspect of who we are. It remains central even if we develop a chronic illness, if a spouse dies, if we are surrounded by friends and family, if we suffer from mental illness, or if we live alone and maintain a solitary life. Fundamentally, we are sexual beings.

Studies of sexuality and aging traditionally have been quantitative in which people recorded a valued response (1 to 5, for example) to questions having to do with frequency, pleasure, availability, and so forth, with sexual intercourse (Clarke, 2006). They provide good baseline information but are incomplete for understanding human meanings and motivations. Early research failed to account for changing definitions of sexuality across the life course. In other words, early research measured a question researchers had but did not measure what their respondents were thinking, or the meaning the respondents gave to sexuality in their lived experience.

More current methods of research have been qualitative in nature in which the *experience* of the person becomes the unit of analysis. Daniluk (2003), in her narrative studies of women's relationship to their sexuality, contends that sexuality is by definition concerned with the individual's personal experience. Sexuality is that which "appears to be most important . . . to the particular woman in question" (p. 7). It is shaped by each person's experience. She therefore contends that meanings given to sexuality are constructed in an interaction between the individual's experience and the culture or society in which they live. That definition might mean sexual intercourse; it might mean touch, intimacy, emotional closeness; or it might be a combination of all of these.

The role models available to us as we grow into maturity are critical. Findings from numerous studies show that greater knowledge about sexuality and later life is associated with more permissive attitudes. Most interesting might be the finding that the simple presence of a relationship with at least one grandparent significantly and positively relates to adult attitudes toward elderly sexuality.

Cohort and culture clearly shape sexual relationship in later life. Modern historical analyses of marriage and marriage expectations provide an example (Botkin et al., 2000; Dion & Dion, 1996). People who were 16–19 from 1937 to 1944 (World War II cohort) grew into adolescence and adulthood with strictly defined gender roles and a solid marriage was considered the ultimate objective of life. The "silent generation" born 1936–1940 is the last modern generation to respect the authority of American institutions and corporate paternalism (Huyck, 2001). Ninety-three percent of women from this generation became mothers. Following this cohort, the women's movement challenged traditional gender roles, including the value to women of marriage and childbearing. Sheehy (1995) refers to the next cohort (1946–1964) as the "Vietnam generation." They represent baby boomers and according to Margaret Huyck's analysis

[Partly] because there are so many of them . . . competition for everything has been fierce and they have responded by emphasizing

individuality (to stand out from the crowd) and hard work to get ahead of the competitors. (p. 13)

In this cohort gender roles were challenged, and cohabitation—living together without marriage—became an acceptable norm (Huyck, 2001). The rules for marriage and sexuality were forever shifted.

Sheehy (1995) has named the next two cohorts the "me generation," born 1956 to 1965, who married later or not at all, and the "endangered generation," born 1966 to 1980, for whom committed relationships of marriage seem particularly difficult. This cohort is more likely to have parents who divorced and to have experienced the turmoil of marriage dissolution in their families of origin.

A significant change in the two decades straddling the year 2000 has been a cultural shift in recognition of same-sex romantic relationships. Driven in part as it is by the "Vietnam generation" with strong identities of individualism and personal rights, challenges to the legal inequalities between heterosexual and homosexual partnerships shape the sociopolitical discourse in the early 2000s.

The cultural context of sexual identity, shaped in no small part on historical trends, is huge and its ripple impacts family composition, the work force, and social policy, to name only a few of the institutions and organizations that respond to shifting cultural values.

Sexual Invisibility

The sexuality of older people is largely invisible. This is partly because for many older people, sex and sexuality are intensely private matters. It also emerges from our cultural focus on youth: that which is not youthful (or new) lacks function and is of no use. The myth that older people automatically lose interest in sex distorts perceptions of older people. It also influences older people's perceptions of themselves, because we incorporate our culture's attitudes,

norms, and values into our self-concepts unless there is some other powerful reason or validation for being "different." As a result, elders in our society become sexually invisible. They are not viewed as being sexual and therefore any suggestion of sexuality is received with derision (what does he think *he's* going to do?), condescension (did you see how *cute* he was around that young woman?), humor (did you hear the one about the old guy who . . .), shock (how could a woman her age *act* like that?), or a deadly combination of all four. Simply stated, people generally do not put themselves in positions to be treated with scorn; therefore, their sexuality becomes hidden. Invisible.

Physical Attraction and Youth

The idea is commonplace that sexual tension is based mainly on physical attraction between the sexes and very young men and women are the most physically or sexually attractive. Advertising, film, television, and stage promote the theme that good looks, youth, and sex go together. Models are generally young, and standards of beauty are set by those no older than their mid-20s.

Many people consider only the young, perfectly proportioned body to be sexually attractive. The signs of age, especially when combined with obesity or socially unacceptable features, are assumed automatically to be unattractive. When people consider older people to be neither sexually stimulating nor desirable, the next step is to assume that no interest in sex or sexual activity occurs in old age. Assumptions can produce self-fulfilling prophecies in which old people internalize those cultural messages about themselves.

This self-fulfilling prophecy might not always apply, however. Wilcox (1997) surveyed 144 men and women age 20 to 80 to determine their personal body attitudes. She found that self-esteem, health, and masculinity (an instrumental orientation) were positively related to body attitudes, but that these relationships did not differ by age or by gender. In other words, men and women with good self-esteem also had good body attitudes, that is, they liked their body. Eighty-year-olds with good self-esteem liked their body as much as 20-year-olds with good self-esteem. Wilcox suggested that **social comparisons** (Heidrich & Ryff, 1993) are an aspect of this body attitude. Social comparisons describe a process by which people compare themselves with people who are having more difficulty along a particular dimension. This is a "downward" comparison and might be a particularly functional adaptation for older adults in maintaining a good sense of body image. For example, I have disfigured fingers from arthritis, but unlike Mrs. Smith, I still enjoy going dancing.

Romantic Love

Although romantic love has been variously defined, it has a number of generally accepted characteristics: idealization of mate, consuming interest and passion, fantasy, and desire for a blissful state of togetherness. Clichéd ideals—love at first sight and love conquers all—are often elements of romantic love. There is no inherent reason why only young people should have exclusive access to romantic love. Yet in all forms of media, romantic love is predominantly an emotion for young people. From Shakespeare's *Romeo and Juliet* to the present, passion and romance is primarily for young lovers. Perhaps the concept came to be associated with young people because they are most often involved in mate selection. However, the increased divorce rate and longer life expectancy provide options for mate selection to continue throughout life. The belief that only young people are capable of strong passionate feeling invalidates the passions of older people.

The concept of **sexual function for procreation,** which disassociates elders and sex, was embodied in Christian Catholicism by the teachings of St. Augustine (Feldstein, 1970). Accordingly, a woman's femininity was measured by her childbearing and mother role. Likewise, a man's masculinity was measured by the number of children he fathered. Because of the association between sexuality and procreation, men and women beyond the usual childbearing years were believed to be asexual—most clearly so for women, whose ability to procreate ends abruptly with menopause. Generally in mainstream North America sex is considered an expression of love and happiness, with the reward being pleasure rather than children. Deeply rooted ideas take a long time to die, however. Some men still fear that a vasectomy will rob them of virility, and some women fear that menopause or a hysterectomy will rob them of femininity.

Homosexual Relationships

Because of the stigma attached to homosexuality many gay and lesbian couples are reluctant to disclose their sexual orientation and consequently demographic data are based on estimation (Hoyer & Roodin, 2009). Stereotypes of older gay men and lesbians are of lonely and predatory outcasts in either homosexual or heterosexual communities. Indeed, aging can be a particularly difficult course for gay men and lesbians. Psychotherapist Joel Frost (1997) states:

> Negative myths and stereotypes about older gay men and lesbians, the dearth of positive older role models, continued discrimination in a heterosexist society, and developmental pressures without a clear model for gay aging make is especially difficult and complex for gay men and lesbian women. Indeed, many older gay men have to now struggle with being the age of the internalized homophobic stereotype of the "dirty old man" they learned to avoid when they were younger. (p. 268)

The current political and social discussion of civil rights of homosexuals and particularly of same-sex partners has helped to highlight some of the discrimination faced by these groups (Boxer, 1997; Kimmel, Rose, Orel, & Green, 2006). Particularly among younger people, there is a good deal of social activism on behalf of issues relevant to gays and lesbians, and this activism will shape social policy in the future. *Outing Age: Public Policy Issues Affecting Gay, Lesbian, Bisexual and Transgender Elders* (2001) identifies the following key social policy inequalities favoring "institutionalized heterosexism."

- Social Security pays survivor benefits to widows and widowers but not to surviving same-sex life partners of someone who dies

- Married spouses are eligible for Social Security spousal benefits which can allow them to earn half their spouse's Social Security benefit if it is larger than their own Social Security benefit. Unmarried partners in life-long relationships are not eligible for spousal benefits

- Medicaid regulations protect the assets and homes of married spouses when the other enters long-term care; no such protections are offered same-sex partners

- Tax laws and other regulations of 401(k)s and pensions discriminate against same-sex partners, costing the surviving partner in a same-sex relationships tens of thousands of dollars a year

- Basic rights such as hospital visitation or right to die in the same nursing home as one's partner are regularly denied same-sex partners

The social and psychological circumstances for most older gays and lesbians are vastly different, however, than those for younger men and women. A classic study in the early 1980s (Dawson, 1982) estimated the U.S. population of gay men and lesbians over 60 to be 3.5 million people. These men and women developed their sexual identity and "came out" in times of social persecution and fear. They had few positive role models and were surrounded by highly negative stereotypic messages about themselves (Frost, 1997). Developmentally and socially, most older gays and lesbians are charting new waters, developing as role models for younger adults in the process. It is a difficult course to chart: in addition to the impediments of social attitude and self-perception, many older people must address the ageism that occurs in gay, lesbian, and bisexual communities just ageism exists in heterosexual communities (D'Augelli & Garnets, 1995).

Later life relationships tend to be strong ones, grounded in maturity and self-understanding, whether they be same-sex or opposite-sex partnerships.

Given the diversity and difficulty of generating representative samples for study, little is known about the older gay and lesbian population from a social science perspective (Michaels, 1996). Kurdek's (2005) review of homosexual households focused on division of labor, relationship conflict and stability, and factors associated with the prediction of relationship quality. They found in general the division of labor to be fairly equally shared, and that same-sex couples appear to have more effective conflict resolution behaviors than do heterosexual couples. Same-sex couples generally have a strongly developed community of friends, and in Kurdek's study between 8 percent and 21 percent of lesbian couples and 18 percent and 28 percent of gay couples have lived together for 10 years or longer in relationship quality similar to those of heterosexual couples. Clearly, research on aging in lesbian and gay male communities lags behind that of research on heterosexual aging, and it is equally clear that the assumptions and theoretical frameworks used for one cultural group must be carefully challenged when applied to another.

Older lesbians and gay men typically live within a self-created network of friends, significant others, and selected biological family members. The Senior Action in a Gay Environment (New York City, 1999) found that 65 percent of 253 gay and lesbian elders reported living alone—nearly two times the rate of all people 65 or older in New York City (Cahill, South, & Spade, 2001). Like the general population older women were two times as likely to live alone than were men (Brown, 2000). Another study with a more culturally diverse population found that 75 percent of older adult gays and lesbians in Los Angeles lived alone (Cahill, 2001).

The lifestyles and life choices of gay males and lesbians are similar to those of heterosexuals. In reports from older lesbians, although the respondents suffered from ageism, sexism, and homophobia, they did not regret their sexual orientation. They were able, as are other groups of older adults, to make a healthy adjustment and live their lives with high self-esteem (Kimmel et al., 2006).

Research on Sexuality

The **Kinsey studies** of the sexual practices of men and women were landmark, classic studies. *Sexual Behavior in the Human Male* appeared in 1948, followed by *Sexual Behavior in the Human Female* in 1953. No one had previously studied the sexual histories of so many individuals in such depth. Although very few of Kinsey's sample subjects were older men and women, Kinsey concluded that the rate of sexual activity gradually declines with age. The men in Kinsey's study were sexually most active in late adolescence (ages 16 to 20), the women in their 20s, and the rate for both genders gradually declined thereafter.

A great deal of research has followed Kinsey's stepping-stone into aging and sexual behavior. A few of the most recent studies are explored here.

Masters and Johnson

Masters and Johnson's studies of human sexual behavior were considered revolutionary for their time period, just as Kinsey's were for his; they studied not just sexual histories but also observed and recorded the physiological responses of couples having sexual intercourse. This research resulted in their first book, *Human Sexual Response* (1966). Their second book, *Human Sexual Inadequacy* (1970), is based on their treatment program for those suffering from sexual dysfunctions. Although their technical language is rather difficult for the lay person to understand, the discussion that follows is taken largely from a very readable book that interprets their work and is endorsed by them (Belliveau & Richter, 1970). One methodological problem relative to aging with Masters and Johnson's study is that most of their "old" subjects were between 50 and 60. They studied few individuals over 60.

Erection is chiefly caused by scuraum, eringoes, cresses, crymon, parsnips, artichokes, turnips, asparagus, candied ginger, acorns bruised to powder and drank [sic] in muscadel, scallion, sea shell fish, etc.

Aristotle, 4ᵀᴴ century BC

Body processes slow down with age, but do not stop. Masters and Johnson found that if a middle-aged husband and wife are aware of these changes and take them into account in their lovemaking, there is no reason why pleasurable sex cannot continue. For men, the body processes affected by the aging process include the length of time needed to get an erection and the firmness of the erection. The ejaculatory expulsion of semen is less forceful, and the erection may be lost faster after ejaculation. In addition, it may take longer to have another erection. If the older man is aware that these changes do not signal the end of his ability to have an erection, he can relax and enjoy sexual activity.

According to Masters and Johnson, men over 50 have many advantages over younger men. Their ejaculatory control becomes better, and they can maintain an erection for a longer period of time without the strong drive to ejaculate. As partners, many women say that longer lovemaking is more satisfying—time allows for greater opportunity to bond emotionally during sex. Sex is, for many women, more an act of bonding and emotional connection than an act of physical performance.

The aging female is subject to all the negative attitudes of our society regarding women and sexual matters: the beliefs that a woman's sex drive is not as strong as a man's, that women are more passive, that women are not supposed to enjoy sex as much as men. These myths become even stronger in terms of the older woman—for instance, the myth that it is unnatural for women to continue sexual relations after menopause. Far less attention has been paid to sexuality in aging women than in aging men, in no small part due to the sexism of the male-dominated medical profession of the nineteenth and twentieth centuries (Morley & Kaiser, 1989).

Masters and Johnson reported that it is natural and beneficial for older women to continue sexual activity with no long periods of abstinence. Older women do, however, experience a slowing process, just as men do. Vaginal lubrication occurs more slowly with aging and less lubrication is produced. While younger women take 15 to 30 seconds for lubrication, older women may take as long as four or five minutes. The clitoris may get smaller with age but it still receives and transmits sexual excitement. There is less increase in the size of the vaginal canal. And, with age, the lining of the vaginal walls becomes thinner. Older women, like older men, generally experience a shorter orgasmic phase. Menopause alters the hormone balance and again, huge individual differences exist, from attitudes and self-concept to an individual woman's biology.

If [a woman] . . . is normally developed mentally, and well-bred, her sexual desire is small. If this were not so, the whole world would become a brothel, and marriage and a family impossible.

J. G. Richardson, MD, 1909

Masters and Johnson reported an increase in the rate of masturbation in older women. Many women are widowed or divorced and are isolated from male sex partners. Particularly among older couples, husbands may be ill and may not be able to participate in sexual intercourse. Masturbation can be an effective and *available* form of sexual pleasuring.

Among women, as men, aging has advantages. Postmenopausal women are no longer concerned about the risk of pregnancy or the side effects of contraceptives. They may also be free of the anxieties and pressures of motherhood. That there is no apparent reason why menopause should necessarily blunt the female's sexual capacity, performance, or drive. Masters and Johnson found no endpoint to female or male sexuality set by advancing years.

The Duke Longitudinal Study

In spite of the small numbers of older people Masters and Johnson studied, their findings are of tremendous importance regarding the sexual *potential* of older people. But Masters and Johnson made no attempt to study just how many older people take advantage of their sexual potential and remain sexually active. The **Duke longitudinal study** tried to answer this question and to also assess the changes in sexual interest and activity that come with age (Verwoerdt et al., 1969). The study, which began in 1954 at Duke University, still continues. Different researchers involved in the study continue to generate and publish findings. One fascinating finding is that one in six of us will be even more interested in sex as we age (Witkin, 1994).

In the Duke study, sexual activity was one of many aspects studied in the lives of 254 older people. The sample was composed of nearly equal numbers of men and women whose ages ranged between 60 and 94. Four patterns of change in sexual activity were observed throughout the 10-year longitudinal study of 1954 to 1965: inactivity, sustained activity, decreased activity, and increased sexual activity.

The researchers found a great deal of variability in sexual behavior among the aging individuals. Approximately 15 percent of the sample fitted into the rising category, whereas unmarried women were almost totally inactive sexually. Ten years did not greatly decrease the number of elders who were sexually active. About half of the men and women who survived into their 80s and 90s reported having sexual interest. Among men surviving into their 80s and 90s, continued sexual activity was not rare; and about one-fifth of these men reported that they were still sexually active.

In summary, the Duke study shows enough sexual activity on the part of the old and the very old to illustrate the sexual potential Masters and Johnson found in older adults. On the other hand, the sexual interest of many Duke study participants exceeded their actual sexual activity. Although not clearly explained, this discrepancy was perhaps caused by ailing physical health, psychological reasons, or lack of an acceptable partner.

Other Studies on Sexuality

The Kinsey study has been criticized for conclusions made from its cross-sectional methodology. To make age comparisons in a cross-sectional study, researchers must compare old people who were raised in conservative times with young people raised in more permissive times. In other words, the criticism holds that Kinsey's report might accurately show patterns of interaction for different age groups at one point in time; however, it reliably says nothing about *change over time*.

Starr and Weiner (1981) studied responses of older individuals to a lengthy questionnaire on sex and sexuality. Older adults, even those in their 70s and 80s, reported being sexually active and stated that sex was as good or better than ever, though somewhat less frequent than previously in their lives. Three-fourths of the elders studied said their lovemaking had improved with time. These researchers challenge the Kinsey studies as to the rapidity of decline of sexual activity with age.

Unhappy marriages reduce sexual activity for older couples just as it does for younger couples. Rates of extramarital sex, though lower than those for younger couples, still occurs for the same reasons: inattentive or uninterested spouses. Some couples have open marriages. Homosexual experiences are reported, some of which take place for the first time in old age. Findings such as these dash stereotypes of sexually inactive and uninterested elders, many of whom have struggled for sexual frankness and openness throughout their lives, fighting taboos that were far more strict than those young people face today. Some older people improve their sex lives over the years by breaking down psychological barriers; some do not (Gott, 2005; Gott & Hinchliff, 2003; Matthias et al., 1997; Mooradian & Greiff, 1990; Peate, 1999).

Older people find various ways to compensate for sexual changes brought about by aging or poor health. They slow down the lovemaking process, emphasize oral sex and manual stimulation of the genitals, and participate in fondling and cuddling. Many report that the sensation of touch is more important, meaningful, and appreciated than in their younger years. Some women use a lubricant to compensate for lack of vaginal moisture. The biggest problem for older, unmarried women is that of having a partner; men their age are in small supply. Older adults demonstrate a strong interest in sex and a large capacity for enjoying life.

A study in the *Johns Hopkins Medical Letter* reported that 70 percent of 68-year-old men and the majority (over 50 percent) of women in their 70s regularly engage in sexual intercourse. One of the most debilitating factors related to diminished sexual activity is the self-fulfilling prophecy of asexual aging: the belief that sexual prowess diminishes with age. Knowing what changes to expect with aging as well as how to deal with those changes is key to remaining sexually active. At any age health problems such as

diabetes, alcoholism, depression, or anxiety can interfere with sexual function.

For the very old, biological studies of changes in the male reproductive system reveal these findings: need for greater direct penile stimulation to achieve erection, slower erection (but only a few minutes slower), briefer stage of ejaculatory inevitability, reduced amount of seminal ejaculation, and reduced need for ejaculation at each and every sexual contact (Zeiss & Kasl-Godley, 2001). If a man becomes impotent, the cause is illness or disease, not aging. New knowledge and techniques are continually being made available in dealing with either impotence or premature ejaculation. In addition, sensations become less genital and are more diffused with age. In the aging female, changes include slower lubrication, greater need for direct stimulation of the clitoris, shorter orgasmic phase, and irritability of vaginal tissue and outer lying tissue with low levels of estrogen. But these issues do not mean the end of one's sex life. Sexuality is among the last of our faculties to decline with age (Witkin, 1994).

Bretschneider and McCoy (1988) studied the sexuality of a healthy population of individuals between the ages of 80 and 102. The most common activities in order of frequency were touching and caressing, masturbation, and intercourse. Although 63 percent of the men had intercourse at least sometimes, only 30 percent of the women did. The researchers concluded that, for men, the frequency of sexual intercourse does not change greatly after age 80 compared to the previous decade. This finding would suggest that physiological shifts have already occurred for the 80-year-old man, and he has an available sexual partner—probably a wife. For either men or women in general there is no decade this side of 100 in which sexual activity is totally absent.

Changes inherent in the aging process need not interfere with sexual activity because compensatory strategies are easily employed, and indeed most older adults have been

We are sexual beings. Warmth, caring, and a deep appreciation helps to bring out the best in both people in a partnership, regardless of age.

shown to adapt easily to such changes (Zeiss & Kasl-Godley, 2001). Physically disabled or terminally ill individuals continue to be sexual beings, too, and sexual counseling is now commonly available through hospitals and clinics to assist couples with physical disabilities or impairments. Sometimes people simply need reassurance that they are sexually attractive and capable, regardless of their physical limitations. It is relatively common for heart attack and stroke victims to report fearing that sexual activity may trigger a life-threatening episode. Information, knowledge, and caring for oneself and one's partner are primary aspects of a continuing and satisfying sexual life.

Most of what we know about aging and sexuality is by studying predominantly healthy, well-off Caucasians (Zeiss & Kasl-Gorley, 2001). Although the information available on elders and sexuality is generally valid, professionals in health care and human services must honor the gap in knowledge about sexuality and nonwhite, nonmainstream groups.

Improving Attitudes About Aging and Sexuality

Ruth Jacobs' book published in the early 1990s, *Be an Outrageous Older Woman,* describes her surprise and dismay at the shock evoked by her writings and seminars on sexuality. Her writings, although not graphic, are open and candid about the sexual desires of older women. She believes that our definition of what is sexual has undermined older adults and needs to be changed. If being sexy is measured by frequencies of orgasms and other quantitative counts, sexuality may show some decrease with age. But if sexuality refers to one's sensuality, then most adults do indeed become more sexual with age.

In one of Jacobs' workshops an older woman who assumed her decreased interest in sex was due to aging discovered that the tranquilizers she had been taking for four years were responsible. Some drugs can severely affect the ability of a woman to have an orgasm, so people must check on side effect of any drugs they take. For example, sexual dysfunction is not an uncommon side effect for men who take certain medications for hypertension.

Once I asked Grandma when did you stop liking it?, and she was eighty. She said, "Child, you'll have to ask someone older than me."

GRANDDAUGHTER, 1998

Sexuality in Special Circumstances

The most common reason for an older adult to report low interest in sex is the lack of a sexual partner because of divorce, death, or illness. Under special circumstances we come face-to-face with cultural taboos and social prohibitions, most especially on the topic of sex in nursing home settings. Many long-term care settings do not allow couples to share a room and provide no locations where couples could have privacy for intimacy with a partner. Early studies in long-term care facilities found strongly negative attitudes about sexual activity among staff (Eddy, 1986). Many facilities have modified their rules and now offer training programs to increase staff knowledge, although the need remains for training among the professionals and para-professionals in gate-keeper positions within long-term and assisted living care models (Bouman & Arcelus, 2001; Walker & Harrington, 2002).

Many old people continue to engage in sexual behaviors, despite the lack of marital partners and living circumstances that range from incarceration-like, to the more affluent life-care communities. We find that sexual patterns and attitudes do not change much over the life course. Those who enjoyed and sought out sexual opportunities earlier in life enjoyed and sought out sex in their elderly years.

It is very probable that cognitively competent older people handle their sexuality in a nursing home setting by private use of masturbation, and by psychological repression. One study compared the sexual attitudes of nursing home staff with those of the residents (Lauman et al., 1994). They found that residents were more likely than *staff* to agree that sex is not needed for women after menopause or for men after age 65 and that sexually active elderly people are "dirty" (Richardson & Lazur, 1995). This form of repression is probably functional in most nursing home settings—it helps people to get past any felt need for sexual expression. It also illustrates, however, the loss of part of the human condition for nursing home residents. The lack of sexual contact hurts, as well as the lost sexuality. Most tragically it is probably unnecessary if alternative opportunities for the expression of sexuality were provided.

Living in the communal circumstances of a nursing home can be difficult. Behavioral and psychological problems may result when residents are ridiculed or prevented from enjoying physical contact. When residents hold hands or say they want to marry, this may evoke laughter and sarcastic remarks, leading to humiliation. Treating an aged man and women like children because they want to hold hands or embrace is insulting and robs the older adult of an important aspect of identity.

According to one study of nursing home life (Richardson & Lazur, 1995), nursing homes can address this issue of sexuality directly in several ways, including

- improving privacy ("do not disturb" signs, respected by staff)
- educating staff about human sexuality in later life

Group living situations are often not conducive to maintaining an intimate relationship. Ideas about age-related role behavior are slowly changing to include respect for the continuing need for expressed sexuality.

- helping to arrange for conjugal visits or home visits
- encouraging other forms of sexual expression, such as hugging and kissing
- evaluating complaints about sexual functioning
- advising the elderly to discontinue medicines that may affect sexual function
- providing information and counseling about sexuality to interested patients

We all need to know that the possibilities for intimacy and enchantment with another continue to the end of our lives. Sex is for life. The frequency and vigor of sex may change with age, but sex in later life remains as important in our later life to self-identity as it was in out youth. Regardless of age, the loving expression of sex says, "I value you. I appreciate you. I like having you in my life. I trust you, and want to touch you and be touched by you."

Chapter Summary

This chapter considers older couples and three categories of older singles: the widowed, divorced, and never married. An unknown percentage of older people live together without being married, but we do not directly address this important and growing phenomenon in this chapter. As baby boomers begin to move into old age, more information on the social significance of unmarried partners will become available.

The myth that older people automatically lose interest in sex distorts perceptions of older people. Unhappy marriages reduce sexual activity for older couples who experienced that just as it does for younger couples. About 70 percent of men and 50 percent of women think often about sex and physical closeness with the opposite sex, regardless of age.

Having sex, feeling sexual, having intimate relationships, and being touched with loving care—caressed—are profoundly important human experiences regardless of one's age. Old age can separate people from marriage partners due to death, divorce, or living arrangements, but old age does not mean asexuality.

Older homosexual adults report satisfaction with their lives—or not—in way as do heterosexual adults. Impediments to maintaining a same-sex relationship are far greater than for opposite-sex couples in terms of public policy, but homosexual relationships into later life do last and are reported as being supportive and loving, as are heterosexual relationships. Regardless of sexual orientation, living arrangements profoundly impact intimacy in coupled relationships although this appears to be slowly changing.

Sexuality is an important part of who we are as humans and who we are as self-identified persons. Our sexual identity develops with us into later adulthood and older adults, especially those with available partners, remain sexually active.

Key Terms

achieved relationships
ascribed relationships
Duke Longitudinal
    Study
grief work
intimacy
Kinsey studies
Masters and Johnson
    studies

occupational regret
relationships
romantic love
sexual function for
    procreation
social comparison
widowhood

Questions for Discussion

1. Look through a selection of birthday cards online. What references do you find to sexuality and aging? Print out some examples to share with the class.

2. Look on TV and in magazines for references to sexuality in advertising. How is old age and sexuality used in advertising? How does that compare to the use of sexuality and youth in advertising?

3. What are some advantages to sexual activity in late life? What are some things that hinder sexual activity in late life?

Fieldwork Suggestions

1. Interview a widow or widower who has been alone for several years. Ask questions about how his or her life has changed by losing his or her partner. Find out what his or her spouse or partner was like.

2. Browse through an online dating site for older people looking for someone. How do their profiles differ from those of younger people? What is the same? Do you find a general theme to the profiles of older people who are dating?

Internet Activities

1. Search for an Internet site with information on widowhood. Develop categories of the material you find: who do you guess would use this site? What does the site market, if anything? What is the intent or purpose of the site?

2. Locate an Internet chat group specifically for older people who want to connect with other older people. What observations can you make about this site?

Work and Leisure

8

Carter Still Standing Against Injustice

Byron Williams

Former President Jimmy Carter may well go down as having the most accomplished post-presidency in history; his commitment to human rights is unparalleled.

Carter recently added to this already distinguished legacy by making a difficult personal decision. After more than 60 years, Carter broke the sociological and theological ties he formed with the Southern Baptist Convention (SBC).

The Nobel Prize-winning former president recently wrote, "Faith is a source of strength and comfort to me, as religious beliefs are to hundreds of millions of people around the world. So my decision to sever my ties with the Southern Baptist Convention, after six decades, was painful and difficult."

The cause for this irreconcilable difference was the failure of SBC to recognize women as equal with their male counterparts. SBC's official statement on women reads:

> Women participate equally with men in the priesthood of all believers. Their role is crucial, their wisdom, grace and commitment exemplary. Women are an integral part of Southern Baptist boards, faculties, mission teams, writer pools, and professional staffs. The role of pastor, however, is specifically reserved for men.

This led Carter to write:

It was, however, an unavoidable decision when the convention's leaders, quoting a few carefully selected Bible verses and claiming that Eve was created second to Adam and was responsible for original sin, ordained that women must be "subservient" to their husbands and prohibited from serving as deacons, pastors, or chaplains in the military service.

At its most repugnant, the belief that women must be subjugated to the wishes of men excuses slavery, violence, forced prostitution, genital mutilation and national laws that omit rape as a crime. But it also costs many millions of girls and women control over their own bodies and lives, and continues to deny them fair access to education, health, employment, and influence within their own communities.

SBC is the largest Baptist denomination, with more than 16 million members, and probably its most conservative. Its origin as a stand-alone denomination dates back to 1845, when it split with northern Baptist over slavery.

In 1995, SBC voted to adopt a resolution renouncing its racist origins, formally apologizing for its past defense of the institution of slavery.

Carter is not the first high-profile elected official to leave SBC. Former president Bill Clinton and former Vice President Al Gore have also left, citing disagreements with a number of SBC positions.

SBC is hardly alone. Many churches, along with organizations outside of religious circles either overtly or covertly, struggle with gender equality. Lest we forget, the 19th Amendment, which guaranteed women the right to vote, was ratified in 1920.

With few exceptions, the American church has traditionally been slow to adapt to change.

It was moderate white clergy and not conservative pastors from SBC affiliated churches that called Martin Luther King's nonviolent tactics in Birmingham, "extreme" in a full-page ad in 1963, which provoked King to write his famous "Letter from Birmingham Jail."

The legacy of the historical black church standing at the vanguard of the Civil Rights Movement has more to do with the few than it does the masses. In 1963, when King led Project "C" in Birmingham of the 500 black churches in the city, less than 20 actively participated.

Moreover, a number of historical black churches still struggle with gender equality today.

In an ironic twist, First Baptist Church of Decatur, GA, a 2,700-member Southern Baptist church called Julie Pennington-Russell to become its senior pastor last month—the first woman to lead a SBC church.

Whether this signals a philosophical change within SBC or Carter's resignation will lead a mass exodus from the denomination, though both are unlikely, misses the point.

If those of us who are not members of SBC, or affiliated with any religious institution, look at this with an elitist eye that narrows Carter's decision to one that holds only internal ramifications we might overlook the injustice that occurs in the organizations with which we do associate.

I applaud the former president for his progressive stand. Carter had to fight 60 years of familiarity and comfort to see an injustice he could no longer tolerate. That is indeed a lesson we can all embrace.

Source: http://www.huffingtonpost.com/byron-williams/
carter-still-standing-aga_b_249522.html.

To work or not to work in old age: that is the question. The answer has implications for both society and the older person. It also implies a choice that is limited or nonexistent for many people. Some people are forced to retire because of age discrimination or because of illness. Others must continue to work, often at menial jobs, because they cannot afford to retire. Further, the economics of supply and demand govern one's presence in or out of the job market. If there is a demand for your work, you stay; if not, you are encouraged to retire.

The issue of work and retirement is a thorny one. Retirees may either feel elated and free or devalued and depressed. Because of changes in routine, personal habits, and opportunities for social interaction, retirement can bring stress even when it is voluntary. Though studies show that most older people generally make a satisfactory adjustment to retirement, given good health and sufficient income, some do not have these benefits; and a minority are not satisfied with retirement, regardless of their health and financial status. *Retirement* is a dirty word to some; to others it represents freedom from a daily grind of work and a doorway to a fresh chapter of life.

In this chapter, we explore the options of retirement, analyze discrimination against the older worker, and examine the difficulties old people have in adjusting to retirement. We also look at the meaning of leisure in American culture. When we are young, our time away from work is called "leisure"; when we are old, it is called "retirement." What is the difference?

As a social institution, retirement has been the major means of redividing the life span so that old age can be identified by a common dimension[:] as a period of inactivity on a pension.

GUILLEMARD, 1980

The Concept of Retirement

The social definition of **retirement** has changed in the past few decades. For Atchley (1976) retirement is an event that occurs when a person definitively stops working and withdraws from the formal labor market. Social scientists have since pointed out that some people withdraw from the formal labor market because they lack the health to continue working, or because of age-related disabilities, or because they have become unemployed and are unable to find a new job. Retirement is quite different than struggling with the inability to work, and consequently being dependent on family or public welfare for survival. In this chapter, retirement is seen from the perspective of a social institution: it is a person's definitive economic inactivity, with income from work replaced by income established for support after employment ends. Retirement is also a stage in the life course, with a set of social rules and with special social meaning.

For individuals, the meaning of retirement ranges broadly. Researchers analyzing data from a study on the problems and suggestions of retirees (Marcellini et al., 1997) summarize:

For some, it is a way of being free from daily routine, making it possible for the person to dedicate him- or herself to more satisfying activities linked to personal interest. For others, instead, it is a difficult period of line in which [they] feel deprived of a social role, and problems regarding finances, health and loneliness can arise. For everyone, though, it is a period of great change in which lifestyle has to be restructured in many ways. (p. 377)

The outcome of retirement depends on individual characteristics of people—their lifestyles and the sociocultural context in which they live.

In 1900, nearly 70 percent of American men over age 65 were employed. By 1960, the figure had dropped to 35 percent; by 1976, to 22 percent; and by 1992, to 15 percent (U.S. Bureau of the Census, 1991). In 1992, the percentage of women in the workforce aged 65 and over was 9 percent; this number has remained fairly constant for the last two decades but has experienced a 1 percent increase over the last five years and will rise further in the years ahead. Most of the labor force growth in the 1990s has and will come from increased participation by minorities and middle-aged and older women. The trend toward early retirement among men appears to have leveled off but among women it has been offset by a rise in midlife careers (Burtless & Quinn, 2002). The impact of the global recession of the 2000s on work and retirement is still uncertain; however, it is expected to shift retirement planning.

Economists and social scientists have collected a huge array of labor market statistics in the past 30 years. Trends in labor force transitions show a precipitous drop in labor force participation among older people, especially among men. The drop in participation rates for the 20-year period of 1969 to 1988 for men aged 60 was nearly 18 percent and for men 65 and over was 10 percent (U.S. Department of Labor, 1989). Women in 1989 had a higher labor force participation rate than in 1968, but the drop in labor force participation rates from 1968 to 1989 was only 3 to 4 percentage points among women aged 62 and older (Peracchi & Welch, 1994). Unlike men's patterns, labor force participation for women is associated with changes in marital status, childbirth, spouse's income, and family background, so a complete picture of retirement decisions for women over this same 20-year-period is less clear in this study. Projections for future retirement statistics are made on past retirement patterns; however, cohort differences are likely to create changes in the actual patterns that emerge. These generational differences have to do with economic and social status at the time individuals prepare for retirement and then actually do so. Current workers in middle age were educated after the end of World War II, and retirement decisions are being made in a time of economic and labor market turbulence. For those who retired in the last 30 years pension plans were less frequently provided, and Social Security benefits were lower than for current middle-aged workers. Fewer retiring men had wives who were also in the work force, and retirements did not have to be coordinated within a family. Because of longevity issues, fewer of the retiring workers in the past had elderly parents needing care. All these factors impact retirement decisions and an individual's adaptation to retirement.

For most people in the United States, retirement is an expected life event. Even though mandatory retirement has been abolished for many jobs, other factors are leading older men to favor retirement. On the other hand, more older women are entering or remaining in the workforce. However, not all workers retire completely. Studies show that more than one of every five older Americans who retire return to work at least part time.

In the agricultural era of the United States, few workers retired. Most workers were self-employed farmers or craftspeople who generally worked as hard as they could until illness or death slowed or stopped them. Stopping work, however, did not have to be abrupt. A gradual decline in the workload could occur simultaneously with a gradual decline in physical strength.

Older people who held property could usually support themselves in old age. For example, a homesteader who began to grow old and

In retirement, some people develop new relationships and discover interests that were not developed when their attention was in the world of work and raising a family.

to experience difficulty in doing heavy work could usually pass the work on to family, retaining his or her authority as the children assumed more and more responsibility. Because he or she remained in charge, the older farmer did not have to quit producing entirely. No one could technically force "retirement."

Before 1900, few people lived past age 65. Those who owned property lived longer than those who did not own property or have helping family members. Older people without economic resources had to do heavy work because of their need to support themselves. When they could no longer physically continue to work, they often ended up in alms houses or died of malnutrition and neglect. The "good old days" were not good for many old people.

The Industrial Revolution brought many complex changes. Increased productivity created great surpluses of food and other goods. More people ceased being self-employed and went to work in large factories and businesses. Government and bureaucracy grew. When civil service pensions for government workers were introduced in 1921, the retirement system began.

In 1935, with the passage of the Social Security Act, all conditions for institutionalizing retirement were met. The law dictated that people over 65 who had worked certain lengths of time were eligible for benefits, and 65 became the age for retirement. Since then, employees in both the public and private sectors of the economy have retired in increasing numbers.

We also need a word about women's work pattern cohorts. Women born in the early 1900s were not encouraged to enter the workforce. A Gallup Poll in 1936 found 82 percent agreement for the statement that wives of employed husbands should not work (Burkhauser, 1996). Not until the 1960s did the concept of work outside the home as other than a temporary role for women develop. Even then, wives' employment was often seen as a supplemental source of money for "extras." Before the 1960s,

the majority of permanently employed women were single. Today, married women tend to have interrupted work histories—entering the labor force and then leaving it upon the birth of a child (the "mommy track"). Thus many women in their old age today (the married ones, at any rate) do not have the extensive work histories that older women in the future will have, even though many of them worked temporarily outside the home during the Depression and World War II.

Now more than 95 percent of all adults expect to retire some day. Although retirement has become an accepted feature of modern life, one must question whether retirement is wise for everyone.

The Trend Toward Early Retirement

A different trend is usurping retirement at age 65. More and more companies are encouraging employees to retire before age 65 without a substantial loss of pension benefits. American industry, which seems to presume that the young have greater vitality, has steadfastly worked for **early retirement** and restrictions on work opportunities for older workers. In industry, retirement at age 55 is not uncommon and 62 is typical. And contrary to popular belief, the trend for early retirement is not among people in poor health. In 1993–1994 the typical early Social Security beneficiary was as healthy and as wealthy as the typical same-age worker who did not retire (Burkhauser et al., 1996).

This trend in industry has characterized the white-collar world of work. Whether blue or white collar, most baby boomers say they would like to retire by age 55. Some (about 40 percent) want to retire between the ages 56 and 65, and some, after 65 or never. The desire for and expectation of early retirement has increased over the last several decades. From 1950 to 1990 the median retirement age dropped from 67 to 63.

In 1961, early Social Security benefits were introduced. In the year previous to these benefits, over 79 percent of men aged 61 and nearly 76 percent of men aged 63 were in the labor force. In 1993 those percentages were 64 percent and 46 percent, respectively—despite improvements in mortality and morbidity (Burkhauser, 1996). In the 1970s the trend toward making pensions and other benefits available before age 65 greatly accelerated, from the government's pension system, to other public and private institutions. One study indicated that those who were offered full pension benefits before age 65 were twice as likely to retire early. In 1992, 38 percent of early retirements from large companies were the result of incentive offers. For the most part, companies were seeking to cut their payrolls and at the same time avoid layoffs.

Studies of companies and educational institutions show that most corporations now have early retirement inducements in their pension plans. Only a handful of organizations offer incentives to continue working beyond age 65.

The following excerpt illustrates the way in which Hewlett-Packard used early retirement to scale down its number of employees when business was off:

Plagued by manufacturing problems that have hurt its earnings for the last two quarters, the company announced a corporate-wide plan to offer early retirement to 2,400 workers in plants across the country. The program is available to workers who are at least 55 years of age or older and have 15 or more years with the company. Those who opt for the program get a half-month's salary for each year of service up to a maximum of 12 months' salary. (Silver, 1990)

Likewise IBM's Japan unit, needing to reduce the size of its workforce, offered an early retirement plan, hoping to remove a significant number of employees aged 50 and over (IBM's Japan Unit, 1993). These early retirement packages have become typical of large corporations that wish to cut costs by downsizing their operations. Such downsizing was especially common in the cyclical high-technology industry, with its recurring booms and busts. Some workers like having work options, grabbing a retirement package at one company, and then getting an equivalent or even better job elsewhere. Others wanted to retire anyway. The unhappy ones are those who feel forced to retire either by company downsizing or for health-related reasons.

Some American businesses have begun to question the policy of encouraging workers to retire at or before age 62. The labor pool of young workers is declining, and they sometimes cost as much to train as to retain the older workers (Cyr, 1996). The value of work differs by cohort, and some executives find that older workers have a stronger work ethic than do younger workers. Some of these companies have introduced **late-retirement incentive programs (LRIPs)**. Labor statistics indicate, however, that although most businesses fear too many workers will take advantage of LRIPs, most workers still opt for early retirement. Incentives for early rather than late retirement are still the order of the day for business and industry. The only change is that many early retirees find the "good deals" not so good after all; buyout plans are growing skimpier, and new jobs are scarce. The trend toward early retirement therefore can bring with it a number of complex issues.

Older Worker Discrimination

Some early retirement incentives are offers that older American workers cannot refuse, and many older workers are wondering whether they are being subjected to disguised **early retirement discrimination**. For example, at the age of 56, a man who had worked for 31 years for a large manufacturing company was hoping to work 6 more years and retire at age 62. But his corporation, like many others across the United States, decided to trim its payroll and abolish numerous jobs, including his. The company offered him half-pay retirement for the next four years, provided that he did not work elsewhere. He accepted these terms out of fear that other jobs were not available to him.

A one-time incentive is not the same as an early retirement option. An early retirement option offers the employee the choice to stay or to go. In contrast, with the one-time incentive, the employee either accepts or gets laid off. For every employee who gladly accepts an early retirement incentive, another feels that he or she is being forced out the door. Under the **Age Discrimination in Employment Act (ADEA)**, all early retirements must be voluntary (Blouin & Brent, 1996). The question remains, what is voluntary? Do you leave of your own free will if your boss drops strong hints that you are not wanted and hints at a big bonus offer if you leave early? Or if you are told that in a year your job will be phased out?

Employers are slowly starting to realize that jettisoning older workers can result in an irreparable loss of skills and expertise.

REDAY-MULVEY, 1996

Along with the multinational web of new global marketplaces, new forms of industrial competition are emerging "precisely at the time the population is aging" (Minda, 1997, p. 564). Downsizing, restructuring, reengineering, outsourcing, reduction-in-force (RIF)—all are common terms to the postindustrial work environment of the twenty-first century. These management shifts were all part of an economic recovery between 1990 and 1998; however, median family incomes remained stagnant and average weekly earnings of most rank-and-file workers fell (Minda, 1997) during the first part of the century. Some economists and social analysts believe the newly emerged definition of work has created social changes as profound as those of the Industrial Revolution 200 years earlier (Minda, 1997; Weaver, 1997).

In the United States, outsourcing and downsizing has shifted much work from permanent workers to non-union, lower-paid, part-time, contingent workers, who often do not receive benefits such as health care and corporate pensions (Minda, 1997). In the early 1980s the profile of the typical unemployed worker was younger, unskilled, and blue-collar workers. By the 1990s, the profile of the unemployed worker included many older, skilled white-collar workers (Congressional Budget Office, 1993). The risk of job loss is rising for workers 45 to 55—the age category that has suffered most as a result of corporate downsizing (Bureau of Labor Statistics, 1996).

Workplace changes accompanying the global recession in the first decade of the twenty-first century have had profound impact on work, retirement, and leisure plans of people. The changes include loss of manufacturing jobs and a subsequent shift to service and knowledge work. This results in a mixed workplace outcome: on the one hand, younger workers have less experience and therefore lower salaries and are more affordable for economically stressed industries. Additionally, most younger workers are far more technologically savvy than are the older workers, and technology is increasingly integrated into management and production. Older workers might require technology training that younger workers already possess. On the other hand, certain professions such as law or medicine require many years of education and apprenticeship. Top managers emerge through the ranks of lower-level management. From one perspective older workers are more valuable to an industry or business; from a different perspective, younger workers benefit that business.

The risk extends internationally. When Congress amended the ADEA to apply to U.S. companies abroad, it neglected to include the entire class of American workers working for foreign employers stationed in the United States (Madden, 1997). With increased globalization of the economy, the protection of all workers, including older workers, becomes more complex yet more essential.

The Taxpayer Versus Early Retirement

People are asking, "Can we afford old age?" Business and industry are pushing for early retirement programs, and these programs are sometimes a boon for the older worker. But here is the irony: life expectancy is greater than ever, and the supply of younger workers is shrinking, creating a larger percentage of older workers. Social Security would be more likely to stabilize if older workers stayed on the job longer, and the government, contrary to many companies' wishes, is attempting to see that they stay longer. The Social Security retirement age will gradually become 67 (in other words, the normal retirement age will be 67 for those reaching age 62 after 2022), and those taking benefits early will receive 30 percent less (they currently receive 20 percent less) than those receiving benefits at the normal retirement age. In this way, federal policy is encouraging workers to extend their work lives. But business and industry give incentives to encourage the older worker out of the market. Economics researchers make it very clear that most people respond to economic incentives in choosing to retire.

The current work place is likely to have a broad mix of ages among workers. Some people cannot afford or choose not to retire as soon as possible.

Early retirement has generally been interpreted positively. It is an indication that society is able to provide economic security for a large number of older people who no longer must be in the labor market. This in itself is an historic development. Additionally, retirement has been a positive force in avoiding generational conflicts between older, employed workers and younger workers, seeking employment (Towers, 2007). Despite its widespread acceptability, however, retirement poses problems for many older workers who find themselves unemployed as well as unemployable.

Denis (1996), in an analysis of early retirement from a corporate perspective, listed six advantages of early retirement plans:

1. They can reduce the need for massive layoffs

2. Payroll costs can be reduced

3. The number of senior, more highly paid employees can be reduced

4. When done across-the-board (not targeted), employee morale may be preserved or at least not unduly disrupted

5. Because of the release component, employment discrimination claims can be minimized, if not avoided

6. Promotion channels can be opened so that other qualified employees can advance within the company

The downside of early retirement, according to Denis, is that employers "may be deluding themselves about the overall effectiveness of these programs in cutting staff and, in particular, ridding employers of their less-productive employees" (p. 66).

When employees seek redress under the law for what they believe to be discrimination, they often do not win. In a landmark retirement case heard by the U.S. Supreme Court in 1995 *(Lockheed Corporation v. Spink)*, the Court supported the corporation's retirement policy. Lockheed had amended its retirement plan to provide financial incentives for certain employees to retire early. The plan required that participants receiving the early retirement benefits must release any employment-related claims they may have had against Lockheed. Employee Spink was eligible for the early retirement package but refused to sign the release. He retired without earning the extra benefits, and subsequently sued Lockheed. The Court held that the Lockheed retirement plan was permissible but introduced the concept of "retirement sham." If the payment of early retirement benefits is purely a sham, a court might find that the plan assets were misused. Just how egregious the transaction must be to be considered a "sham" was not defined by the Court (Denis, 1996).

The economic incentives to business for introducing early retirement packages are strong. To keep older workers on the job and to thereby save Social Security funds, the government will need to strengthen its measures—to increase the amount of money older workers can earn without reducing their Social Security benefits, for example. And business and industry should try harder to find a place for the older worker, who may well offer a gold mine of experience, wisdom, and loyalty. Unfortunately, workers near retirement seem expendable to most large-scale employers. But as the number and percentage of younger workers shrink, private business may finally be motivated to retain and retrain older workers. It will take collaboration between the private and public sectors to reach the proper integration of policies that accomplish (1) an extension of the working life to manage human resources and to finance pensions and (2) the development of well-protected and regular part-time work to sustain policies of gradual retirement (Reday-Mulvey, 1996). As the twenty-first century begins, a new definition of *work* is being forged, and hopefully it will represent a balance of the public interest with that of the private.

Age Discrimination in Employment

Age discrimination in employment starts long before the traditional time for retirement. Even those still in their 30s have trouble getting into some training programs and schools, such as flight school and medical school. The problem of job discrimination is severe for older workers in spite of federal laws that prohibit it. The 1967 Age Discrimination in Employment Act (amended in 1974 and 1978) prohibits the following:

1. Failing to hire a worker between age 40 and 70 because of age
2. Discharging a person because of age
3. Discrimination in pay or other benefits because of age
4. Limiting or classifying an employee according to his or her age
5. Instructing an employment agency not to refer a person to a job because of age, or to refer that person only to certain kinds of jobs
6. Placing any ad that shows preference based on age or specifies an age bracket. *Exceptions:* the federal government, employers of less than 20 persons, or jobs where youth is a "bona fide occupational qualification," such as modeling teenage clothes

Prior to 1978, the ADEA protected employees from age 40 through 64. The amendments adopted in 1978 extended that protection to age 70. Though all upper limits were removed for federal employees, the law does include exceptions that allow the federal government to retire air traffic controllers, law enforcement officers, and fire fighters at younger ages. Nevertheless, the U.S. Supreme Court in 1985 ruled that the city of Baltimore could not force its fire fighters to retire at age 55, despite federal regulations requiring retirement at 55 for most government fire fighters. Other cities are easing their rules to allow fire fighters, police officers, and other public safety employees to stay on the job until 65 or older.

Interpretations in the law can be subtle. In 1996 the U.S. Supreme Court upheld the suit of a 56-year-old worker who had been replaced by a 40-year-old worker, both ages of which are within the protected category of ADEA. In *O'Connor v. Consolidated Coin Caterers Corporation*, the Court ruled that the case was discriminatory because an older person was replaced with a substantially younger worker (Blouin & Brent, 1996).

The Older Workers Benefit Protection Act of 1990 (OWBPA) amended the ADEA to prohibit employers from denying benefits to older employees. This amendment was designed to reduce the disincentive to hire older workers when the cost of some benefits to older workers is greater than the cost of those benefits to younger workers, as for most health care programs for example (http://www.eeoc.gov 2009). In limited circumstances, however, an employer is permitted to reduce benefits based on age so long as the cost of providing the reduced benefits to older workers is the same as the cost of providing benefits to younger workers. Most benefits packages covering health plans now coordinate retiree health plans with eligibility for Medicare. This illustrates only a small portion of the complexity of health care reform.

Some companies required older workers to sign a waiver agreeing not to sue before they could qualify for an early retirement bonus. Now any waiver and release signed by an older worker going through a job transition must meet certain requirements. The agreement must be clear and concisely written, and the employee must have 21 days to decide whether to accept. Additionally, the agreement must clearly specify that the worker is waiving all rights under the ADEA (Blouin & Brent, 1996).

The protection of older workers has become necessary in light of downsizing and demand for high corporate profit margins, even in a time of economic growth and especially in a time of economic recession. When layoffs are for legitimate business reasons rather than to eliminate those employees who are highly compensated, they are legal. When actions are not legal, the onus is on the older worker to take his or her case to court—at enormous economic and emotional cost to the "retired" or otherwise laid-off worker. It is estimated that for every case that even gets to court (much less every case in which the worker wins the suit), there are 50 cases in which the unhappy worker did not sue (Abel, 1998).

The current economic recession has dramatically shifted the American myth of the golden age of retirement. Under the presidency of Bill Clinton a proposal to "fix Social Security first" was received with very different generational responses. When asked hypothetically what should be done with a budget surplus, two-thirds of the full-time working public felt it should be used for domestic programs rather than for tax cuts or debt reduction, but generational differences were stark. Those aged 50 and older would fix Social Security. Those under 50 would put it to other public programs, such as education or environmental concerns (http://www.people-press.org/report 1998).

By 2009, 37 percent of full-time employed adults of all ages said they have thought about postponing retirement, and this proportion swells to 52 percent among workers 50 to 64 (Morin, 2009). The Pew Research survey of American workers found in 2009 that workers approaching retirement age (the Threshold Generation) were inclined to delay retirement based not so much on income but on the extent of retirement income loss they experienced in the recession with mutual funds, stocks, or retirement accounts such as a 401(k).

Barely two in ten (21 percent) of those aged 50 to 64 reported being "very confident"

they have assets to tide them over into retirement. This compares with 37 percent of full-time workers younger than 30 (Morin, 2009). Investment losses and financial confidence are closely related. More than 82 percent of working members of the Threshold Generation who lost money during the 2009 recession say it will be harder for them to meet their financial needs in retirement. This compares with 66 percent of those ages 18 to 49 (Morin, 2009). This finding is not surprising: younger workers have more time to recoup lost asset value than do workers preparing for, or thinking about, retirement.

Age Discrimination in the Job Search

Objectively, age should not be a determining criterion in finding work. One's ability to do the job is what counts. Age discrimination in the labor market makes finding work more difficult, however. Because older workers have difficulty finding jobs, changing fields is even more difficult. The skills and energy of older people can be greatly underutilized. More older people want to work than is commonly believed. One study found the following (Commonwealth Fund, 1993)

- As many as 5.4 million older Americans (55 and over)—one in seven of those not currently working—report that they are willing and able to work but do not have a job.

- More than one million workers aged 50 to 64 believe that they will be forced to retire before they want to.

- More than half of all workers aged 50 to 64 would continue working if their employer were willing to retrain them for a new job, continue making pension contributions after age 65, or to transfer them to a job with less responsibility, fewer hours, and less pay as a transition to full retirement.

Social service programs can help older adults through a paperwork maze, or to understand a legal complexity that impacts the elder's life.

Older persons who feel that they have been denied employment or have been let go because of their age can take steps to get their job back or to receive compensation by making a formal written charge of discrimination against the employer to the Wage and Hour Division of the Department of Labor. Individual cases of age discrimination are difficult to prove, however. In the investigation of the employer's hiring, firing, and retiring practices, the government tries to reveal a pattern of discrimination. For example, if an employer has hired a good many people over the past several years—but nobody over age 30—a pattern of discriminating against older workers in favor of younger workers may be documented. The older worker will have a better chance of winning if he or she

hires a lawyer. A class-action suit may be filed if a number of job applicants feel they have been denied employment because of discrimination. This action, of course, requires financial resources as well as personal energy and many workers simply sigh and quit. The law protecting older workers has, however, made it unlawful to retaliate against an individual for opposing employment practices that discriminate based on age.

Resources that work diligently to help the older worker find employment are increasing. For example, employment agencies formerly violated the ADEA by stipulating upper age limits or using words such as *junior executive*, *young salesperson*, or *girl* in advertising. Employment agencies were also found guilty of failing to refer older workers to potential employers. Such violations, unfortunately, are still common practice; however, employment agencies such as "Kelly Services: Encore" cater to people over 50 years of age. They advise older applicants that they do not have to give their date of birth on a resume, and remind them that potential employers may not legally ask a person's age in an interview.

Other employment resources are aimed specifically at older adults, and the Internet has become a central source for sharing this information. Job sites for older workers include websites such as DinosaurExchange.com for retirees with experience. Enrge.us is a network for retired government federal, state, and local government employees. RetiredBrains.com offers nationwide job listings that can be searched by industry or state (Brandon, 2007).

Women made up 47 percent of the total labor force in 2003 (Department of Labor, 2004).

More women are in the labor force more continuously than ever before, yet 60 percent of these women are segregated into three occupations: sales, service, and clerical. Of the 27 percent in administration, management, and professions, many are teachers or nursing administrators. Further, women's earnings peak at age 35 to 44, whereas men's peak later in life at ages 45 to 54. One major reason is that women have the kinds of jobs that do not get rewarded with big raises. In the very near future, many employers will need to change their attitudes toward these women or risk not only legal action on the basis of age and gender discrimination, but possibly understaffing as well if the predicted labor shortage occurs.

On-the-Job Discrimination and Ageism in Layoffs

Ageism in the labor market is deeply ingrained institutionally and culturally. Ageism in employment occurs in job categories of every kind. Studies of professionals, engineers, and scientists who are unemployed show age to be a significant variable in explaining their layoffs. Ageism is a factor in both blue-collar and white-collar occupations and for both sexes. Occupations in sales and marketing may suffer the most age discrimination, because they place more emphasis on youth and a youthful image than other occupations, as do occupations requiring physical stamina, such as construction. Once out of work, older workers are likely to remain unemployed much longer than their younger coworkers and to find job hunting a nightmare. If they do find work, it is usually at a much lower salary and with fewer fringe benefits.

Though protected by law, age discrimination is a tough charge to prove. In the last several years, thousands of cases filed with the Equal Employment Opportunity Commission (EEOC) have led to only several hundred lawsuits, and not a high proportion of those have been won. However, in spite of mishandling many cases, the EEOC has obtained some very large settlements involving millions of dollars in back wages and pension benefits:

- Blue-collar workers at the Kraft General Foods plant in Evansville, Illinois, sued for bias when the plant closing led to the firing of workers (Geyelin, 1994).

- Ceridian Corporation and Control Data Corporation paid nearly $25 million to more than 300 former employees who claimed they were fired because of their age. The 1997 court ruling favored the employees, who accused the corporation of a pattern of discrimination by laying off older workers and replacing them with younger workers. The suit was begun in 1989—the proportion of the settlement required to cover legal costs will probably be one-third of the total amount (Ceridian Control Data, 1997).

- Westinghouse Electric and Northrop Grumman Corporation settled two age-discrimination lawsuits in 1997 (Westinghouse Electric, 1997).

- A 78-year-old physician for the medical unit of the Los Angeles County Sheriff's Office was returned to his job after the County settled his claim that he was forced to retire (Jail Doctor Wins, 1998).
- A news correspondent got a settlement from CBS on age discrimination (Lambert & Woo, 1994).
- Apple Computer has been sued for age discrimination in a "wrongful termination" suit (Fired Executive Sues, 1993).
- An ex-AT&T manager won just under $2 million in an age-discrimination verdict (Sandecki, 1993).

The cost in time, dollars, and personal stress required to pursue legal action is enormous. In addition, the line between discrimination and decision making for corporate well-being has become less clear. Some factors creating this lack of clarity include conservative court decisions, stock-holder pressure for corporate high profit margins, and changing technology. In the past, courts have supported age-bias claims in which the layoffs had "disparate impact" on employees over the age of 40. Four federal appeals courts, however, have recently questioned the idea that people can claim age bias merely because an employer's actions had a harsher effect on older people. Instead, the courts have said, "older workers must now show that their employer intentionally discriminated against them" (McMorris, 1997, p. B1). In a job market in which technological skills are important, more recently trained people probably will have the most current skills unless the company provides appropriate employee training. If a company lays off workers based on up-to-date skill levels, or otherwise makes decisions that they feel further the economic stability of the corporation, is it age discrimination?

One particularly ominous struggle is taking place in the UK between nearly 200 Intel workers and Intel Corporation, the microchip manufacturer established both in Britain and in the United States. The workers have taken the unusual step of forming a pressure group (Former and Current Employees of Intel—FaceIntel) and publicizing their argument on the Internet. The concerns have to do with the firm's use of "ranking and rating" to assess employees. This ranks employees on "equal to" or "slower than" other employees, according to a moveable line determined by the company. Anyone receiving two out of three, or two consecutive "slower thans," is automatically put on a disciplinary corrective action program. Those who do not reach a target set by their manager face discharge.

The employees claim that a disproportionate number of people receiving "slower thans" are older workers and workers with disabilities—two groups protected by anti-discrimination laws in the United States. Intel denies the group's allegations, saying that it remains as flexible as possible in the face of the astounding competition in the microelectronics business. This competitive environment requires the company to have employees who can work under demanding circumstances. A random sample telephone survey by FaceIntel of the 1,400 laid off workers reportedly found that 90 percent were over 40 years old, and 40 percent had some form of disability (Welch, 1997).

The following age-discrimination complaints are typical of those reported by older workers on the job:

- Position terminated: older people are told their jobs are terminated; when they leave, younger workers are hired in their places
- Sales force downsized: the sales force is trimmed, and older workers are the ones eliminated
- Retirement credits refused: older workers reach a certain salary level, and there is no more potential for salary (or pension) increases
- Dropped for medical reasons: older people are told they are being dropped for medical reasons

In some cases older workers are passed over for promotion or are the first fired, sometimes to protect a company's pension funds, sometimes to save salary costs. The reasons for and means of discrimination against older workers vary. Employers frequently prefer younger workers—sometimes because of age prejudice, sometimes because younger workers will accept a cheaper wage, and sometimes because the company feels it will receive more years of work before it must pay retirement benefits. Employers in a tight job market can get by without giving substantial salary increases, because the older worker, afraid of unemployment, will settle for a low salary. The older worker is highly vulnerable in a tight labor market.

Health and life insurance fees, which tend to increase with worker age, constitute a supposedly inordinate cost of retaining older workers. The older, tenured worker also tends to receive a higher salary and more vacation time. Surely, companies could deal more creatively and tactfully with older employees. For example, many older people would gladly contribute toward their health and life insurance costs if it meant keeping a job; in turn, companies could be more realistic—and less paranoid—about the older workers' health concerns.

Myths About the Older Worker

Negative stereotypes of the older worker still persist. The older worker is thought to be accident- or illness-prone, to have a high absenteeism rate, to have a slow reaction time, and to possess faulty judgment. Stereotypes contribute greatly to on-the-job discrimination. Common **myths about the older worker** include the following:

- Older workers cannot produce as much as younger workers
- Older workers lack physical strength and endurance
- Older workers are set in their ways
- Older workers do not mix well with younger workers—they tend to be grouchy
- Older workers are difficult to train—they learn slowly
- Older workers lack drive and imagination—they cannot project an enthusiastic, aggressive image

Retention of older employees in the labor force requires some changes in attitude. In some instances older workers have a younger supervisor. Both employee and supervisor must be able to feel comfortable with this situation. A Harris Interactive survey in 2006 found that while 90 percent of employed U.S. adults stated people over age 50 age "with-the-times," 70 percent think their company does not value older workers (http://www.seniorjournal.com/). Most older workers, however, believe their companies are impartial to employees' ages. Researchers suggests, however, that age discrimination is so institutionalized, and so unconscious, that younger supervisors might have lower expectations with older workers, and actually exaggerate or over-reward their good performance. Older supervisors, on the other hand, choose to perceive themselves as valued members of the organization and do not identify personally as an "older" employee.

Businesses need employees, and they need employees who are honest, are dependable, have a good work ethic, and are mature. Online business zine *Entrepreneur.com* examined some benefits of hiring older workers and concluded that the unique skills and value of older workers make them an added value to a company. Hiring older workers, the author suggests, may be a matter of rethinking the costs of high turnover in a more youthful workforce versus the benefits of experience and mature standards of older workers (Bastien, 2006).

Age is not necessarily a determinant of the capacity to do well on the job. The fact that the age force is increasing means that opportunities for a wide variety of interactions with older workers will increase. The more frequently younger people are around older people, the less negatively they see the elderly as a group, and older people are seen more positively—as individuals, some of whom are winners and some of whom are not.

Older Worker Performance

On-the-job studies reveal individual variations in the ability of older workers. Most importantly, however, such studies generally show that older workers are as good as—if not better than—their younger counterparts. One study indicated that older workers have superior attendance records (less absenteeism); that they are likely to be stable, loyal, and motivated; and that their output is equal to that of younger workers (Commonwealth Fund, 1993). The work ethic of older people tends to be very high, which leads to high job satisfaction.

Experience can often offset any decrements that come with aging. The ability to do heavy labor does decline with age, but this decline is gradual and jobs vary greatly in the physical strength they require. Furthermore, an older person in good physical condition is quite likely to outwork a younger person in poor physical condition. Studies of work loss due to illness show that workers aged 65 and over have attendance records equal to or better than that of most other age groups. The U.S. Bureau of Labor reports that workers aged 45 and over have better safety records than younger workers: the highest overall accident rates occur in the 18 to 44 age group (U.S. Bureau of Census, 1991).

Older workers may take somewhat longer to train, but considering their careful work, the investment in time should be worthwhile. Some firms who recognize this say the problem is not "How can I get rid of the older worker?" but "Where can I find more?"

Adjustment to Retirement

Just how difficult is adjustment to retirement? Though one myth claims that people get sick and die shortly after retirement, studies do not confirm this. Retirement in and of itself does not lead to poor physical health. Research shows, however, that adjustment to retirement can be difficult.

Studies of retirement adjustment show varying results in terms of adjustment, satisfaction, and happiness. A general finding is that a minority of people have serious problems with retirement, whereas the majority adjust reasonably well. The message here seems to be that of continuity. The kind of person one has been does not change significantly just because of retirement. This man is typical of those who have no problem adjusting to retirement.

I turn 80 next February. It's always a milepost, I suppose, but no big deal. I retired about 75 percent at age 65 and then fully retired at age 70. It was easy, very easy, for me to retire. I like the freedom to do what I want, go where I want. I never understood these people who are restless and unhappy in retirement. It seems to me they haven't got much imagination. (Colburn, 1987)

Financial problems top the list of reasons for unfulfilled retirement expectations. Those at a marginal or lower-income level are affected most adversely by retirement. In fact, about one-third of retired men end up going back to work and many retirees worry that they will outlive their retirement funds (Morin, 2009).

Clearly, if one's retirement income is adequate, retirement has a much greater chance of success. Good health is also a factor. If the retiree has the money and physical mobility to pursue the lifestyle of his or her choosing, adjustment comes more easily. The importance of these two factors suggests that adjusting to the role of retiree depends more on physical and monetary resources than on mental set. Lifestyle is another factor: if a person is involved with family, friends, and activities, then adjustment is usually more successful. Conversely, if a person has nothing to do and no associates, then adjustment may be poor.

Willingness to retire and attitude toward retirement are also important factors. Naturally, reluctant retirees show negative attitudes toward retirement. Those who retire voluntarily are healthier mentally and physically than those who did not feel control over the decision to retire (Gall et al., 1997). A positive attitude toward retirement also depends on an individual's expectations. In general, those who expect to have friends and social activities, and who expect that their retirement will be enjoyable, usually look forward to retirement.

Gerontologist Guillemard (1993) identified five types of retirement patterns:

1. Withdrawal—Extreme reduction of social activity; long "dead" periods exist between actions performed to ensure biological survival.
2. Third-age retirement—Professional activities give way to creative activities (artistic creation, hobbies) and cultural improvement. (This term is commonly used in France. Life has three stages: childhood, adulthood, and **"third age."** With each stage a new positive phase of living may begin.)
3. Leisure retirement—The focus is on leisure activities (vacations, trips to museums and other exhibitions, theater shows) with an emphasis on consumption.
4. Protest retirement—Characterized by political activism; much time is devoted to associations of the elderly to protect the interests of the retired.
5. Acceptance retirement—Acceptance of traditional retirement values; lengthy time periods spent in daily exposure to television and other forms of mass communication.

According to Guillemard, those with professional backgrounds tend to cluster in groups two and three. Those with working-class backgrounds are overrepresented in the other categories. Studies of white-collar/blue-collar differences in adjustment to retirement reveal various findings. Former top managers and executives can have difficulty adjusting because they feel a loss of power and status, on the whole, white-collar workers generally adjust fairly well because they have more resources at their command. Professional types, such as educators, often show a good adjustment to retirement. Because they are well educated, they tend to have interests and hobbies, such as community affairs, reading, travel, and art, that lend themselves well to retirement. Some professional people write or act as consultants. They seem to have many ways to spend their time fruitfully.

Blue-collar workers indicate a greater readiness to retire because their work may bring them less satisfaction. For example, a man or woman with a dull, routine factory job may look forward to leaving. However, studies show that those with the least education and the lowest incomes have the most trouble adjusting to retirement. Socioeconomic status has been found to have a major influence on retirement rates. The lower one's socioeconomic status, the earlier one retires. Individuals with less education who are employed in occupations characterized by low skills and labor oversupply tend to retire early.

Retiring persons face several adjustments that relate directly to retirement: loss of the job itself, loss of the work role in society, loss of the personal or social associations that work provides, and loss of income. In addition, events such as declining health or the loss of a spouse may coincide with retirement. When all these events occur at the same general time, adjustment is stressful and difficult.

Retirement Preparation Programs

Retirement itself has become a life stage that people look forward to with the idea of pursuing travel, family activities, hobbies, and education. This image is widely accepted;

Maintaining stable and meaningful goals throughout life is critical to happy adjustment to changes brought about by aging.

however, it is a transition requiring personal adjustment. **Retirement preparation programs** can aid in this adjustment. Retirement preparation programs are now common in industry and business, helping people facing retirement understand benefits, Medicare and other health plans, and the role of other assets in retirement. Outplacement companies have been created to assist with layoffs and to offer assistance to laid-off employees (Marcellini et al., 1997). As noted previously, the laid-off older workers are less likely to find re-employment than are the younger workers.

Most outplacement organizations are offered by government agencies and by companies whose workers are covered by a private pension. These retirement programs fall into two categories: (1) limited programs that explain only the pension plan, the retirement timing options, and the benefit level associated with each

option; and (2) comprehensive programs that go beyond financial planning to deal with topics such as physical and mental health, housing, leisure, and the legal aspects of retirement. Individuals exposed to comprehensive programs have a more satisfying retirement.

Successful adjustment to retirement from a psychological perspective depends on the ability to maintain stable and meaningful life goals and sense of purpose. Planning can help prepare the individual by addressing systematically the issues that will emerge in retirement (Price, 2000). Having knowledge of one's financial security, health status, timing of a partner's retirement, and other family responsibilities such as caring for grandchildren of aging parents are all factors to be thought through and addressed for successful adaptation in this life transition.

People need to be socialized into post-work roles just as they are socialized into other life roles. Anticipatory socialization, which prepares a child for adult roles, is also necessary to prepare an adult for successful retirement in old age. We can prepare for retirement in several ways: by saving money, by deciding what our goals are, by beginning to care for and improve our health, by forming meaningful relationships with a sense of permanence, and by expanding our interests so that work is not our primary focus.

Work and Leisure Values

Even though all workers are expected to retire, our traditional value system gives higher social esteem to those who work. In the past, leisure was an accepted lifestyle only for the extremely wealthy. Adherence to the traditional work ethic is not as complete as it once was. What value older people assign to leisure is a research question getting some interesting answers.

Work

Sociologist Max Weber (1958 [1904–1905]) proposed that before the Protestant Reformation in the sixteenth century, work had been regarded as a burden and something to be avoided. With the Reformation, religious reformers like Martin Luther and John Calvin suggested that all work had an inherent dignity and value. In fact, diligence in the performance of work was part of the highest form of Christian behavior. This ethic has remained through the centuries: although no longer is work considered to be the direct glorification of God, the Judeo-Christian culture still respects work for work's sake. Current evidence suggests that a 1990s shift in the meaning of the work ethic includes both high values in hard work *and* negative and disparaging views of others who may lack strong work ethic (Mudrick, 1997).

The **work ethic** is a long-standing American tradition between religious belief and work helped to form the firm conviction that it is im-

moral not to work. "Idle hands are the devil's workshop," a basic precept in the early American character structure, still influences our attitudes today. Work has high value not only because of its moral quality but also because of its practical and personal value. Many people truly enjoy their work and derive pleasure from it. Work can foster interest and creativity as well as a feeling of pride and accomplishment.

To some, work is not a value in and of itself but is, rather, a means to identify one's social standing. It is not work that matters but having a job. Indeed, everyone knows that the general question "What do you do?" requires a very specific answer: to have any status, one must name some type of occupation. Work provides a means of achieving self-identification and placement in the social structure. This very fact has made many American women second-class citizens, because work at home has not traditionally been considered an occupation.

A common theme in the literature—both popular and academic—is that the work ethic has declined. A nationwide study in 1955 and again in 1980 reported a significant decline from 52 percent to over 33 percent in full-time workers in the USA who said they enjoy their work so much they have a hard time putting it aside. Additionally, workers who said they enjoyed hours on the job more than hours off the job declined from 40 percent to 24 percent (Glenn & Weaver, 1983–1984). This finding, however, might reflect an increase in the cultural value of *leisure* rather than a decrease in the work ethic. A follow-up to the study found that there were few fundamental changes in the work ethic in the United States from 1972–1978 to 1988–1993. The study found, in fact, that among white workers of both sexes in white-collar jobs, there is evidence for *increased* work ethic values (Weaver, 1997). Gender and ethnicity changes that did emerge included: more white females in blue-collar jobs and more white men in white-collar jobs said they would continue to work if they were to get enough money to live as comfortably as they would like for the rest of their lives. More women in white-collar jobs felt that people get ahead by their own hard work than by lucky breaks or help from others; fewer men and fewer women from all ethnic groups reported being "very satisfied" with their jobs (Weaver, 1997).

Work matters to Americans. One problem with having a strong work ethic is that it may be difficult to reconcile it with retirement. Work attitudes can be so ingrained in retirees that they carry over to nonwork activities. A strong work ethic tends to be related to low retirement satisfaction. The least satisfied retirees tend to be those with high work values who do not perceive their retirement activities as being useful. It has been found that retirees with strong work values are not as active in social activities because they have a hard time enjoying them. Although most people eventually make a satisfactory adjustment, a 1993 survey found retirement to be one of the most difficult transitions in life: 41 percent of retirees said the adjustment was difficult. The younger the retiree, the more likely that retirement was a difficult adjustment (Weaver, 1997).

Perhaps it is because the line between one's job and "work" has grown less distinct than it once was. Extreme devotion to the work ethic does seem to be waning, and acceptance of retirement as a legitimate stage in the life course is increasing. As technology advances, the United States needs fewer workers to produce necessary goods and services, thus lowering the value of long work hours and increasing the value of other activities. With shifts in longevity, retired people have more company—role models for a healthy, active life following retirement are now increasingly prevalent.

Leisure

The work ethic is alive and well. The traditional conceptualization of the life course is organized around realms of preparation for work, work, retirement from work; this is overlaid on the realms of marriage, childbirth, childrearing, grandparenthood, death. Traditionally, the male "specialty" has been in the work realm, and the female "specialty" has been that of home and children, with the understanding that movement between the two realms takes place in the process of living. This model is changing. Shifts in gender role expectations, work demands, longevity, and the needs for continuing education and training to update skills in a changing technology-driven marketplace have combined to shape this change. The model for the twenty-first century seems to be one in which work comes and goes throughout the adult life course, as does recreation, as does education. The new model brings changes in attitudes toward the concepts of work and leisure activity.

A Gallup Poll revealed that 8 out of 10 adults of all ages think time is moving too fast for them. One benefit of retirement is having more time. The older the person, the more content they seem to be with the amount of time they have. Only 44 percent of those younger than 50 said they have enough time, whereas 68 percent of those 50 and older believed they had time enough for their tasks. According to the poll, all Americans wish they had more time for personal exercise and recreation, such as aerobics, hunting, fishing, tennis, and golf (47 percent); hobbies (47 percent); reading (45 percent); family (41 percent); and thinking or meditation (30 percent). Given more time, most Americans would relax, travel, work around the house or garden, or go back to school. **Leisure values,** that is, the acceptance of leisure pursuits as worthwhile, are becoming stronger.

Sociologists have observed that for the past several decades we have treated leisure as another commodity that is produced and consumed. The leisure industry is one of the largest segments of the American economy, and it keeps hundreds of thousands of persons employed. Ironically, Americans who become overly engrossed in working to attain the symbols and commodities of leisure must work harder to obtain them and often end up with less time to enjoy them.

In some segments of American society, older people are becoming a visible and contented leisure class as they accept the new lifestyles of the retired. The retirement community concept has, no doubt, made the role of retiree more legitimate. Elders who choose the traditional symbols of leisure—the boat, recreation vehicle, or home in a warm climate with access to a golf course and swimming pool—seem to go beyond merely owning the symbols to enjoying the lifestyles that the symbols represent.

In the past, older Americans were considered "invisible consumers," a segment of the population to be ignored by marketers. Their presence is now acknowledged by virtually all institutions in the economy. On the positive side, recognizing elders as consumers gives them greater visibility and legitimizes their role as people enjoying life. On the negative side, consumerism directed at older people as a group often does not recognize that they are a diverse group whose needs and wants vary dramatically.

The sheer size of the over 65 age group market, however, has grabbed the attention of the commercial world. News that the mature market is growing and that older adults have greater discretionary income from improved pension and retirement planning is showing up in bold print on marketing magazine covers.

Expanding Work and Leisure Opportunities

In spite of age discrimination and retirement eligibility, a substantial number of older adults are employed. Many middle-aged women are entering the job force for the first time in their lives. And, although men have retired in larger numbers since the turn of the century, women have not. Women's retirement tends to be more closely related to factors such as whether they have raised children as a single parent, and the level of work responsibility they hold. Currently as previously, women workers' experience of the relationship between family and work is closely

tied to their retirement decisions—the options available in addition to economic choices.

More older people want to be employed full time than are. Still, a more popular choice for older people is part-time work. But part-time jobs are relatively uncommon in many sectors. For example professional, technical, administrative, and other high-paying jobs tend to be full time. Full retirement age is based on one's year of birth: 65 is retirement age for those born before 1937. For people born in 1943, full retirement age is 66, and full retirement age will increase gradually each year until it reaches age 67 for those born in 1960 or later. According to federal law, people who receive Social Security benefits and who have not reached full retirement age earn benefits if they make less than $14,160 annually. For every $2 over the limit, $1 is withheld from the benefits (http://www.ssa.gov/pubs/10003.html, 2009). These amounts become guidelines for many older adults. Not surprisingly, those who lack private pension coverage are more likely to continue working full- or part-time in their later years.

A nationwide study of 11,000 workers examined the work choices of those who retired. More than one-fourth of those workers who leave full-time employment will take up new employment within four years. Most of the remainder will remain retired. The most likely work change is to take a new full-time job (11 percent of the sample); second, to take a part-time job (10 percent); and third, to work part time at the job from which one retired (just under 5 percent). Factors leading a retiree back to full-time employment included low pensions and personal savings and, in contrast, the offer of high wages. Those who were eligible for good pensions tended to retire completely (Bureau of Labor Statistics, 2004).

In another large study, workers of all economic status decreased their work hours gradually as their retirement approached (Choi, 1994). Not unexpectedly, the extent of work reduction was based on preretirement income, since reduced work reduces earnings. The study revealed the fluid relationship between work and leisure, especially during the early years of retirement. "For many workers," the authors concluded, "the transition from work to leisure is not a very finite process, at least in the early years of retirement" (p. 6).

Interest in work following formal retirement from a job seems to be related more to people's individual characteristics and their life goals than to postretirement income among those whose income is 50 percent above poverty level and higher (Morin, 2009). Attitudes toward future employment among some retirees are very positive and among others, quite negative: people age as they have lived. Those who look forward to retirement and have planned their leisure activities for retirement do not want work and resent if they feel compelled for financial reasons to work again. Others plan for continued work, either part time or full time, and find themselves frustrated to be rejected in job searches.

Where and how do older people find jobs they want? Although jobs are hard to find for any age group, they are especially so for elders. If an elder's original goal is to have leisure and a good-paying part-time job as well, opportunities for leisure may seem more limited when the actual time comes to get that part-time job. Older part-timers are on track for jobs that can seem quite disappointing for some people. For example, some retail outlets recruit elders, but the pay is minimal. Fast-food chains might offer a better wage, but the working conditions can be difficult—socially, physically, and psychologically—for some older people who are more accustomed to work in which they have a valued skill or knowledge base. The common occurrence of being paid less per hour as a part-timer for the same work as a full-timer can lead to low morale for part-time workers of any age.

Full-Time Work Opportunities

An obvious and far-reaching way of expanding work opportunities for seniors has already been mentioned; that is for private business and industry to keep their older workers and discourage, not encourage, early retirement. More older workers should be retrained and motivated to stay in the labor force.

Older workers who want to change jobs or branch out on their own need support in achieving these goals. Rather than retire, able elders sometimes shift their focus or their employers. A nuclear engineer may become an expert witness on nuclear power. A lawyer may leave a large law firm set up practice on his or her own. A retired professor may become a lecturer, conduct workshops, or become a consultant. Consulting is a common job for "retired" managers, employers, and engineers.

Business leaders who look at demographics (shrinking young population, growing elders) have taken steps to hire older people. Days Inns of America currently employs many older persons; in fact, they are more than 30 percent of its labor force. Days Inns analyzed its Atlanta Center by comparing costs and benefits of hiring older versus younger workers. It found the following:

- After a half-day of computer familiarization training, older workers can be trained to operate sophisticated computer software in the same time as younger workers—two weeks

- Older workers stay on the job much longer than younger workers—an average of three years compared with one year. This resulted in reduced average annual training and recruiting costs per position

- Older workers are better salespeople: they generate additional revenue by booking more reservations than younger workers, although they take longer to handle each call for the reservations center

- Older workers are flexible about assignments and willing to work all three shifts

Many older people turn to self-employment as a way to earn money. Creativity, initiative, capital, and sometimes past business experience are needed to start a business in old age.

Older adults are sometimes hired for jobs not easily filled by a younger person, such as that of a Park Service docent. Salaries for such jobs are often stipends intended to acknowledge the contribution of the worker.

Still, self-employment is often easier than getting hired to work for someone else. Slightly over 40 percent of all employed elders are self-employed. One "retired" man at age 62 started a travel agency that eventually expanded into three more. At age 60, another started an equipment leasing business. A 60-year-old woman started producing how-to videos on more than 150 subjects ranging from watercolor painting to foreign language instruction to being a magician; the videos are now distributed nationwide (Duff, 1994). Internet sites are rich sources for marketing, and older adults have begun using this resource. Some re-

tirees have businesses in arts and crafts, such as real grandfathers who make grandfather clocks, and these items are marketed through individual or collective sites.

A way of assisting older people in finding work is to encourage self-employment, whether by helping them, say, for example, to start new businesses or by providing the means for advertising and selling the goods they make. One urban senior citizens' center established a shop to provide local artisans with a sales outlet for quality handmade goods. Rural towns, too, offer stores selling quilts, stitchery, wooden carvings, and other crafts made by local elders. A hobby can become a source of income as well as venue for socialization, as hobbyists meet one another as well as other people at such craft sales.

Part-Time Work Opportunities

Some older people welcome the opportunity to work part time and companies such as McDonalds, Wal-Mart, and Walt Disney World, all of which actively recruit older workers, although the jobs are minimally challenging. Some companies, such as Travelers Insurance at Hartford, Connecticut, and Aetna Life and Casualty, offer an early retirement program to the long-term workers who command the highest salaries. But Travelers hires back pensioned workers part time when they are needed. Retirees are recruited regularly at "unretirement parties" hosted by the company. Open to all retirees is a job bank, which offers flexible part-time temporary work. The company profits by not having to train new workers and by not having to create new pension funds for them. Thus, one place to look for work is in the company that you retired from.

One company that hires older workers for new assignments is the F. W. Dodge Company, a data-gathering firm headquartered in Kansas. Until recently, the firm hired temporary, part-time workers to assist in filling out forms. Because these temporary workers were not always dependable, the company redesigned the jobs as part-time permanent positions and recruited retirees. The response by oldsters was extraordinary: 90 percent of the positions were filled. Employees work several days every month. The older employees have proven to be more reliable and the accuracy of the data has increased (American Association of Retired Persons, 2005).

The federal government has several small-scale work programs designed to aid retired persons by providing work to supplement their Social Security benefits. For example, the Green Thumb Program pays older people living near or under the poverty level minimum wage to plant trees, build parks, and beautify highways. The Foster Grandparent Program also employs old people near the poverty line to work with deprived children, some mentally retarded and physically handicapped, others emotionally disturbed. The older workers try to establish meaningful relationships with the children. For their efforts as foster grandparents, the elderly receive an hourly wage, transportation costs, and one meal a day.

Work opportunities could expand if employers were more flexible in scheduling the hours and days older employees worked per week. Management consultants have suggested that workers aged 55 and over be allowed gradually shrinking workweeks, with final retirement not occurring until age 75 or later. Another suggestion is that companies allow employees the option of reducing their work load to two-thirds time at age 60, one-half at 65, and one-third at 68. Still another alternative would allow the older person near retirement to work six months and to "retire" for six months, alternating work and "retirement" periods until permanent retirement. Flexible work/retirement plans enable older persons to gradually retire and at the same time remain productive on the job even into their 70s and beyond, if they desire.

Part-time work is sometimes possible through **job sharing.** Job sharing has generally been found not only to please workers by giving them more options, but also to benefit employers. One employer stated that it increases productivity: "When people know they're working only two or three days a week or only up to four hours a day, they come in all charged up. You can bet they're much more productive." Job sharing increases job interest, lowers absenteeism, gives the company decreased turnover, and increased morale which results in greater productivity. Employers' arguments against job sharing are (1) people may have less commitment to part-time work than to full-time work, and (2) training and administrative costs might increase because of the greater number of people involved. Neither of these points is borne out in the literature (Morin, 2009; Turner, 1996).

Polaroid has a program offering part-time work to employees aged 55 and over. Permanent part-time opportunities are available through job sharing. Varian Associates, a high-tech firm in California's Silicon Valley, offers reduced work schedules before retirement. As a rule, the reduction entails a four-day work week the first year of transition and a three-day schedule in the second year. A third year of half-time work is possible. At Corning, Incorporated, the "40 Percent Work Option" enables people to retire, collect their pension, and work two days per week at 40 percent of their preretirement salary. To qualify, they must be age 58 or older and have been with the company 20 years or more.

Polaroid also has a "rehearsal for retirement program" allowing individuals to take unpaid leaves of absence. Leaves can run as long as six months. At Kollmorgen Corporation in Massachusetts, employees may reduce their time on the job to do volunteer community service while receiving full pay and benefits 12 months before retiring (American Association of Retired Persons, 2005).

Retraining

If new technology makes older workers obsolete, they should have options for updating their knowledge and skills. In the past, people assumed that professionals and their skills naturally and unavoidably became obsolete. Now, however, **retraining** and mid-career development programs have proven effective; and national training programs have trained older workers as successfully as younger workers, despite the fact that these training programs could work harder to consider the unique problems of adult retraining.

Career development programs for middle-aged and older workers are far too few. In our society, which generally punishes those who make job changes, the older worker who changes jobs loses most of the retirement benefits from his or her previous job. People would have much more career flexibility if retirement benefits were transferable. They would feel freer to change jobs, and companies would be more inclined to hire the middle-aged. For those whose careers require working so intensively that one becomes hardened and emotionally drained, spending a few years in a related field would actually increase one's effectiveness in the original career, if one chose to return to that field of work.

Under Title V of the Older Americans Act and with the help of the Job Training Partnership Act (JTPA), older workers are receiving some job training. The JTPA offers special training programs for economically disadvantaged older workers aged 55 and over; 3 percent of the funds allotted to the act must be spent on such training. Some businesses are offering more than severance pay to laid-off workers: a change to start a new teaching career. Pacific Telesis, Chevron, AT&T, Rockwell, IBM, and Kodak, as a part of their severance program, pay tuition and expenses for retirees to go back to school to become teachers in kindergarten through twelfth grade. At PacTel, managers with undergraduate degrees in math or science who are bilingual are given priority (Severance Offers Chance to Teach, 1991).

Other diverse businesses offer continuing education and retraining. General Electric and AT&T have classes to teach employees the latest changes in technology. Pitney Bowes has a retirement educational assistance program in which employees and their spouses over age 50 are eligible for tuition reimbursement up to $300 per year per person, continuing for two years after retirement for a maximum of $3,000 per person. These same retirees often become teachers for classes at Pitney Bowes (American Association of Retired Persons, 2005). The examples given here may be a "drop in the bucket," but they serve as examples of how older people can be retrained, reeducated, and encouraged to stay in the workplace.

Leisure Opportunities

As we discussed previously, Americans generally use leisure for keeping busy with something, so that they rarely *relax* or refresh themselves. Time away from work at any age should be a time for *being*, not doing. We all need conditioning to learn to be, to express our individual talents and interests, and to find fulfillment in the pleasure of self-realization.

Boredom does come with nothing to do, but real leisure is self-satisfying. Meditation, reflection, and contemplation are fun when divorced from the necessity to do them. Playing with ideas, trying to resolve puzzles, or brainstorming new inventions can be relaxing and pleasurable. A group of older citizens could meet daily for lunch to share their ideas and experience: "We solve the problems of the world, but no one listens." All of us have undeveloped personal resources, talents, and abilities that could be realized. We need not limit our self-development only to what is necessary to hold a job. For example, a particular college professor likes to sit quietly and speculate. Students coming to her office say, "I see you are not doing anything." Much that can be enjoyed cannot be seen!

We should try to see the role of retiree as a valid and legitimate one, well deserving of leisure in its real sense. Our work values need to be examined and put into perspective. Passively watching television, a common activity of both young and old, is hardly a form of self-realization. Our educational system, both formal and extended, should prepare us not only to work, but also to find fulfillment as human beings.

Chapter Summary

Retirement has not always been characteristic of American life. Ceasing work at a given age and receiving a pension has become more and more common for Americans after the Great Depression when Social Security was enacted. Most now expect to retire and many take an early retirement. Orientation to work and leisure are important factors influencing the other person's decision to retire. The older person with a strong work ethic who enjoys status and friends on the job may continue working. Age discrimination operates against older people in employment. There are negative stereotypes of older workers. Employers may want more years of work than the older person has to give. "Youth cult" values encourage hiring the young instead of the old. Finding work in old age can be difficult for anyone.

Many factors affect one's adjustment to retirement, and the capacity to enjoy leisure is not a given. Sufficient retirement income appears to be of great significance in retirement. Work and leisure opportunities for elders in our society could be greatly expanded. More part-time options and gradual retirement programs would give older persons more choice, and hence, satisfaction with post-career plans.

Key Terms

age discrimination in employment

Age Discrimination in Employment Act (ADEA)

early retirement

flexible work/ retirement plans

job sharing

late-retirement incentive programs (LRIPs)

leisure values

myths about the older worker

retirement preparation program

retraining

work ethic

Questions for Discussion

1. If a person on Medicare develops a cancer or heart disease the hospital treatment is completely covered. If that same person develops Alzheimer's disease and needs help at home or long-term care in a living facility, there is no coverage at all. Is this fair?

2. Medicaid planning is a form of "spending down" and transferring assets to become eligible for Medicaid benefits. Medicaid will cover long-term care costs, but Medicare will not. Would it be ethical to spend down your money to become eligible for long-term care if you or your spouse had Alzheimer's disease?

3. What are the issues unique to elders living in rural versus urban areas? What could be done to improve life for each group?

Fieldwork Suggestions

1. Get information on Medicare. Write a simplified description of the plan as if you had to explain it to an elderly person.

2. Find out what services are available for elders in the smallest rural town near you. Compare those services to what is available in the largest city near you, or to where you live.

3. Research on the Internet the concept of "spending down" to become eligible for Medicaid. What are the pros and cons?

Internet Activities

1. Locate a website with current age-discrimination cases in the state in which you live. Who sponsors the site? How accurate do you believe the information is? How easy or hard was it for you to find it?

2. You are a career counselor, providing assistance to a 57-year-old male who was recently laid off by a nationwide hardware chain. His former job was inventory control for six stores in a certain region of the country. What websites on government programs and services will you access for information on his behalf? (Hint: begin with websites for Pension Benefit Guaranty Corporation and the Social Security Administration.)

9 Finances and Lifestyles

Childhood Sweethearts Reconnect after 85 Years

Georgia Garvey and George Houde | Chicago Tribune

The bride wore blue. She carried a new pink rosebud with a touch of baby's breath, and on her wrist dangled a silver bracelet lent to her by a family friend. But she wasn't worried about completing the tradition of being married with "something old, something new, something borrowed and something blue." "I'm the something old," she said with a laugh.

The bride, Lorraine, 92, and groom, Roland "Mac" McKitrick, 93, said their vows 85 years after becoming childhood sweethearts in Wisconsin. In the interim, they'd moved away, lost touch and built families, only to fall in love again decades later. Not letting cobwebs gather, they moved their three-year relationship from engagement to wedding in three days, tying the knot at Community Presbyterian Church in Mt. Prospect on Friday.

Lorraine McKitrick said she wanted to avoid the hassle of planning a complicated wedding. Instead, her new daughter-in-law Pat Ridenour and friend Kathy Graham witnessed a simple ceremony in which the vows were sealed with a kiss and a song. "There's no other music, so I figured I could contribute something," said Roland McKitrick, a barbershop quartet and church choir singer, who serenaded his bride a cappella with "I Love You Truly," made famous in "It's a Wonderful Life."

The look in their eyes just before they kissed for the first time as husband and wife echoed the innocent gazes in the picture on the altar. In the faded photo, a boy in well-worn overalls stands in a field, beside him, a girl with bobbed hair in a simple dress. The photograph was taken in the 1920s, near the time the McKitricks met at English Ridge Public School in Wisconsin.

"His father was a farmer," Lorraine McKitrick said. "My father was a cheesemaker." They walked to school together, played in the countryside and became friends. But eventually, they went separate ways.

He went to the University of Wisconsin and majored in music. She moved to Georgia, went to business college and got a job. Both married, had children, and were later widowed.

The two lost contact over the years, but about three years ago they reconnected through siblings who had remained in touch. On Wednesday, Roland told Lorraine he needed to talk to her.

"I thought, 'What have I done wrong?'" she recalled. "He just said, 'I want to ask you to marry me,' and I said, 'Go ahead.'"

Though their reunion story may be sweet, it wasn't completely smooth. Shortly after they reconnected, they wound up at a dinner with family members. But Roland McKitrick brought along a female friend, a former co-worker. The future Mrs. Lorraine McKitrick was annoyed.

"He leaned over the table and asked if I remembered him telling me I was his sweetheart in 3rd grade," she said. "I said, 'No.'" Only later did she recant. "I sent him the photo of us as kids," she said. "I wrote 'I lied' on the back."

Source: http://www.chicagotribune.com/news/local/chi-old-sweethearts-26-jul26,0,5294865.story

In 2008, 36 million people were over age 65 in the United States—about 12 percent of the population. At the turn of the century, 420 million people in the world were 65 or older—about 7 percent of the world's population. This number is projected to be around 974 million by the year 2030 (He et al., 2005). These proportions of older adults have a huge impact on health, financial well-being, and therefore lifestyles. The social and economic implications of the aging baby boom will require modifications to Social Security, Medicare, and disability and retirement benefits, and one of the outcomes of longevity will be altered family structures and plans. The decline in personal income brought about by the global recession of 2007 creates a yet more complex picture of the financial well-being of older adults (Lenze & VonKerczek, 2009).

In the early 1900s, older Americans were almost all poor or near-poor. They faced a variety of unjustified economic hardships, given the generally high standard of living among the rest of the population. With the advent of Social Security and other pension programs, the financial status of the elderly improved throughout the twentieth century. Changes in Social Security in the 1960s and 1970s provided more comprehensive coverage. Still, the bulk of elders live relatively modest lifestyles. To cope with inflation, a big problem in the 1970s and early 1980s, the U.S. government raised interest rates. In the early 1990s when a recession hit, interest rates were dramatically lowered. Interest-sensitive monies are savings accounts, bonds, and utilities stocks—those very institutions in which elders have invested their savings and pensions.

Financial Status

The financial status of the *average* older American is rather ordinary. The household income of those over age 65 is not much different from the average income for all adult households. Job losses, lower interest rates, and smaller corporate dividend payments accompanying the recession impacted all Americans, younger as well as older. Before 1974 people aged 65 and older were more likely to live in poverty than any other age group. Since then, children have been more likely than working-age or older adults to be living in poverty.

On average, the poverty rate for people aged 65 and older declined from 15 percent to 9 percent between 1974 and 2006. The percentage of older adults living with low income declined from 35 to 26 in that period (FIFAS, 2009). The poverty rate is somewhat higher for older adults than for younger adults, but is still below the poverty rate for children.

The "greatest generation" cohort, which reached adulthood in the 1940s and 1950s, had high marriage rates, high birth rates, and low divorce rates. More women remained married throughout their lifetimes, and more children were raised by two parents. The post–World War II era is exemplified by improved opportunity, stable family life, and upward mobility. The boomer generation, the first of whom who turn 65 in 2011, is a radically different population.

They are more racially diverse and better educated than previous generations. Women are more likely to have worked full-time for several years and to qualify for Social Security and other retirement benefits than were their mothers. Increased longevity and life expectancy among this population increases their likelihood of needing long-term care and exhausting financial resources. They are more likely to be divorced, reducing the likelihood of available spousal caregivers in later life. Although more women in this generation have worked full- or part-time than the previous generations of women, boomer women have lower pension income than do boomer men (He et al., 2005). The reconciliation between family life and work impacts women's work in the twenty-first century as it did in the twentieth.

Numbers derived from population trends, however, can be misleading. Social class, race, ethnicity, and gender all combine to unequally distribute power and wealth in the country. The circumstances over an individual's lifetime have financial consequences in later life. The poverty rate for women living alone, for example, is more than two times the average rate for all people, whereas that for African American women living alone is nearly 50 percent—more than four times the national average (U.S. Census Bureau, 2005). Eighty-five percent of people over 65 today belong to the non-Hispanic white majority; however, the minority population is growing and aging. It is projected that by the middle of this century non-Hispanic whites will constitute about 67 percent of the older population (Moody, 2000).

Near-poverty rates are equally descriptive of the circumstances of older adults. These older people have been variously described as near-poor, deprived, or **economically vulnerable.** Such an income is inadequate to allow most people to lead a full life in the broader society. This economic state creates a cohort of socially excluded older adults for whom even available social services are beyond their grasp (Vaux, 2004).

A major reason for the increased income (reduction in poverty) of elders over the past two decades is that Social Security payments are higher because of automatic cost of living adjustments. Additionally, more women are now covered under pension plans and Social Security. The number of young and middle-aged women in the workforce who pay into the system has steadily increased since the 1960s, and as these women retire they receive greater benefits than did women from previous generations. Major sources of income for older people in 2006 were Social Security (reported by 89 percent), asset income (primarily the value of a residence, 55 percent), private pensions (reported by 29 percent as primary income source), and earnings from investments (reported by 25 percent). Proportions of retirement income sources are shown in Figure 9.1. Note the relative size of Social Security as a source compared to private pensions. In 2007

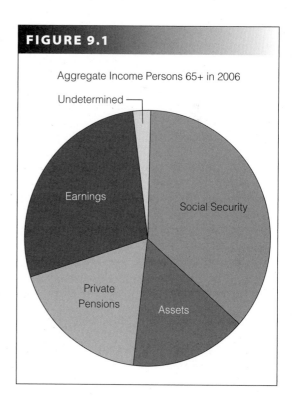

FIGURE 9.1

Aggregate Income Persons 65+ in 2006

Undetermined

Earnings

Social Security

Private Pensions

Assets

Social Security constituted 90 percent or more of the income received by 35 percent of all Social Security beneficiaries (AoA, 2008).

Aging is not a unidimensional phenomenon. The income of the head of the household tells only a small portion of the story. Generally elders live in families, families consist of generations, and family patterns include both personal and financial interactions across the generations. The financial and personal caregiving exchanges that take place over time powerfully impact the financial status of all generations in a family. Another point to emphasize in studying finances in later life is that the upcoming cohort of baby boomers will age much differently than their grandparents did. The grandparents of boomers are their model for aging since, generally speaking, our parents and grandparents become our role models for aging. Cutler and Devlin (1996) discuss a "wealth span"

that corresponds with the life span. Middle age is a time in one's wealth span when a variety of emotional, biological, social, and financial changes begin to emerge and are likely to form the basis of new sets of expectations and concerns and therefore of plans and financial decisions within the family unit (p. 25).

Wide Diversity in Financial Status

Census data suggest that the spread of wealth is great among older people, meaning that those 65 and over have a broad range of income. Income, as we have pointed out, is not equally distributed among ethnic groups. One of every 11 elderly whites was poor in 1996, compared to one-fourth of elderly blacks and almost one-fourth of elderly Hispanics (AARP, 1997). The population of minority older adults is projected to increase from just under 5.7 million in 2000 (16.4 per-

Older people have widely differing financial resources. These men may be making a choice to sit on the park bench and visit or they might have no place else to go.

cent of the elderly population) to 8.1 million in 2010 (20 percent of the elderly population), and to 12.9 million in 2020, or 23.6 percent of the elderly.

The moral economy of a society is its set of beliefs about what constitutes just exchange: not only about how economic exchange is to be conducted. . . but also. . . when poor individuals are entitled to social aid, when better off people are obligated to provide care, and what kinds of claims anyone—landowners, employers, government—can legitimately make on the surplus product of anyone else.

DEBORAH STONE, 1988

Disadvantaged people—the poorly educated, women, and minorities—have not shared equally in the increase in retirement income (AoA, 2008). The equalizing effects of Social Security are more than outweighed by private pensions and asset income (interest on savings), which are received mainly by those in the upper-income brackets. Inequality throughout life is cumulative and is magnified in old age as accumulated disadvantage. In other words, the advantages of good education and/or good jobs lead to better pension coverage and savings as well as other assets in the later years.

Women, ethnic minorities, those who live alone, and the oldest-old constitute 90 percent of the elderly poor. Their financial distress is a function not only of aging but also of education, race, and past employment. Single, widowed, and divorced women have a particularly difficult time, because living alone is the most expensive way to manage. Because women live longer, and because many have not paid as much into Social Security, older women are many times more likely than older men to live alone in poverty.

Poverty has declined over the past several decades, but not extraordinarily given the state of the economy. Gerontologists are concerned that

older Americans are being unfairly blamed for the country's huge economic deficits. The myth that all elders are down-and-out has transformed into the myth that elders are well-to-do and depriving the young of their fair share. In reality, most older people are struggling in the lower- to middle-income ranges. If the benefits of Social Security were withdrawn from that income, many more of America's elderly would move into poverty and beyond, into destitution. The new trend toward decreased government support for social welfare places the well-being of this population in a precarious and insecure position.

Home Ownership

The net worth of a family unit or individual is the total value of all assets, minus debts. The overwhelming majority of the American population saves little for old age, does not maintain significant savings accounts, and does not hold other financial assets. However, around 78 percent of the elders owned their own homes in 1996, a figure that is 12 percent higher than a decade before. Although more older people own their homes than do younger adults, the problem for older home owners is to remain in their homes and stay financially solvent. High energy costs and the generally high cost of living can lead to difficulties. The percentage of income spent on housing, including maintenance and repair, in 2000 was higher for older persons (34 percent) than for the younger consumer population (27 percent); however, about 80 percent of older home owners in 1995 owned their homes free and clear (AARP, 1997). In the past many older people have been "house rich and cash poor"—that is, they have minimal monthly income, but many thousands of dollars in equity in their home. The greatest asset for most elders is the value of their home. In 1988, the federal government introduced a plan of **reverse mortgages,** whereby a loan is made against home equity owned by the borrower. A borrower (the home owner) increases indebtedness while drawing down the equity in the home. The loan does not require monthly repayment and becomes repayable when the borrower ceases to use the home as a principal residence, which may be when he or she moves (to long-term care, e.g.) or dies. At that time the home is sold, and the lender (government and/or its agent) receives the value of the loan plus interest. In another approach to home equity conversion, a bank guarantees a monthly income to the home owner for the remainder of his or her life and claims ownership of the home upon the home owner's death (Vitt, 1998).

Just previous to 2000 it was estimated that more than 600,000 elderly home owners living in poverty could be raised above the poverty line if they obtained a reverse mortgage (Kutty, 1998). One study sponsored by the U.S. Department of Housing and Urban Development estimated that the poverty rate of elderly households could be reduced by three percentage points by means of reverse mortgages, and 68 percent of older home owners owned their homes free and clear in 2007 (Kutty, 1998; Sawyer, 1996; American Housing Survey, 2007). Full ownership of the property could make a reverse mortgage desirable.

The potential income resource of reverse mortgages is dependent on the equity—the resale value—of a home. In previous decades home ownership has been the primary asset of older adults, and currently this continues to be so. Boomers, however—the next generation of older people—generally do not own their homes outright. Coupling this partial ownership with deflated property values brought about by the recession of 2007, many people at the cusp of "old age" owe more on their homes than their mortgages. For the upcoming generation of elders, reverse mortgages as they are now configured will not be a practical option.

Reverse mortgages, when they work, are of value only for those older people who choose to age in place—that is, those who wish to live in their present home until their death (or a move). If an older person were willing to move,

selling the home and investing the proceeds in an income-generating asset would make more sense from a strictly economic perspective (Kutty, 1998). Most older people, however, are strongly motivated to remain in their own homes for as long as possible. Houses are memory holders: one's lifescape is contained in a home. Familiar surroundings, including the familiarity of a neighborhood—even one that has changed over time—are comforting. For many elders, a reverse mortgage option can provide exactly the right payoff: the investment in a home over all those years is paying off for them in their old age, assisting them to remain solvent in the space of their choosing. For others, it is an option they choose not to take; still others have fewer options than this.

A number of elders own mobile homes, which are typically cheaper than houses. Many people have found the mobile-home-park lifestyle to be an affordable one. A summary of mobile-home facts and figures reveals the following:

- Six percent of older Americans live in mobile homes.

- Almost half the residents of manufactured housing are age 50 or older.

- More than one-third of manufactured home residents are retirees or part-time workers.

- More than 90 percent of older residents own their manufactured homes; 23 percent own their lots.

- More than half the older residents of manufactured homes live in mobile-home parks.

- California and Florida are the states that have the most manufactured homes.

- *Double-wides*—composed of two sections—and even larger mobile homes are growing in popularity, accounting for half of all units produced in 2000.

These facts and figures speak for themselves as to the popularity of mobile homes and their affordability. A modest lifestyle of the owners is also implied. Residents, however, must pay rent for the space on which the mobile home is placed, and these rents have risen in recent years. In many areas of California, older adults are called "prisoners of paradise," because the rents of these spaces have escalated at a frightening pace. Mobile-home owners and even condominium owners who pay maintenance fees are subject to cost increases over which they have no control. Older people would benefit from legislation controlling these costs.

Home Rental

A renter is more subject to inflation, enforced living conditions, involuntary moves, and other forms of control by others than is a home owner. In 2007, of the 2.9 million households headed by older people, 20 percent were renters. The median household income of older home owners was $29,899 whereas that of older renters was $15,130 (American Housing Survey, 2007). Some cities have adopted rent-control ordinances to help such people as elders living on fixed incomes, but such policies are rare.

Social Security or Social Insecurity?

The United States has had a strong work ethic since its inception. In 1929, the U.S. stock market began a steep decline into a crash; subsequently, in 1931, many large financial institutions failed, and the United States entered the Great Depression. This was perhaps the first time as a nation that Americans were faced with the reality that hard work, diligent savings, and good citizenship were not enough to keep one from economic disaster. Good people—admired political and community leaders, as well as ordinary, hard-working blue-collar folk—were economically ruined overnight. The collapse was like a house of cards: businesses failed; factories shut down; people lost jobs; banks repossessed properties for which loan payments were not

being made; then many of the banks failed, taking the savings of ordinary and extraordinary people alike down with them. The poverty rate in the United States was tremendous, and older people were among the very least competitive applicants for whatever jobs were available. The extent of economic interdependency was clearly driven home to the average U.S. voter: it indeed might take more than hard work, good values, and trust in the institution of banking to protect oneself economically, and society as a whole had an obligation to help out those who were most desperate.

Social Security began in 1935 to counteract the effects of the Great Depression. It was designed to provide, in the words of President Roosevelt, "some measure of protection. . . against poverty-ridden old age." The intent of the first Social Security Act was to provide a safety net for those who were most vulnerable to desperate poverty. It was assumed that pensions and personal savings would support people in their later years; however, in 1994—about 60 years after the act was first passed—40 percent of all income received by older people was from Social Security (Quadagno, 2005).

Social Security was designed to supplement pensions or retirement savings, rather than to provide the total income of aged people. Although everyone seems to agree that Social Security should not be and was never designed to be the sole source of retirement income, for many elders it continues to be just that. If it were not for Social Security benefits, nearly half of our aged would live in poverty. In general the lower one's income level, the more important Social Security becomes as a component of the household budget. It is the most common source of income for all races. For either blacks or Hispanics, Social Security is much less likely to be accompanied by other forms of income. Obviously, for the individual with no income from earnings, savings, stock and bonds (asset income), or pensions, Social Security is a much higher percentage of the total income. Under these circumstances, there is no universal agreement on the adequacy of the benefits or the appropriate level of benefits.

In 2004, for one-third of Americans over 65, Social Security benefits constituted 90 percent of their income (Social Security Administration, 2004). Presently, with the global economic downturn, Social Security is of even greater importance as a safety net to older adults living in poverty or near-poverty. Americans of all generations, however, report that political leaders are out of touch with the public's views on Social Security. The media is also criticized for poor analysis of the benefit and proposed changes to it. In 1998, 82 percent of people aged 50 and older felt that Social Security needed to be a primary policy focus for the nation (Young, Old Differ, 1998). Lack of adequate, calm, accurate public analysis and discussion of these issues impedes and politicizes the development of policy and planning.

At the end of 2008, almost 51 million people received Social Security benefits: over 35 million retired workers and their dependents, 6 million survivors of deceased workers, and 9 million disabled workers and their dependents. Total benefits paid in 2008 were $615 billion; total income was $805 billion, plus assets held in special U.S. Treasury securities valued at $2.4 trillion (OASI, 2009). These combined assets will continue to rise; however, annual costs will exceed the tax income by 2016, requiring the use of the Treasury securities to fund the assets (OASI, 2009).

The United States was one of the last industrial nations to introduce a major public system of social insurance against old-age dependency. By 1935, 27 other countries already had well-developed national retirement systems. The Social Security Act of 1935 required that employer and employee each pay into the Social Security fund 1 percent on the first $3,000 of the worker's wages. (In 2007 workers paid 6.2 percent of their paychecks, up to $97,500, into Social Security.) Congress eventually amended the act to extend benefits on the basis of social need in addition to earned right. Amendments adopted in 1939 made certain dependents of retired workers and survivors of deceased workers eligible, too. In 1960, amendments extended benefits to all disabled persons. Over the years, the amount of money that a retired person may earn before forfeiting part of the benefit has increased.

Fewer Workers to Support More Retirees

Social Security operates on a principle of sharing: some individuals die before they are able to collect benefits, whereas others live long enough to get back much more than they contributed. The ledger supposedly balances. However, the increased percentage of older adults and their increased life expectancy is one reason for concern about Social Security solvency.

In 1945, for instance, 42 workers paying into Social Security supported one retiree, whereas today just three active taxpayers support one retiree. In about 30 years, the baby-boom generation will begin to retire, creating even greater strains. In the year 2035, only two workers will support every retiree. Most citizens believe that their payroll taxes go into a kind of insurance pool, which is held and invested until the day they retire. In fact, Social Security taxes are transferred almost immediately to retirees: the system takes money from workers and hands it to retirees. With longevity, Americans have been receiving more than they and their employers ever paid into the system.

Social Security Adjustments

Social Security benefits rise with the cost of living. This change was brought about by Social Security amendments instituted in 1972. Under the current law, this **cost-of-living adjustment (COLA)** occurs automatically whenever the Consumer Price Index increases in the first quarter of the year (in comparison with the same period in the previous year). Effective January 1995, for example, Social Security benefits increased 3 percent based on the 1994 increase in the Consumer Price Index from the previous year. However, even with the COLA provision, many who now receive Social Security are still living below the poverty line.

How to keep Social Security solvent is a major social issue. Some changes in Social Security have been made to restore the system's fiscal health. In 1983, compromise amendments were painfully worked out by a bipartisan committee and were subsequently adopted by Congress. For example, since 1988 college students between the ages of 18 and 22 can no longer draw benefits on a deceased parent who paid into Social Security. The practices designed to protect the Social Security system include:

1. Accelerating Social Security tax rate increases.
2. Gradually raising the eligibility age for full benefits (in the year 2000, the retirement age began its increase from 65 to 67). This change affects people born in 1938 or later. For example, those born between 1943 and 1954 will not receive full benefits until age 66.
3. Substantially increasing tax rates for the self-employed.
4. Levying an income tax on higher-income Social Security beneficiaries.
5. Expanding coverage to newly hired federal employees.
6. Creating sharper decreases in starting benefits for early retirees—a 30 percent reduction at age 62 began in 2000.

Changes in the system are necessary to keep Social Security from defaulting by 2037. Adjustments in place include a gradual increase in the age of full retirement. The initial age of retirement, for people born before 1937, was 65. For people born in 1943 the full retirement age is 66, and it will increase gradually each year until it reaches age 67 for people born in 1960 or later. In 2008 and 2009, people retiring at

The social responsibility for older and younger generations is different. This girl will attend public school for many years to come; her grandfather's income is supplemented by Social Security.

full retirement age had no limits on the income they could earn in addition to Social Security. Those choosing to receive their benefits before reaching their full retirement age are entitled to receive all their benefits if their earnings are below a specific amount at which time for every $3 above the limit, $1 is deducted from the benefits. In 2008 the maximum earnings was $36,120, and in 2009 it was $37,680 (OASI, 2009).

The question of how close to bankruptcy the Social Security system is is more complex than it appears to be. Solvency depends on future income and expenditures based on the size and characteristics of the population receiving benefits, the size of the workforce and workforce earnings, and the level of monthly benefits amounts (OASI, 2009). Social Security is said to be heading for bankruptcy because it uses earmarked taxes and has a trust fund or Office of Old-Age and Survivors Insurance (OASI). As Figure 9.2 indicates, in 2008 the fund was adequately financed for the next 10 years, although annual costs will exceed tax income beginning in 2016. Much of the baby-boom generation will be receiving payments by that date. People are living longer, and so payments are made for more years than previously. Additionally, after the 1960s couples began having fewer children—a smaller generation of workers to pay Social Security benefits (OASI, 2009). The Social Security Administration projects the cost rate to rise rap-idly from about 2012 through 2030 because of baby-boom generation retirements, which will cause the number of beneficiaries to rise more quickly than the labor force. Thereafter, however, the cost rate will rise slowly for about five years, and then it will remain fairly stable for the next 25 years (OASI, 2009). Long-range planning is clear and critical—and is a central political issue. Social Security, however, was designed to be free from year-to-year political winds, and therefore designed as a "pay-as-you-go" system.

Social Security will become bankrupt and unable to pay benefits only if taxpayers are unwilling to raise taxes to pay for those benefits—the benefits that "spread the cost." An alternative to raising taxes would be to reduce benefits by increasing the number of years

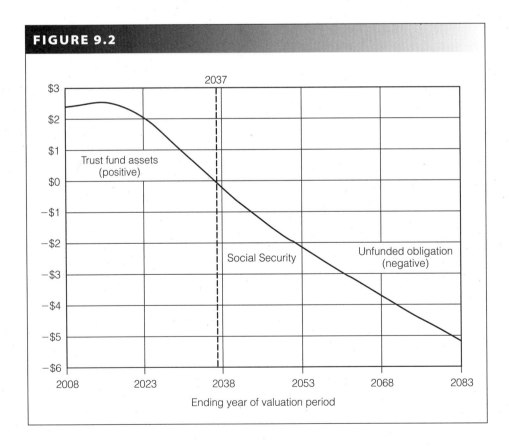

FIGURE 9.2

people must work before receiving payments, for example. Yet another alternative would be to introduce means testing. An *affluence test* would reduce benefits to higher-income individuals to allow for more adequate benefits to lower-income people. This model appeals to many Americans, but economists fear that it would discourage savings—for "income" includes all an individual's assets. Consider, for example, a situation in which both persons 1 and 2 earned the same salaries over their lifetimes, but their standards of living were very different. Person 1 was the last of the big-time spenders, and person 2 saved and scrimped and had a substantial savings account. Through means testing, person 1 would qualify for Social Security, whereas person 2 would not.

A more radical suggestion is to eliminate Social Security all together and instead require mandatory savings: to privatize Social Security. This plan would replace today's pay-as-you-go program with a funded system of individual accounts invested in the stock market (Moody, 2000). Proposals to privatize only a portion of Social Security have also been designed. The fundamental argument for privatizing is the belief that investment in the private sector will exceed the value of future Social Security benefits. Following the economic crash of 2007, however, the public has shown less interest in having no government-guaranteed safety net for average middle-class Americans.

The unknowns are enormous: to completely privatize is to tell citizens, "You're on your own," rather than to spread the risk through an insurance program model (Social Security) and help one another. Regulations controlling sales commissions on investments by workers might also reassure the public.

The United States is not alone in its dilemma to provide old-age security for its citizens. While the ratio of people aged 65 and over to those 20 to 64 will rise from 21 percent in 1990 to 36 percent in 2030 in the United States, it will rise in Germany from 24 percent to 54 percent, and France is close behind that proportion (Feldstein, 1997):

> *The Organization for Economic Cooperation and Development projects that government retirement benefits (excluding health costs) will exceed 16 percent of the GDP of Germany, France, and Italy by 2030, compared with about 7 percent of GDP in the United States (Feldstein, 1997, p. 25).*

The bottom line in the emerging debate is what sort of social contract the nation wants and to what extent does a nation consider itself responsible for the well-being of its citizens.

Inequities in Social Security

Under Social Security, people under age 66 and between ages of 66 and 69 are penalized for working, although after age 70 there is no penalty on wages earned. Earned income is subject to both income and Social Security taxes, although Social Security payments are subject to neither. Nor is there a penalty for income earned through capital gains or interest on savings. In the end, the system is more beneficial to those who have money than to those who do not; however, it is far more critical to the economic quality of life for those with fewer resources.

Salaried workers are penalized the most by the present system. Older workers who need money often settle for hourly part-time jobs at minimum wage or even less, so that their wages will not be docked. In fact, most federal government employment programs for elders provide income only as a supplement to Social Security and pay minimum wage or less. Military, federal, and state government and other pensioned retirees, however, generally face no penalty against their pension earnings after retirement, because salaries and wages are the only income subject to the penalty.

Adequate income for later life is critical, but savings and assets are unequally divided in society. For those with adequate resources old age can be a profoundly rewarding time.

Before complete retirement, some employees receive retirement benefits (other than Social Security) from one job while receiving full salary from another—a practice called "double dipping." The end result of one of the changes in Social Security—an income tax levied on higher-income Social Security beneficiaries—is that about 20 percent of people who get Social Security have to pay taxes on their benefits. This provision affects the more well-to-do, who typically have income from a variety of sources such as salaries, interest from savings, and dividends from stocks and bonds.

Congress has considered reforming the Social Security law to eliminate the ceiling on earnings. However, this change has never been enacted. The reluctance of Congress to remove all limits on extra benefit income is due in part to uncertainty as to the effects of such a move. Congress has reasoned that an older person who could earn more than the yearly amount of unpenalized income probably does not need full Social Security benefits, thus reducing the strain on the system for the benefit of others. But this ruling works a hardship on those making a very modest living. Further, if there were no ceiling, it might reduce the number of people opting for early retirement by keeping them employed full time. All these would ease the strain on the system.

Still another criticism of Social Security involves the two-career family. The law favors one-earner households. Working couples pay up to twice as much in Social Security taxes as single-earner families, yet they seldom collect twice as much in benefits. A woman who has contributed throughout her working years may find that, because of her lower earning capacity, she may fare better with 50 percent of her husband's benefits than with 100 percent of her own. (She is not entitled to both.) In essence, many working women receive none of the money they paid into Social Security. Another problem is that divorced men and women, particularly those who have not worked outside the home, may be left totally unprotected. To draw on an ex-spouse's benefits, a man or woman must have been married to that spouse for 10 years. For example, a woman married to one or more husbands, each for nine years, who has not paid into Social Security, is not eligible for any benefits. A man or woman married 10 years to one person and 10 years to another would not qualify for benefits from both. Divorced women suffer more than divorced men because they are more likely to have worked at home and to have not paid into Social Security.

Assistance from Poverty

The definition of *poverty*, for purposes of social policy, is framed in terms of inadequacy, yet what is adequate might be *absolute* (having less than an objective minimum), *relative* (having less than others in society), or *subjective* (feeling one does not have enough to get along) (Hagenaars & de Vos, 1988). Poverty, then, must be understood in its political context, which is dynamic and changing over time. At one point in time, keeping citizens from starving to death might have been the bottom-line social objective (absolute adequacy) for a society. At another time, providing nutritious food, adequate safe shelter, and access to information for decision making is the bottom-line objective (relative adequacy). The difference in these two values is primarily economic—economic strength and stability drives social policy, and to some extent, social values. It is an important concept to bear in mind while making the difficult trade-offs in cutting the budget pie.

The poverty threshold in the United States was set in the 1960s by estimating a minimally adequate food budget and multiplying it by a factor of three, because research from the 1950s indicated that families spent about one-third of their income on food. Since the 1960s the threshold has been modified in prices, but not for standard of living (Meyer & Bartolomei-Hill, 1994). Since the 1990s the standard of living has included a broad range of American families planning to send children to college. This aspiration, based on *subjective* adequacy, has been supported by government programs because of the value of a well-educated populous to the greater good.

A federal program formerly known as Old Age Assistance, which was intended to aid those who were not covered adequately by Social Security, was renamed **Supplemental Security Income (SSI)** in 1974. Instead of being managed by local offices, it is run by the Social Security Administration. Funds, however, are not provided by Social Security, but by general U.S. Treasury funds. The program provides a minimum income for elders, the blind, and the disabled by supplementing Social Security benefits if they are below the amount stipulated by SSI. In essence, it is a guaranteed annual income, which is very important for those who have little or no Social Security. The monthly amounts vary from state to state.

The federal SSI program makes no allowances for special circumstances. The amount is clearly defined, calculated by subtracting the income of an individual or a couple from a guaranteed level. In addition to income, individuals must pass an asset test: couples have countable assets (those that "count" in the defined allowances) of less than $2,500, and singles must have less than $1,500 to be considered eligible. The value of an owner-occupied home, a car needed for transportation to medical treatment or work, life insurance valued at less than $1,500, and personal property or household furnishings valued at less than $2,000 (McGarry, 1996) are excluded in this test.

The SSI program is designed to be a safety net for the poorest segment of the population. Although it is called "supplemental," benefits decrease substantially if the person or couple has other sources of income. This program has been strongly criticized because it does not consider differences in the cost of living between states and because it allows for no flexibility for special circumstances. It has been estimated that only 56 percent of older people who are eligible for SSI presently receive benefits (McGarry, 1996).

Retirement Winners and Losers

Social Security payments are determined according to detailed guidelines and legislated formulas and are not based solely on need, unlike SSI payments. Depending on a person's personal circumstances or work history, the amount of a monthly Social Security check can be adequate or inadequate. Here are some hypothetical examples of people who fared differently in retirement:

Once relatively well off, Ed earned $75,000 per year as a chemical engineer. But he never stayed long enough at one job to qualify for a private pension. He spent his full salary every year on his wife and four children, hobbies and travel, and their home. He retired at age 65, with a Social Security benefit of $1,733 per month, making his annual income $18,396. How will he handle the giant financial step downward?

Jim, a former construction worker, earned an average yearly salary of $30,000; so did his wife, Jane, a teacher. Both retired in 1998. Their combined monthly Social Security check is $2,028; their union pension checks provide another $1,500; they also have a joint savings account that yields $600 in monthly interest. Now they earn almost as much as they did while working. (Their combined working salary after taxes, was $50,000). They are now making $49,536 annually and are moving to a condo in Hawaii.

Former Senator Calvin Bigg was in the Air Force for 20 years before joining a law firm and then entering politics. He has an annual $25,000 military pension; for his years as a lawyer, he qualifies for Social Security; and he qualifies for a civil service pension for his years of public service as senator. His three combined pensions total more than $80,000 a year, not including interest on his savings and dividends from real estate holdings, which bring his yearly total to $130,000.

Wilona still misses her husband who died in 1980. She was age 61 at the time and eligible for Social Security benefits as his widow. In 2000, she was 81 years old, she is healthy, and she wants to go on some boat cruises. But her yearly income, in spite of cost-of-living adjustments (COLA), is less than $6,000 a year—income hardly sufficient for living, let alone traveling.

At its inception, Social Security was meant to provide a stable future and a safety net for all citizens. It was never meant to be the sole source of income, which unfortunately it is for many older Americans.

Medical Expenses in Old Age

Medical expenses rise with age. Although elders represent only 13 percent of the population, they spend 25 to 30 percent of the money used for health care. Contrary to popular belief, Medicare and Medicaid do not cover all medical expenses when one is over age 65. Actually, these programs do not cover many chronic conditions, leaving the individual to pay for treatment. The *Medicare Handbook* lists the following expenses as not covered:

- Dental care and dentures
- Over-the-counter drugs and most prescribed medicine
- Eyeglasses and eye examinations
- Hearing aids and hearing examinations
- Routine foot care
- Most immunization shots
- Custodial care in the home
- Custodial long-term care (nursing homes)

Examine the list carefully. The program's coverage excludes most parts of the body that tend to change with age and require care. As medical costs rise, one's standard of living declines.

Medicare and Medicaid

Medicare is a part of the Social Security system; those who are eligible for Social Security benefits are also eligible for Medicare. The vast majority of those people are older adults.

Medicare is the main insurer of people over 65 years of age: 95 percent of older persons are covered by Medicare hospital insurance (Part A), and 98 percent of those have supplementary medical insurance (Part B) as well (Cowan & Hartman, 2005). Medicare requires deductibles and 20 percent coinsurance for most services, which must be paid for by **Medicaid**, if the individual meets the financial criteria; supplemental private insurance (Medigap) purchased by the individual; or the individuals themselves. Medicaid has traditionally provided medical coverage for poor and disabled people.

Two primary interacting factors cause concern for the solvency of Medicare: the growth of the over-65 population and medical costs. Total number of Medicare beneficiaries in 2007 was just under 45 million. Medicaid recipients increased by nearly 300 percent (including an increase in dependent children). The number of **dually entitled** persons (people covered by both Medicare and Medicaid) amounted to 21 percent of those 45 million Medicare recipients (State Health Facts, 2006).

Steadily rising medical costs threaten both the Social Security system and the older adults. Some examples are as follows: skilled nursing facility payments for Medicare increased from $7.1 billion in 1994 to $9.1 billion in 1995—an increase of 28.2 percent, and due in no small part to longevity. Home health agency benefit payments for Medicare grew from $12 billion in 1994 to $15.1 billion in 1995—an increase of 25.8 percent. Medicare hospice expenditures have significantly increased, from $1.4 billion in 1994 to $1.9 billion in 1995—an increase of 35.7 percent (Health Care Financing, 1996).

In 1967 national health costs, including Medicaid, were $51 billion, or 6.3 percent of the gross national product. By 1995, total Health Care Financing Administration (HCFA) outlays were $248.9 billion—16.4 percent of the federal budget. Although older people are increasingly considered to be the cause of the national budget deficit, nearly 20 million

dependent children under 21 received Medicaid benefits in 2000. Dependent children represented 51.5 percent of the Medicaid budget in 1995. The five million people over 65 receiving Medicaid represent only a little over 12 percent of all recipients.

Medicare has two parts. Part A, called Hospital Insurance, helps pay for inpatient care you get in a hospital, skilled nursing facility, or hospice, and for home health care under certain conditions (CMS, 2008). Most people on Social

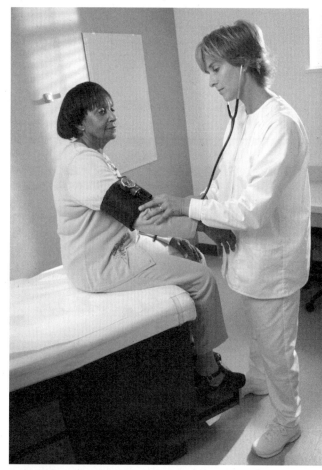

Adequate medical care is crucial throughout life. Medicare helps older adults with acute conditions but chronic health care needs remain unaddressed by policy.

Security pay no premiums for Part A because they or their spouse paid Medicare taxes while employed. It covers hospitalization for up to 90 days plus 60 "lifetime reserve days" that can be used to extend the 90-day covered period. When the days have expired, a per-day charge is assessed. All premiums, deductibles, and co-insurance amounts change every year on January 1st.

If inpatient skilled nursing or rehabilitation services are necessary after a hospital stay, Medicare pays for up to 100 days. Medicare does not pay for "custodial" care given in a facility that lacks appropriate medical staff—it is designed for a "spell of illness," or to assist with acute illness, not chronic conditions. Some hospice care is also covered. If a recipient is confined at home, Medicare can also pay for limited visits from home health workers.

Medicare Part B, called Medical Insurance, consists of optional major medical insurance for which the individual pays a monthly fee. Over 98 percent of Medicare beneficiaries select Part B as well. Medical insurance helps pay for necessary doctor's office visits, outpatient services for physical and speech problems, and many medical services and supplies that are not covered by Part A. It also pays for some preventive services such as flu shots. Most people pay the standard monthly Medicare Part B premium, which was $96.40 in 2009 (CMS, 2009).

To qualify for Medicare payments under Part A, hospital care must be "reasonable and necessary for an illness or injury." Congress established Medicare to pay a restricted portion of a citizen's medical expenses resulting from only the most serious illnesses. That is why the most common health needs of seniors, such as drugs and dentures, are not covered. In addition, the program places both a dollar limit and a time limit on what is covered. Recall, for instance, that Part A has a deductible that the patient must pay. Other restrictions and limitations get rather elaborate. Suffice it to say that many frail, sick older people pay thousands of their own dollars for medical care. Medicare does not usually cover all expenses, especially for long-term, chronic conditions, and private insurance plans such as Blue Cross, Kaiser, and EmCare cover only portions of either office visits or a hospital stay. Therefore medical expenses can get costly, and an older adult may be just a major illness away from financial disaster. Over the past several years, Medicare has "tightened" its system covering less, not more, of hospital and medical costs. The patient or some other source pays the difference between what the doctor charges and what Medicare pays the doctor.

Of the nearly 45 million Medicare beneficiaries, about 90 percent are using a traditional **fee-for-service** model. They pick their doctors, and Medicare pays a set fee for each service. Most beneficiaries also purchase private supplemental insurance to pay for uncovered costs. It can become an expensive proposition, even with Medicare assistance. The Public Policy Institute of the American Association of Retired Persons found in 1998 that over 2 million elderly people at or below poverty line were using 50 percent of their yearly income or more for medical expenses that were not covered by Medicare (Stamper, 1998). These expenditures went for such things as prescription drugs, outpatient care, and premiums for private supplemental insurance. Additionally, adults whose incomes already place them in the poorest 20 percent of the population are two-and-a-half times more likely to become disabled in any one year than those in the top 20 percent (Vaux, 2004).

One alternative to a fee-for-service plan is the **health maintenance organization (HMO),** also called managed care. Older adults are attracted to the HMO model because they often pay drug benefits that other plans do not cover. On average people 65 and older spend 19 percent of their income on health care although HMOs have much lower out-of-pocket costs (Crystal et al., 2000).

In the managed-care model, elders can use only the doctors and facilities on a limited list, but there are often additional benefits such as a more broad range of coverage for pharmaceuticals and the convenience of avoiding a confusing bureaucratic billing system. Concerns with the HMO model have been that access to needed care will be restricted as a method of controlling costs. Some people fear that more costly services will be rationed, resulting in unequal access. The authors of one major study on health provision models concluded that access to home health services is particularly problematic in the managed-care model and that HMOs need to address the needs of frail and more vulnerable Medicare beneficiaries more effectively (Nelson et al., 1997). As the United States struggles to define its position on the provision of health care for its citizens, fears about the fairness of services, the role of profit, and accessibility of care have become loudly expressed.

A third option for Medicare services is called **point of service**, in which beneficiaries can visit doctors outside the network but must pay an additional cost. Another option that is growing in popularity is the **preferred provider organization (PPO)**, in which the beneficiary can visit any doctor in the health care network without a referral or see doctors outside the network at an additional cost. Medicare pays the PPO a set fee for all services for a set period of time.

The **provider-sponsored organization (PSO)** is quite new. Owned and operated by doctors and hospitals (as opposed to health insurance companies in the HMO model), PSOs are organizations similar to managed-care plans in which Medicare pays the health plan a monthly fee for each recipient. In the **Medicare medical savings account**, which was not put into effect until around 1999, a high-deductible insurance policy provided by Medicare is combined with a tax-free medical savings account to help pay the deductible. At the end of the year, the beneficiary keeps any unused Medicare money.

The final option for Medicare services is the **privately contracted fee for service**, in which the beneficiary may visit any doctor or purchase

any health plan, but pays extra for uncovered or expensive services.

Beyond question, those with greater financial resource have a broader range of options for health care services in later life. Because Medicare is an **entitlement** (provided categorically, not need based), people with higher income pay the same as those with limited income in terms of dollars, but the proportion of their income is much less.

Extremely low-income individuals who qualify for SSI are also eligible for Medicaid in most states in the nation. Two groups of people qualify: the categorically eligible and the medically needy. The categorically eligible are people with certain characteristics (aged, blind, disabled, or a member of a family with dependent children that is headed by a female or in certain cases by an unemployed male) (Cowan & Hartman, 2005). The medically eligible must have a strictly limited income including savings but excluding one's home and automobile, household goods, and $1,500 in burial funds. For those who qualify, Medicaid helps pay for a wide variety of hospital and other services. Medicaid often pays for services that Medicare does not cover, such as eyeglasses, dental care, prescription drugs, and long-term nursing home care.

Medigap Policies

"Medigap" insurance policies are popular among older people who can afford them. In spite of continually rising policy costs, a majority of those on Medicare are estimated to have a **Medigap policy** as well. Such policies are sold by private companies to help cover the "gaps" in health care protection for which Medicare does not provide. Prior to 1992, insurance carriers could sell any benefits they chose as Medigap so long as the minimum benefit requirements were met. The range and costs were enormous. In 1992, federal legislation was implemented to standardize policy benefits. Any company that sells Medigap coverage must offer policy

A (a specified package of core benefits). Policy A is the starting point for policies B through J, with policy J having the most comprehensive coverage.

The required core benefits include:

- Daily Medicare co-payment of a specified amount for hospital days 61 through 90, adjusted upwardly for days 91 through 150
- Full cost of up to 365 additional hospital days during your lifetime
- Twenty percent co-payment of Medicare's allowed amount for doctor charges

The policy level, as well as the company providing the policy, needs to be carefully chosen by the consumer because the range of choices can be confusing. For example, policy C appeals to many seniors, but the cost from most carriers is often greater than the benefit. Policy H might be best if significant prescription drug expenses can be anticipated, but the cost of the policy must be balanced with its benefit. For the first six months after signing up for Part B of Medicare, a person can buy *any* Medigap policy sold in their area. After that time, the insurance provider has the option to turn down an applicant.

Consumers will find wild price differences for this insurance, even within the same region. For example, the cost of policy A ranged from $436 to $1,032 for a 65-year-old man in California in 1998 (Roha, 1998). Advice to people shopping for a Medigap policy includes:

1. Comparison shop—review policies from at least three insurance carriers before purchasing.
2. Ask for current premiums for 70-, 75-, 80-, and 85-year olds—some policies increase faster with age of beneficiary than others.
3. Inquire about a Medicare Select plan— a hybrid managed-care/fee-for-service system offering premiums reduced from 10 percent to 20 percent in exchange for giving up some choice of health care providers, especially hospitals.

No Place to Go

The aging population is the major determinant of increasing nursing home admissions: the number of elderly requiring long-term care is expected to double by 2025. The proportion of nursing home residents over 65 is 46 people per 1,000; and that increases to 220 per 1,000 among those 85 and over (Hicks et al., 1997). In the next 30 years, if those proportions double to meet the growing needs, a substantial, if not alarming, increase in costs and need for facilities will result. The out-of-pocket costs for nursing home care have also dramatically increased. In 1980, the nursing home industry consumed 8.1 percent of all spending for personal health care services, a total of $17.6 billion. In 1994, those costs came to $72.3 billion and accounted for 8.7 percent of all personal health care expenditures (Levitt et al., 1996).

For many older Americans, the cost of long-term health care is insurmountable. In a few months a couple can use up all their assets. If their income is sufficiently low, they can apply for Medicaid to cover nursing home care. This is a devastating experience, and one that more commonly occurs to the oldest-old (those aged 80 and over), to women, and to those living alone.

In terms of medical coverage, the forgotten minority are those elders whose income is just above the Medicaid cutoff, yet too low to cover the cost of nursing home care. Medicaid is administered by states; eligibility and dollars per month coverage vary widely. In Delaware, one of the best states for coverage, nursing homes are well reimbursed; therefore, good homes are available and willing to take even "heavy-care" patients. In contrast, a study of working- or middle-class Florida residents just above the Medicaid level of eligibility found that these elders were experiencing severe emotional, physical, and financial hardships trying to cope at home. Nursing home care, unavailable or unaffordable, was urgently needed for many (Quadagno, 2005).

With no adequate government programs to pay for nursing home care, older patients suffer, as do hospitals and frustrated care providers. Many low-income elderly patients who cannot get into nursing homes remain in hospitals even if they no longer need the expensive hospital care. Hospitals that keep older patients for longer periods than Medicare or Medicaid fund rack up huge losses. The "boarder elderly" problem affects about 25 percent of the nation's hospitals, with urban hospitals hurt the most. Seventy-five percent of hospitals say that finding nursing homes for their "boarder" patients is difficult. Although "buyers" regularly visit some hospitals shopping for "profitable" elders on Medicare or Medicaid to fill their nursing homes, they have trouble finding them. A patient like 70-year-old Lucille, who suffers from seizures and is on a ventilator, does not interest them. The reimbursement for heavy-care patients does not meet the true cost of care, so nursing homes limit the heavy-care patients they take. Lucille is left to fend for herself; there is no real safety net for her.

The message here for older people is this: do not stay too long in a hospital or nursing home unless you can pay for it yourself. If you are poor and have no government subsidy, nobody wants you and there is nowhere to go.

Private Pensions

A pension is money received upon retirement from funds into which an individual usually has paid while working. Some employers have plans to which the employee contributes; others have plans to which the employee does not have to contribute. An employee may have both a pension plan and Social Security or either of these programs. On the average, **private pensions** provide a much lower percentage of the income of elders than does Social Security.

Trends in Private Pensions

Private retirement plans created in the 1980s, such as Individual Retirement Accounts (IRAs) and 401(k) pension plans, have become popular. In both plans, the benefit amounts are proportionate to the amount of money the employee contributes (and the employer matches in the case of a 401(k) plan), as well as by how long the money is invested in the plan at the bank or with the employer. These plans transfer more of the retirement costs directly to the individual employee. For cost-cutting reasons, U.S. employers have, in droves, eliminated traditional pension plans in which they footed the bill. These plans guaranteed retirees a specific income based on salary and length of employment. Many employers offer no substitute; others offer 401(k) plans or similar types of individual plans (IRAs), which may or may not have a limited contribution by the employer. And they do not guarantee a monthly income for one's retirement life. Such studies show that the current generation of workers is not going to have large retirement benefits unless they save their own money. There is also a definite shift away from fully paid health care for retirees by private employers.

Tax incentives further encouraged this avenue of retirement planning. However, the Tax Reform Act of 1986 eliminated some tax advantages, and uncertain interest rates (on which IRAs are based) have made these alternatives less reliable over the long term. When interest rates were high in the 1980s, the IRA and 401(k) (along with other savings plans, such as certificates of deposit) were seen as good sources of supplementary income to Social Security or employer pension plans. When interest rates came down in the early 1990s, these saving plans did not seem as good or as reliable as they were in the 1980s. When interest rates as well as stock market prices dipped to extremely low levels in the deep recession of 2008, many people's retirement investments lost up to 40 percent of their value. For those on the cusp of retirement (the boomer generation), those investment losses have caused careful reconsideration on the timing of retirement (Morin, 2009).

In any case whatever is in the account when the worker retires is what he or she gets. And if it runs out before the retiree dies, too bad. One good thing is that a worker who changes jobs can take the IRA or 401(k) along—they are portable—unlike the company-sponsored pension.

Problems with Pensions

Numerous gaps exist in private pension plan coverage. Temporary or part-time employees do not typically have a pension plan offered by their employers. Some companies simply have no pension plans. And to be eligible for pension benefits in many companies, the individual must have worked for a considerable number of years. Thus, the person who changes jobs every seven years might work for 40 or more years but accrue very small or no pension benefits from any job. The worker who changes jobs several times will typically get less in total pension income than a long-term worker, even if both earned the same pay, worked the same number of years, and were in similar pension plans.

In spite of a pension reform bill passed in 1974, many loopholes in pension plans continue. Before the Pension Reform Act of 1974, a person who lost a job before retirement might lose all the money he or she had paid into the company's pension fund. The Employee Retirement Security Act of 1974 established minimum funding for private pension plans and offered termination insurance to protect employees if their company ended the plan for any reason.

Congressional investigation of private pension plans has revealed that many workers, after a lifetime of labor rendered on the promise of a future pension, found that their expectations were not realized. In some instances, people were laid off in their late 50s or early 60s, just previous to their retirement, losing their pensions in the process. In other instances, corporations changed ownership—a small chain of independently owned grocery stores, for example, rotated ownership among key family members every eight years or so. Each "sale" provided a legal loophole for the company to eliminate or otherwise alter the pension fund. As more giant companies fall on hard times, retirees face a wave of broken promises. When Pan Am Airlines entered Chapter 11 bankruptcy, the pension plan was terminated. The federal government stepped in to cover employees according to the Pension Reform Act of 1974, but determined that they were obligated to pay workers only $596 per month of the $1,000 per month they were expecting. Pan Am and many other financially strapped companies have taken the option to "underfund" their pension programs, intending to make up for it later. One such company is General Motors, one of America's oldest and most venerable businesses. GM laid off many thousands of workers in the deep recession of 2008, and its bankruptcy restructuring proceedings put retirement benefits for workers in an uncertain position. For those companies which go out of business—and there have been many—later never comes.

Typically, if a person leaves a job before 30 years of service or before reaching the retirement age the company specifies, benefits are drastically cut. There are other reasons for getting small pension benefits. Sometimes, a company will stipulate that an employee's years of service must be uninterrupted. For example, a truck driver who belonged to the Teamsters Union for 22 years and retired anticipated that he would receive pension benefits of $400 per month. Upon retirement, however, he was told that he was ineligible for a pension because he had been involuntarily laid off for three and a half months during one of his work years. Denied a pension he thought was rightfully his, this Teamster sued the union, claiming that he was never told of the pension plan risks. He won his case.

When not completely denied, benefits from a pension fund may be small because the recipients are either widows or widowers. Most private pension plans require the wage earner to sign a form, if he or she wishes to allocate survivor benefits—usually considerably less than those the employee would receive—to his or her spouse. An employee who refuses to do this need not inform his or her spouse, and the spouse may have no way of learning the details of the pension. The federal government and private companies, because of privacy laws, may not divulge pension information to anyone but the employee. A widowed homemaker, consequently, cannot always count on getting something from her husband's pension.

Some retirement plans base pensions heavily on wages earned during the final three years of the worker's employment. Thus, in spite of pension plan reforms, loopholes prevent many elders from receiving equitable benefits. For example, a teacher who worked full time for 30 years, but who worked on a half-time basis for the final three years, might draw a pension largely based on the salary received during those last three years. Thus, the teacher's pension benefit would be considerably less per month because he or she had worked part time rather than full time—a severe penalty for trying to ease into retirement. Teacher retirement differs from state to state, and teachers must investigate those provisions and restrictions. Anyone employed by an organization that offers a pension plan should become fully acquainted with the provisions of the plan.

Even after reform, current pension laws and provisions *do not* encourage job change. Yet, data from the field of occupational psychology indicate that with advancing technology, college graduates today will *have* to change careers, on an average, three times in their lives. What can be done? Young people, eager to get a job, rarely question or quibble about retirement policies, and too many young people still cannot imagine themselves getting old. However, young people *will* become old; and they must concern themselves with the retirement benefits of the jobs they accept. Perhaps, in time, all private plans will merge into one federal plan of Social Security, with each individual's account showing total earnings regardless of occupational mobility.

A potential problem with private pensions is that they do little to contend with inflation. How can companies offering pensions take inflation into account? This is a grave challenge for financial planners. An opposite problem in the 1990s was the recession and the dramatic lowering of interest rates. Many elders use the fixed interest on savings accounts to pay for living expenses. When interest rates are cut from say 10 percent to 5 percent, the income from the accounts is cut in half. Today's times bring fear and anxiety to retirees, which can only be alleviated by a certainty of financial security.

Gender differences are sharply evident when postretirement income is analyzed, underscoring the finding that old women are more likely to be poor than are old men. Whether a pension is received and in what amount play a major role in differentiating the life chances of old people. Opportunities and decisions made earlier in life ripple into late life. Each of these choices women make—to marry, to withdraw from the labor market, to raise children—increases her likelihood of having limited pension resources in her old age (DeViney, 1995).

Lifestyles of the Poor

What does poverty mean to you? You might say it is low spendable income, and that would certainly be true, but how does income translate to lifestyle? It is the housing in which you cannot live, the food you cannot afford, the stores in which you cannot shop, the medical care you cannot get, the entertainment you cannot enjoy, the items (color TV, VCR, car) you cannot possess, the clothing you cannot buy, the places

A Profile of Poverty

Rather than a sudden displacement from an upper- or middle-class economic situation to a lower class, the overall picture of the aged poor is typically a descent from a lifelong lower economic class to the lowest, culminating in total dependency on a government paycheck that is far too small.

The study of poverty among elders is more a woman's story—70 percent of the aged poor are female. Though averages show that the older poor female is typically over 75 years old and white, the poverty rate among African-American females is three times higher than white females.

A discussion of the poorest poor—the homeless—is given next.

The Homeless: An Urban Dilemma

Close to three-fourths of all older U.S. residents live in urban areas. Of these urban seniors, nearly half live in central cities. Although younger people are more likely to abandon a deteriorating city area, older people are more attached to their area; for both emotional and financial reasons they are not as likely to move to the suburbs. Urban issues such as crime, pollution, transportation, housing, and living costs in a large city are the issues of older people. The bulk of the older population in a central city lives in neighborhood communities, often racially and ethnically homogeneous—ethnic enclaves constituting people of European, Hispanic, or Asian ancestry.

The United States Interagency Council on Homelessness (USICH) reported in February 2008 an unprecedented documented decrease in street and chronic homelessness across the nation, based on locally reported data. Their report indicated that from 2005 to 2006, 20,000 people had moved from the streets (http://www.ich.gov/library 2008). About 1.6 million people used a homeless shelter or lived

A larger proportion of older adults live in urban settings, however not all urban settings are by any means blighted. The value of services, libraries, and opportunities for socialization, cannot be over estimated.

you cannot go, the gifts you cannot give, the holidays you cannot celebrate. These and other deprivations are the essence of poverty.

Both inner cities and rural areas contain disproportionately high shares of the older poor, yet the lifestyles of older adults can be more constrained than that of younger people because of health and mobility, as well as income-generation possibilities.

in transitional housing between October 2007 and September 2008: about the same number as the year before, and in January 2008 about 664,000 people were in homeless shelters or in the streets. That's a drop of about 7,500 from the year before, but the count was made just before the nation's economic woes were beginning (HRSA, 2008).

An American tragedy hit large cities with a jolt in the early 1980s: homelessness. The problem grew even faster in the 1990s. High inflation in the 1970s, economic policies favoring the rich, corruption in the savings and loan industry, budget crunches at all levels of government, and lack of low-cost housing all worked to swell the number of homeless people. The noose tightened around the desperately poor in the "greed decade" of the 1980s, during which time the dominant social paradigm became the corporate profit model. A strictly capitalistic profit model has no room for those who neither participate in production nor have an economic or familial safety net.

A new feature of poverty in the 1980s and 1990s became the large number of homeless women and children. Though homelessness received considerable media attention, formal studies of the homeless focus more on the young than on the old. Crane, in her review of three studies on homeless elderly people, found agreement that catastrophic causes for homelessness were rare. Risk factors were more commonly discharge from an institution or release from prison and exploitation and abandonment by family members. Each of the studies profiled older people who had "lifelong difficulties with other people, lack of a supporting family, poor education and lack of job skills, and personal problems such as alcohol abuse and criminal behavior" (Crane, 1994, p. 633). Additionally, among homeless elderly people in New York City, psychological trauma following a relationship breakdown or bereavement was listed as cause of homelessness.

The nation focused on the tragedy of homeless children and families, and by the turn of the century homeless elders were mostly a forgotten population. This decade has added older adults facing homelessness for the first time, or at risk of becoming homeless, to the population of chronically homeless adults in the midst of the country's economic recession. Today's housing crisis and job losses will inevitably play out in the nation's homeless shelters, and a certain proportion of those newly homeless will be older adults.

Older people who are homeless are more likely to have multiple medical problems. They generally lack social supports and are prone to depression, dementia, and other mental health problems. Although services exist for the homeless population, older adults find the barriers that homeless people face to be more difficult to overcome. In addition to their homelessness, older people are faced with physical challenges that all aging people have: poor mobility, diminished hearing or sight, various chronic health conditions. Older homeless people are at a greater risk for victimization and injury than are younger people. They are less able to defend themselves and frequently have monthly income sources such as Veterans' benefits, SSI, or Medicare. They constitute a tragically vulnerable population.

A Homeless and Elderly Report from Health and Human Services included the following description of older homelessness in America:

In New York City, he is a 78 year old man with a history of depression who has lived in one of the flophouses in the Bowery for over 20 years. On a cold winter night in Danbury CT, he is a 58 year old man who is sleeping on the streets. His left foot was recently amputated after suffering from frostbite. In Philadelphia, he is a 65 year old veteran who sleeps above a steaming manhole in the middle of the sidewalk in order to keep warm. And in San Francisco, she is an 82 year old woman who is preparing to leave her home of 22 years after the landlord, with plans to convert eh building into condominiums, illegally evicted her and others. She does not have family or friends from whom to seek help and will likely go to the local homeless shelter. (HSRA, 2008, p. 1)

Whether portrayed as derelicts, as victims of misfortune, or as people burdened by structural forces beyond their control, the image of homeless people as reflected in most court opinions is one of weakness, helplessness, and despair.

DANIELS, *BUFFALO LAW REVIEW*, 1997

A 1999 report on homelessness identified causes of homelessness for older adults as social isolation, living alone, lacking economic stability, substance abuse, and no family or social support (NCH, 1999). In Boston, a study of homeless elders revealed an average age of 77. Many of the homeless people in this study had lived on the streets an average of 12 years—they were chronically homeless before old age. This population had an average of four chronic diseases and suffered from chronic alcoholism, severe mental illness, or both. Most resisted all efforts to place them in nursing homes (Knox, 1989).

Poverty and homelessness, of course, are associated, and poverty is disproportionately an issue for women, African Americans, and Hispanic Americans. Older men and women suffer homelessness in essentially the same way. Often labeled "bag ladies," homeless women cannot afford even the most inexpensive shelter, often shun the confinement of an institution, and carry their few possessions with them, eating from garbage cans or wherever they can get a free meal. Their very existence says something profoundly tragic about the role of elders in our society. For older, homeless men "skid rows" are common living environments. Estimates show that one-fifth of the homeless

live in such areas. A skid-row resident is essentially homeless, even though he or she may not live exclusively on the streets. When they can afford it, they get cheap accommodations in run-down hotels or boarding houses. Typically, the men live on the street, periodically abuse alcohol or other drugs, may be physically or emotionally impaired, lack traditional social ties, and are poor.

An early study of 281 homeless men aged 50 and over in the Bowery area of New York City provides us with a look at survival strategies (Cohen, 1988; Cohen & Sokolovsky, 1989). The men were paid $10 upon the completion of a two- to three-hour interview. Fifty may not be very old, but in the Bowery men looked and acted 10 to 20 years older than they are. Of the final group interviewed, 177 lived in flophouses, 18 in apartments, and 86 on the street. The number of the oldest men living on the street was 78.

Perceptions of "home" are important. . . Some individuals living in marginal accommodation develop a sense of attachment and rootedness to their housing and perceive it as "home."

CRANE, 1994

The men had developed skills that enabled them to survive in a treacherous environment. One-half of the non-street men had spent two or more years at their current address, suggesting considerable stability. The median number of years residing in New York was 13. Approximately half of the street men had been living on the streets for three years. One had been on the streets for more than 25 years. Most had very little income. One man bragged, "Sometimes I live on two bucks a month." Their lives revolved around getting food, cigarette butts, some wine, a bench to sleep on, a bit of hygiene, and a degree of safety. One man said, "I'm really tired; I'm worn out. You walk back and forth from 23rd to 42nd Street. I must walk 30 miles a day." Income sources were welfare or SSI, friends or relatives, panhandling and hustling, and odd jobs. Quite a few had long work histories. Some sought food from garbage cans or obtained scraps from restaurants. The men had their friendships with one another, counting on one another for help in times of trouble. Crimes such as muggings were their biggest fear. Though many of these men were eligible for government aid of some type, they did not apply for it, relying first on themselves, second on one another, and last on local charities such as missions. The researchers suggest that flophouses offer a better chance of survival than streets and that supporting and strengthening local service groups would help these men. Project Bowery in lower Manhattan has been partially successful in providing assistance—food, clothing, job and housing referrals—to older residents, and has become a national model of accessibility to support services for older homeless people.

Marginal and poor elders also live in single room occupancy (SRO) hotels. The Tenderloin area of San Francisco fits the stereotyped image of the decaying central city. Although unique in some respects, it characterizes sections of all large cities. The Tenderloin is primarily a residential neighborhood where most inhabitants occupy cheap single rooms in apartment buildings, rooming houses, and hotels. The area seems to house society's discards; it is a dumping ground for prisons, hospitals, drug programs, and mental institutions.

Many residents are transients who drift in, stay awhile, and drift on. The most numerous and yet least visible of those who live in the 28-block area are the elderly; most of them live alone in small rooms. During the day, they go out on short walks and for meals. A few never make it past the hotel lobby, where they sit. Others barricade themselves in their rooms and are not seen for long periods of time. Very few leave the buildings at night.

Higher-priced rooms usually include a private bathroom. The cheaper rooms may include a sink and toilet, but more often the bathroom is down a dimly lit hallway. Many hotel managers take no responsibility for meeting special needs of older people; their job is simply to operate the hotel at a profit. Many of the buildings do not have cooking facilities. Some residential hotels maintained specifically for elders provide food on the "continental plan," which includes three meals per day, six days per week. On Sundays, only brunch is served.

Some of the residents are alcoholics; others are lifelong loners, detached from larger social networks. Both men and women, reluctant to trust, tend to present a "tough" image and protect their privacy. The death rate is high in the Tenderloin, with common causes of death including malnutrition, infection, and alcoholism, all of which hint at neglect, isolation, and loneliness. In spite of the negatives, some elders seem to like the independence and privacy that a hotel room in the Tenderloin can offer, much preferring their SRO quarters to an antiseptically clean room in a nursing home.

A survey of 485 aged SRO residents in New York City had similar findings. Many of the elderly residents had weak family ties, strong preferences for independence, and longstanding attachments to central city neighborhoods. The SRO hotel, although run down, was an acceptable solution for housing because more "standard" accommodations were not affordable (Crystal & Beck, 1992). Some are able to find meaning and purpose in life despite rough circumstances; others are not.

Many downtown urban areas have undergone urban renewal. The older hotels in these marginal areas have been torn down, and the residents, with their low incomes and limited housing options, face a crisis. For example, by the late 1980s the skid row in Chicago no longer existed. **SRO hotels** were eliminated. Although an alcohol treatment center was built, no affordable housing remained; and the residents left to seek cheap shelter in other areas.

Eliminating the skid row destroyed a community and a way of life. The area had been a settled community, in that 30 percent of its residents had lived there four years or more. And, counter to the skid-row aging-alcoholic image, 37 percent drank no alcohol. About 60 percent had the meager economic means to make independent life in an SRO hotel a viable living arrangement, and the inhabitants needed the proximity to employment that the urban downtown environment provided. They might have liked a more pleasant room and perhaps better food, but if the hotel had a TV, lobby, working elevator, housekeeping services, and was reasonably clean, they were satisfied with the convenient location, the autonomous lifestyle, and the affordability. But they had no power to resist redevelopment (Hoch & Slayton, 1989).

Jason DeParle (1994) provided an excellent summary of the history of SROs in the United States:

> They sprung up in downtowns during the early part of the century to house railroad workers and other transient laborers. Over time, many of them deteriorated into slum housing for drifters, drinkers, and the mentally ill. Then came developers, who, enticed by cheap land, razed the skid rows and replaced them with galleries and office towers. Over the past several decades, as many as a million cheap hotel rooms have been destroyed. (p. 53)

Some cities have begun to view their SROs as a community resource for the prevention of homelessness, rather than as a community liability that should be demolished (Shepard, 1997). Residents of SROs are, after all, only a short step away from living on the street. Perhaps the specter of people sleeping in doorways and camped out in alley-ways will become too great an ethical burden for the public-at-large, and society will see fit to fund social programs to address homelessness.

Rural Elders

Rural is defined by the U.S. Census Bureau as territory outside places of 2,500 or more inhabitants. It includes ranches, farms, other land, and towns that have lower than 2,500 people. The term *nonmetropolitan* is much more general and refers to counties that are not metropolitan. A metropolitan county (or counties) contains an urban area of 50,000 or more and the surrounding counties totaling at least 100,000 people.

Elders in rural areas represent all social classes. Some are owners of large farms or ranches; others own oil fields, factories, businesses, or have been professionals such as doctors, lawyers, and teachers. Their only commonality may be a love of the rural lifestyle, which in itself varies from the South to the Midwest to the North and East coasts and varies by population density. Some rural areas contain small towns; others, acres and acres of farms and ranches with no towns for miles.

In 2000, the 10 states in the United States with the highest poverty rates for people over 65 were Mississippi (20 percent), Tennessee (20 percent), the District of Columbia (19 percent), Arkansas (19 percent), South Carolina (19 percent), Louisiana (18 percent), Alabama (17 percent), North Carolina (16 percent), Georgia (16 percent), and South Dakota (15 percent). Note that with the exception of South Dakota, the rates are highest in the South. All areas except the District of Columbia are agricultural states and have large rural populations. Individuals who lived a marginal lifestyle as farm workers rather than owners tend to exist on Social Security as their only source of income when they retire, if they even have that.

The lifestyle of the rural poor is potentially more isolated than that of the urban poor. We imagine the rural elders living in comfortable homes close to neighbors and family in small towns, shopping at nearby stores, and busily involving themselves in the social activities of the community; but this is not necessarily true. The older person may live miles from town or from neighbors. Housing in rural areas is often more run down than in urban areas and more likely to be substandard. Lack of a formal transportation system and lower income impacts the older rural person's access to services, as well as to specialty health care services, friends and family members, and participation in social activities in more distant regions (HSRA, 2008).

Incomes for rural elders are often lower than those of their urban counterparts. Compared with urban elders, a greater percentage of rural elders are poor and an even higher percentage are **near-poor.** This is especially true in the South. Yet income in and of itself may not be the most appropriate indicator of this group's well-being. The rural elderly have a somewhat lower cost of living and tend to be more satisfied and better able to make do with what they have. Rural elders are likely to have lived 20 or more years in their communities; they have commonly developed low expectations with regard to services, both social and medical.

Rural and urban elders construe health differently. Evidence shows that education and marital status are apparently less related to a sense of life satisfaction among elderly rural women than are social factors over which they have no control, such as their standard of living or community changes (Butler & DePoy, 1996). One study concluded that rural women seem to feel less entitled to be satisfied with their lives (my lot in life) than do urban women of the same age. Expectations and comparisons, in other words, shape people's perception of quality of life, and female rural elders in this study, expected less.

Rural African Americans generally tend to be particularly disadvantaged economically. They have substantially lower incomes than their rural white counterparts. The disparity becomes even greater among those in the oldest-old category (HSRA, 2008).

The rural crisis of the last decade continues: small farmers are being forced to abandon farming because they cannot make a living. The cost of supplies and equipment has far outpaced the price increases for beef, pork, and grain. Young people have been leaving the farm life for years. Many towns and counties in the Midwest have been slowly losing population for the last 50 years.

Government-sponsored low-income housing is less available in rural areas. Rural residents are likely to migrate to cities simply to improve their housing circumstances. Yet retirement complexes in large cities often do not offer a viable way of life to longtime rural dwellers. They do not really want to sacrifice the advantages of rural living, such as peace and quiet and longtime friendships, for better medical care and a more modern home. Studies of rural life tend to document a marked resilience and strength. Interviews reveal rural elders to be self-sufficient, proud, and able to survive on small Social Security checks.

The devil wipes his tail with the poor man's pride.

JOHN RAY, ENGLISH PROVERBS

It is life near the bone, where it is sweetest.

H. D. THOREAUX, *WALDEN*

The Culture of Poverty

Location is only one way in which the lifestyles of the poor can vary. Poor people's degree of independence, pride, dignity, and happiness, and their ability to meet their physical needs and provide meaning in their lives, can also differ.

Most people in the United States find it difficult to think of poverty as a stable, persistent, ongoing condition. The **culture of poverty** refers to the abject hopelessness, despair, apathy, and alienation in poverty subcultures. This concept, which has been used to describe minority subcultures, has application to poor elders who are hopeless and despairing about their situation. Some have trouble living life with dignity, hope, happiness, and meaning. This tends to be more the case in inner cities where elders are fearful and alienated. Elders in rural areas seem to be more content with their lifestyle.

Relative Deprivation

According to the theory of **relative deprivation,** a person is deprived if his or her resources come up short in comparison with another's. This theory, which has been applied to financial adequacy, holds that the older person who compares his or her income with that of another and feels relatively deprived is not satisfied with the income regardless of its actual amount. Applied in another way, the older person who looks back on past income and considers the former a better income than the present one may also feel relatively deprived and therefore dissatisfied.

One study hypothesized that not only real income, but the older person's subjective interpretation of that income, affects financial satisfaction. The theory ties in with the symbolic interactionist definition of a situation: if people perceive a situation as real, its consequences become real. An older person can examine relative deprivation both in terms of comparison of oneself with others and in terms of comparison between one's present and past financial circumstances. Objective income influences financial satisfaction only indirectly: feelings of relative deprivation directly affect subjective financial well-being. Subjective feelings allow some to be satisfied with low incomes and others to be dissatisfied with high incomes.

A Piece of the Good Life

In reflecting on what old people deserve in later life, we must conclude that they deserve a fair share of the "good life"—respect, dignity, comfort, and resources for experiencing new or continuing pleasures. Actually, a select number of aged in our society's upper echelons are doing quite well, having amassed great assets over their lifetimes. But they account for only 5 to 6 percent of the population aged 65 and over. Many elders, especially those at the very bottom of the economic ladder, struggle to enjoy—or to simply get—their piece of the American pie.

One obstacle to attaining this good life is that older people typically have less income in retirement than when they were working. A rationale for this is that the older person is not doing anything; therefore, little money is necessary. However, this rationale is based on a negative image of nonactivity, which often becomes self-fulfilling. Actually, the retired person with plenty of leisure time needs an adequate sum of money to enjoy it. During our working years, we spend eight hours per day on the job; we spend more time traveling to and from work; and we spend still more time resting up from work. After the working years, one may justifiably need more money for pursuing leisure activities, such as travel, school, and hobbies. Elders without money most commonly fit the negative stereotypes of aging.

Businesses in cities and towns recognize the modest means of many older persons and offer discounts of all kinds. Local programs provide seniors with token assistance such as free banking service and discounts in theatres, restaurants, and transportation. In the Richmond, Virginia, area, for example, local merchants initiated a **senior discount program.** All elders who wanted to participate were issued photo identification cards. After nine months, over 600 merchants and over 19,000 older persons were participating. The AARP has a national discount directory, and some cities have a "Silver Pages" section of the Yellow Pages that lists special services and shops that cater to the "senior citizen."

Under senior discount programs, older people receive discounts from 10 to 15 percent on drugs, groceries, baked goods, taxi fares, haircuts and hairstyling, TV and radio repairs, auto tires, shoe repairs, restaurant meals, banking services, movie and theatre tickets, miniature golf, cassettes and compact discs, jewelry, hearing aids, dry cleaning, electrical work, furniture and appliances, hardware, flowers, books, clothing, auto repairs and supplies, art supplies, services at health spas, and pest extermination. Merchants, who offer a variety of explanations for their voluntary participation in such programs, often join because they believe such programs will promote their goods and services. In short, such programs indicate the awareness of the need to ease the financial burden on older persons. Maggie Kuhn, founder of the Gray Panthers in the mid-1970s, described contemporary programs for older people as "novocaine treatment" that dulls the pain without really changing anything. Other people are more optimistic, however. They feel that discount programs are a step in the right direction and are greatly appreciated by older people.

We might easily wonder why our society seems so unwilling to allow its members a comfortable passage through middle and old age. In *Critical Perspectives on Aging* (1991), Vicente Navarro, professor of health policy at Johns Hopkins University, posed the following question: Why is there no debate about the morality of an economic system that does not provide security, joy, and relaxation for those citizens who have built the country through their sweat and toil? Navarro's perspective is that most Western populations would like to see their countries' resources distributed according to need and that the unrestrained capitalism of the United States is failing in this area.

He poses another question: How can the United States believe that it is the society of human rights when basic rights such as access to health care are still denied? This question is the tiger in the closet for America in the twenty-first century.

Chapter Summary

The poverty rate for people aged 65 and older declined from precipitously 1970 to 1996, in no small part due to the benefits of Social Security. Women, ethnic minorities, those who live alone, and the oldest-old constitute 90 percent of the elderly poor. Most old people, however, are not desperately poor, but struggling in the lower to middle income ranges. If the benefits of Social Security were withdrawn from their income, many more of America's elderly would move into poverty.

Under Social Security, people under age 65 and between the ages of 65 and 69 are penalized for working, although after age 70 there is no penalty on wages earned. An income tax levied on higher-income Social Security beneficiaries resulted in about 20 percent of people on Social Security paying taxes on their benefits. SSI provides a minimum income for elders, the blind, and the disabled by supplementing Social Security benefits if they are below the amount stipulated by SSI. It is designed to assist the very poor.

Food, shelter, personal safety, health, and the opportunity to be socially interactive are minimum requirements for a good quality of life. It is the responsibility of the society as a whole to provide for all of its citizens, including poor elders whose fortunes have changed and desperately poor elders who have never experienced comfort and security.

Key Terms

cost-of-living
    adjustment (COLA)
culture of poverty
economically
    vulnerable
Medicaid
Medicare
Medigap insurance
near-poor
private pension

relative deprivation
reverse mortgages
senior discount
    programs
Social Security
SRO Hotels
Supplemental
    Security
    Income (SSI)

Fieldwork Suggestions

1. Get information on Medicare. Write a simplified description of the plan as if you had to explain it to an elderly person.

2. Find out what services are available for elders in the smallest rural town near you. Compare those services to what is available in the largest city near you, or to where you live.

3. Research on the internet the concept of "spending down" to become eligible for Medicaid. What are the pros and cons?

Questions for Discussion

1. If a person on Medicare develops a cancer or heart disease, the hospital treatment is completely covered. If that same person develops Alzheimer's disease and needs help at home or long-term care in a living facility, there is no coverage at all. Is this fair?

2. Medicaid planning is a form of "spending down" and transferring assets to become eligible for Medicaid benefits. Medicaid will cover long-term care costs, but Medicare will not. Would it be ethical to spend down your money to become eligible for long-term care if you or your spouse had Alzheimer's disease?

3. What are the issues unique to elders living in rural vs. urban areas? What could be done to improve life for each group?

Internet Exercises

1. Locate Quotesmith (www.quotesmith.com) on the Internet to determine the range of Medigap policies in your state, and specifically in your region. If you were seeking **Medigap insurance** for your parent or grandparent, in what way would you use this resource? Find three other sources of information on Medigap insurance (hint: check out AARP, HCFA, the insurance commissioner's office in your state).

2. You represent a 75-year-old man whose wife is in long-term care. They own their own home and have a savings account of approximately $120,000. Locate a site or sites with information on Medicare, Medical, Social Security, and/or other programs that would help you give your client some primary points in making economic plans.

Living Environments

10

It Gets Better, or So They Say

By Maggie Scarf

"I'm so glad that I'm not young any more," sang 70-year-old Maurice Chevalier in the film "Gigi." Quite right, says Maggie Scarf, an American journalist who has written a succession of books on family and marriage. In this one, she turns her attention to couples in their 50s and 60s, and finds older marriage is full of unexpected pleasures.

Is it just a sense of the shortening future that makes married couples grow more comfortable together with age? Perhaps. But, as she points out, the future for older people is not nowadays so short: in 1900 life expectancy at the age of 65 in the United States was 11.5 years for men and 12.2 for women; by 2003 that had risen to 16.8 and 19.8 respectively, and is still growing.

This is a journalistic book—a collage of interviews with half a dozen couples and descriptions of research reports—and the material is sometimes stretched. But the clearest message, both from the research and from the couples, is upbeat. Those later years are, of course, a time of adjustments. The nest empties. Retirement approaches. The body may start to hint that it is not immortal. Lust languishes, in spite of Viagra. As in late adolescence, argues Ms. Scarf, people once again have to forge an individual identity. Without a growing family or a career to provide self-definition, older people must answer anew the teenage question, "Who am I?"

This time around, however, the answers are more manageable. The popular image of these late years may be of crusty and decrepit ancients. The reality is that, as people age, their conflicts grow less acute and their ability to draw pleasure from the more agreeable aspects of life increases. In the words of one group of researchers whose work Ms. Scarf examined: "Older couples, compared to middle-aged couples, expressed lower levels of anger, disgust, belligerence and whining and higher levels of one important emotion, namely affection." The parts of the brain associated with anger and aggression gradually shrink as people cross the threshold to old age. Emotional stability steadily improves.

Should unhappily married people split up? Ms. Scarf addresses this question in a chapter called "Does Divorce Make People Happy?" The short answer seems to be, rarely. She quotes a study of people in unhappy marriages which followed up the couples five years later. Its conclusion: unhappily married adults who divorced or separated were no happier, on average, than unhappily married adults who stayed married to the same partner. Only one in five of them was happily remarried. More surprisingly, a majority of those who remained married pronounced themselves happy at the end of the five-year period.

Among Ms. Scarf's interviewees are several couples who went through a rocky patch but stayed together. What word, she asked one couple, would describe the later years of their lives together. "The new beginning," said one partner. "Peace," said the other. So be it, if you can make it to that point. Perhaps Chevalier was right.

Source: www.economist.com/books/displaystory.cfm? story_id=12000969

Think of the one place where you can be yourself, keep the things that mean the most to you, live in the way you prefer, and shut the world out when you feel like it. There are lots of names for it, but they all mean the same thing—home. In this chapter, we will discuss the problems of housing, which to some may mean a home or to others may mean only a shelter. Psychologically, there is a big difference between *housing* and *home*.

One of the most important challenges facing the nation now and in the near future is providing safe, adaptable, quality, affordable housing for the growing elderly population. Quality-of-life concerns are never simple, and the issue of "adequate" housing options for America's elders is an example. The expanding population of elders will be quite diverse: the housing needs of a 65-year-old newly retired person will probably be quite different from those of an 85-year-old person. Between now and the year 2050, the population of people over 85 is expected to quadruple, from 5 million to 20 million. Such dramatic shifts in the age distribution of a population require that a responsible society work to understand the social implications of that shift and prepare for the well-being of its citizens. Housing is one area that profoundly affects the quality of life of an individual, and among special-needs populations it requires study and policy-level response. This chapter will consider issues of income and affordability; competence and opportunity; the need for adaptability (of houses, people, and social policy); the emotional power of the meaning of home; and options, both current and future, for housing for older people.

Living Environments

Housing—where a person lives—is a primary key to his or her quality of life. Housing is comfort, safety, community, memory, connection to family and the larger community. This is especially true for older people living on fixed incomes and those who are becoming frail. The diversity of the aged population underscores the need for many housing options. Compared with other industrialized nations, the United States has been slow to acknowledge the housing needs of its older citizens and to offer the needed options. The housing market for older people is continually changing as the older population ages and grows. Elders need appropriate housing that offers safety and comfort in a convenient, desirable location and at a cost within their budgets. Gerontologists are concerned with the effect of the **living environment** on older people—not only in terms of the house structure, but also in terms of the environmental context. The features of a residential area and its surrounding community help determine whether an older person is going to be happy. Because older people are typically in their homes many more hours per day than are younger people, their living environments must make a positive, meaningful contribution to their lives.

Diversity in Living Arrangements

The kinds of problems older people face in housing depend to an extent on the location and type of housing they have and on their personal circumstances, such as their health and marital status. The majority (67 percent) of older noninstitutionalized people live in a family setting, according to the U.S. Census Department of Health and Human Services data (He et al., 2007). About 81 percent of older men and 57 percent of older women live in families—usually with their spouse (Figure 10.1). The number of people living in a family setting, of course, decreases as widowhood increases. Single people tend to choose to live independently until there is a specific reason not to do so, which generally occurs when they become more physically frail with age. In fact, the elderly, whether single or married, are the least likely to change residence than other age groups (AoA, 2006). Living alone or in a

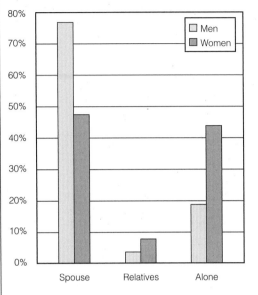

FIGURE 10.1

Living arrangements of persons 65 and over

Source: Adapted from the Administration on Aging, Department of Health and Human Services, 1997.

Affluence at any age allows for a wide range of choice in living environments. Most older adults prefer to live independently in their own homes.

large home is difficult and can be unsafe if, for example, an individual is visually impaired or is unable to regularly negotiate steps.

> *In the home are objectified both self-identity and social identity. Patterns of living, social norms and values, together with personality and biographical experiences, are hardly more clearly expressed anywhere else than in the home.*
>
> EKSTROM, 1994

Options and Choices for Living

In the United States and other developed nations older people typically live with their spouse until widowhood, and then live alone until health and care needs make living alone untenable. In developing countries people typically live in multigenerational households. In both developed and developing countries, one's primary source of care and support is his or her spouse (Kinsella & Velkoff, 2001).

The National Older Adults Housing Survey, completed in 2005, has provided one of the most complete studies of aging and older American adults' housing preferences to date (NAHB Research Center, 2005). The study was a computerized survey sent to a national sample drawn from the Harris Poll Online panel of respondents and as such represents a particular group of aged and aging adults—those who are computer literate and had an interest in responding to the survey. A total of 1,814 people participated, including those in the baby-boomer cohort. Inclusion of this age group helped provide an indication of the future direction of housing preferences. The basic findings of the report included:

1. The older the person, the more interested they are in having features in their homes to increase accessibility and support aging-in-place efforts.

2. Most respondents had not made specific plans for where they would move if it became necessary.

3. Most respondents desire to age in place, but many do not believe their current home will continue to meet their needs.

4. Most respondents were not interested in moving to age-restricted communities; they lacked a clear understanding of what age-restricted communities offer.

5. Safety; convenience; proximity to medical facilities, places of worship, and shopping areas were important to most of the respondents.

6. People living in age-restricted communities were attracted by amenities for recreation and social activities.

7. Housing amenities included bedroom and full bathrooms on the main-entrance level, one-story living, exterior requiring little maintenance, and energy efficiency.

8. Desired modifications to increase safety included wide hallways and doorways, nonslip flooring, adequate turn space in the bathroom for a wheelchair, grab bars and curb-less showers in the bathroom, and shower separate from the bathtub (NAHB Research Center, 2005, p. 2).

One objective of the survey was to identify amenities most important to older people who chose age-restricted living and mixed-age housing. Table 10.1 shows some of the features for both groups. Note that some features are the same on both lists; however, the order is different (NAHB Research Center, 2005).

The report concluded that most older adults prefer to live in a house that is of the same size or smaller than their current home; community centers, planned activities, and recreational facilities were seen as the primary attractions in age-restricted communities. Boomer respondents in the sample, however, had difficulty projecting

TABLE 10.1

Features important in retirement community selection

Age-Restricted	Rank	Mixed-Age
Lower crime	1	Lower crime
Reduced home maintenance	2	Lower cost of living
Weather/climate	3	Proximity to hospitals/health service
Lower cost of living	4	Weather/climate
Maintenance costs included in fees	5	Less expensive than previous home
Community amenities	6	Proximity to shopping
Home safety features	7	Reduced home maintenance
Less expensive than previous home	8	Proximity to children
Proximity to hospitals/health services	9	Proximity to grandchildren
Community safety features	10	Home safety features

Source: NAHB Research Center (2005). *National Older Adult Housing Survey 2005*, p. 26.

themselves into a future of more limited functioning in which a move to new housing would be beneficial. This finding is consistent with that of previous research on housing preferences indicating that people—even those with economic flexibility—avoid making a decision to change housing until there is little planning time remaining (NAHB Builder Survey, 2003; AARP, 2000, 2003). Among the boomer-generation adults, 75 percent expressed an optimistic sense that they will age in good health and would (or did) find retirement rewarding (AARP, 2004). Three-quarters of the boomer respondents believed they would be able to remain in their homes, and only 51 percent believed they would need to make modifications to their homes as they aged. This finding is probably related both to a reluctance to visualize oneself as frail in the future and to examples set by parents or relatives who have aged in place.

An array of housing options has developed with the burgeoning older population. State and county directories to assist people in making the best possible housing choices for themselves or a family member describe an array of local choices, depending on one's economic and physical circumstances. Some brief definitions of terms that are described later in this chapter are the following:

- *Active adult community*. Large residential developments for those 55 or older; include active recreation programs and provide the benefits of home ownership such as tax advantages.

- *Assisted living or residential care apartments*. Independent apartments including supportive services such as meals, services, transportation, and other personal care. The facilities vary greatly in size and cost, and some are a component of a continuing care retirement community (CCRC).

- *CCRC or life care*. Provides a wide range of health and supportive services within the same facility. Residents can move from one type of housing to another as their needs change.

- *Residential care facilities*. Sometimes known as board and care homes; can be privately operated or large public facilities. Both provide room, meals, housekeeping, and personal care services.

- *Shared housing*. Two or more unrelated people share a dwelling for security, help with living activities, companionship, or financial benefits. Some shared housing is assisted through "match-up" programs in the community; other housing might be owned by an organization and rented to a number of older people, appropriate to the size of the space.

- *Skilled nursing facilities (SNFs) or convalescent hospitals*. California licenses SNFs to provide 24-hour custodial or maintenance care for chronically ill people who are medically dependent. Most nursing homes are certified for short-term stays through Medicare and long-term stays through medical-assistance programs (Medicaid, e.g.).

The older population is scattered throughout the United States, although some areas and states have disproportionately high numbers. In 2006, just over half (51 percent) of people over 65 lived in nine states: California (nearly 4 million); Florida (3 million); New York (2.5 million); Texas (2.3 million); Pennsylvania, Ohio, Illinois, Michigan, and New Jersey each had over 1 million (AoA, 2006). The 65-plus population increased by 14 percent or more of the total population in eight states in 2006: Florida (16.8 percent); West Virginia (15.3 percent); Pennsylvania (15.2 percent); North Dakota (14.6 percent); Iowa (14.6 percent); Maine (14.6 percent); South Dakota (14.2 percent); and Rhode Island (13.9 percent) (AoA, 2007). Inner-city areas with high percentages of elders typically house those who have "aged in place." In such areas the people have aged along with their houses.

Older people tend to reside in ordinary houses in ordinary neighborhoods, rather than in institutions or age-segregated housing. Thus, the housing needs of elders are obscured because many are dispersed and invisible as a distinct group. Those who live alone tend to have fewer financial resources than married couples, and their housing tends to be of poorer quality. Single men have the poorest housing, in spite of having higher incomes than single women, who, after fulfilling roles as homemakers, often seem to have more of an emotional investment in their housing.

America's Older Homeless

Some elders have no home at all. Homelessness was addressed previously in more detail, but this chapter on housing addresses the problem once again.

Some older adults have no home. Others have homes that are neither physically nor psychologically safe.

Although the *proportion* of homeless men is declining because of the increased number of women and children in this situation, the *number* of homeless men is increasing. It is estimated that over 3 percent (5.7 million) people in the United States were homeless between 1985 and 1990, whereas in the one year of October 2007 to October 2008, 1.6 million people were homeless. Nan Roman, president and CEO of the National Alliance to End Homelessness, in Washington says, "More people are becoming homeless when they're older, which is new. The programs that inoculated older people against homelessness are not keeping up" (quoted in *AARP Bulletin Today*, 2008). This condition is a reflection in part of the economic downturn experienced in 2007, in which many people lost their homes as well as a large portion of their retirement portfolio value when the stock market fell dramatically.

An emerging concern, related to the high costs of medical care coupled with limited resources for housing options, is of people who become homeless while in the hospital. The vignette below illustrates a case study:

> *Having lived a year in a residential home, a 79-year-old man developed a chest infection and was admitted to this hospital. He was given intravenous antibiotic treatment and supportive care and recovered over the next week, returning to his previous state of mobility. The residential home was contacted to arrange return transfer, but its staff asked to reassess him first. The assistant manager visited, and the next day the nursing staff was told that the manager had found the patient's mobility to be inadequate and therefore he could not to return to the home. He was homeless in hospital. The ward physiotherapists reassessed him and considered him to be safe and fully recovered. He considered the residential home to be his home and he wished to return there (www.bmg.com/cgil/content; downloaded 2009).*

The population of homeless Americans has an "unacceptably high risk for preventable disease, progressive morbidity, and premature death" (Plumb, 1997). A study of 55 chronically homeless older men in Seattle found them to have serious health care needs, to be subject to violence and crime, and to be facing loneliness and frail health brought about by a history of poor nutrition, an overconsumption of street drugs and alcohol, and/or the general fragility of inefficient networks of social and emotional support (Goering et al., 1997).

Public homeless shelters provide the homeless a temporary sanctuary, but the rigid routines and the loss of self-esteem they suffer in such places are not conducive to staying. Many homeless people have a life-long pattern of being asocial "loners," with limited social skills and frail mental health. Older homeless women, increasing in number and proportion, like the men, often suffer from histories of asocial behavior and/or mental health problems (Daniels, 1997). Outreach agencies such as Project Rescue in New York and the Hostel Outreach Program in Ontario have been successful in providing respite and health care to the homeless. But little funding is available for such programs, and special-project funding is more often targeted to the severely mentally ill or homeless families—groups that can be systematically categorized—than to the larger issue of homelessness and poverty (Daniels, 1997; Buhrich & Teeson, 1996).

We on the street have long recognized that Social Services operates under the policy of "don't make it too easy for them." . . . A county caseworker told me that before she could process my application I must bring in receipts from all the people I'd panhandled from so she could total my monthly income. But my personal all-time favorite is "provide proof of no income."

"KIM," HOMELESS VOICES, 1997

The 1996 welfare "reform" bill called the Personal Responsibility and Work Opportunity Reconciliation Act was designed to force people to work; its underlying assumption is that poverty and homelessness is an individual responsibility that can be addressed by getting a job. The reform bill of 1996 was accompanied by cuts in social programs that have left holes in the safety nets available for people living on the streets—both the chronically homeless and the "temporary" homeless whose number has largely increased in the country since 2007. In much of the country new jobs do not exist for those who have lost theirs. Joblessness is not a choice but is forced on some workers as a result of a deep economic recession. Additionally, in parts of central and south Florida, low- and middle-income workers are increasingly occupying shelters. Properties lost to foreclosure and abandoned in Cleveland have become shelter for many of the homeless who are trying to stay off the streets, according to the executive director of the Northeast Ohio Coalition for the Homeless (AARP, 2008). In Santa Barbara, California—an extremely affluent oceanfront city—an option is provided for the homeless who have cars, gas, and insurance (and often retain gym membership for a place to shower). Specified private and municipal parking lots throughout the city become outdoor lodging each evening for 55 of the estimated 300 homeless people who sleep in their vehicles. Social workers circulate the lots each night; the project is run by New Beginnings Counseling Center and is funded by the city (AARP, 2008).

In old age, one's own home fosters the maintenance of one's personhood and integrity. A home not only provides shelter but also is a symbol of permanence and identity. The loss of home is devastating to one's being. Along with the loss of self is often the loss of friends and neighbors, support, and a platform from which to reach out and restructure one's life. When bodies age, the desire to remain settled and safe is strong. Homeless elders symbolize the tragic nature of a competitive society in which the less fit are ignored and neglected. Lack of a place to call home in one's old age is tragically incomprehensible.

Dissatisfaction with Housing

Compared to the homeless, those who have a place to live are doing well and counting their blessings. Yet older people can be dissatisfied with their housing situations for a number of reasons.

Personal Changes

It is generally not until people are in the oldest-old age grouping (85 and older) that the more profound effects of aging begin to affect the ways in which they can live comfortably in their homes. However, for a significant minority of elders, acute or chronic health conditions will necessitate a change in their housing (Frolik, 1996). Among those people, a decline in strength and vigor makes it inappropriate to remain in their current living situation.

Changes in one's life may lead to a housing situation that does not match one's needs. First, as the older persons' needs change, a house may become too large. A large proportion of older individuals are widows or widowers living alone. When a house that once was the appropriate size for a family is unnecessarily large and empty, a smaller house or an apartment may be more suitable.

Those with physical disabilities may not want to put their energies into keeping up a big house. Limited mobility and agility make it hard for older people to climb stairs, stand on stepladders, use bathtubs, and reach high cupboards. The person who loses a driver's license due to failing health or as a result of forgoing car ownership may be unhappy if the home is also isolated from friends and stores.

Financial Changes and Increasing Maintenance

Shrinking personal finances lead to difficulties in paying required housing costs. The older person who retires may receive only half the income he or she received when working. For home owners, property taxes and insurance premiums take ever-increasing chunks out of fixed income; sizable rent increases may force them to move to less adequate dwellings. Of the 2.2 million households headed by older people in 2005, 80 percent were owners and 20 percent were renters. About 68 percent of older home owners owned their homes free and clear, although 43 percent of owners and 82 percent of renters spent more than 25 percent of their income on housing costs (AoA, 2007). The housing costs of taxes, insurance, and repair will increase, but property resale, over which the elder has no control, will not result in higher monthly payments for the home owner.

The home ages with the person. Older people may have lived in their homes 20, 30, 40, or more years. The older house requires more maintenance, and hiring others to perform the work is often costly. Leaking roofs must be repaired, appliances made usable, lawns mowed, and windows washed.

The amount of money spent annually on the upkeep of a home is less for home owners age 65 and older than it is for younger home owners. The average home maintenance cost for people 60 and younger is about $466 annually, which drops to about $145 for home owners 75 and older (Kausler & Kausler, 1996). Housing that requires little upkeep is ideal, but the risk becomes that older people not do necessary home maintenance. An estimated 10 to 15 percent of elders live in dwellings that are substandard, with rural areas having the highest rates.

Urban Blight

The calm, quiet residential neighborhood in which inner-city residents choose to grow old often becomes a run-down area of high crime. Older people may be afraid to go out at night or to stay in their own homes. What was once an attractive residential area can deteriorate or become a "concrete jungle" of commercial

One person's castle is another's prison. Housing is generally appropriate if the older adult wants to be there.

buildings or factories, with extensive air and noise pollution. Drug trafficking, burglary, theft, muggings, and assault as well as rape occur with greater frequency in these areas.

Desire to Pursue Leisure

Old people who have no financial or health problems may wish to relocate closer to recreational amenities. Some want milder climates; others want to live among those in their own age group. Those who choose to migrate, who must go through the difficult decision of whether to uproot, often select a retirement community or center. Many such places exist, especially in Florida, Arizona, and southern California, where the climate is mild; but others are scattered across the United States.

Problems in Relocating

According to an AARP study (2004) the most common reasons older people gave for relocating are they want to be closer to family, they can no longer afford current housing, they desire more appropriate housing, and/or they are moving to a retirement community or area.

A factor not included in this list, but a very important one in the decision to relocate, is the physical inability to continue to live in the present environment. Whatever the reasons for deciding to relocate, a number of issues are associated with making the move. Four of these issues are discussed here.

Shortage of Appropriate Housing

The availability of adequate housing for elders is woefully inadequate in some communities. A shortage exists both in low-cost housing and in housing that meets the special needs of the oldest-old.

One of the biggest shortages is in low-cost housing. Demand far exceeds supply in all areas of the United States. Those with low incomes—for example, those who live entirely on Social Security or Supplemental Security Income—often live in substandard housing. Government housing programs for older adults lost their funding in the 1980s due to budget cuts. The National Affordable Housing Act of 1990 has some provisions to help older home owners. For example, a home equity provision allows use of home equity to make home repairs (Redfoot & Gaberlavage, 1991). The equity provision is available to all home owners, regardless of their income; however, a number of older people—especially the young-old—live in homes that are not yet paid for. In the present economy, housing and property values have dropped by substantial amounts in some areas of the country. Many people who counted on the continued appreciation (increase) of property values have found themselves in houses worth less than they paid for them, and indeed worth less than they still owe in mortgages. This makes equity loans for repairs and other purchases out of the question. Additionally, many Americans used equity loans to make purchases such as home remodeling, automobiles, helping adult children get into homes of their own, and so forth. This depletes the equity available for the individual to borrow against for simple home repairs.

Budget constraints at the federal, state, and local levels offer little hope that other funds for housing will be forthcoming. In 1997 the status of Section 8 housing was considered fair game in the battle for budget balancing, as lawmakers struggled with the ideological issues of the degree to which government was responsible for providing its citizens with safe housing. Section 8 housing is federally subsidized rental assistance for income-eligible seniors and people with disabilities. They are specific properties, many of which are housing complexes, and all complexes generally have waiting lists.

A suggested alternative model is that project-based subsidies be replaced with tenant-based assistance—that is, the federal assistance is provided to the individual who then locates his or her own housing, rather than subsidizing builders and housing project management. The weakness in this argument is that the private sector will have no incentive to build inexpensive housing. Sister Lillian Murphy is president of the nonprofit Mercy Housing, which operates about 3,000 units for low-income elders and families in eight states. She fears that housing for the poor will virtually disappear if no incentives are provided for the for-profit industry. "The market has never served the poor, [and] it never will," she said. "If [for-profit companies] are not getting the subsidies, they will walk away" (Wells, 1997). To allow elders to remain in their original homes, some states provide property tax relief by not taxing property after the owner is a given age. When the older person dies, the tax bill is deducted from the estate. A second option, discussed at greater length in Chapter 9, is the Home Equity Conversion Mortgage Insurance program or reverse mortgage plan (FHA Program, 1988). Both tax savings and reverse mortgages, however, require that the individual own their own home.

Ideally, the physical structure of a house for the older person would include the following:

- Wide hallways and doors to allow wheelchairs to move freely
- Protective railings and dull-surfaced floors to reduce the probability of falls
- Low-hung cabinets for easy access
- Increased brightness in lighting fixtures to accommodate reduced visual acuity
- Acoustical devices to increase the volume of doorbells and telephones

The oldest-old have little physical tolerance for extreme temperatures; thus, the thermal environment must be carefully regulated. The house should have no change of levels, not even a step down into a living room. Walk-in, sit-down showers rather than bathtubs are a must. Nearly as important as services available to and facilities of the living environment is the **surveillance zone**—the visual field outside the home that one can view from the windows or glass doors inside. A surveillance zone provides an important source of identity and participation in one's environment. The concept has implications for location, design, and landscaping (NAHB Research Center, 2005).

Housing designed without consideration for the special needs of the elderly is just as inadequate as housing built on an unstable foundation, or housing with inadequate plumbing.

Uprooting

Relocation is difficult whether it is voluntary or involuntary, because it involves leaving close friends and associates as well as a familiar environment. It also poses the uncertainties of change. Older people, like everyone else, have quite varied emotional experiences to forced moves; however, making a move that one has not fully chosen—moving with regrets—can have a particularly powerful emotional impact on elders.

Most studies on the emotional impact of voluntary and involuntary moves made by elderly people have focused on stress. Stress has to do with loss of control. It is the emotional experience of being unable to manage a situation or changes that occur, and of the struggle to reshape those changes to gain a sense of competence. A complex blend of emotions, or feelings, are present in the stress of changing one's home, most especially if that change is not fully voluntary. Issues having to do with security, attachment, belonging, estrangement, anxiety—all are connected to the emotional meaning of *home*.

Mats Ekstrom (1994) developed categories of emotional states that people in his study experienced when they were forced to move from the place they called home (see Table 10.2). Coping with stress requires a reinterpretation of the *meaning* of the stressful situation; and it requires the development of a plan of action designed to remedy the situation. The person must *reinterpret* feelings of mistrust, self-estrangement, guilt and sorrow, and—most centrally—violation, or adjustment to the stress of the move will not happen. Ekstrom provides the poignant illustration of a man who was required to make an involuntary move:

For Mr. Peterson his home was intimately associated with who he was. In spite of the fact that his wife had been dead for several years, little had been touched. Her bed was in the same place as before, the furniture arrangement had not been changed, the ornaments had not been moved, and in the kitchen and linen-cupboard things were ordered in the way that she had ordered them. In conjunction with the move he had to dismantle all this, and he never felt able

TABLE 10.2

Emotions described by an elderly population being forcibly relocated

Emotional Experience of...	Definition
Stress	Feeling the lack of resources to cope with, and possibilities of escaping from, considerable problems, demands, and/or threatening situations. Exposure to potentially stressful situations, and also the capacity to handle them, is dependent on social context and dispositions and patterns of actions developed over time
Trust and security	Related to the degree of dependability, continuity, stability, and anonymity of the social environment in which one lives
Belonging, self-estrangement, and Meaninglessness	Related to the capacity to set our imprint on the environment through creative, self-expressive action; realize our thoughts in various objects; and realize our possibilities of living in a place, and in the presence of objects, to which, over time, we have established a deeper relation
Guilt, shame, pride, and dignity	Indirectly, also joy and sorrow; created when we regard ourselves through other people's eyes and reflect on our actions in light of other people's values and expectations
Violation	Direct exercise of power, insult, and infringement of social identity; vulnerability to such actions is socially determined

Source: Adapted from Ekstrom, Mats. (1994). Elderly people's experiences of housing renewal and forced relocation: Social theories and contextual analysis in explanations of emotional experiences. *Housing Studies*, 9(3), 369–391.

to put it together again. Several months after the move what had been his home was still largely packed in boxes that stood all over the new flat. The furniture was scattered here and there, and only in the bedroom were there curtains and pictures on the wall. His relationship to his new flat was an alienated one, at the same time as he felt that his life had been encroached upon by the landlord, for which reason he felt deeply violated. Mr. Peterson finds it hard to hold back the tears as he speaks: "They took my home away from me. . . I was forced to take apart the home we had, and I'll never be able to put it back the way it was. . . I live in a furniture depository and I'm never really happy." (p. 388)

Mr. Peterson's story demonstrates the importance of the home environment—the relationship of things, their placement, and space to memories. *Home* has deep emotional meanings for all of us, and managing those emotions might take more resources than any one individual—especially one who is experiencing the losses of the physical self—has to give it.

Migration

More than 65 percent of older people living in their own homes have lived there 20 years or longer (AoA, 2004). Living in an area for a long time fosters a feeling of neighborhood integration and security and closeness to friends and family. Yet many older persons can and do move, some very happily.

Retirement communities have sprung up throughout the country, advertising their amenities for the "golden age lifestyle." It must be kept in mind that most older people do not move at all until it becomes absolutely necessary for them to do so because they are unable to thrive in their present housing. This is projected to be particularly true for the newest cohort of older people, born between 1925 and 1942 (Brecht, 1996). By the end of the 1990s, almost all the marketing efforts and retirement services had been directed to those born before 1924. The needs and interests of this new generation of elders are different from those of their predecessors (Clark et al., 1996). The boomer generation, those born from 1946 through 1964, will represent yet another different cohort of older adults whose physical and psychological needs are different from those of the previous generation. The critical need for both personal and policy planning can be seen when we take a longer perspective view of aging in the country.

Amenity relocation (moving to an age-restricted community for one's retirement pleasure) is a concept made real by an industry specializing in building living. Most housing built in the 1960s and 1970s offered no nursing care or assistance with daily living. The communities focused on hospitality services such as meals, housekeeping, transportation, and activities. They targeted the newly retired (Brecht, 1996). In the 1980s and 1990s, the ages of those moving into independent living units were typically late 70s and early 80s rather than the late 60s and early 70s (Newbold, 1996). That population requires a different approach to facility amenities than did the first wave of retiree-relocators.

Among those who do move, a first move typically occurs around retirement for the young-old. In any five-year period, about 5 percent of the population over age 60 make a long-distance move. Long-distance moves are more popular among "sixty something" couples who have both the financial resources and the desire to relocate during their early elder years. Some retirees have planned their move for years and have vacationed in the spot many times. Though the reasons for relocation vary, they typically involve the attractive leisure amenities retirement communities offer—arts and crafts, music, golfing, boating, fishing, tennis, or social activities. Elders from the Northeast and Midwest still migrate toward the Sunbelt of the West and South.

Migration may be either permanent or seasonal (Newbold, 1996). For several decades now, Sun City, Leisure World, and other large retirement communities of the Sunbelt states have offered housing and amenities at reasonable prices. Over half the popular retirement communities are in southern states or in the southern parts of states: Florida, Texas, Arkansas, Arizona, and California. The northern and western states of Michigan, Wisconsin, Minnesota, Oregon, and Nevada also have clusters of retirement communities. Florida still has the largest percentage of elder residents of all states.

Some older persons are **snowbirds,** living in the North for several months during summer and going south for winter. The opposite, those who live in the South and move north in summer, have been labeled **sunbirds.** Sunbirds have been understudied, yet the tendency for older Arizona residents to leave in summer is comparable to the tendency of older Minnesotans to go south during winter. Researchers find a trend toward snowbird eventually making a permanent migration to the South, then becoming a sunbird. Seasonally migrant elders compared to nonmigrants are more likely to be married and retired; they also have slightly higher levels of education and backgrounds as white-workers. Studies of Sun City and similar retirement communities show that the stereotype of the older person moving away from family and friends to a foreign and unfamiliar world is largely false. Retirees generally hear about such places from relatives and friends and visit the community before moving to it.

Some people do move, but not that many. The popular belief that Snowbelt seniors will flee to warmer climates if they can afford to obscures the reality that only 5 percent actually move away from their community. Most projects attract residents from within a six- to seven-mile radius (Olson, 1998).

The pressure for a second move occurs when older people develop chronic disabilities that make it difficult to carry out everyday tasks. The presence of a spouse is helpful and may act to postpone the move, which, when made, may place the disabled elder nearer to adult children who can offer assistance or to medical facilities. With urbanization and industrialization, many grown children have moved to cities for jobs. Older parents from rural areas, if they are to be near their children, must move to the city. An example is that of a middle-aged Iowa farmer who gave up farming after years of drought and moved to the city to secure a more financially stable livelihood for his family. In the process, he left his parents, both in their 70s, behind on the farm. With increasing age and/or declining health, these older parents will need to move closer to their displaced son and his family or make arrangements for local services (if any are available). Sometimes the second move is to some type of congregate or assisted housing, in which the older person can maintain a degree of independence.

The 2000 census repeated the 1990 census finding that more older persons live in the suburbs than in central cities. Declining health and the need for health care are often associated with migration to suburban areas. Given the closing of rural hospitals and the inadequacy or nonexistence of rural health care systems, worry about well-being often acts as a strong stimulus for an unwanted move to the city. The young-old who do not have to worry as much about health care tend to migrate to rural areas. The old-old tend to come to the suburbs or city from rural areas for necessary social and medical services.

Integration versus Segregation

The issue of **age-integrated housing** versus **age-segregated housing** concerns whether one would rather live entirely among age peers or in a mixed age environment. Age-restriction might threaten an elder's ties with the larger society, or it might be psychologically and physically more comfortable. As we saw in Table 10.1, people who have made the choice to live in age-restricted communities tend to be older and frailer than people who live in age-integrated communities, and both groups have slightly different priorities regarding their communities. For example, people with very strong ties to children and grandchildren are more likely to want to live in age-integrated communities whereas older adults who are more independent from extended family, or who have physical needs best met by special housing and living circumstances, might prefer the pace and protection of age-restricted communities.

Age-concentrated retirement housing has strong support from middle- to high-income elders. Interviews with elders in three apartment buildings with low, medium, and high densities of older adults revealed that residents associated high-density senior citizen housing with larger numbers of friends, more active friendships, and slightly better morale (NAHB

Research Center, 2005). However, the idea that age-concentrated housing is the best and most desirable situation for older adults appears to be more individually determined, rather than something understood by gerontologists and geriatricians. Living environments with high densities of elders seem to have little effect on morale, either positive or negative.

Many elders live in age-concentrated housing because they like the accommodations and amenities. Others are simply in "old" neighborhoods: age concentration is high because everyone has lived there for many years. They do not, by choice, limit their interactions to other old persons. Researchers conclude that, in general, though age concentrated housing is a desirable alternative for some, it should not serve as a guide for housing policy at large.

Age-concentrated retirement has strong support from some older adults. The opportunity for socialization and meaningful activity is seen as a positive outcome of living in "adult communities."

Some sociologists predict more age integration in the years to come, particularly if the young-old stay in the job market longer and continue their activities of middle age. Some social critics view age-segregated housing as a means of insulating older people from ageism in the larger society. It can be comfortable for elders to live among other people who understand their aches and pains, have experienced the death of a spouse, are patient in the need to speak directly and more loudly to someone with a hearing deficit, and can share the anxiety related to a heart condition or an impending hip operation.

Housing Options

This section will review some of the housing options that appeared in a previous section to greater extent. Types of housing range from independent to dependent on a continuous scale. Living in one's own home is the most independent lifestyle; living in a hotel-type residence is semi-independent; and living in an institution is the most dependent lifestyle.

Generally speaking, the more dependent the new lifestyle, the more difficult the adjustment. Sometimes the problem in institutionalized group housing is not the quality of care provided, but the fact that the older person really does not want to be there. Many elders are dedicated to a lifestyle of self-reliance and self-direction and have little tolerance for the regulations of a nursing home. They would like better food and shelter, but not in exchange for the freedom to decide their life's course. A truly independent person might choose to live on a menu of eggs, beer, and ice cream in a cockroach-infested hotel rather than in a nursing home.

On the other hand, an isolated, frail, disabled elder might welcome the safe haven of institutionalized care. Acknowledging the diversity in age, personality, and health of those age 65 and older opens up more and more housing alternatives for them. The needs and requirements of

the young-old in good health are much different from the frail elders in poor health. In this section, housing options for both the young-old and the oldest-old are considered.

Nearly every study affirms the conventional wisdom that older people strongly prefer to age in place, to grow old within familiar territory that has provided a context for their lives, whether they live in single family dwelling, elderly housing complexes, or naturally occurring retirement communities.

SYKES, 1990

Aging in Place

Most people do not move from their own homes until they must do so for reasons of safety. The image of retiring and moving to Golden Age Village is just that: an image. Some people do retire and move; indeed, in some states elderly migration is a growth industry. Retirees are viewed as the ultimate "clean industry" (Clark et al., 1996). Moving does not reflect the first choice of most new retirees these days, however. Most people go to great lengths to remain in their own homes—a home not associated with being old, frail, or having special needs. That home is the home of memories and is often associated with the self-concept of mature and responsible adulthood. **Aging in place** has become a popular phrase reflecting people's general preference to stay in their own home for as long as possible despite increasing frailty and its associated problems. The surroundings are familiar; memories are associated with the rooms, the garden, and around the kitchen table.

Older Americans have moved less frequently each year since World War II (Golant & LaGreca, 1994). About 84 percent of people age 55 and older say that they want to stay in their longtime homes rather than move to

Aging in place has profound psychological meaning for some elders. It is sometimes difficult for families to balance their needs for the elder to be safe and care for with the elder's needs to age in place/die in place.

senior housing. When they are forced to move, 63 percent prefer to remain in the same city or county and only 11 percent prefer to move to a different state. "Their dominant preference is to stay in their homes, and not move, never move—stay there forever," said Lea Dobkin, a housing specialist with AARP. "[For developers], it means use caution. It's a more competitive environment."

Many older people already live informal retirement communities known in the senior housing industry as **naturally occurring retirement communities (NORCs)**. These are buildings or neighborhoods where the residents have aged over the years. There are many cases in point, in the suburbs, in the rural Midwest, and in inner cities (Newbold, 1996). The neigh-

borhood, like the individual, goes through a life cycle of young to middle-aged to old.

Aging in place is not always easy. As people age into their 80s and 90s and beyond, more tasks pose difficulties, even for those in good health—writing checks, driving a car, shopping, or bathing. Some public and private agencies are available to provide services that older people themselves and their families and friends cannot provide. Home-nursing services and homemaker services, providing bathing and grooming assistance, housecleaning, some transportation, are available. Home-delivered meals and senior centers with meals and activities also provide services. Social services might also be needed for emotional health and support. Increasingly, communities have volunteer programs to make calls to older people once a day to check on their health and safety. These calls are especially important for those living alone.

Downsizing

The move to a smaller home, say a condominium or a mobile home, if it is in the same town or general area, is a form of aging in place. The reasons for such a move were discussed earlier in this chapter: the upkeep is easier, taxes are lower than those for a large house, and the location may be more convenient.

Mobile homes can provide a comfortable living environment and be a viable housing option for older adults. Perhaps the most significant factor in opting for a mobile home is related to its affordability. Mobile homes provide older adults with a sense of home ownership, privacy, and security.

Mobile homes are generally purchased on-site, and residents pay a monthly site rental fee, which is likely to increase over time. The quality of mobile homes, however, is now assured by the Housing and Urban Development (HUD). Most counties have mobile home parks designed exclusively for older residents, age-restricted to over 55.

Enough people live in their recreation vehicles (RVs) the year around for the term **full-timing** to have evolved. This is one more choice that older people have. Some fulfill a lifetime retirement dream when they "hit the road" in their RVs and see all the places in Canada, the United States, or Mexico that draw them. It is, of course, most viable for younger-old adults with the physical strength and competence for driving and making various automotive repairs. It is a lifestyle that has not gone unnoticed by sociologists and gerontologists.

Adult Day Care

Adult day care offers medical and social services at a center to older adults who commute from home. Many service providers call their programs "frail elder programs" to avoid the comparison and association of child-care centers with programs for adults. Centers may be nonprofit organizations, government-run centers, or profit organizations run by insurance companies such as Elderplan, Kaiser, SCAN, and Seniors Plus. Multipurpose senior centers provide health care facilities in which older persons are brought in the morning and taken home in the afternoon. These might be day hospitals that offer medical or psychiatric treatment or day-care centers that provide social and recreational services. Studies show that persons who can avail themselves of adult day care reduce their risks of hospitalization.

Some nonprofit organizations provide multiple services to elders. Through the Older Americans Act, the federal government now funds a number of multipurpose senior centers. One such organization, the Minneapolis Age and Opportunity Center, largely government-funded, provides such services as daily meal deliveries, laundry, and transportation. A Touch of Home in Rochester, New York, provides respite and socialization opportunities for frail elderly and their caregivers. The program is an example of other small nonprofit agencies that provide "therapeutic recreation that emphasize opportunities for success in a structured, safe and fun environment" as well as transportation, sometimes a meal, personal care, and transportation (Tracy, 1996).

Frail-elder programs are especially critical for people who choose to continue living in their own homes, or who are living with a spouse or adult-child caregiver. The socialization and care provision offered by these programs can make the difference in whether an individual can continue to live independently. The Touch of Home program in Rochester grew from a small program in the basement of a church in 1990 to services at three locations in the region, providing services to more than 60 people by 1996. The need is undeniably there; however, the funding to support programs and subsidize program costs for individuals and families is stretched thinly between competing social services programs. Program managers for A Touch of Care concluded that the greatest hurdles facing adult day services are

---

funding streams and private market acceptance. As government funds diminish and decision-making is decentralized to local governments, identification of local community-based funding streams is critical. Partnerships with employers, other service providers and public elder service agencies, as well as education and awareness-raising of local government officials, will be critical to the survival of non-profit, community-based adult day-care programs. (p. 41)

Insurance programs are now being offered to assist the older person in staying at home. Such plans are becoming affordable and workable. A variation of the health maintenance organization (HMO) is the social health maintenance organization (SHMO). Elderplan in New York, Kaiser in Oregon, and Seniors Plus in Minnesota offer health insurance programs that provide personal care and household assistance services at home or in day-care settings.

A new variation of health insurance policies offers Life Care at Home (LCH). Insurance agencies are tailoring LCH insurance policies that cover medical and personal costs for the older person who wishes to stay at home. One must proceed with caution in purchasing an LCH plan because the LCH involves a large initial entry fee, a monthly fee, and at least a one-year contract. Such private long-term care insurance policies are expensive and may have restrictions that limit coverage.

The On Lok Health Center in San Francisco, patterned after the day-care system in England, provides day-care services to elders in Chinatown. A van with a hydraulic lift transports the elderly to the center, where each day begins with exercise and reality orientation. People are introduced to one another in both English and Chinese. On Lok assumes responsibility for providing all services needed by the functionally dependent. If not for On Lok, the participants would need institutional care. Though the cost per person for this service is about the same as that for institutional care, the psychological benefits and the social facilitation make the service well worth the price.

Granny Units and Shared Housing

An accessory apartment, also called **a granny unit**, is an independent unit contained within a single-family home, generally with an outside entrance. It is often attached to the home of an adult child. It is appropriate for an older home owner wanting additional income or a person wanting to live closer to others (MCCA, 2007). Older people can also install granny units in their backyard. They could potentially rent their home or their granny unit for income or trade for services such as nursing or homemaking. Although a relatively new concept to the American public, it is an old concept to the Japanese and Australians, who refer it to the granny flat or elder cottage.

Advocates of granny-unit housing, such as the American Association of Retired Persons (AARP), must fight zoning laws in some towns and cities. Areas zoned for single-family homes often do not welcome added granny units. The primary desirable feature of granny units is that they allow the older person a choice to live independently and still be around other people. If the older person can live in a granny unit rather than in the adult child's house, there is more privacy and freedom for everyone. The "granny flats" experience in Great Britain suggests that "they have an important but limited role" in enabling people to live close to their families (Tinker, 1991). But despite cost and regulations this housing alternative is gaining support in more countries worldwide.

Shared housing that involves some type of group living is another concept that is relatively novel in American awareness. In Great Britain, there are more than 700 Abbyfield homes—homes in the community that house several elders and are run by specialists in aging. This kind of program is being developed in the United States. Another alternative is renting a room of one's house to another person and sharing such rooms as the kitchen and living room. Such an arrangement usually predicated by the need for financial relief, services, and companionship. Many variations are possible. An older person could rent a room of the house in return for income or services.

The difficult part is to find a good tenant–home owner match. More and more community agencies, such as Share-a-Home in Memphis, Tennessee; Independent Living in Madison, Wisconsin; and the Marin Housing Authority in California, are helping to coordinate shared housing. Again, it can be anticipated that as the boomer generation moved into old age, attitudes about housing will change. Shared housing will probably be more available by then, and it will also be more acceptable to the older person whose social experience has included living with other people—in college dorm rooms, in shared apartments before marriage, in commune settings as young adults for some.

Adult Communities

Many towns and cities have **adult communities** for those over age 50. Typically, children under 18 are prohibited from living in these places; court cases have upheld the legality of the age limit. Adult mobile-home parks with age limits on younger persons are also quite popular. An adult community may have a grocery store, a bank, a medical clinic, and convenience stores within its boundaries. The sizes of adult communities within cities vary. A recent trend is for older folk with interests in common (e.g., artists or gay retirees), to form their own communities.

Retirement Cities

Some older people seem to like the age-segregated **retirement cities**, because such communities offer so much. Sun City, Arizona, after which many other retirement towns have been patterned, has a population of over 46,000 and a minimum age of 50 (others, such as Youngstown, Arizona, have set the lower age limit at 60). Located 20 miles northwest of Phoenix, Sun City is a large complex on 8,900 acres with a number of social and leisure activities, such as tennis, bridge, quilting, wood carving, and language classes. The lure is the affordability and the weather. In 1991, prices for homes started at $65,000 in Sun City West, a development started in 1978 when all of the original Sun City units had been built and sold. Residents tend to be upbeat about cities made up of all senior citizens. They get involved more easily because of the many activities shared by people in their own age group.

Several million older people have bought or leased property in the retirement cities that have mushroomed around the country since 1950. One of the most successful retirement community developments has been Leisure World, with locations in several sections of the country. Most retirement villages have swimming pools, clubhouses for dancing or crafts, symphonies, golf courses, and free bus rides to shopping centers and entertainment. One community excludes dogs and puts a three-week limit on visits by children. Such features make age-segregated housing more desirable to some older persons.

Assisted Living

Assisted living (or assisted housing), an alternative to an SNF (nursing home) residence, typically refers to community-based residential housing of some type (apartment, condominium, or semicommunal living), including supportive services to assist in the activities of daily living. It is designed to serve the needs of the frailer person who wants to live in a homelike setting, but requires some assistance to do so. This level of care is a fairly new concept. Older adults requiring assistance with their daily living tasks formerly had only the option of hiring home assistants or moving to SNFs in which by law, "all needs" are to be provided for. The assisted living philosophy is one of individualizing and maximizing the elder's independence, privacy, and options. It is a clear hands-off approach: people will be provided whatever assistance they need to maximize their independence. This is in stark contrast with the philosophy of skilled nursing care, which is to be hands-on, providing people assistance with all their needs, including medical needs. Assisted living services generally include meals and the option of eating in a dining room setting or eating in one's room, housekeeping, transportation, and help with everyday tasks such as bathing and grooming.

Growth in assisted living has come about because developers and investors can see the emerging market and the potential for profit. It is, in fact, currently the hottest senior housing market. In 1996 construction for assisted living accounted for 54 percent of all senior housing built in the United States (Murer, 1997). Assisted living development reached approximately $33 billion in revenues by 2000, with over four million beds (NAHB Research Center, 2005).

In the mid-1980s hospitality corporations such as Marriott and Hyatt made substantial investments in long-term care, which had previously been seen as a health care or hospital arena. This somewhat legitimized the industry for other investors. In a recent analysis of assisted living facility development, Olson (1998) reported that "in the past 24 months there has been a tremendous increase in the money available to develop senior housing . . . A large part of this capital comes from Real Estate Investment Trusts (REITs) seeking to diversify their portfolios." Wall Street and bonds also provide development funds, but the majority of financing comes from conventional sources such as banks (Olson, 1998). Financing provides

the structures; the cost of the services is paid, separately from rent, by the resident. Not surprisingly, most of the facilities generating the greatest amount of excitement are designed for upper- and upper-middle-income elders.

Funding is also available for housing for the truly low-income elderly—those with annual incomes of about $15,000 or less. Section 42 of the Internal Revenue Tax Code provides for a low-income housing tax credit to corporations investing in low-income elder housing. Some communities use Community Development Block Grants to build low-income housing; still others look to other federal programs such as the HOME Program, established under the National Affordable Housing Act of 1990 (Nolan, 1997).

Problems arise in the options for building low-income assisted living housing for elders because of the need for *assistance*. Services must be funded separately, because for low-income elders those costs cannot be passed on to the resident. Some developers have used a new Medicaid home- and community-based waiver that allows for the financing of health care services in an assisted living setting (Nolan, 1997). Other areas simply compete with local service providers for the limited funding available to provide living assistance to the poor.

In its current state of development, assisted living is a Cadillac: nice to have but affordable to a relative few. Unlike a Cadillac, though, assisted living is more than 'nice to have'; it is becoming a necessity for many elderly Americans who are, nevertheless, unable to afford it. Aging but still living independently, in need of support services but a long way from needing skilled care, they are natural customers for assisted living. Is gaining access to it beyond hope for them?

NOLAN, 1997

Board and care homes are similar to the old mom-and-pop operation in a large family home where unoccupied bedrooms become available to boarders. Relatively little attention has been paid to board and care facilities, which are variously referred to as residential care facilities, adult congregate living, rest homes, adult foster care, personal care, care facilities, and group homes (Garrard et al., 1997). Generally the board and care facility offers a bedroom in a private home, whereas an assisted living facility offers self-contained housing units. Operators of board and care homes are typically middle-aged women who live in the homes with their spouses, and many of the operators have worked previously as nurses or aides in health care settings (Kalymun, 1992).

Sunrise Retirement Homes and Communities is the country's largest operator of assisted living centers, operating more than 30 such communities (Estrada, 1994). When a husband-and-wife team started their first Sunrise Center by renovating a nursing home, the term *assisted living* did not exist. In 2000 more than 3,000 elders lived in their centers. One example of their efforts is the Sunrise Retirement Home near Washington, D.C., a three-story "Victorian mansion" with 42 one- and two-room units. The building has the feeling of a country inn with a fireplace, large bay windows, and a sun room filled with wicker furniture. Residents' rooms are grouped around common living rooms and lounges. Units rent for $2,500 to $4,000 a month, including housekeeping, meal services, and minor medical care.

Marie Morgan, 73 years old, was withdrawing and deteriorating in a nursing home when she opted for assisted living and moved to Rackleff House in Canby, Oregon (McCarthy, 1992). Despite incontinence and a heart condition, she is thriving in her assisted living quarters. She gets out and about using her walker to circle the grounds, does her own laundry, and is able to keep her dog, a Lhasa Apso, in her private apartment. For about $2,200 a month in 1990 she got housekeeping, meals, laundry facilities, transportation, social activities, and regular visits from nurses. These costs are currently greater, with cost variation depending on the facility itself; the area of the country; range of services provided; and proximity of the living quarters to shopping, cultural events, medical care, and so forth.

Some residents in Rackleff House have mild to moderate Alzheimer's disease, like the person who loiters by the copy machine in the office. She was a bookkeeper most of her life and finds the office bustle comforting. The staff understands and works around her. In Oregon assisted living centers have become so widespread that nursing home population has been reduced by 4 percent even though the over-65 population rose by 18 percent. More than 20 assisted living units are available in Oregon.

Rackleff House opened in 1990 and was created by a professor at Portland State University motivated by her mother's needs, not by profit. Renting a room in Rackleff House is affordable to lower-middle-income elders. The average age is 89, and half are on Medicaid. They can furnish their rooms with their own things. Rackleff looks like a big yellow farmhouse. It is not lavish, yet it has a cathedral ceiling in the dining room and a secure enclosed courtyard. A fireplace flickers in the front parlor. Traditional nursing homes have a staff–patient ratio of about 10 to 1: Rackleff's is one worker per three residents.

Did you know . . .
More Americans are concerned about financing long-term care than about paying for retirement (69 percent to 56 percent)?
NATIONAL COUNCIL ON AGING AND JOHN HANCOCK MUTUAL LIFE INSURANCE COMPANY, MILLER, 1997

Continuing Care Retirement Communities

The **continuing care retirement community** (**CCRC**) represents an upscale form of congregate housing that requires a strong commitment on the part of the resident. CCRCs, also known as **life-care communities**, require a large entry fee (from $50,000 to $1,000,000), with a monthly fees of $500 to $5,000 (depending on the area of the country, quality of housing, and number of services available), and commitment to at least a year-long contract. In return, the entrant is provided with housing, health care, social activities, and meals. Residents must be capable of independent living on entry but can remain in the facility even if they become incapacitated in later life. The model represents a range of housing and services, from independent living through long-term care.

Housing options develop as the market is perceived to be there. For example, some universities are now investigating CCRCs for alumni and faculty. According to a plan developed for Davidson College in South Carolina, residents will be able to audit classes, use campus libraries and sports facilities, and become involved in other ways with the university campus life (Olson, 1998). Part of the exchange, aside from providing a profitable and valuable service, will be that the increased investment in the university's well-being by the CCRC residents' involvement will result in an increase of bequests to the university.

The University Retirement Community (URC) in Davis, California, another nonprofit community, was begun by a group of University of California retirees and other people interested in having a CCRC in Davis. The vision was to have a place to retire, populated by like-minded people who enjoyed learning and doing new things—living the life of the mind so to speak. In 2000, the facility was opened through a unique partnership between a retirement services corporation, the city of Davis, and local citizens. The facility has garnered architectural awards and was accredited by the Continuing Care Accreditation Commission (CCAC). Of approximately 2,100 CCRCs in the country, only about 341 were accredited in 2005. Accreditation is based on three areas: governance and administration; financial resources and disclosure; and resident life, health, and well-being. The URC provides an excellent model to other locales for developing and building a nonprofit retirement community based on unique aspects of a particular community and/or interests.

The entire life-care industry has grown dramatically in the last few years. Dozens of big-name developers and corporations have built or plan to build CCRCs and other assisted living senior residences. The options are marketed in very sophisticated ways, and there is a growing concern for the need to protect older consumers. Legislation has been passed in many states to protect the older consumer. Contracts and rental agreements are regulated, and most states have certification and registry requirements. California and Florida, for example, require extensive disclosure statements prior to certifying facilities; Indiana and Colorado require that the potential resident receive copies of such disclosure statements. Those considering life-care communities must protect themselves as much as possible by examining the facility and contract very carefully before signing.

Nursing Homes: What Do They Offer?

Meeting the needs of the infirm is the responsibility of society, not just the responsibility of service providers, the government, or their families. Here we will consider what nursing homes can offer. How well they meet needs ultimately depends on how well the public informs itself concerning the problems of nursing homes and the rights and services to which residents are entitled.

Who Needs a Nursing Home?

Nursing homes are becoming large-scale operations. Since 1985, the number of nursing homes decreased by 13 percent while the number of beds increased by 9 percent (DHHS, 1997). The number of nursing home residents was up only 4 percent between 1985 and 1995, however, despite an 18-percent increase in the population aged 65 and over. Before this 1995 finding by the Department of Health and Human Services, the utilization rates had kept pace with the increase in the elderly population (DHHS, 1997). People are *not* aging in the same way they did 25 years ago.

The typical nursing home patient is female, white, widowed, and over age 80. Most of the residents come to the facility from another institution, usually a general medical hospital, rather than from their own homes. Most nursing home residents have multiple problems such as arthritis, heart trouble, diabetes, or vision/hearing impairment. In addition to physical impairments, there are often mental health problems, such as disorientation (confusion), impulse control (anger), and emotional affect disorder (depression). People living in nursing homes generally require assistance in dressing, eating, toileting, and bathing.

Making the Decision to Move

Whether an older parent, relative, or friend will need nursing home care can often be anticipated some months before the individual actually enters the facility. The limitations in behavior brought on by chronic illness or age may be progressive as well as irreversible, usually allowing the individual and the family time enough to make a thoughtful and careful selection of homes and to ease the transition from one living arrangement to another.

Most people are unwilling to entertain the possibility of a nursing home and simply wait for circumstances to force a quick, hurried decision, usually one filled with emotional trauma for all concerned. The typical nursing home patient enters the facility from a hospital rather than from a private home, signifying that nursing home care is a necessity rather than a choice. Families often burden themselves for long periods of time trying to care for an elderly parent who could receive care as good or even better in a nursing home. The phrase "never put me in a home..." rings in the ears of most people whether the words were spoken to them or not. The idea of "old folks' home" conjures up scenes of unclean, smelly facilities with old people strapped in wheel chairs, nodding off, perpetually waiting—for something, someone. It's sad, depressing, and generally an uninformed stereotype.

Families can anticipate an eventual need for a frail family member, and all members can be involved in the decision-making process. All concerned can discuss the advantages and disadvantages of nursing home care and together inquire about and visit potential facilities. A grandparent living with adult children may not be happy to be in the midst of bustling young people or to be the cause of unusual alterations or deviations in a family routine. Families can and should discuss these problems in an atmosphere of mutual love and concern. The older person "dumped," without discussion or plan, in a nursing home after a sudden medical crisis may well feel deserted and hurt. The American Health Care Association has published three brochures for those considering nursing home care: "The Nursing Home Dilemma," "Thinking About a Nursing Home?" and "Welcome to Our Nursing Home, A Family Guide." These brochures are available on request; they detail the many aspects of the decision-making process and can be the basis for beginning conversations about future needs (American Health Care Association, 1994). American Health Care Association websites are excellent resources for information on nursing homes and decision making.

Finding a Nursing Home

Over 30 percent of the population in one national-level random-sample study would rather die than move to a nursing home (Olson, 1998).

Some professionals and family caregivers might add that the other 70 percent simply do not know about daily life in a nursing home, cost notwithstanding.

Many nursing homes provide fine patient care and make special efforts to meet the psychosocial needs of their residents. Others are skilled in the delivery of medically related services but lack the foresight, knowledge, and skill to address the psychological and emotional needs of their residents. The better facilities generally have long waiting lists, so early planning is essential to eventual placement.

Unfortunately, most families spend little time in visiting a variety of nursing homes to secure a satisfactory placement. However, resources can assist in the search for the right nursing home. The local Council on Aging and the local chapter of the state Health Care Association can provide the locations of licensed facilities in the area and the factors to consider in choosing a home. The local Ombudsman program maintains a list of all formally reported complaints made by or on behalf of people who are residents of long-term care facilities. The Office of Nursing Home Affairs can provide information on characteristics of good homes. Finally, the local chapter of the Gray Panthers typically has a nursing home committee; various other watchdog groups publish information on how to choose a nursing home. The following was compiled by social workers and nursing home professionals as preliminary ideas about how to select a nursing home.

- Make at least two visits to a particular nursing home, once at a mealtime and again in mid-morning or mid-afternoon. Visit several other facilities, so you can make comparisons.

- Make sure you know what the basic rate covers. Investigate extra charges for professional services and medications.

- Take time to observe interactions among staff. Does there seem to be an easy relationship, or do you pick up any tension? The more staff seem to like one another, the greater the probability that the environment is a happy one for residents, too.

- Extend your observations to interactions between staff and residents. Are you comfortable? Do staff extend beyond courtesy to each resident and seem to have a genuine regard for each person (you, after all, are "company" and "company manners" might be on display)?

- Ask to see a copy of the month's menu. Is it varied and interesting? Does the food on the menu actually get served? How is it served? Have breakfast, lunch, or dinner in the facility. How was it? How did it make you feel?

- How many hours a month does a registered dietician spend in the facility? Experts believe four hours a week should be an absolute minimum.

- Is there a registered nurse on duty on the afternoon and evening shifts? This is required by law for facilities of 100 beds or more. All facilities with more than six beds should have a registered nurse on the daytime shift.

- Is there an activities director (required by Medicare-certified facilities)? Talk to him or her. What are the reasons for specific activities? What is the range of activities, and how often do they occur? Are there activities for bedridden residents? What are they? Do they seem interesting to you?

Adequate resources for later life can make a huge difference in a person's quality of life at the end of life.

> *Death is a dramatic, one-time crisis while old age is a day-by-day and year-by-year confrontation with powerful external and internal forces, a bittersweet coming to terms with one's own personality and one's own life.*
>
> ROBERT BUTLER, 1994

Nursing Homes: How Bad Are They?

Virtually all of us are either directly or indirectly affected by nursing home care. Although at any given time only 5 percent of the nation's elders are in long-term-care institutions, this figure is deceptive. About 1.5 million residents were receiving care in 16,700 nursing homes in 1995. Almost 90 percent of those residents were aged 65 or older, and more than 35 percent were aged 85 and older (DHHS, 1997).

Families are also affected. Nine out of ten children can expect that one of their parents (or their spouse's parents) will spend time in a nursing home. Nursing home populations are projected to increase from the 1990 level of 1.8 million to four million within the next 24 years—an increase of well over 50 percent (DHHS, 1997). That means that a lot of people are going to be directly and indirectly affected by the quality standards of nursing homes.

The question remains: are all homes as bad as we imagine them to be, or is a bad nursing home relatively rare? It turns out that this question is not an easy one to answer. For starters, we can easily identify truly negligent care, but can we identify good care—"quality" care? The very concept is laden with personal judgments—good quality in Mr. Jones' estimation might not be good quality from the perspective of his wife. Bad care, on the other hand, can be obvious to the observer or more subtle and not immediately apparent. Nursing home violations cover a broad range of conditions, just a few of which include residents being tied in chairs for hours at a time; lack of an activities program; bedridden elders lying in urine-soaked bedding; unappetizing and possibly undernourishing meals; and the use of styrofoam drinking cups and plastic utensils, which impose a hardship on those with tremors and arthritis. Where is the line between *bad* and *good-enough, all things considered?* What about *quality?*

Identifying the specific dimensions of nursing home quality in a way that makes sense to professionals, consumers, and policy makers is a relatively new research endeavor. In a major qualitative study, Rantz and associates interviewed people with a variety of experiences in providing nursing home care to determine the variables they used to assess nursing home quality (1998). The people interviewed represented a range of responsibilities in long-term care, including administrators, nurses and activity directors, social workers, ombudsmen, state regulators, professional home care staff, and hospice and mental health personnel.

The two key indicators of quality that these experienced nursing home professionals identified were *interaction* and *odor*: caring interactions between staff and residents and the absence of the odor of urine—an indication that residents are adequately toileted and kept clean. Table 10.3 summarizes the Rantz study findings on specific components of good and bad nursing homes. Note in the table that many of the identifiers of poor quality are subtle and might not be immediately obvious to overwhelmed family members or to the frail elder himself or herself.

The decision of when to move, where to move, and whether to move is a loaded one for all family members. Add to that burden the uncertainty of what kind of care a facility truly will provide for the person you love (or yourself), the load on that decision is enormous.

It seems apparent that quality nursing home care is something that cannot be legislated: the difference between "good-enough" and "high quality" takes an organizational philosophy—a dedication to providing *care* in the nursing home. *Caring* incurs meeting individual needs, and meeting individual needs incurs a good deal of knowledge and understanding about the individual being cared for. This, in turn, requires a low ratio of residents per staff person and an organization dedicated to the mental as well as physical well-being of its guests. Unfortunately in old age care as well as earlier in life, money holds quality of life hostage of America's older adults. As a nation, the United States has not made elder care a priority.

> **Did you know . . .**
> *Medicare will provide only skilled nursing and rehabilitative services, and only in Medicare-approved nursing homes—fewer than 10 percent of existing facilities. Medicare Nursing Home Benefits provide*

- First 20 days paid 100 percent by Medicare
- Additional 80 days, patient pays $92/day; Medicare pays remainder
- Beyond 100 days, patient pays all nursing home costs

The economics of nursing home management has another whole range of problems. The government reimburses nursing homes through two systems, both of which can be corrupted by dishonest owners. On the flat-rate system, the nursing home is paid so much per patient. The operator who wants to make a large profit on this system keeps costs as low as possible by providing cheap food, having as few registered nurses on the staff as possible, and providing no physical therapists or psychiatric counseling.

TABLE 10.3

Factors in nursing home quality

Nursing Home Care Components	Good Care Quality	Bad Care Quality
Environment	• Clean • No odor • Maintained • Bright/good lighting	• Odor of urine/feces • Shadowed/poor lighting
Care and treatment	• Attentive, caring, residents listened to • Treated as individuals • Restorative care, ambulating • People up, dressed, clean, and well-cared for • Food is good	• Residents unkempt, exposed, not clean, unshaven, disheveled, clothes dirty, poor nail care • Complaints from residents about care, lying in urine, not being taken to toilet
Staff	• Knowledgeable, professional • Busy interacting and working with residents • Open and listens to family • RNs involved in care • Education encouraged • Low turnover	• Interact inappropriately or ignore residents • Visitors can't find staff
Milieu	• Calm but active and friendly • Presence of community, volunteers, animals, and children • Residents engaged in age- and functionally appropriate activities	• Chaos • Residents screaming and no one paying attention • Unfriendly atmosphere • An institution, not a home
Central focus	• Residents and family	• Survival of agency • Leadership void • Financial gain without regard or understanding of services needed by residents

Source: Adapted from Rantz, M., et al. (1998). Nursing home care quality: A multidimensional theoretical model. *Journal of Nursing Care Quality,* 12(3), 30–49.

Healthy residents are the most desirable ones under this system.

The second system is called the cost-plus system. Here, the nursing home is reimbursed for its costs, plus a "reasonable" profit. The way to make money under this system is to pad the bills—that is, the government is billed for more goods and services than the facility received or billed for goods and services never delivered. Doctors and nursing homes gain if doctors perform "gang visits," stopping just long enough for a quick look at residents' charts and perhaps to visit briefly with people. The government is then billed as though each resident had a separate appointment—a task that would take days rather than hours.

Who speaks for the frail, sometimes confused residents who are not objects of a television news expose? Ombudsman programs are an independent nonprofit agency, federally and state-mandated to be an advocate for long-term-care residents. Employing a staff of well-trained volunteers, this agency investigates and resolves complaints from and on behalf of residents of nursing homes. The agencies, though seriously underfunded by federal and state sources, provide a critical watchdog role. In one northern California county in 1997, the three-staff-person, 20-volunteer program handled 1,052 cases, including over 200 abuse cases. Volunteers made 1,429 visits to long-term-care facilities in the region and contributed 5,155 volunteer hours (OPSC, 1998).

The debate over whether proprietary (for-profit) homes offer a lower quality of patient care than nonproprietary (nonprofit) homes goes unanswered. One study using strict quality control measures found no difference in the quality of care between profit and nonprofit facilities (Duffy, 1987). Other analyses disagree, observing that not-for-profit facilities generally have greater concern for the psychological and spiritual development and well-being of their residents (Rosenkoetter, 1996). Close to three-quarters of all nursing homes, however, are operated on a for-profit basis.

Those in the business who are really concerned about elders and provide quality care deserve credit. Credit must also be given to all the honest doctors, pharmacists, nursing home inspectors, social workers, and therapists. A number of nursing homes are excellent by any standards. The Jewish Home for the Aged in San Francisco is considered to be the best among others. The living quarters resemble rooms in a nicely furnished college dormitory. Virtually all residents are out of bed and dressed. The home has a beauty shop, and there are many activities and opportunities for therapeutic exercising on stationary bicycles and other equipment. "They take care of me from head to toe," said one resident.

New Trends in Nursing Homes

The majority of nursing home residents have disabling mental health conditions, such as depression, Alzheimer's disease, or a history of mental illness. As a result, many nursing home residents present behavior problems, such as agitation, abusiveness, or wandering. Nursing home care presents an increasing challenge for staff, family, and residents alike. In the past, the health care system provided scant rehabilitation for frail elders and held low expectations for older people in nursing homes. Now, however, there are positive new trends for both the physically and mentally disabled.

Those over 65 are more likely than any other group to suffer strokes, lower limb amputation, hip fractures, and heart disease. With rehabilitation, many can achieve a level of independence that allows frail elders to live in their own homes. Rehabilitation units represent a departure from the traditional nursing home. One nursing home with a rehabilitation unit revealed that 57 percent of its residents were discharged after an average stay of three months in the unit. The unit had a team composed of a geriatrician, psychiatrist, physical therapist, social worker, nurse, occupational therapist, and nutritionist. The conclusion was that patients discharged "quicker and sicker," without rehabilitation, are likely candidates for relapse and readmission (Adelman, 1987).

It is estimated that 65 percent of all nursing home residents have at least one mental disorder (Lantz et al., 1997). The mentally ill, who present far more behavioral problems than other nursing home residents, need rehabilitative services. Such services have been found to be generally effective. According to Shea et al. (1994), however, fewer than 25 percent of nursing home residents with a mental disorder actually received such services in the year of the study. Furthermore, only about 5 percent of those residents with a mental disorder received *any* mental health treatment in the last one-month period. One-half of the services they received was provided for by general practitioners who have very little training in geriatric psychiatry. Obviously, although services for those who receive them tend to be somewhat effective, much more must be done to provide the quality and quantity of services for those who need them. It has been argued that nursing homes, in fact, should be mental health facilities. A more in-depth preparation of staff is necessary, and a wider range of mental health professionals must be utilized to meet these needs.

Family and community involvement can improve life for elders in a nursing home. Both resident-only and resident/family counseling groups reduce anxiety and increase feelings of internal control for those in institutions. Family members who strive to keep family connections, offer optimism for recovery, and help the resident to maintain dignity can do much to uplift spirits. Family members tend to judge a nursing home not so much by its technical care as by its psychological care. The views of involved family members provide the staff with good feedback that hopefully can lead to improvements. Families of nursing home residents are often unfamiliar with ways to assist their relatives. Nursing homes that establish a "partnership" with families in sharing responsibilities for the nontechnical aspects of care enhance the psychological and emotional welfare of the resident as well as that of family.

The Omnibus Budget Reconciliation Act of 1987

With the enactment of the **Omnibus Budget Reconciliation Act of 1987 (OBRA)**, the nursing home industry became one of the most highly regulated environments for health care delivery in the United States. In addition to tightening survey and certification procedures, the act strengthened patients' rights in planning their care and making treatment decisions. Other requirements, such as written care plans, mandated nursing assistant training, and the employment of certified social workers, were designed to upgrade the quality of patient services and insure their participation in planning their care.

OBRA also intensified requirements for survey teams making routine and unscheduled audits. As a result, many past procedures, such as placing patients in restraints without medical authorization, denying patients full disclosure of their medical records and diagnoses, and failing to assure patients of their rights, may become grounds for censure or, ultimately, the closing of a facility by regulatory agencies. Although OBRA does not purport to correct all problems related to the quality of care, it represents a concerted effort to upgrade and enrich services to institutionalized elders.

Enforcement of the provisions of OBRA is generally left in the hands of state health departments that are generally woefully understaffed, which results in only occasional audits of nursing homes to ensure that OBRA mandates are being carried out. Although it is hoped that the enactment of OBRA has led to a higher quality of care for residents, empirical evidence does not confirm that those outcomes have been achieved.

Chapter Summary

Theoretically, many acceptable kinds of housing are currently available to those who must relocate:

- Home of a relative
- Public housing
- Deluxe high-rise apartment
- Retirement village
- Mobile home
- Apartment complex or duplex
- Hotel-type residence with meals or a board and care home
- Intermediate day care/own home at night
- Assisted living (a residential complex with many services and facilities)
- Small group residential home
- Institution

Reality, however, does not keep pace with theory. Old people are often forced into housing that is inadequate or that needlessly increases their dependence.

Most prefer to live in their own homes where possible, but others enjoy the stimulation of living in retirement communities, shared housing arrangements, community care retirement centers, or similar living environments. For older adults with limited incomes, public housing or mobile homes may provide a practical alternative. Although the percentages are declining, many continue to live with relatives—usually a daughter or son and their families. Regardless of the type of living situation, most prefer to remain as independent as possible.

For those who decide to relocate (either voluntarily or involuntarily), problems may arise with respect to housing shortages, downsizing, decisions over whether to live in an integrated or segregated environment, or finding a community in which to live that is compatible with their lifestyles. Relocation is never a simple matter, although under optimal conditions it can enhance both the lifestyle and the life satisfactions of those who elect to do so.

For a small percentage of older adults—those who are debilitated—the nursing home is often the only viable housing option. The lack of personalization often creates dependency and the loss of personal freedoms for the resident. OBRA legislation was enacted to increase the quality of life in nursing homes by providing for resident input and offering a broader range of higher-quality services.

Everyone needs a home. The older we get, the more important a safe harbor becomes. Home is not just a place to exist, but a place to be. More needs to be done to meet this most important psychological requirement. We all deserve to live out our lives in dignity, pride, and comfort.

Key Terms

adult community
adult day care
age-integrated housing
age-segregated housing
aging in place
amenity relocation

assisted living
board and care home
continuing care retirement communities (CCRCs)
downsizing

full-timing
granny units
life-care community (LCC)
living environment
mobile-home park
naturally occurring retirement communities (NORCs)

Omnibus Budget Reconciliation Act of 1987 (OBRA)
public housing
retirement city
snowbirds
sunbirds
surveillance zone

Questions for Discussion

1. List some of the changes that can occur late in life, which affect the living arrangements. How many are positive changes? In what way? How many are negative changes? In what way?

2. How do you see yourself living in old age? What would be your ideal living environment?

3. Imagine you have an elderly relative who can no longer live alone due to a stroke. She can still get around and feed herself and she is mentally competent, but she needs assistance with cooking, driving, and doing chores. This person had no money for daily care at home, but she does have a small monthly income. What could you do for her?

Fieldwork Suggestions

1. Research the costs of improving the safety of an elderly person's home. Often simple changes in a home can allow a person to age in place safely when they no longer have good vision, balance, and strength. For instance, removing loose rugs and adding hand-holds to doorways and grab-bars in the bathroom can all prevent falls. Increasing light levels can also improve safety. What other safety measures can you list?

2. Find out what services are available in your community to help elderly persons with home safety, food, and transportation.

3. Visit a skilled nursing facility and ask for a tour. What do you think of the facility? How do you think it would be to live there?

Internet Activities

1. Locate the home page for the American Association of Homes and Services for the Aging. What resources do they have that would give you information if your private nonprofit agency were interested in building an assisted living complex for poor elders in your community?

2. Your parents live 500 miles away from you, their closest adult child. Your mother requires extensive caregiving, and your father is much weakened from a recent heart attack. You fear that his exhaustion in trying to care for your mother will kill him. You feel that the family needs to consider placing your mother in a nursing home. Use the internet to search out the information you need to bring to a family gathering that is being held to discuss, "What will we do with Mom?"

The Oldest-Old and Caregiving

11

Alone Time with Pets Helps Seniors

Dogs apparently need no help lifting the spirits of lonely people. A study has found that nursing home residents felt much less lonely after spending time alone with a dog than when other people joined in the visit. The residents shared their problems and story in "intimate conversations" with the visiting dog, researchers said.

"It was a pretty surprising finding," said Dr. William Banks of Saint Louis University, who co-authored the study with his wife, Marian Banks, a postdoctoral fellow in nursing at Washington University at the time.

"They were happier with the one-on-one . . . bonding with the animal. It suggests human interaction is not value-added, and might be slightly detrimental."

Residents at three St. Louis nursing homes who said they wanted dog visits were divided into two groups. One group received one-on-one visits with a dog; the other group shared the dog visitor with several other residents.

Researchers suspected the visiting dog would prompt socialization—and reduce loneliness—but the residents who shared dog visits with other people reported only slightly less loneliness. The big winners were the residents who had exclusive visits with dogs. Their loneliness decreased substantially.

The research will be published in the March issue of *Anthrozoos*.

"The study also found that the loneliest individuals benefited the most from visits with dogs," William Banks said.

An earlier phase of the study, conducted by Marian Banks in Mississippi in 1997, found that nursing home residents who received one to three dog visits a week had substantial decreases in loneliness, as measured in a psychological test instrument known as the UCLA loneliness scale.

Those without dog visits had no change in loneliness.

The next phase of the study, at Saint Louis University, will look at whether robotic dogs popular in Japan have a similar effect on lonely seniors. Researchers will measure both residents' loneliness and ability to attach to the robotic dog.

Marian Banks said a Japanese study showed that the robotic dog, Aibo, elicited smiles from Alzheimer's patients.

Marian Banks said she used an academic mentor's golden retriever to conduct the first phase of the study in Mississippi.

But for the second and third studies in St. Louis, she's used Sparky, a mixed-breed dog she found four years ago in the alley behind her house.

She said his sweet disposition won her over, and she adopted and trained him. She said Sparky sits next to nursing home residents on their bed, listens to their stories, and lets them groom him. "He sits there very nonjudgmental," she said. "When they go to a nursing home, they lose all their possessions. They need to belong, love and be accepted. The dog gives unconditional love. They say the most incredible things in the presence of a dog."

As the population of the old-old has mushroomed, caregiving to an older person within a family setting has become more common. Informal caregiving usually precedes and sometimes accompanies or replaces the formal caregiving offered by hospitals and other institutional settings. We have no clear-cut norms or customs dictating care for older family members who need help in caring for themselves. This chapter examines the oldest-old population and the caregiving that is offered in a family context.

The Oldest-Old

Recently, researchers have begun separating old-age groups in their studies. As early as 1974, Bernice L. Neugarten separated the older population into the young-old (age 65 to 74) and the old-old (age 75 and over). Before that, studies considered all the elderly together, obscuring important differences and offering little insight into the social realities of the **oldest-old**. Other terms for the oldest-old have also come into use, and they may have slightly different meanings: the very old, the extreme aged, the dependent elderly, and the frail elderly. These terms, often used interchangeably, need to be clarified when they are used.

There is no clear-cut time at which all gerontologists agree that one joins the "oldest-old" and clearly, how old is "old" is relative to one's own age. The six-year-old sees the 10-year-old as being very old and sees himself as being old in terms of his baby brother. In 2009, the Pew Research Center found generation gaps in the perception of "old" to continue to exist. Survey respondents of age 18 to 29 believed the average person becomes old at age 60. Middle-aged respondents put the threshold closer to 70, and those 65 and older responded that old age began at 74 (Pew Research Center, 2009). The discussion in this chapter reflects that fact. Here both age 75 and over and age 85 and over are used at different times to identify the oldest-old,

TABLE 11.1

United States aged population, projected (in millions)

Age in Years	1995	2000	2025	2050	2075	2100
65 and over	34.2	35.4	60.8	74.1	83.7	89.9
75 and over	15.1	16.8	25.1	39.3	45.9	50.6
85 and over	3.8	4.4	6.3	14.7	16.9	20.1

Source: Social Security Administration, Office of Programs: Data from the Office of the Actuary, Washington, D.C.

depending on the studies cited. As longevity and the proportion of very old elders increase, however, "oldest-old" more commonly refers to 85 and over.

Table 11.1 shows the projected number of people aged 65, 75, and 85 and older for approximately the next 100 years. People aged 85 and older are the fastest growing segment of the older population, currently representing 15 percent of the older population (all Americans 65 years or older). Note that the number of people in this age group will be more than five times greater in 2100 than in 1995; the number will be nearly one and a half times greater in 2100 than in 2050. The table shows that there were 3.8 million people aged 85 years and older in 1995 in the United States—a huge number relative to existing social policy—and that number is expected to grow to nearly 15 million by 2050.

Figure 11.1 shows the percentage distribution of older Americans by age group. Nearly 12 percent of the older population was in the 75-to-84-year-old range in 1998. By the year 2050, those age 75 and over will constitute 40 percent of the aged population. The needs of older adults from 75 and older will weigh more heavily in the future as their number and the percentage of their population increase. As a society

FIGURE 11.1

Percentage of older Americans by age group

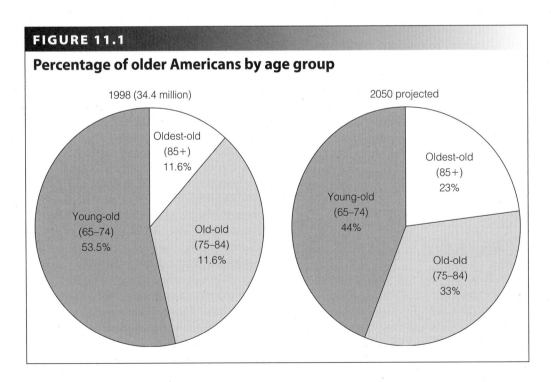

1998 (34.4 million)

Oldest-old (85+) 11.6%
Young-old (65–74) 53.5%
Old-old (75–84) 11.6%

2050 projected

Oldest-old (85+) 23%
Young-old (65–74) 44%
Old-old (75–84) 33%

we will need to provide a viable lifestyle for this burgeoning group. People do not necessarily decrease in functional ability with age, however.

Robust aging is sometimes called "successful aging." It refers to aging while maintaining good mental and physical status. The most robustly aging people have certain characteristics in common: they report far greater social contact than those who are not aging well; they have better health and vision; and they have experienced fewer life events in the past three years than their less well compatriots (Bauman et al., 2001). This finding is a global finding: healthier, more socially integrated people whose later lives have not been disrupted by huge personal, family, or social trauma remain strong mentally at oldest-old ages (Kinsella & He, 2009).

Physical Health

The 75-to-84 age group is, by and large, healthier and happier than stereotypes would have us believe. In fact, the general consensus seems to be that physical losses in old age do not begin taking a heavy toll until after age 85. The 85-and-older group uses approximately 10 times as many hospital days as those age 45 to 64, whereas those 75 and over use about 4.5 times as many (U.S. Census Bureau, 2000). An 85-year-old is 2.5 times more likely to enter a nursing home than is a 75-year-old.

In one of the first longitudinal studies of the health and well-being of elders (median age just under 89 years) conducted in the 1980s, only moderate declines in mental, physical, and daily living functions were found over a 10-year period (Palmore et al., 1985). The findings contradict the gloomy view at that time that the old-old experience rapid decline. In the 1990s, a mere 20 years after the first wave of the study cited just previously, it became clear that traditional views of aging need rethinking. It is now recognized that people who do not suffer from illnesses or severe impairments can exhibit a broad range of healthfulness and well-being. People in their 90s and older are often very healthy and robust.

> *Did you know . . .*
> *Among people 65 to 79, women seem to have a slight advantage over men in cognitive abilities. A gender crossover seems to occur between the ages of 80 and 89, however, after which men are on average higher than their female counterparts. This occurs because men who are cognitively impaired generally die earlier than do women, leaving mainly mentally intact men to live into later life.*
>
> Thomas T. Perls, 1995

The health of the oldest-old is a basic concern. A critical measure of health is whether a person can manage daily activities alone or requires the help of others. A measure of **functional ability** is the ability to do **personal activities of daily living** (PADLs or ADLs) without help. These activities include bathing, eating, dressing, toileting, transferring oneself in and out of a bed or chair, and walking or getting around inside the home (see Figure 11.2). A second measure is the ability to do **instrumental activities of daily living** (IADLs), such as shopping, housework, money management, and meal preparation (see Figure 11.3). Less than half of the oldest-old (85 and over) need some help in performing ADLs, and about 60 percent need help with IADLs.

The oldest-old have a number of common chronic disabling conditions. The most common problems are bone and joint problems, heart disease, vision and hearing problems, mental impairment, drug intoxication (from prescribed drugs), falls, and urinary incontinence. In spite of common health issues, many of the oldest-old

---

FIGURE 11.2

Activities of daily living (ADL) index

Physical Self-Maintenance Scale (PSMS)

Subject's Name _____ Rated by _____ Date _____

Circle one statement in each category A–F that applies to subject.

A. *Toilet*
    1. Cares for self at toilet completely, no incontinence.
    2. Needs to be reminded or needs help in cleaning self, or has rare (weekly at most) accidents.
    3. Soiling or wetting while asleep more than once a week.
    4. Soiling or wetting while awake more than once a week.
    5. No control of bowels or bladder.

B. *Feeding*
    1. Eats without assistance.
    2. Eats with minor assistance at meal times and/or with special preparation of food, or help in cleaning up after meals.
    3. Feeds self with moderate assistance and is untidy.
    4. Requires extensive assistance for all meals.
    5. Does not feed self at all and resists efforts of others to feed him/her.

C. *Dressing*
    1. Dresses, undresses, and selects clothing from own wardrobe.
    2. Dresses and undresses self, with minor assistance.
    3. Needs moderate assistance in dressing or selection of clothes.
    4. Needs major assistance in dressing, but cooperates with efforts of others to help.
    5. Completely unable to dress self and resists efforts of others to help.

D. *Grooming* (neatness, hair, nails, hands, face, clothing)
    1. Always neatly dressed, well-groomed, without assistance.
    2. Grooms self adequately with occasional minor assistance, e.g., shaving.
    3. Needs moderate and regular assistance or supervision in grooming.
    4. Needs total grooming care, but can remain well-groomed after help from others.
    5. Actively negates all efforts of others to maintain grooming.

E. *Physical Ambulation*
    1. Goes about grounds or city.
    2. Ambulates within residence or about one block distant.
    3. Ambulates with assistance of (check one) a. another person _____ b. railing _____ c. cane _____ d. walker _____
       e. wheelchair _____
    4. Sits unsupported in chair or wheelchair, but cannot propel self without help.
    5. Bedridden more than half the time.

F. *Bathing*
    1. Bathes self (tub, shower, sponge bath) without help.
    2. Bathes self with help in getting in and out of tub.
    3. Washes face and hands only, but cannot bathe rest of body.
    4. Does not wash self, but is cooperative with those who bathe him/her.
    5. Does not try to wash self, and resists efforts to keep him/her clean.

Source: Adapted by M. Powell Lawton, Ph.D., Director of Research, Philadelphia Geriatric Center, Philadelphia, from *Older Americans Resource and Assessment Multidimensional Functional Assessment Questionnaire.*

Instrumental activities of daily living (IADL) scale

Name _____ Rated by _____ Date _____

1. Can you use the telephone
   without help, — 3
   with some help, or — 2
   are you completely unable to use the
   telephone? — 1

2. Can you get to places beyond walking
   distance without help, — 3
   with some help, or — 2
   are you completely unable to travel unless
   special arrangements are made? — 1

3. Can you go shopping for groceries
   without help, — 3
   with some help, or — 2
   are you completely unable to do any
   shopping? — 1

4. Can you prepare your own meals
   without help, — 3
   with some help, or — 2
   are you completely unable to prepare any
   meals? — 1

5. Can you do your own housework
   without help, — 3
   with some help, or — 2
   are you completely unable to do any
   housework? — 1

6. Can you do your own handyman work
   without help, — 3
   with some help, or — 2
   are you completely unable to do any
   handyman work? — 1

7. Can you do your own laundry
   without help, — 3
   with some help, or — 2
   are you completely unable to do
   any laundry at all? — 1

8a. Do you take medicines or use any medications?
   Yes (If yes, answer Question 8b.) — 1
   No (If no, answer Question 8c.) — 2

8b. Do you take your own medicine
   without help (in the right doses at the
   right time), — 3
   with some help (if someone prepares it for you
   and/or reminds you to take it), or — 2
   you are completely unable to take your own
   medicine? — 1

8c. If you had to take medicine, could you do it
   without help (in the right doses at the right time), — 1
   with some help (if someone prepared it for you
   and/or reminded you to take it), or — 2
   would you be completely unable to take your
   own medicine? — 3

9. Can you manage your own money
   without help, — 3
   with some help, or — 2
   are you completely unable to manage money? — 1

The IADL Scale evaluates more sophisticated functions than the ADL Index (see Figure 11.2). Patients or caregivers can complete the form in a few minutes. The first answer in each case—except for 8a—indicates independence; the second, capability with assistance; and the third, dependence. In this version the maximum score is 29, although scores have meaning only for a particular patient, as when declining scores over time reveal deterioration. Questions 4–7 tend to be gender-specific: modify them as you see fit.

Source: Adapted with permission from M. Powell Lawton, Ph.D., Director of Research, Philadelphia Geriatric Center, Philadelphia, from *Older Americans Resource and Assessment Multidimensional Functional Assessment Questionnaire.*

have no *functional* limitations: that is, their chronic diseases do not keep them from living alone and doing their own personal and household chores. A comprehensive study of people over 85 years of age and living in home settings found them to have good to excellent functioning in mental and social domains, with the largest functional impairments having to do with physical functioning (Krach et al., 1996). Common chronic diseases in this group were arthritis, hypertension, and cardiac problems. The elders in this study reported that they needed visiting nurses, home health aides, and help with shopping and transportation to maintain themselves.

The implications for society of robust longevity are vastly important ones. As a result of better ways of life and medical advances, the "new" model of living into very old might work to compress morbidity, mortality, and disability into a shorter period of time at the very end of the life span. This means that as the old become very-old, the massive drain on the economy to care for a growing population of frail, disabled older people that some people predict and fear will not happen.

As a result of . . . demographic [changes] and work-force trends, the long-term care industry has boomed over the past few decades, from just over 500,000 nursing home beds in 1963 to 1.7 million today.

MARY H. COOPER, 1998

Living Arrangements and Marital Status

Only a small percentage of the oldest-old are institutionalized. Of the oldest-old, those age 85 and over, more than 50 percent still live at home. More than 35 percent live alone at home; about 75 percent of those 85 and over are widowed. An estimated 25 percent of the oldest-old live in the household of an adult child or other person (usually a relative), and another 21 percent reside in nursing homes. For the oldest-old, the most critical need in the future will be for programs and policies that reduce the risk of dependence and promote self-determination. If housing were more affordable and home help services more available, more people could live in their own homes.

Most people live in their own homes. Not until elders reach age 90 are there more than 50 percent who cannot live independently. A 2005 report by the Census Bureau reports that the most noticeable change in living arrangements since 1980 occurred in the number of women aged 85 and older who live alone. This proportion increased from 42 percent in to 57 percent in 2005. This shift reflects better health status and more financial resources available to assist the older person to remain at home. However, much of the help needed for the oldest-old to live alone is not covered by Medicare.

For example, if a medical diagnosis is osteoarthritis of the knees, the primary problem is trouble with walking. Although the problem is medical, and some help involves pain medication (to control but not cure), major help is needed with shopping and self-maintenance. Such categories of help are not "treatments" and therefore are not covered by Medicare.

The marital status of men and women at the upper age levels afford a dramatic comparison with the oldest-old women much more likely to be widowed and the oldest-old men, married. Compares the percentages of single, married, widowed, and divorced men and women under 75, and 75 and over. The number of widowed men in the age group 75 and older more than doubled (from 10 percent to 24 percent) compared to the number in the 65 to 74 age group. For women in both age groups, the percent of widows are dramatically higher than the men. Nearly 36 percent of women aged 65 to 74 are widowed. The majority (65 percent) of women 75 and over are widowed, whereas the majority (70 percent, in fact) of men in the same age group are married. Widowhood is the predominant lifestyle for women 75 and over.

Evaluation of Life

A study of the oldest-old revealed that at some point between 80 and 90 years of age, individuals became willing to describe themselves as "old" (Bauman et al., 2009). A clear awareness emerges that the spirit is willing but the body is unable to cooperate. Malcolm Cowley (1980), who wrote of his personal experience with aging in *The View from Eighty*, describes this reality:

Everything takes longer to do—bathing, shaving, getting dressed or undressed.

Travel becomes more difficult and you think twice before taking out the car.

Many of your friends have vanished.

There are more and more little bottles in the medicine cabinet.

You hesitate on the landing before walking down a flight of stairs.

You spend more time looking for things misplaced than using them.

It becomes an achievement to do thoughtfully, step-by-step, what you once did instinctively.

Evaluations of reality vary with one's age and life view. Among people age 85 and over who are in good health, living at home, and still driving, the absence of limitations keeps them feeling lucky in life. A study of 12 centenarians found a sense of comparison to be an important part of the evaluation of life: "I'm doing pretty good for my age" was a commonly heard remark (Pascucci & Loving, 1997). **Life satisfaction** is the sense of well-being experienced and identified by the individual, not something recorded by measures external to the individual. To a certain extent, satisfaction in later life seems to depend on an elder's sense of self-efficacy—the extent to which a person is able to master the environment effectively and feel a sense of control in life (Jacob & Guarnaccia, 1997).

Most life satisfaction studies have focused on the young-old and show that life satisfaction does not particularly change over time for an individual. This implies that life satisfaction is more a personal way of seeing the world than something that is affected by current circumstances. Measuring life satisfaction, however, is difficult to do—some researchers believe that self-report measures of life satisfaction are superficial and do not actually measure an individual's subjective experience of well-being. Gerontologists generally believe that phenomenological methods (interviews in which the subject's own words are used to define well-being) are more useful in studying quality of life and life satisfaction among older people.

A greater number of longitudinal studies of the old-old would help us to better understand this age group. Life may or may not be different as one progresses from age 75 to 85 to 95 to 100. Among the "one hundred over 100" (Heynen, 1990) interviewed over 20 years ago for a book about centenarians, full and meaningful lives were the norm. No single factor predicted who would live to be over 100—some were smokers, worrywarts, or nonexercisers. Some were homebodies, others travelers; some were religious fanatics, others skeptics. Old people are as varied a group as are younger people—there are just fewer of them, and their activities and therefore their interests are generally more limited. Studies of the very old continue to identify themes of good health, good coping patterns, and a strong sense of well-being throughout life (Pascucci & Loving, 1997; Aquino et al., 1996; Bauman, 2009).

These qualities do *not* imply the jolly, twinkly-eyed old person of popular stereotype. Common qualities of the very old also include being completely willing to voice an opposing opinion if need be and being an assertive voice for self-care. The very old are a group of people who have survived because of the combination of (1) their genetic makeup; (2) the lifestyles they have both chosen and been blessed with throughout their lives; and (3) personality characteristics having to do with taking care of themselves physically, psychologically, socially, and spiritually. They are generally people who are able to experience joy in surviving and appreciation for what is available to them.

Informal Caregiving

Among those older adults requiring assistance in living, a great number receive in-home care from relatives. Although the numbers vary depending on definitions, at least 20 percent of all noninstitutionalized elders are in need of help. This percentage more than doubles for those age 85 and over. For every disabled person who resides in a nursing home, two or more equally impaired aged live with and are cared for by their families. An even larger number of people provide help to old people who do not share their households.

A myth that the extended family, especially the three-generation family, was common in the 1700s and 1800s is widely accepted. In fact, the three-generation family is more common in the 2000s than it was then. The primary reason is due to demographics: more older adults are alive today.

The preferred living arrangement of older people, however, is to live independently from adult children. The trend over the last several decades has been to encourage such independence with more care being provided in the homes of elders, not in the homes of their adult children.

In 2006, more than half (55 percent) of noninstitutionalized older adults (65 and older) lived with their spouse. This represents roughly 72 percent of older men and 42 percent of older women (AoA, 2007). Just less than 30 percent of women aged 75 and older lived with a spouse. But half of women (48 percent) aged 75 and older live alone, a proportion that increases with advanced age.

In 2006, more than 670,000 grandparents aged 65 and over maintained households in which grandchildren were present. In addition, over 800,000 grandparents over 65 years lived in parent-maintained households in which their grandchildren were present, and roughly 450,000 of these grandparents were the people with primary responsibility for their grandchildren living with them (AoA, 2007). It appears that families are there for their older members as the need arises, despite the preference for independence that American older people retain.

Informal caregiving comes from relationships that exist naturally in a person's environment, such as family, friends, church, and organizations that are not professional or financed by the government (e.g., members of a garden club). Informal care is given by family, friends, or other community members whereas formal care is provided by professionals and paraprofessionals. **Formal caregiving** includes health care professionals, hospitals, and day-care

Informal caregiving is often provided by social organizations like the church this man attended for many years. His friend visits three times a week to share and help with lunch.

centers and nursing homes. Staff are professional or paraprofessionals and are paid for their caregiving.

The stress of caring for a frail, disabled person is very costly in human terms among both formal and informal caregivers, but particularly for informal family caregivers, whose psychological relationship with the care recipient is generally intense and complex. It is not uncommon for an elderly caregiving spouse to die, leaving the "frail" spouse behind. Informal caregiving is most commonly carried out by a spouse, and when a spouse is not available, by a daughter.

From a statistical perspective, the typical caregiver is a 46-year-old working woman who spends 18 hours a week caring for her chronically ill 77-year-old mother who lives nearby. Forty-one percent of caregivers also care for their own children under the age of 18 (Cooper, 1998). Most caregivers get help—usually from a son, daughter, or other relative—in providing services such as transportation, grocery shopping, and household chores for their relative (Adams, 1998). Despite the myth of the nonfunctional nuclear family, research has found that families generally work hard together to provide care for a frail member. This care adds a heavy toll on family caregivers, but it occurs nevertheless.

Obligations of Adult Children to Parents

The **modified nuclear family** concept describes the typical American situation. There is a great deal of family interaction, for example, visiting and exchanging gifts and services, but there is no extended family household. Most elders live in separate households, but they are not cut off from their families. However, the standard indicators used to measure family interaction do not measure emotional closeness; nor do they measure how responsible the family feels for frail, older parents.

It's just really, really hard. It was a year and a half of unbelievable sandwich-generation problems. With the best will in the world you aren't necessarily doing the best thing because you don't know what that is. What worked yesterday won't work today, but how you're supposed to figure it out, I don't know.

ADULT CHILD CAREGIVER, REPORTED IN COOPER, 1998

Gerontologists observe that no clear cultural guidelines and no specific norms for behavior exist in the area of intergenerational relationships between elderly parents and adult children. Some older parents expect that (1) married children should live close to their parents; (2) children should take care of sick parents; (3) if children live nearby, they should visit their parents often; and (4) children who live at a distance should write to or call their parents often. Indeed, that is why the phrase "modified extended family" has been used—it describes families that keep close ties even though they do not share households. These norms or expectations of close relationships are adhered to in the typical American family.

He doesn't know us anymore. But every time I go to say goodbye it kills me. I'm so sick of saying goodbye. . . . One time when the doctors told us he wouldn't make it through the night, my mother and I actually made funeral arrangements. . . . It was so hard. I grieved. I mourned. He didn't die— and I actually got angry. I was like—when is this going to end? Of course I felt guilty.

DAUGHTER, QUOTED IN MCCARTY, 1996

Extreme financial or physical dependence of aged family members tests the limits of adult

Vignettes to elicit sense of responsibility of adult children for dependent parents

Vignette 1

A 64-year-old woman had a major stroke 8 months ago, causing paralysis of her right side. Because of this weakness, she is unable to get out of bed and into her wheelchair or onto the toilet without physical assistance. Her 71-year-old husband, who has some minor health problems, finds it very difficult to do this for her and is only just managing. Also, he is obviously having trouble coping emotionally with seeing his wife in such a weak condition.

 Their only source of income is the government old-age pension, so they are unable to afford private help.

Vignette 2

A 76-year-old man has had Alzheimer's disease for the last 4 years. He presently lives with his wife in an apartment, where a community nurse visits him every week. Besides poor memory, his major problem is that he has no understanding of his illness. He gets quite upset, and at times aggressive, when anyone tries to get him to do something against his will. He tends to get more agitated at night and has trouble sleeping. He has reached the stage where he requires supervision for most activities and so cannot safely be left alone for any length of time.

 His wife, who is 72 years old, is in reasonable physical health but is becoming exhausted with having to take care of her husband. She is unable to get a good night's sleep and cannot rest during the day because she is afraid her husband will get into trouble.

 They are living on their old-age pension supplemented by a small amount of savings.

Vignette 3

An 82-year-old widower lives alone in a small second-floor apartment with no elevator. He has quite bad arthritis, causing pain in his knees and deformity in his hands. He can get about his apartment safely with specially designed canes but is unable to get outside without help. Even though he has few visitors he claims not to be lonely, but the visiting social worker thinks that he must be.

 Two years ago he had an operation to remove a cancer from his bowel. Unfortunately, he required a permanent colostomy, which means that he has to wear a plastic bag on the side of the abdomen to collect his stool. Because of the arthritis in his fingers he needs help to empty and maintain the bag.

 He receives the old-age pension along with a small pension from his former employer.

Source: Wolfson C. et al. (1993). Adult children's perceptions of their responsibility to provide care for dependent elderly parents, *Gerontologist,* 318.

children's sense of responsibility. In seeking to determine whether there is a solid moral underpinning for filial responsibility, Canadian researchers used the vignettes in Figure 11.4 to elicit adult children's sense of responsibility for physically ill or disabled parents (Wolfson et al., 1993).

 Respondents in the Canadian study were first asked what levels of assistance adult children *should* provide in each case. Second, adult children were asked to imagine those described in the vignettes as their own parents. They were asked what levels of financial assistance, emotional support, and physical assistance they *could* provide if the elders described in the vignettes were their own parents.

 The overwhelming majority of people interviewed (almost all) felt that emotional support *should* be provided by adult children in general and *could* be provided by them for their parents. Not quite as many, but still a substantial majority, believed that adult children *should* provide physical assistance; a somewhat lower percentage *could* provide physical assistance. And a majority believed adult children *should* help out financially. The lowest figure in the study was the degree to which adult children *could* help financially, and even these scores were reasonably high. The researchers concluded that adult children feel a strong moral obligation to provide care for their disabled parents.

Another study in a similar vein asked respondents to agree or disagree with statements such as the following (Mangen & Westbrook, 1988):

If an old man has a medical bill of $1,000 that he cannot pay, his son or daughter is morally obligated to pay the debt.

The purpose of the study was to deeply understand the **intergenerational norms**, which are the standard, expected behaviors of one generation toward another. Interestingly, on the earlier item older generations were more likely to disagree than younger generations. Among Americans, the strong value of independence may well prevent older parents from expecting their adult children to take on any economic responsibility for them. In contrast, the majority of adult children agreed with the earlier statement.

Not all studies show similar expectations of aid to older parents from adult children. Intergenerational conflict was revealed in a classic study of three generations of women, using a hypothetical story about a widowed mother to ask each generation what adult children *should* do for an elderly parent (Brody et al., 1984). Respondents from each generation recommended in general that adult children not share a household with the mother. The youngest group of women was most in favor of sharing the household, a feeling the researchers interpreted as a youthfully idealistic view of the caregiver role.

Older parents judged female family members as acceptable to provide such personal care as meal preparation, housework, and grocery shopping, whereas male family members were judged unacceptable. Middle-aged daughters reluctant to change family schedules or give up jobs outside the home were somewhat unwilling to become caregivers, yet the oldest mothers in the study were the most likely to expect caregiving from their middle-aged daughters.

A study recently released by Stanford University Medical Center found that the physical cost to middle-aged daughters who care for a frail parent can be actually greater than that to the spouse of the person requiring care. Daughters experienced a significantly greater increase in heart rates and blood pressure than did wives during social interactions with the ailing loved one, and the daughters recorded significantly more distress in interpersonal interactions (Evans, 1998). Divided energies, divided loyalties, and feelings of guilt for being unable to meet all the demands of caregiving exact a heavy toll.

Did you know . . .
The number of extended family households in the United States steadily decreased from the turn of the twentieth century until the 1980s, when the downward trend reversed. This shift resulted from increases in horizontally extended households among immigrants—primarily Mexican, Guatemalan, and Salvadoran. It is related to young, single adults living with relatives and in increases in poverty rates among immigrants from these countries which precludes independent living among the newest immigrants.

JENNIFER E. GLICK ET AL., 1997

Weakened Family Support Systems

As far back as 1977 Treas predicted that the family would be more and more strained by caring for its oldest members. She cited three factors that contribute to the **weakened family support system**, which remain relevant today: (1) demography, including fewer offspring and longevity; (2) women's changing roles; and (3) changing intergenerational relationships.

Demography

Long-run trends in mortality and birthrates have startling consequences for the kin network. At the millennium, the aging parent, having raised fewer children, will have fewer offspring to call upon for assistance than did his or her own parents. Middle-aged children without siblings to share the care of the elderly parents will feel increased strain, and those adult children will themselves enter old age along side their own parents, who will be then in the oldest-old category.

Kin networks offer fewer options and re sources when the younger generations have fewer members. Generally speaking, an aging couple will fare better when several children can contribute to their support. And having a number of grown children increases an aging widow's odds that at least one will be able to accommodate her.

With more and more people living into their 80s and beyond, more of the old will be frail elderly, or the old-old. The burden of their physical, financial, and emotional support may be considerable, especially for the young-old children of old-old parents, children whose own energy, health, and finances may be declining.

Women's Changing Roles

Traditionally, providing older parents with companionship and services has produced a sexual division of labor. The major burden for physical care and social activity has traditionally fallen on female relatives' shoulders, whereas the burden for instrumental care such as tax completion, medical bookkeeping, and the like, has become the responsibility of male relatives. In a study of the division of caregiving responsibility among siblings, Wolf and associates (1997) found that the presence of siblings does reduce the workload on any one adult child; however, it is far less than a one-for-one basis. They further found that the primary predictor of parent-care hours is the presence of sisters: when one's siblings include sisters, the workload on any of the siblings is less.

Changing Intergenerational Relationships

In the past, children have been expected to tend to their aging parents in return for inheriting the family farm or business. This exchange created a pattern of economic interdependence. Parents had greater certainty of care in their later life through their ownership of the greatest family asset.

Children are now less likely to take over their parents' farm or business. They are freer to take jobs elsewhere or establish careers independent of their parents, and indeed the more well educated a person the more likely he or she will move a longer distance from his or her hometown location than those with no more than a high school education (Cohn & Morin, 2008). Emotional ties, however, remain strong in the American family. Affection, gratitude, guilt, or a desire for parental approval still motivates adult children to care for their aging parents. Thus, the basis for helping is now less for economic and more for psychological reasons.

We may view these changes as either positive or negative. On the one hand, the power of older people to ensure family support is reduced. On the other hand, most parents—and certainly adults in the present old-age category—hope for upward mobility for their children and that often requires taking a job and building a life elsewhere for those children. Social Security and other governmental programs have also reduced the dependence of the old on family-support systems. In the postindustrial society and the welfare state, both grown children and aged parents have been somewhat liberated from complete economic dependence on each other.

> *It seems that society is so enamored with the notion of "the family," that we overlook the fact that parenting can be a very painful experience. Too often, we deprive our older people of the chance to express the guilt, grief, uncertainty, fear, and uneasiness that are tied to their concern that they have not been effective parents.*
>
> ROBERT BUTLER, 1977

Changes in the family, demographic shifts, and changes in the economic structure have necessitated greater levels of governmental and formal intervention in the care of older people. Programs such as day-care centers, hot meals, and housekeeping services are essential to help relieve overburdened family-support systems. However, these programs still suffer from high cost and limited availability.

Spousal Caregiving

Spouses provide a large percent of caregiving, with wives being more likely to be caregivers than husbands. About 40 percent of all caregiving is provided by spouses, 14 percent by husbands, 26 percent by wives (AoA, 2007). Living to very old age as a married couple represents a biological and social achievement. Nevertheless, the consequence of marriage in later life is not always blissful. If illness and chronic impairments loom large, heightened anxiety, interpersonal difficulties, and economic strains may result. Older married women often become part-time nurses for their husbands, who tend to be older and in worse health than themselves. Of noninstitutionalized elders in the community, some are homebound, and of those, some are bedridden. A husband or wife,

if available, is most often the primary caregiver, and if the couple is lucky, paid helpers and children provide additional assistance.

In addition to family structure, both gender and race differences shape the caregiving experience. A recent study comparing male and female Anglo- and African American caregivers showed that men of either race are less likely to refer to caregiving in terms of emotional work, and men and African American women caregivers are less likely to acknowledge a difference in the caregiving role between males and females (Miller & Kaufman, 1996).

Spouses are first-line responders when it comes to caregiving. Relationships of mutual caring provide the strongest context for satisfaction in the difficult task of assisting with basic needs.

Sometimes the caretaking role is assumed for years and years. In the following example, the husband has Alzheimer's disease:

For nearly a decade, Hildegarde Rebenack, 69, has watched her 78-year-old husband, Robert, deteriorate from Alzheimer's disease. Robert was a bank examiner, a man proud of rising above his eighth grade education. Now, he spends his days staring at a collection of stuffed animals in their Louisiana home. . . . Robert was diagnosed in 1982. By 1984, he could no longer be left alone, and, two years later, Hildegarde had to put him into a nursing home. . . . Last March, the latest price hike forced her to bring him home. "We had 37 good years together," says Hildegarde, her voice breaking. "But the last six years have been hell." (Beck, 1990)

For the spouse of an impaired person, decline challenges long-established patterns of interaction, goals, and behavior. The loss of functional capacity in one partner has reciprocal emotional effects on both partners, and those effects are highly variable, depending on the couple. Spouses meet with circumstances that bring up the most intense emotional issues: life and its meaning, freedom, isolation, and death. Different styles of caregiving bring different problems and issues. Some women feel "burned out" by giving too much of themselves over the years, both before their husbands' health began to fail and after. Both men and women sometimes need to redefine their later life roles as spouse, and they may require assistance in establishing an identity as an individual as well as someone who is part of a marriage as the marital role changes with the changing health of the frail partner. In a study of husbands who were caregivers, Harris (1993) named the following four types:

1. *The worker*: He models his role after the work role. He plans his work schedule every day. He reads everything he can about Alzheimer's disease and has a desk to organize the insurance papers and other materials.
2. *Labor of love*: He provides caretaking out of deep feelings of love. He often holds his wife's hand and embraces her. He cares out of devotion, not out of duty.
3. *Sense of duty*: He provides care that stems from commitment, duty, and responsibility. He says, "She would have done the same for me. I will never abandon her."
4. *At the crossroads*: He is a typical new caregiver who hasn't oriented to the role. He is floundering and in crisis.

Harris concluded with a variety of support recommendations: (1) educational groups led by a male caretaker and nurse clinicians; (2) support groups limited to men, because men are often not comfortable discussing personal matters in predominantly female groups; (3) computer networking programs for men who would not join groups; and (4) more quality and affordable in-home respite services.

These conclusions are consistent with a second study that provided a comprehensive psychosocial intervention program available to all family members, male and female, over the entire course of an aged family member's disease (Mittleman et al., 1995). By providing options for helping interventions, spouse caregivers were able to use services most suited to their unique situation and relationship with their spouse and their larger family. The results showed significantly lower rates of depression among the treatment group than among those in the control group. Supportive intervention can work to help ease the tremendous emotional and physical burden of caring.

Adult Children as Caregivers

Now is the first time in history that American couples have more parents than children. The number of years of caregiving that a married person with children might plan for is around

18 years for children, and 20 years for elderly parents. These numbers are particularly dramatic for women: nearly 30 percent of aging people who need home care receive it from adult daughters; 25 percent from wives; and 20 percent from more female relatives such as sisters or sisters-in-law or female nonrelatives. This compares with 10 percent of home care from adult sons. Adult sons, however, are likely to provide *instrumental* support: communication with the physician, completion of insurance forms, and the like.

The greatest strain seems to be upon "women in the middle": middle-aged women who have children and jobs and are also responsible for the home care of parents. The demand of multiple roles on the caregiver can be profoundly stressful. This stress is mediated by the history of the relationship with the parent, the extent of social support the caregiver has, her coping skills, and her self-image (McCarty, 1996; Adams, 1996). Some adult child caregivers, for example, have long histories of conflict with the very parent for whom they now feel responsible, and these histories shape the caregiving experience for both the caregiver and the care recipient. One care-providing daughter (reported in Brackley, 1994) wrote:

> When I was young my mother manipulated me until I could get away from her. To have to take responsibility for her now is quite distressing. Everyone thinks she's "cute," or "quite a character," but don't understand that these qualities are not endearing on a lifetime basis. In fact, they are outright distressing when they seem to take over your life. I am trapped in a circle of anger and guilt, guilt and anger, with a smattering of denial and grief.... (p. 15)

Unmarried adult children who share a household with the ailing parent provide the most care. Being married and employed decreases the amount of help given. Living apart further reduces the amount of help elders receive. Adult children caregivers most commonly offer help with the following activities: getting out of bed and going to the bathroom, shopping for food, traveling, doing laundry, preparing meals, doing housework, bathing, taking medicine, dressing, getting around the house, and providing personal and supportive communication.

Family size, socioeconomic level, and ethnicity are all determinants of family caregiving patterns. In one study of middle- and upper-socioeconomic neighborhoods, non-Hispanic whites with no children were more likely than are Mexican Americans to report other family or nonfamily as their available caregiver. Among those with five or more children, however, non-Hispanic whites were *less* likely than Mexican Americans to report nonfamily as their caregiver (Talamantes et al., 1996). These findings imply that Mexican Americans have access to a more extensive support system, especially informal support, than do non-Hispanic whites and highlight the need for services and policy to address ethnic differences.

Parent care involves a constant tension between attachment and loss, pleasing and caring, seeking to preserve an older person's dignity and exerting unaccustomed authority, overcoming resistance to care and fulfilling extravagant demands, reviving a relationship and transforming it.

ABEL, 1991

Unlike caregiving for children, who become more independent with age, caregiving for the impaired older parent requires more effort as the years go by, and the ultimate outcome of the caregiving task will be that parent's death. Coping skills and management strategies are stretched to the limit. Providing care for a frail parent promotes compassion and personal growth among some, or is overwhelming for others, with a negative outcome for both the adult child and the frail parent.

An early study of 30 women caring for a chronically ill parent or parent-in-law along with the person receiving care observed two kinds of parent-caring roles: (1) care provision and (2) care management (Archbold, 1983). A care provider was one who personally performed the services an older parent needed, whereas a care manager was one who identified and obtained needed services, then supervised their provision by others.

The care-providing individuals were generally more strained, had lower income and less knowledge of available resources, were more involved in the heavy physical work of daily care, and were less likely to provide stimulating activities and entertainment in their caregiving role. Care providers are far more likely to be women. In contrast, though care managers, the predominantly male family members did not totally avoid strain and stress, they experienced less than that of the care provider. The financial costs care managers incurred were greater, but expenses were high for care providers as well.

We might note here that the Tax Reform Act of 1986 entitles anyone who works and pays for the care of a dependent to a federal tax credit. However, care must be for a child or other relative for whom the taxpayer provides *more than half* the needed support, and the maximum credit allowance is only 30 percent of care expenses per dependent. Therefore, the act provides very little actual relief.

Childlessness

Children were once described as one's old-age insurance. Today, however, 20 percent of the women over age 44 have no children, and this trend will accelerate with increased longevity among the baby-boom generation who experienced decreased fertility rates and increased divorce rates (Zezima, 2008). This rate is double the level of 30 years ago, and women who do have children have fewer than ever—an average of 1.9 compared with the mean average of 3.1 children in 1976 (AoA, 2007). This drop in fertility rate will impact care and caregiving for adults in their later years.

Most studies on childlessness have addressed psychological and emotional health rather than actual care provision. The general profile of the childless individual, compared with the elderly parent, emerges as someone who is more financially secure and in better health; more reliant on siblings, nieces, and nephews, and on hired care providers; and experiencing no greater loneliness, isolation, or unhappiness than their counterparts with children (Choi, 1994; Zezima, 2009). Patterns of help seem to be shaped more by marital status than by parental status.

Other researchers have found that childless elders follow the "principle of substitution" by turning to extended kin for help, but these sources are less available than are children. Childless married couples tend to rely primarily on each other and remain otherwise independent from extended kin. Unmarried older people, having established lifetime patterns for seeking assistance, seem to be more resourceful in using a variety of people and social resources to meet their needs. Having no necessarily built-in family support system, the childless elderly require more highly available substitute supports.

Elder Caregiving to Adult Children and Grandchildren

The number of older parents who care for dependent adult children and grandchildren is increasing. More than six million children live in households headed by grandparents (4.5 million children) or other relatives (1.5 million children). In most of these homes grandparents and other relatives take on primary responsibility for the children's needs (Fact Sheet, 2007).

Reduced mortality in the second half of the twentieth century among children with mental and physical disabilities has placed many older parents in the role of being a care provider throughout their lives. It is no longer unusual for an 85-year-old parent in good health to be caring for a 65-year-old offspring with developmental disabilities, or a grandchild who has been abandoned by parents. In those instances, grandparents often assume this responsibility without either of the child's parents in the home.

Sometimes older parents assume the caregiver role because no one else is available to care for their dependent children or grandchildren. Support groups for elderly parents have become increasingly available in recent years to help them with their caregiving, especially in times of stress and crisis. A study funded by National Institutes of Health, Child Welfare Division, reports that between 1983 and 1993 the number of abused and neglected children had increased nearly 135 percent. The Project Healthy Grandparents was implemented in 1995, when children with grandparents represented 50 to 70 percent of the children in kinship care. Few assistance programs existed at that point.

The study based its model on the assumptions, based on previous research, that mental health, social support, physical health, legal services, educational support services, financial benefits, and substance abuse referrals were areas of special need for older adults returning to the parenting role.

The impact of later-life caregiving on 105 mothers of adult children with mental illness and 208 mothers of developmentally disabled adult children was assessed to determine their experienced levels of stress (Greenberg et al., 1993). It was found that elderly mothers had the most problems with mentally ill adult children. The mentally ill children posed greater behavior problems and were less likely to have day care outside the home or be employed. The mothers' social network of friends and relatives was smaller than the mothers of developmentally disabled adult children.

The emotional cost of caring for a disabled child is high. Strawbridge and colleagues (1997) found differences in mental health, but not in physical health, between elderly people caring for an ailing spouse and those caring for an adult child. In that study, grandparent caregivers experienced decreases in physical health as well as mental health.

Solving Caregiving Problems

The stresses and strains of caregiving have drawn considerable attention over the last decade. The discussion that follows summarizes some of the solutions that have been offered.

Stress and Stress Management

A fierce tangle of emotions comes with parenting one's parents: anguish, frustration, inadequacy, guilt, devotion, and love. Stress comes not only from these emotions, but also from work overload. "Caregiver distress" indicates the negative stresses of caretaking, including role strain, subjective burden, depression, anxiety, hostility, and other troublesome emotions. Scales such as the Zarit Burden Interview (ZBI) have been developed to measure degrees of caregiver distress (Knight et al., 1993). One specific emotion that has been widely studied is depression. Gerontologists seek to understand how much of the depression precedes the caregiving distress and how much is solely the result of the new added role. Those who are not depressed to begin with cope better than those who are. Personality factors have been examined, using personality inventories, for their role in one's ability to cope (Hooker et al., 1994).

Coping has been conceptualized as a response to the demands of specific stressful current situations. Coping techniques and abilities vary from person to person. These variations in ability need attention and explanation so that those with poor skills can be helped. In one study the specific stressors imbedded in caring for severely impaired elders were specified. The most commonly identified characteristics of the person with Alzheimer's disease that cause stress to the caregiver are memory defects, loss of ability to communicate, and unrelenting decline (Williamson & Schulz, 1993). The process in which a caregiver travels has been identified by Opie (1992) as that of "taking on the responsibility for caregiving, its content, the relationship between various family members, and the interface between informal and formal care" (quoted in Adams, 1996, p. 705). Opie's process model does not include the well-documented grief response that accompanies caregiving, which is a heavy load.

Problems in caregiving arise from (1) the strain of responsibility for direct personal care of the elder, (2) the caregivers' own current personal and health problems, (3) role strain from the demands of other work and the need for leisure, (4) inter-sibling problems and other strained family relationships, and (5) arranging outside help and coping with bureaucratic mix-ups. A social support network is very important to caregivers.

Although helping parents can be rewarding for many adult children, it can be accompanied with enormous burdens, both financial and emotional. The principal caregiving adult children are "women in the middle," pulled in many directions from competing demands on their time and energy. Brody (1990) reports that three-fourths of daughter caregivers said that

The ability to appreciate life's small graces as well as to laugh at life's absurdities is often the outcome of a long life well lived.

they felt they were not doing enough for their mothers. Adult children often express feelings of guilt as caregivers because of a deep commitment to "repay" their parent for care given in their childhood and infancy. This is impossible, which results in guilt.

Ellen McCarty's intensive interviews with 17 adult child caregivers (1996) found that daughters were less likely to experience stress if they could:

1. Restructure her parents' former identities and her own prior filial relationship

2. Seek out helpful management strategies such as taking time for herself, taking it "day by day," receiving support from siblings, using humor, perceiving support from formal caregivers, and "having someone to talk to"

3. Perceive and respond to available support from family and significant others

4. Seek support from professional caregivers

5. Grieve in response to perceived losses and changes in self and others

6. Engage in positive restructuring of life beliefs in the face of reality (p. 800)

Psychological Interventions and Respite Care for Caregivers

The 1980s was a time for drawing attention to the burdens of caregiving. Gerontologists have documented the problems again and again and still study what kind of help is most useful. A controlled study of psychosocial interventions (psychological help from paid professionals) and respite programs (programs that allow the caregiver time off) showed them generally to be moderately effective (Knight et al., 1993). The kinds of **psychosocial interventions** for the caretakers were varied: individual counseling, family counseling, support groups, educational groups, problem-solving groups for the caretaker and patient, social worker visits, and family consultants.

Support groups are typically available in larger communities at mental health clinics and also from individual counselors. They usually meet once a week for eight weeks or so, in which individuals share problems and help one another with emotional support, friendship, and ideas. A paid professional leads the group. We should keep in mind that such group interventions work for some, but not for everyone. Counseling, either individual or family, can also be helpful if families can afford it; although for some families suffering from poor relationships, no amount of professional help can "fix" the problem. In this case, emotional distancing may be the only stress-management technique that would alleviate guilt and despair. Organizations are available in most communities to offer services to caregivers or refer them to individuals who do.

In the study cited earlier by Knight et al. (1993), respite care was somewhat more effective than psychosocial interventions. **Respite care** can mean placement of the dependent elder in a nursing home for two weeks or so, or it may involve bringing a hired home care worker into the home for one to two weeks or more while the caregiver goes on a vacation. Time off for the caregiver is very important in relieving distress. Intervention and respite care should not measure its success only by the alleviation of caregiver distress. Other important outcomes have been found: improved functioning of the dependent elder, reduced use of hospital days for the elder, and improvements in patient mortality.

Future research should direct attention to how much and what kind of intervention is necessary to achieve a desired effect (Knight et al., 1993). If interventions and respite care work, which they often do, the next step is to determine what works best at what levels of strength, with which kinds of caregivers for elders with which kinds of impairments. A broad range of ethnic differences in real and perceived family support and in coping styles exist, and social service programs must identify and address these cultural differences to adequately support the provision of family-based care (Talamantes et al., 1996; Strom et al., 1997).

The Home Care Crisis

The percentage of elders is increasing, and more family members work outside the home. Is a crisis emerging? Who will care for the very old in our society? Many gerontologists believe that a trained paraprofessional labor force for long-term care of the aged and disabled must be a national priority. Others believe that the next cohort of elders will be more healthy and less disabled than their parents, and that medical and lifestyle advances will compress their morbidity and morality into the very end of their lives, as opposed to living for decades with disabling diseases. Regardless of which scenario plays out in the 21st century, the number of Americans needing long-term care reached around 9 million by 2000 and is expected to mushroom around 24 million by 2060 (Cooper, 1998). Home health care personnel are now the fastest-growing segment of the long-term care industry, and the demand for their services will continue.

Home care workers are generally poorly paid, overworked, and poorly trained. They are disproportionately middle-aged ethnic minority women. Wages, job conditions, and opportunities for caregiving work have typically been at the lower end of the employment scale, with no chance for advancement. Outstanding **paraprofessional caregivers** in such job categories as nurse's aid, home health aide, personal care attendant, chore worker, and homemaker should receive prospects for increased responsibility and advancement. Poorly paid **paraprofessional home care workers** are a transient work group with a high turnover rate. Paraprofessionals at present have little training. In some states they must undergo about 75 hours of training and pass a competency test. In many states, however, the requirements include fewer hours of training and no test. Given such easy entry into the field one might expect a large labor pool, but this is not the case. Many states report having trouble delivering state-funded home care services because of worker shortages and high turnover. Increased wages and better working conditions will reduce caregiver turnover, and continuity of care will be improved.

Paraprofessional home care wages lag behind those with similar jobs in nursing homes and hospitals. Homemaker rates in Massachusetts are at the higher end of the pay scale, running from $11.50 to $12.50 per hour. However, large cuts in state home care funding have resulted in the loss of 500 paraprofessional homemaker jobs in that state. Another problem in the industry is the poor job image associated with the long-term care of chronically disabled elders (MacAdam, 1993b). More attention needs to be paid to upgrading the job image and recruiting, training, and retaining quality home care workers for elders (Eustis et al., 1993).

Many families do not qualify for state-funded home care workers. Nearly 46 percent of all Medicare users of home health care must purchase services in addition to those Medicare offers (Kane, 1989). Only 25 percent of the disabled elderly population getting home assistance receives skilled nursing care at home. When they have to pay out of their own pockets, many go without care. Those in need who have enough income and access to services do seek paid help (Stoller & Cutler, 1993).

Although private insurance should be available at affordable prices to cover home care, it typically is not. The gerontologist Nancy Kane states that, "Society should be compelled to come up with better ways to both pay for home care and provide the services" (Kane, 1989, p. 24). She proposes a managed long-term care insurance program focused on maintaining older people in their homes, a program similar to most HMOs. She estimates that 40 to 50 percent of the very old could afford such an insurance policy. The federal government could assist in a number of ways, such as subsidizing low-income elders who want to buy into the program.

The private sector has a stake in caring for elders. There were 33 million Americans aged 65 or older in 1990; 70 million will reach old age in this century when life expectancy may increase to 90 years. Currently, about 80 percent of all in-home care received by frail elders is being provided by family members. This elder care costs business billions of dollars per year in absenteeism, lost productivity, and increased turnover (Scharlach, 1994).

Businesses are run better if caregiving adult workers receive outside help in caring for aged parents. To date, only about 3 percent of U.S. companies have policies that assist employees who care for the elderly; but many more are considering such programs. Such programs need to develop, and quickly; for the growth in the labor force in the 1990s will come primarily from women aged 35 to 54, Brody's stressed "women in the middle."

Researchers suggest that employers could provide more policies, benefits, and services for employees who are caregivers. They recommend (Neal et al., 1993):

1. Part-time job options such as job sharing, voluntary reduced time, or shared retirement
2. Flexible work hours, such as a spread during the day and evening, or a compressed workweek with a day or two off
3. Policies encouraging working at home
4. Paid sick, vacation, and personal leaves
5. Parental/family leave, paid or unpaid
6. Medical/emergency leave
7. Dependent-care reimbursement plan in which the employer directly subsidizes a portion of the employee's caregiving expenses
8. Long-term care insurance that includes home care by paraprofessionals and respite for family caregivers
9. Education about caregiving
10. Resource and referral
11. Counseling
12. On-site day care
13. More tax relief for caregivers

Options for care provision in later life must be developed now, before the tidal wave of the boomer generation requires care. In some families, home care with occasional paraprofessional assistance works to provide the best care for the elder; in others, broad ranges of outside assistance are necessary; the needs of elders in other families will be best met by living in assisted living housing or long-term care facilities. Some companies such as IBM, Stride Rite, and Travelers Insurance Company have elder care programs. They offer flexible work schedules to allow employees time to care for older relatives and further pay for leave up to 26 weeks for elder care. Adult day-care costs can be lower than bringing in home health care aides, depending on how much care is needed. The elderly are a very diverse group with widely diverse needs. Services must address those needs.

Chapter Summary

This chapter extends information in Chapter 6, providing an emphasis on the oldest-old and caregiving. Norms governing intergenerational relationships are not clear-cut, which leaves room for elders to feel disappointed when adult children do not live close by and act in a supportive manner. Aged women far outnumber aged men, especially at advanced ages. The existing studies show the oldest-old to be healthier and more active than stereotypes would have us believe. Most manage to live outside nursing homes, either in their own homes or with adult children. Chronic conditions are common but most people are not totally disabled by them. Caregiving by adult children is becoming more and more common as more elders live into their 80s, 90s, and 100s. The largest percentage of caregiving is provided by older wives and middle-aged daughters. A smaller but substantial percentage of caregivers are older husbands and sons. Caregiving places a great deal of stress on families because more women are employed outside the home, and families have fewer children than they once did. In spite of the burdens, the family in the United States remains strong and a major source of aid to its members. More paraprofessionals are needed in the home care field to offer quality care at a reasonable price to the oldest-old.

Key Terms

caregiver distress
coping
formal caregiving
informal caregiving
instrumental activities
    of daily living
intergenerational
    norms
modified nuclear
    family
oldest-old

paraprofessional
    caregiver
paraprofessional
    home care worker
(personal) activities
    of daily living
psychosocial
    interventions
respite care
weakened family
    support system

Internet Activities

1. Beginning with the home page for Asset and Health Dynamics Among the Oldest Old (AHEAD) locate current information on health factors among the oldest-old in the United States. What ethnic differences are evident? (Use www.umich.edu/~hrswww/index. html for your initial address in this search.)

2. Locate three of the following resource home pages on the Internet, and prepare an outline highlighting each resource that is made available to the consumer:

   - American Association of Homes for the Aging
   - American Association of Retired Persons
   - American Heath Care Association
   - Assisted Living Facilities Association of America
   - Health Care Financing Administration
   - National Citizens' Coalition for Nursing Home Reform
   - National Council on the Aging

Special Problems

Growing Old Behind Bars

Frank Green

The number of older prisoners in Virginia has more than doubled in the past 10 years, creating new issues for the state's prison system.

Winter sunshine slices through a narrow security window and falls on Aloysius Joseph Beyrer's white hair, slight shoulders and the linen covering his fractured hip.

Like the rest of the country, Virginia is coping with a growing number of aging inmates. Beyrer, 84, is the state's oldest and his home, the Deerfield Correctional Center, focuses on geriatric inmates.

In 1999, Virginia had 2,015 prisoners 50 or older. Today, there are almost 4,700, and by 2011, state officials expect there to be 5,057.

A drop in the number of paroles granted to inmates who remain eligible is a factor in Virginia's increasing number of older inmates. Truth-in-sentencing reforms that in 1995 led to stiffer, no-parole sentences for violent crimes are expected to contribute to Virginia's aging prison population in coming years.

At Deerfield, wheelchairs and walkers line aisles in the secured assisted-living dormitory, where it would be easy to confuse the frail residents with those in nursing homes. But it would be a mistake to do so.

Beyrer, a veteran of prisons in Virginia and elsewhere, thinks Deerfield, "is pretty good," though security comes first there, even for octogenarians like Beyrer, who is serving 100 years for sex crimes. The prison's goal is to provide older inmates care and some dignity, not freedom.

The warden, Keith W. Davis, who has a master's degree in social work, makes it clear he is not running a spa for the golden years. "This is not a perfect world. We do not have unlimited resources," he said.

Even with a blank check to meet all their medical and mental-health needs, Davis said no one wants to grow old or die in a prison. "That's a big challenge for the staff . . . . We do what we can do, but we can't cure oldness," he said.

"Offenders are like the rest of us. We get old, we get ill, we die," he said. Deerfield provides a continuing-care community, he said, "so they can reach what we believe is their fullest potential—body, mind and soul."

• • •

Experts say substance abuse, little or no health care before imprisonment and the stress of living behind bars can leave a 50-year-old inmate physiologically 10 to 15 years older than his chronological age.

In general, older inmates require more supervision and medical and mental-health care, as well as special diets, mobility aids and special housing.

Deerfield, Virginia's only prison dedicated to geriatric inmates and inmates with special medical needs, accommodates 1,080 inmates, 90 of them in wheelchairs and 65 percent over the age of 50.

Other older inmates and older female inmates are in prisons such as the Fluvanna Correctional Center for Women and the Greensville and Powhatan correctional centers.

Critics point out that many older inmates are far less likely to commit new crimes and could be released at great savings. Prison officials, however, believe their care would largely be at public expense in or out of prison.

And though older people are less likely to commit crimes, some still do. Beyrer was 67 when he was convicted in Virginia Beach of statutory rape, aggravated sexual battery and forcible sodomy.

Deerfield's head nurse, Bonita Badgett, said 800 of the inmates there have at least one chronic medical condition such as diabetes, high blood pressure or asthma. The prison psychiatrist, Dr. Amit Shah, said the major problem he treats is depression.

In October alone, the prison handled 5,200 prescriptions.

Badgett has a staff of 14 registered nurses, 25 licensed practical nurses and 21 nursing assistants. Two physicians are at the prison three days a week and the psychiatrist visits once a week. At least one registered nurse is on hand at all times.

• • •

Deerfield was selected 10 years ago as the site for older offenders. An expansion opened in 2007 and there is now an 18-bed infirmary, a 57-bed assisted-living dorm, a larger ancillary-care dorm, a dorm for diabetics and a dorm for other special-needs inmates.

More than 75 percent of Deerfield's prisoners have violent records and nearly 30 percent are sex offenders. Security measures are complicated by health-care needs, said Maj. Stanley Mayes, chief of security for the prison.

"These guys have a lot of serious medical needs and . . . there are a lot of unusual [and potentially dangerous] pieces of equipment or property that we will allow them to have that you typically wouldn't see in another prison," he said.

Officers must be sensitive to prisoners who are gravely ill, suffering a heart attack or a stroke. "But not be deceived by someone who is faking to get an advantage to facilitate an escape," Mayes said.

Not everyone at Deerfield is happy. More than 200 inmates signed a letter to Gov. Timothy M. Kaine last year complaining about the parole board's low grant rate. One inmate claims staff stole his pain medication as he recovered from an injury.

Parole issues aside, inmates interviewed during a recent tour said they liked Deerfield.

James Henry Tinsley, 59, and partially paralyzed, has been there since 2003. "I been locked up 26 years," said Tinsley, convicted of 55 felonies, including capital murder, robbery and burglary.

"You've got some good people here. I ain't got nothin' bad to say about 'em. . . . As far as the medical, I give it a double A plus," he said.

Another well-traveled inmate, William H. Glazebrook, 74, has been in the state system for 25 years and at Deerfield for a year and a half. "This is Boy Scout Camp compared to the rest of 'em. This is a hell of a lot better," he said.

The Rev. Lynn Robinson, the prison chaplain, says, "These guys here, man, this is a special group of fellows." He said the inmates recently arranged to have Thanksgiving food baskets sent to five families and raised $500 for breast-cancer research.

"The one thing, I think, the community can be aware of is that . . . they need support when they come home," Robinson said.

Also, he said, "Saturdays and Sundays are visiting days, and some of them have family in the general area, and for some reason they don't come to see them. They need to stay in contact with [their] children."

...

Beyrer, Virginia's oldest inmate, was a resident of Deerfield's infirmary in November. Aside from six 1992 felony convictions, little information was available about Beyrer because of privacy rules.

He says that he was once a prisoner at California's San Quentin State Prison. California authorities could not confirm they had ever held him, but New York state archives show he was released from Attica Correctional Facility in 1956.

Dawn Mosena, the nurse manager of the infirmary, said inmates are held there for observation and treatment before and after hospitalization, in addition to long-term care and special-needs inmates such as Beyrer.

She said the staff is planning how to make room for what is expected to be more long-term patients such as Beyrer.

Last year, an inmate's mother was allowed to be with her son in the infirmary when he died. "We want the patient to feel comfortable and the family to feel comfortable and know that they can be with them in those last hours," Mosena said.

"We want to get a hospice program going," she added.

Davis said another problem is that, "a lot of these guys have outlived their families. . . . We could open the door to let them go, and where would they go?"

Badgett, Deerfield's head nurse, agreed. "Some of them we had to keep beyond their release date because we couldn't find a placement for them. There was no family out there, no home, there was nowhere to send them," she said.

Sex offenders, particularly, are difficult to place. Most nursing homes do not want them, and families often reject them because of their crimes, or, "the families simply cannot take care of the needs and medications."

...

At Deerfield, younger and healthier inmates—dubbed "pushers," short for wheelchair pushers—assist the older inmates and perform a wide variety of essential jobs for 45 cents an hour, primarily janitorial and in health care, that help keep the prison running.

One "pusher," James Lee Wainwright, 47, imprisoned in 1990 for armed robbery, helps in the infirmary. He said he has also assisted with health care at another prison before arriving at Deerfield.

"I plan on taking it up when I get out of here," he said.

An infirmary nurse said, "We couldn't function without these guys, literally, without their eyes and their help."

William Robison, chief psychologist at Deerfield, said some inmate helpers perform odious jobs, peculiar to hospitals and rest homes, for infirm inmates. A program has been set up to help the helpers, Robison said.

"The caregivers support group is . . . for guys who are caring for other guys here. You know, if that isn't therapeutic education, what the hell is?" Robison asked.

"It's a little different here, the way we even think of mental health. We try and redeem a guy."

Said Robinson: "What we do is to find them a purposefulness in living in prison and maybe dying in here." He is familiar with programs in other states and said, "I think we're light years ahead."

"We're not soft on crime. Tough love 'em, and they could still die here with some atonement . . . with a sense of humanity and self worth."

Source: www2.timesdispatch.com/rtd/news/local/crime/article/PRIZ04_20090103-212111/168231/

Right at this moment, an older person is being robbed or mugged. Right at this moment, an elder is being abused by caretakers, cheated of his or her savings, or victimized by medical quackery. Some abuses against elders are committed by street thugs, others are perpetrated by presumably reputable business people and professionals, and still others are committed by adult children or paid caretakers. And some of the abuse is self-inflicted. This chapter will review special social problems related to older adults: depression and suicide, crimes enacted against older adults, aging criminals, elder abuse, and fraud. We will look for model social programs and community-based behaviors in which people can engage to benefit the older population.

The actual commission of crime is only one aspect of the problem; the fear of crime brings its own set of problems. Law enforcement and social agencies recognize the crime patterns that most likely affect elders. Older people themselves are devising ways to fight back, and they are taking initiatives, both as individuals and as a group, to protect and defend themselves.

As we develop our understanding of what it means to be old in our society, we must understand that it brings a vulnerability due to ageism and, for some, further vulnerability due to physical and financial limitations. Here we discuss some special problems that, if not unique, occur in high frequency within the aged population. And let us not forget that older persons are not always victims; they can also be perpetrators. This topic will be discussed briefly.

Suicide

Suicide is the ultimate reaction to hopelessness—the acting out of a belief that there is no promise to the future and no reason to live in it. It is inextricably linked with depression, and in 2006 it was the 11th cause of death in the United States (NIMH, 2009). Of every 100,000 people ages 65 and older, more than 14 percent died by suicide in 2006. This is higher than the national average of just under 11 percent per 100,000 people in the general population (NIMH, 2009). Elder suicide is particularly understudied, although the suicide rate of older adults is more than 50 percent higher than that of the general population (McIntosh, 2008).

For four decades, the suicide rates among older Americans declined, but from 1980 to 1992 they climbed by almost 9 percent. During this time, men—predominantly white men—completed 81 percent of those suicides (Centers for Disease Control and Prevention, 1996). In 1933, the first year the National Center for Health Statistics kept suicide statistics, the suicide rate among elders was an alarming 45.3 per 100,000 compared with a national average of 15.1. Between 1950 and 1980—years in which Medicare, housing, and other social programs targeting elders were initiated—suicide among elders declined by 40 percent (Marrone, 1997). By 1981 the rates had dropped to 17.1, compared with a national average of 12 per 100,000 people.

Differences by Gender and Race

Suicide in old age is statistically a man's issue—particularly older white men. All women 65 years and older, despite reporting higher depression rates, have a disproportionately lower rate of suicide. In 2005, the rates in old age for white and black women combined were about half of that for white men over 85 years of age (NIMH, 2009). On the other hand, the risk of suicide for men in the United States begins in the teen years and continues to increase with age, making an alarming increase in very old age. The proportion of **successful suicides** to **attempted suicides** is far greater in old age than in younger age. Older people express and demonstrate a firm determination to complete suicide (Silberman et al., 1995).

TABLE 12.1

Death rates from suicide among people aged over 65 by race, ethnic origin, and gender, 1989 to 1991 (deaths per 100,000)

	Male	Female
White	43.7	6.5
Black	16.0	1.9
Native American	11.4	3.4
Asian and Pacific Islander	18.5	8.9
Hispanic	25.9	2.5

Source: National Center for Health Statistics (1994).

Table 12.1 shows death rates from suicide among older adults by race and gender from 1989 to 1991. Note the predominance of white non-Hispanic older men who died of suicide in this period.

The high suicide rate among aged American men is not an international pattern: in some countries suicide peaks in middle age (Lester & Savlid, 1997). World Health Organization (1991) statistics, however, show that suicide rates throughout the world are generally higher for people over 65 years of age—often by two or three times that of the national average. The highest international suicide rates were found in Hungary, Sri Lanka, Denmark, and Finland, and the lowest rates were found in Colombia, Ecuador, Greece, and Venezuela. Suicide rates among older people in the United States lie somewhere in the middle (Schaie & Willis, 2005). International gender differences in suicide also exist: in Japan, for example, suicide rates for older women have risen steadily over the past three decades—possibly a reflection of the progressive loss of the traditional importance of older women in that culture (Shimizu, 1992).

Examining numbers and proportions of suicides can be misleading. For example, women outnumber men in unsuccessful attempts at suicide by about three or four to one (Schaie & Willis, 1996). Men's attempts at suicide are simply more successful; if *intent* is considered, older women outnumber older men. Men are more likely to kill themselves by violent and more certain means—guns were chosen by 74 percent of men who killed themselves in 1994, compared with 31 percent of women. Women are more likely to overdose with medications or poisons, which can be less certain (Marrone, 1997). Among young adults, more people attempt suicide than actually succeed. By about the age of 50 that trend reverses itself and the suicide completion ratio grows larger. Even considering the ratio of suicide attempts to completion by women, when compared to younger adults, older adults are more likely to kill themselves when they attempt to do so.

Note on Table 12.1 the comparatively low rates of suicide for older black men and women. The reasons for the low rates are not entirely clear, but they are thought to be related to the positive views of aging, a developed tolerance for suffering, strong religious and cultural prohibitions against suicide, and the lack of a giant step downward in status that white males experience in retirement.

Did you know . . .

A study in the early 1970s subjected dogs to minor but discomforting electrical shocks that they could not escape. The animals became apathetic, seemed sad, and were unable to think or move quickly. This behavior led psychologist Martin Seligman to the concept of learned helplessness as a source of human depression. After a series of inescapable losses, if people feel they have no mastery or control of their environment they develop a feeling of futility, even toward events they can control. The resulting emotional state is apathy, hopelessness, and depression.

SELIGMAN, 1975

Causes of Suicide

Some experts speculate that medical technology has introduced a quality of life that older adults cannot accept. Many medical advances give people longer lives but not necessarily a *better* long life: there is a greater *quantity* in terms of longevity, but the *quality* of life suffers.

Suffering and loss are factors known to be related to depression and suicide and take place on psychological, physical, and social levels. Table 12.2 outlines some risk factors related to suicide, including the psychological risk inherent in the process of becoming more self-reflective, physiologically mediated risks of chronic illness and pain, the social risks of isolation through repeated loss, and the behavioral risk of alcoholism—an outcome of a behavior promoting denial (Marrone, 1997). The most significant factor, subsumed in the table under "multiple losses," is the loss of loved ones by separation or death. When children leave home, when friends and relatives die, and especially when a spouse dies, the survivor is often beset with the desperate feeling that there is no reason to continue living. Social isolation has been associated with suicide—and elders are all too often isolated both socially and psychologically through society's attitudes and practices of ageism.

Perhaps most important, physical illness, pain, or disability can be profoundly depressing. Aversion to the bodily changes that illness brings and worry over medical bills, coupled with the possibility of becoming a burden to one's family or society, are typical reasons for older people to take their own lives. Older depressed people tend to use health services at high rates, and indeed about 75 percent of older adults who died by suicide visited a physician within one month before their death (Conwell, 2001).

Despite the disinformation of many health professionals, persistent depression is not an acceptable response to other health issues and the financial hardships that often accompany illness. This belief is often shared by older adults themselves, whose depression/health cycle begins to feel like a hopeless loop. These findings point to the urgency of improving detection and treatment of depression to reduce suicide risk (NIMH, 2009).

TABLE 12.2

Risk factors associated with suicide in later life

Risk Factor	Description
Interiority	A turning in toward ourselves in search of personal meaning; has both healthy and unhealthy aspects: can promote self-knowledge, spirituality, and increased satisfaction with life; or result in unhealthy self-absorption and increased social isolation.
Chronic illness	Organic and psychological decline along with chronic illness and pain contribute to depression. Seventy percent of older suicides visited a physician within 1 month of their death.
Multiple losses	Job loss, death of friends, death of a spouse, the divorce of a child, loss of health and social status—all are cumulative experiences, and can occur so rapidly that resolution of the inherent grief of loss does not happen; associated depression of recurring losses can become chronic depression.
Alcoholism	More important among males 35 to 64 than any other age group; alcohol abuse enhances depression and exacerbates negative live events. Bereaved adults report significant increase in alcohol consumption, psychological distress, and decline in physical health.

Source: Adapted from Marrone, Robert. (1997). *Death, mourning, and caring.* © 1997 West Publishing Co. by permission of Brooks/Cole Publishing Co.

> *Existential philosopher Soren Kierkegaard (1843) argues that the price we pay for not coming to terms with our own death is despair. . . In not coming to terms with death, the person in despair avoids the insights that life is precious, that existence is delicate, that a life filled with vibrancy, choices, and risks is the truest antidote to loneliness and despair, and that the dignity with which you live your life is the dignity with which you die.*
>
> MARRONE, 1997

Living with the imminent reality of dying can be a difficult and lonely walk.

Choosing to Die

The Hemlock Society was founded by Derek Humphry in 1980 to help people make choices about how they would deal with terminal illness. The motto of the society was "Good life, good death through control and choice." Humphry gained international fame after writing *Jean's Way*, the personal account of how he helped his terminally ill wife kill herself in 1975.

In 2003, the Hemlock Society ended as a name, but not before the concept had been sourced in medical and legal texts and made accessible to the larger public by its very language of right to die. The Society believed that suicides were not necessarily the result of poor mental health, but rather of a sound decision based on good mental health in what is called **rational suicide**. Thus, the relationship between suicide and mental health was not as clear-cut to some as it is to others.

According to Humphry (2005), the Society's principal achievements were to:

- Educate and advise thousands of dying people to know how to bring about their peaceful ends when dying, trapped in a ruined body, or just plain terminally old, frail, and tired of life.

- Give help through specialized literature and moral support. The organization never provided hands-on assistance to people to die because it needed, then, to stay within the law, although many of its members quietly, unofficially, did see one another off.

- Draft and launch the first model law governing euthanasia and assisted suicide in America in 1986, from which many others that were to follow were refined.

- Financially and physically back referenda in California, Washington State, Michigan, Oregon, and Maine. Currently only Oregon and Washington have laws permitting doctor-assisted suicide. Those laws remain under Federal Government threat of extinction via the courts.

- Founded and launched a national "Caring Friends" program so that maximum personal support and assistance in dying, within the law, could be ensured to every member. No member needs dies alone or in agony. This is a free program. Credit for this program goes to Faye Girsh, Lois Schaffer, Sally Troy, and Richard MacDonald. Its more business-like name is now "Client Services."

Therapeutic intervention by skilled professionals is always recommended for a suicidal person. Depression can be treatable whereas suicide is not. Many communities have established suicide-prevention centers that maintain 24-hour crisis lines for helping people in distress. The National Center for Studies of Suicide Prevention has assigned a high priority to the problem of suicide among those 65 and older. There is, however, a shortage of skilled therapists to treat self-destructive elders, and people over 65 make only 3 percent of all calls to suicide hot lines; even though suicide is more common in the elderly than in other age group, it is still a relatively rare event.

Why some people can cope with stress and loss and others cannot is not fully understood, nor is the line between "rational suicide" and that emerging from hopelessness and depression very clear (Lester & Savlid, 1997). Barbara Haight (1996) found that older adults being relocated to nursing homes who expressed suicidal ideation reported a strong sense of self-esteem and described suicide as a way to retain control of their lives. We do not sufficiently understand the mixed picture of the hardy survivorship qualities of old people and the isolating effect of relentless loss of peers (Kastenbaum, 1992). All in all, the viability and allegiance to life of elders in general should not be underestimated.

Self-Destructive Behavior

Suicides of people 65 and over leave a very large number of survivors who grieve and mourn. The solitary act of suicide has a ripple effect, affecting many lives—all the way down to yet-unborn grandchildren and great-grandchildren.

In some instances, couples who fear leaving or being left by their partner conduct **double suicides**. The most common pattern of a double suicide is the "mercy killing" of a frail and dying wife, followed by the husband's suicide. Family grief following such a tragedy is huge: what might we have done to prevent this? Did they both truly choose this way to die, or did one person choose it for the other?

Healthy Place: How to Help Suicidal Older Men and Women (2008) lists the following "clues" to look for to determine possible suicidal thoughts in a suicidal person: changes in eating/sleeping patterns; unexplained fatigue or apathy; crying for no apparent reason; inability to feel good about themselves or unable to express joy; withdrawal from family, friends, or social activities; loss of interest in personal appearance as well as work, hobbies, etc. The author points out that a suicidal person might talk about death or not wanting to live—and this should be taken seriously. Some people contemplating suicide take unnecessary risks or give away prized possessions. Being involved, showing interest and support, being nonjudgmental, and taking some actions to remove easy methods they might use to kill themselves are essential to prevent them from committing suicide.

Future Trends

Suicide is twice as prevalent in western states as in the East and Midwest. Older persons in the West are more likely to have moved away from friends and family, and the divorce rate is higher in western states. Although no direct relationship has been found between suicide and divorce or mobility, the trends are important. The trends imply that meaningful social interaction might be more difficult to come by in some areas of the country, and meaningful personal relationships are the most potent antisuicide remedies—failures and losses are burdens easier to bear in the context of close family and friends.

Health experts anticipate an even higher rate of suicide as the baby boomers reach age 65. Baby boomers have had depressive disorders at significantly higher rates than their predecessors—a trend that seems to be steady in subsequent age cohorts. The implications of this projection for mental health professionals are awesome regarding programs and services for a very diverse and large emerging portion of the population.

Crimes against Older People

Though attacks on older people draw huge media attention, national surveys consistently show that older adults are less likely to be victimized than younger adults. However, the older adults report a higher fear of crime than do younger adults (Lanier & Dietz, 2009). The fear, blamed on **perceived vulnerability**—a belief that one is vulnerable—has serious psychological, physical, and financial consequences for the elderly. This fear can result in older adults not wanting to leave their homes, spend money unnecessarily on security systems, and/or generally be more vulnerable to fraudulent safety marketing (Lanier & Dietz, 2009).

Wide variations in victimization rates. Surveys show that a majority of victimizations of older people occur in or near their homes. The converse applies to younger adults, who are more likely to be attacked when they are away from their homes. Unlike younger adults, older people tend to avoid places of danger and restrict their use of public streets.

The concept of perceived vulnerability may well be a reason for attacks on older individuals especially those who are poor, living on fixed incomes, and dependent on public transportation. They may live in neighborhoods with relatively high crime rates. Older people are often not physically strong enough to defend themselves, so their perception of physical vulnerability is not imagined. Most crimes against older people, however, are directed at property: burglary, auto theft, and the like. There is a consistent negative relationship between age and violent crime victimization (Bureau of Justice Statistics, 2005).

A 33-year-old woman decided to experience victimization by "becoming old" herself. Donning a gray wig, wearing semiopaque glasses, and putting splints and bandages on her legs, she created, at various times, three characters: a bag lady, a middle-income woman, and an affluent woman. She traveled through 116 big cities and small towns across the United States, walking the streets, eating at restaurants, and living in motels. While some people were kind to her, she was overwhelmed at the abusive and neglectful attitudes of others and by the constraints she suffered because of transportation systems inadequate for those who have difficulty walking or seeing. She was mugged twice in New York City. Today she lectures to designers and gerontologists on the needs of elders (Ryan, 1993).

Some elders defy the stereotype of being weak and easily threatened. A 73-year-old woman in New York City battled an intruder and foiled a rape attempt in her apartment (Woman, 73, Battles Intruder, 1991). In an elder housing project, the tenants got involved when gang members had been arrested and charged with attacking 12 older women over a three-week period, choking them from behind and stealing their money. Older tenants were instrumental in helping police set up a stakeout and came forward as witnesses to identify those arrested. Anger won out over fear as the residents who "didn't want to take this" came forward (Hevesi, 1994).

These examples are not meant to advocate fighting back in every circumstance. It is not widely recommended that anyone fight an armed assailant. These accounts merely remind us that not all old people are easily intimidated. If this fact were more widely known, the number of attempted attacks might decrease. The stories do illustrate, however, the power of an individual's belief in myths: old people are victims; I am old; I might be a victim.

After decades of inadequate funding for mental health programs, many old people have lived a lifetime of marginal existence. For some, hopelessness is a way of life.

Fear of Attack

The National Opinion Research Center (NORC) has conducted national surveys on the **fear of crime**. When asked, "Is there any area right around here (your home) that is within a mile where you would be afraid to walk alone at night?" more than 50 percent of those over age 65 answered "yes." The image of frightened older persons barricaded behind locked doors, however, is overplayed in the media. A study in Dade County (Miami), Florida, found that older adults did not have a statistically higher fear of crime than did younger adults (McCoy et al., 1996). This suggests that fearfulness might be less evident in relatively age-homogeneous neighborhoods or that older people are capable of making context-based judgments of community safety. The Florida study found that dissatisfaction with neighbor-

hood and physical vulnerability were important correlates of fear of crime, but that actual victimization experience was not the main determinant of fear.

Some studies have suggested that older people are more fearful than younger people, which has to do with the measure used for "fear of crime" (Ziegler & Mitchell, 2003;). The question used by National Crime Survey (NCS) was "How safe do you feel or would you feel being out alone in your neighborhood?" The words "safe" and "feel" can be interpreted in various ways. I might not feel safe in my neighborhood because it does not have flat sidewalks, and I have had a hip replacement. If I were to fall, I would be more vulnerable to someone with ill-intent who might come along. Additionally, one study identified that most of the respondents knew someone who was a crime victim. Of these people, 72 percent said that this knowledge increased their own fears of crime (Joseph, 1997). Clearly, the more life experience one has, the more likely he or she is to know someone who was a victim of crime.

Most literature and research finds, however, that older people are more afraid of crime than younger adults. The literature consistently reports that (a) women are more fearful than men, (b) African Americans are more fearful than whites, (c) those with less money are more afraid of crime than those with more money, and (d) residents of large cities are more fearful than people in smaller towns and rural areas.

The manner in which fear of crime should affect social policy is not entirely clear. On the surface, it would appear that one way to reduce fear of crime among elders would be to segregate them in walled and guarded retirement communities. But age segregation in our society could have undesirable consequences. Studies show generally that the more integrated one is in community activities whereas the less one fears crime. Related to this is the finding that one feels less threatened by crime if he or she knows and trusts his or her neighbors. The social implication is that developing cohesive, close-knit communities reduces fear of crime. These two alternatives are opposites, but either may be needed depending on the possibility of developing a close-knit, friendly neighborhood.

A consistent finding over the years has been that although older women fear crime more than older men, they are less likely to be victims. And though older people are more fearful than younger people, they are less likely to be victims. One explanation for this paradox may be that women and elders, considering themselves to be more vulnerable, do not as often expose themselves to risk. An additional explanation may be that they associate more minor offenses with more serious ones. For example, an older person, thinking that begging is a pretext for mugging, might be more afraid of a beggar than a younger person is. A woman might fear burglary more than a man because of the threat of rape in addition to theft. Thus, the possible consequences of the criminal act can be as frightening as the fear of victimization itself.

Fighting Back

In many cities, new programs aimed at preventing crimes against the elderly, helping those older persons who have become victims, and teaching them what they can do to help themselves are receiving priority. For example, one common approach is the use of police units trained in the problems particular to older people. Such units tip off the elderly to the latest trends in crime, help them to be on the alert for suspicious activity, and instruct them how to be effective witnesses against criminals caught in the act. Programs such as **Neighborhood Watch**, which emphasizes crime awareness in residents of all ages, have resulted in crimes being spotted while in progress.

In many cities, "granny squads" of older people patrol neighborhood blocks and give lectures on how to avoid being raped, robbed, and burglarized. Granny squads, organized under titles such as Heaven's Angels and Gray Squads, recommend such precautions as the following:

- Report all suspicious people and all crimes to the police.
- Use automatic timers to turn on radios and lamps when you are away from home.
- Do not carry a purse. Make a band to wear inside clothing to carry money.
- Have Social Security and pension checks mailed directly to the bank.
- Join a Neighborhood Block Watch, in which neighbors in a block meet one another and watch out for one another's person and property.

Some cities provide escort services to older people when they are the most vulnerable to attack—such as on trips to stores and banks. For example, one Chicago police district, on the day that Social Security and relief checks arrive, supplies a bus and driver to pick up people from two housing projects, take them to a bank and a grocery store, and then return them safely home. New York City police, hoping to bridge the gap between young and old, use teenage volunteers to provide escort services for elderly persons. In Milwaukee, Wisconsin, neighborhood security aides patrol the streets in high-crime areas that house many older residents, walking in pairs to offer safety in numbers. Although they have no power to arrest, they carry two-way radios to call police or firefighters if they encounter suspicious people.

Efforts to protect elders from crime exist at the state level as well. Many states offer reimbursement programs for crime victims, and some of these programs give priority to older persons. A New York state law makes a prison sentence mandatory for anyone, including a juvenile, who commits a violent crime against the elderly. Many states have extended child-abuse laws to include elders. Other states are currently considering legislation that does not allow probation for those who commit crimes against older persons.

Aging Criminals

Although studies find that younger adults are at least 10 times more likely than elders to commit crimes, elderly crime is very much a reality. Most crimes committed by elders are misdemeanors—petty theft, sleeping on the sidewalk, alcohol violations, and traffic violations. Shoplifting is a frequent misdemeanor charge, and most shoplifters are white females. But felonies also occur, most frequently in the form of grand theft and narcotic charges.

There is not necessarily a relationship between economic hardship and crime. Typically, older people are caught stealing lipsticks, perfumes, night creams, and cigars. They are not necessarily stealing to eat. Shoplifting among elders represents the combined influence of stress, age, and merchandising; fear for the future may compel some to shoplift in order to conserve money for anticipated expenses. Some steal to ease fear; others do it to get attention. From a psychological perspective stealing may reflect feelings of deprivation in human relationships (McAdams, 2005). Only a minority of older shoplifters have been engaged in criminal activity throughout their lives.

In general, the older perpetrator accounts for only a minority of all crimes committed. The average age of incarceration is 30.7 for men and 32.1 for females, with only about 6.6 percent of the U.S. prison population above the age of 50 (Camp & Camp, 1996). The proportion, however, is expected to increase, and by 2005 the population of inmates 50 and over in federal prisons is projected to be between 11.7 percent and 16 percent (Smyer et al., 1997).

Fear often accompanies frailty because being frail means having less personal control over life's events. Being frail, alone, and frightened are the bleak circumstances that some elders experience.

Their number more than doubled between 1981 and 1990 (Morton, 1992). The criminal activities most likely to be engaged in by older inmates are (a) violence against a family member, (b) white crimes such as fraud, (c) drug sales, and (d) alcohol-related crimes such as vehicular manslaughter.

Older Professional Criminals

Professional criminals tend to remain active because crime represents their life's work. The longevity of some professional thieves is incredible. Joseph (Yellow Kid) Weil, a prototype for Paul Newman's role in *The Sting*, lived to be over 100 years old and was last arrested at age 72. Willie Sutton, professional bank robber, was into his 80s and still on parole when he died (Newman et al., 1984). Most leaders of organized crime are in the upper registers of the age scale. Organized crime is age-stratified, and the heads of "families" tend to be well over 50. Vito Genovese died in prison at age 71, still commanding his organization from his cell. Older offenders also play an important role in white-collar crime. And, as the percentage of older adults increases in the total population, their percentage of criminal arrests will also probably increase.

Growing Old behind Bars

More people are being sent to prison and receiving longer sentences, increasing the number of older prisoners. For example, from 1990 to 2005 overall number of older prisoners in California rose three times faster than the overall adult population. Younger adults accounted for a declining portion of the prison population (from 20 percent in 1990 to 14 percent in 2005), while the share of adults age 50 and older nearly tripled, from 4 percent to 11 percent (PPIC, 2006). This increase is in part because although young people are most likely to commit crimes, older adults are increasingly

being admitted to prison and more inmates are aging in prison because of tougher sentencing laws dictating the length of time served. This trend can be seen throughout the country: Florida, New York, Virginia, Montana—all show similar patterns of an aging prison population.

Older convicts tend to be chronic offenders who have grown old in a steady series of prison terms, offenders sentenced to long mandatory terms who have grown old in prison, or first-time offenders in their old age. California's three-strikes law of mandatory prison sentencing on the third conviction explains some of the current overcrowding and high recidivism rate. This law mandates life sentences without parole for certain repeat felons, who will inevitably grow old and die in prison. Another concern shared by most states is the adoption of the federal Truth in Sentencing program, which produces stiffer, no-parole sentences for violent crimes. These adaptations in law by fearful citizens throughout the country will contribute to an explosion of older prisoners. A 2008 survey by the U.S. Bureau of Prisons found that there were nearly 125,000 inmates aged 50 or older but fewer than 10,000 beds in facilities dedicated to older inmates (Green, 2009).

California operated the first prison for older inmates in the late 1950s, but closed it in 1971 when the prison population dipped. Since then old felons have been incarcerated throughout the system even though federal studies show that mainstreaming older, sicker inmates with younger ones cost more than establishing specialized units (Kobrin, 2005).

In all states vocational programs for prisoners are cut when budgets are tight. People who have spent time in prison are less likely to find employment once they are released and oftentimes have disrupted family relationships on their release.

The routine of prison life starkly separates the old from the young and middle-aged. The young men tend to be muscled and heavily tattooed, have an "in-your-face" attitude, and form groups, even gangs, for friendship and safety. The old men tend to keep to themselves trying to find quiet. The gang tensions, the chaos, the constant chatter, the televisions going from early morning to late at night take their toll over the years. There is virtually nothing in the construction of prison life that acknowledges the needs of older prisoners.

The greater threat, beyond illness or victimization is despair. Many prisoners. . . "put their lives on hold." As they age, they begin to face the fact that they may never get out. "The dream of getting out, you equate with heaven. Dying in prison, you equate with hell."

PRISON INMATE, QUOTED IN ANDERSON, 1997

Taxpayers as well as prison officials and humanitarians have reason to be alarmed by the increasing numbers. Prison inmates are not eligible for Medicare or Medicaid. Age and illness are associated more strongly behind bars, given a life of chronic stress, idleness, cigarette smoking, and heavy food. The lives lived by older people in prison have often been harsh, exacerbating existing health issues and creating others. A 55-year-old prisoner is likely to have the physiological age of a 65- or 70-year-old person who has not experienced prison life.

Prison costs for eyeglasses and dentures and for treatment for heart surgery, emphysema, prostate problems, strokes, and other age-related needs are rapidly growing and will continue to do so. The average annual maintenance with medical costs for inmates 55 and older in California was estimated to be $69,000 in 1997—triple that of younger prisoners (Smyer et al., 1997). An Iowa study of 119 male inmates 50 and older found increased rates of incontinence and sensory impair-

ment, as well as 40 percent with hypertension, 97 percent with missing teeth, 42 percent with gross physical impairment; in addition, 70 percent were smokers (Colsher et al., 1992).

Tish Smyer and associates (1997), on reviewing the mental and physical health of elderly prison inmates, concluded that new models of incarceration are emerging, and changes must continue to be made.

The use of medical parole/release, community placement (secure skilled nursing facilities. . .), and growing use of prison hospice programs are being implemented as an alternative to placement of the chronically ill or elderly inmate in the general prison population. Those with functional disabilities who require help with activities of daily living may require more intensive contact with skilled professionals and counselors. Their dependent status merely intensifies the meaning of violence and powerlessness to these elderly inmates within the prison environment. (p. 16)

In 2007, the state of New York had the first prison to specialize in dementia-related conditions. The Fishkill state prison housed 1,700 inmates who could receive medical care from basic throat problems to long-term care. The Alzheimer's disease and dementia unit holds inmates from all around the state. All workers on this unit, when hired, go through a training course to learn how to work with these prisoners, whose conditions range from psychiatric or other medical disorders to dementia such that they cannot even remember their crimes. Such programs are essentially becoming necessary for the penal justice system, but prior research estimates the costs of housing, transporting, and caring for elderly inmates to be two to three times higher than those for other prisoners (PPIC, 2006). Certainly following the national recession in 2008, greater funding for shifting prison needs seems unlikely in the foreseeable future.

Elder Abuse

Most abuse of elders is done by the person with whom they live, most often an adult child. Though we are familiar with the "battered baby syndrome," evidence increasingly indicates a corresponding syndrome at the other end of the age scale—**battered parent syndrome**—in which parents are attacked and abused, sometimes fatally, by their own children. Factors that increase the risk of a caregiver's becoming an abuser include alcohol and drug abuse, cognitive impairment, economic stress, caregiver inexperience, a history of family violence, a blaming personality, unrealistic expectations, and economic dependence on the elder (Allan, 1998).

For the incarcerated elder, "aging in place" has a particularly poignant meaning. Re-entering society for the older person can be a huge mental and physical task.

Among community-dwelling elders in the United States and other Western countries, it is estimated that between 3 percent and 6 percent of individuals over the age of 65 report having experienced abuse, usually at the hands of family members (Lachs et al., 1997; Ogg & Bennett, 1992; Podkieks, 1992). Most caregiver abuse, however, goes unreported. It is embarrassing; it might make the caregiver yet more angry; the abused adult is fearful and isolated.

Abuse can be intentional (e.g., willfully withholding food or medications) or unintentional (ignorance or the genuine inability to provide the necessary care; Allan, 1998). On a national level, neglect is by far the most common form of abuse, followed by physical abuse, then financial and material exploitation (My Health Care Manager, 2008). Several measures of abuse have been identified:

1. *Physical abuse* is the willful infliction of pain or injury and may include beating, choking, burning, inappropriate medication, tying or locking up, and sexual assault.

2. *Psychological abuse* includes threats, intimidation, and verbal abuse.

3. *Financial/material abuse* means taking financial advantage of frail or ill elderly. It is the misuse of an elder's money or property: theft, deception, diverting income, or mismanagement of funds.

4. *Violation of rights.* Old people have the right to vote and the right to due process. A conservator, for example, may take away all the rights of an older person.

5. *Neglect* occurs when a caregiver's failure to provide adequate food, shelter, clothing, and medical or dental care results in significant danger to the physical or mental health of an older person in his or her care.

6. *Self-abuse and neglect.* Some old people do not adequately care for themselves. Sometimes it is intentional; other times they simply cannot adequately provide for themselves.

Abandonment, or **granny dumping**, is an active form of neglect. This term is used by emergency workers at hospitals to refer to the abandonment of frail elders. It has been most recorded in Florida, California, and Texas. A Tampa, Florida, emergency room staff found a woman sitting in a wheelchair with a note that said, "She's sick. Please take care of her." More commonly, the older person is brought to the hospital by family members or a nursing home. When the patient has recovered, there is no one available to take the patient home. Elder abandonment has increased over the last few years, as medical and other costs of elder care have accelerated.

The abuse of elders, despite the cruelty of the crime, is not a well-studied topic; therefore, systematic ways to address the problem have not been developed in social policy. Table 12.3 presents a summary of existing theories of elder abuse developed by Dr. Rosalie S. Wolf, president of the National Committee for the

TABLE 12.3

Theoretical explanations of elder abuse

Theory	Description
Psychological model	Relationship of violent behavior to aggressive personality traits; at the most severe end of physical aggression, large proportions of elder abusers have histories of mental illness and/or substance abuse. Intervention is individual, focused on the caregiver.
Situational model	Abuse as a problem of caregiving; high levels of stress and burden of caregiving increase mistreatment and neglect of the frail elder. Intervention is on stress reduction of the care-provider.
Symbolic interaction model	Each person approaches in interaction with personal definitions and expectations; if behaviors match the roles expected, the interaction continues, otherwise conflict arises. The caregiver's subjective sense of stress related to the dependency tasks creates stress, not the tasks themselves. Assessment of the caring situation by caregivers is necessary for abuse prevention.
Social exchange theory	When each individual contributes equally, a fair exchange results; when one party is unable to reciprocate, the exchange is seen as being unfair. Unequal distribution of power in caregiving situations can result in abuse. In this model, intervention should focus on the abuser.
Feminist theory	Identifies violence as a tactic of entitlement and power that is deeply gendered, rather than a conflict tactic that is personal and gender neutral. This model has been more successful in explaining wife abuse than child or elder abuse.
Ecological model	Focuses on behavior within the social context and on the accommodation between the person and the environment. According to this model, mobilization of the community is the intervention strategy.
Political economy model	Elder abuse arises from the way older people are marginalized in society. These experiences are a product of the division of labor and structure of inequality in society, rather than a product of relationship and aging. Intervention focuses on change in social policy.

Source: Adapted from Wolf, Rosalie S. (1997). Elder abuse and neglect: Causes and consequences. *Journal of Geriatric Psychiatry,* 30 (1), 153–174.

Prevention of Elder Abuse. This table shows that resolution of the problem of elder abuse depends on the theoretical perspective taken: psychological models address caregiver personality and relationships in the family setting, while other models look to the community and the larger society (Wolf, 1997).

The primary policy implication for the issue of elder abuse is that strained families must have more help available to them—help from social services, home care services, protective services, personal care services, and family counseling. Increasingly, training programs for police and other first responders focus on identifying an older adult who might need protection from abuse, neglect, exploitation, or abandonment. Regional Area Agencies on Aging that receive reports of abuse investigate the circumstance and can make recommendations to correct the situation. In most cases the agency must have the older adult's consent to provide protective services unless ordered by a court. This is oftentimes difficult to obtain for the reasons mentioned previously.

The profile of the "typical" abused elder has traditionally pointed to the frail, widowed female who is cognitively impaired and chronically ill. Subsequent studies, however, show wider variation in the at-risk older adult. In one nine-year study of more than 2,800 elders linked to abusive victimization through adult protective service records, the predictors of mistreatment were found to be poverty, race, functional and cognitive impairment, worsening cognitive impairment, and living with someone (Lanier & Dietz, 2009). Gender did not emerge as a profile in this study. Indeed, Elder Law Answers (2008) reported an AARP study of 800 fraud victims age 50 and over that found most were educated, held jobs, went to church, and had family and friends. Older adults generally experience comparatively lower crime rates, but they are accompanied by a greater risk for injury—both physical and emotional.

A common predictor of abusive behavior is economic dependence. In one case, an older woman lived with her son and supported him even though he drank all the time and abused her. In another case, an older mother cared for her epileptic daughter who stole money from her. In some cases, wives were abused by the severely disabled husbands they were caring for. Abusers are rarely stable individuals brought to the brink by the excessive demands of an elderly dependent; rather, these caretakers suffer emotionally and are barely able to meet their own needs, much less the needs of another person. The older person who stays in an abusive situation can see no better alternative. The abuser's lack of power becomes a factor in the abuse. The caretaker's dependency, especially his or her financial dependency, rather than the victim's dependency, correlates most strongly with abuse.

Nursing Home Abuse

Abuse can and does take place in nursing homes by staff. Older people are more at risk in institutions than in their own homes because of their exceptional frailty and danger of retaliation by caretakers. Abuse in institutions is not necessarily reported or reflected in national statistics. A survey revealed that 36 percent of the staff at several nursing homes had observed other staff psychologically abusing patients and had witnessed physical abuse that included employing excessive restraints; pushing, grabbing, kicking, or shoving patients; or throwing things at them. Over 70 percent of the psychological abuse included yelling and swearing at patients; this abuse also included isolating patients, threatening them, and denying them food and/ or privileges (Pillemer & Moore, 1989). Although existing regulations monitor the treatment of institutionalized elders, the need for vigilance still remains. Penalties have been stiffened, county and state agencies are being given more power to investigate and prosecute, and people are being encouraged to report suspicious happenings. Signs of abuse include abrupt negative changes in physical appearance;

inappropriate behavior, such as extreme fear, asking to die, or extreme anger; a bad attitude on the part of the caregiver; and deteriorated or isolated living quarters. As more people live to extreme old age and, thus, to the point of becoming physically dependent on someone else—adequate social service programming for elder care is critical.

Fraud

Many of us believe that a person is as good as his or her word. We can be trusted and therefore we trust others. All of us want to believe. Older adults tend to be more trusting, having a lifetime's experience of giving other people the benefit of the doubt—the person trying to sell the product, after all, is simply trying to make a living. Furthermore, when talking with someone it is very difficult for anyone to tell the difference between a good salesperson and a crook.

Fraudulent crimes differ from violent crimes in three primary ways: swindlers use no weapons, never threaten or physically harm their victims, and do not rely on force. Rather, they use persuasion and emotional influence (Identifying and Preventing Frauds [no date]). Violent criminals succeed with physical threat or harm; swindlers succeed by creating mental images of prizes or wealth.

Any person with money is a potential victim of hucksters and con artists. Older people make good targets for several reasons: many have assets, most everyone wants to increase retirement savings (particularly since the 2008 recession), and many older people live alone and are isolated or lonely. Additionally, many older people are overdependent on one caregiver or advisor (Elder Fraud in Denver, 2006; NCL, 1999; Health Fraud downloaded, 2009).

Elder fraud is believed to be vastly underreported. Some research concludes that about 80 percent of abuse cases go unreported—including being a victim to a fraud (Elder Fraud in Denver, 2006)—which is a clear and concerning issue that is attracting regulatory attention. In 2008, for example, more than 500 communities throughout the country held seminars for law enforcement, family and social services programs, and general citizenry on how to identify and prevent fraud against older people (Elder Law Answers, 2008).

Reporting a crime takes courage. It's embarrassing; when someone has fallen for a con artist's line it raises concerns about competence. If I report this, will my daughter and son think I'm not competent? Might the system recommend that I be "put into a home?" Additionally, the criminal justice system feels unsafe for some people who have grown into older adulthood lacking a trust in larger systems of social control.

The following are some descriptions of fraudulent activities that frequently target older adults, who are more likely to be at home when a salesperson stops by and less likely to hang up the telephone on telemarketers. Victimization of older people is especially devastating. It is more difficult to recoup losses by returning to the workforce, for example.

Social Referral

Some dating and marriage services seek out the lonely and the widowed to lure them into paying big fees for introductions to new friends or possible mates. A victim may pay hundreds of dollars for a video to be shown to prospective dates (and then it is never shown) or for a computerized dating service that never generates any dates. Religious cults also have targeted elders and even entire retirement communities as recruitment areas for new members (Collins & Frantz, 1994).

Land and Home Equity Fraud

In one type of **land fraud**, real-estate developers may offer lots for sale in a still-to-be-built retirement community, promising attractive facilities, such as swimming pools and golf courses, and describing the site in glowing phrases. None of the descriptions and promises may be accurate.

Home-equity fraud has left some elders homeless. Swindlers pose as financial experts who offer help in refinancing a home. The swindler ends up with the cash from the home-equity loan or actually gets the owner to unwittingly sign papers transferring title. "Easy financing" fraud often offers home repairs or improvements that carry the highest interest rates and are secured by the home, which the elder will lose if payments are not made (My Health Care Manager, 2007).

Mail Order, Television, and E-mail Fraud

Mail-order catalogs and television advertising can be misleading and result in **mail-order fraud. Phishing** is the sending of e-mails that look legitimate but are not actually from the bank or other reputable agency. Swindlers often ask for information such as bank or credit card numbers or passwords, which are used to steal one's identity.

Current ads, which may be hoaxes or wild exaggerations, tout nutritional miracles, breast developers, and weight-reducing and exercise devices. Mail ordering and phone ordering from television ads are particularly attractive for those who no longer have transportation. Although many firms are honest, the mail-order and television industries have yet to eliminate the racketeers in their midst.

Mail-order health insurance has, in the past, proved to be worth little. Fraudulent hospitalization policies can be filled with small-print exceptions—ifs, ands, and buts that will eliminate the policyholder when a claim is made. Coverage for the person may not exist at the very time of failing health and hospitalization.

The "contest winner" fraud is ever present in our mail system. In some, the "winner" of a vacation, car, stereo, or other prize must either send money for postage or registration or pay to make a long-distance telephone call for more information. Twice in four months, an 88-year-old man flew, confirmation letter in hand, to Florida to collect his $11 million "winnings." The letter read, "Final results are in, and they're official. You're our newest $11 million winner." The fine print, however, fully informed the "winner" that a specific winning number was required. In a statement to Ann Landers, the attorney general of Florida was quoted as saying, "In their zeal to sell magazines, American Family Publishers and their high-profile pitchmen have misled millions of consumers. They have clearly stepped over the line from advertising hype to unlawful deception" (Ann Landers, 1998).

Telemarket Fraud

Online and telemarketing scams rank among the top complaints filed by older consumers. For some, telephone contact is the primary source of interaction with other people. In New York, an 80-year-old woman reported that she had lost thousands of dollars to telemarketers because "I've been a widow for 19 years. It's very lonely. They were nice on the phone. They became my friends." A 77-year-old man said, "I wasn't a victim—I was a sucker. I lost $200,000" (*New York Times*, 1997). The typical story is that the elder has won a wonderful prize—a vacation in Hawaii, a new car, or substantial cash. They are then asked to send a check by overnight mail to cover taxes, postage, and handling of the winnings. Sometimes they are told a courier will pick up the payment to expedite the winnings process. The same people often fall prey to more than one scheme.

Telephones are now the vehicle of choice for committing fraud. It is hard to tell if a caller is legitimate, for starters. It is also difficult to hang up: the caller is a nice young man or woman who's trying to make a living (They Can't Hang Up, 1999). Telemarketing fraud accounts for as much as $40 billion annually, of which a minimum of 80 percent is estimated to target older people (My Health Care Manager, 2007).

The FBI reports that an estimated 14,000 illegal telemarketing transactions take place *every day*. It's a big business. Southern California alone has as many as 300 **boiler rooms** (offices from which the telephone calls are made) functioning at any one time. A popular boiler room product is the vacation scam. Telemarketers tout fantastic-sounding Hawaiian packages for an unbelievably low price—$279 for two people for a week, for example. The "lucky traveler" needs only to provide their credit card number for reservations. But the trip either is postponed indefinitely or uses sleazy hotels with untold extra charges.

The U.S. Justice Department criminal fraud investigation team recommended the following to older people: (1) beware of requests for money to prepay taxes, (2) beware of requests to send a check by overnight delivery or to allow a courier service to pick it up, (3) beware of requests for credit card numbers to show eligibility, (4) beware of a rush for action, and (5) to fight back, just hang up (*New York Times*, 1997). We might add that unreported fraud becomes invisible to regulators. The following are a few resources available for reporting a fraudulent activity or concern:

- National Crime Prevention Council. Publishes Seniors and Telemarketing Fraud 101. http://ncpc.org/topics/by-audience/seniors
- National Consumer League's Fraud Center. Advises and provides referrals for consumer problems. www.fraud.org
- Do Not Call Registry. www.donotcall.gov
- Federal Trade Commission. Advises how to opt out of credit card offers and junk mail. http://ftc.gov/bcp/edu/pubs/consumer/alerts
- Securities and Exchange Commission. www.sec.gov
- North American Securities Administrators Association (NASAA). Verify investment offers. www.nasaa.org
- AARP. Links to examples of investment scams. www.aarp.org/money/wise_consumer

Because some elders experience loneliness and isolation or reduced judgment, they are a group targeted by unethical businesses. Companionship, advocacy, and legal vigilance can help protect the more vulnerable of the country's elders.

Credit Card Fraud

Credit card fraud, in which swindlers find ways to get your card number and charge items on your account, has become widespread. Identity theft occurs when criminals use personal information to access accounts or apply for credit cards in the victim's name. Older adults are less inclined to make a trip to the mall when they want an item. They order items via mail or on the telephone, both circumstances requiring credit card verification.

Door-to-Door Sales

Peddlers of various kinds of merchandise often target the homebound because their loneliness makes them eager for conversation. They are often home during the daytime when neighbors are at work. Obviously, some salespeople are honest; others sell shoddy goods or offer useless services. For example, a salesperson might scare a person into contracting for unnecessary home repairs and then flee with the down payment. Or one salesperson may make a pitch while the other robs the house (Cronk, 1998).

Investment Fraud

Many "get-rich" schemes are used to fleece older people. Older people eager to invest their savings as a hedge against inflation may become victims of numerous investment frauds involving bogus inventions or phony businesses. One California scam artist was convicted of stealing close to $700,000 from mostly older investors who were invited by mail to investment seminars (Brignolo, 1998).

Financial scams are expected to boom with the aging of the boomer generation (Chu, 2005). The sheer number of these people approaching older adulthood and wanting places to maximize their assets is causing "a frenzy of aggressive sales tactics" (Chu, 2005). In 2005 it was reported that boomers have more than $8.5 trillion in investable assets, and over the next 40 years they will be inheriting at least $7 trillion from their parents (Chu, 2005). Lottery scams, sales of promissory notes, and unregistered securities are all areas in which regulators are concerned. "It's the topic of the next few decades: Senior investments and senior fraud," said Patricia Struck, president of the North American Securities Administrators Association (NASAA) (quoted in Chu, 2005).

Estate Planning Fraud

Estate-planning seminars have become common sources for marketing to people seeking high-dividend sources for their investments. Fortunately they are also drawing more regulatory attention because they target areas with high proportions of older people. The seminars are generally free, offer a meal at a nice hotel or conference room, and frequently promote "special circumstance opportunities" that must be purchased immediately.

North Dakota Securities Commissioner Karen Tyler called these seminars a bait-and-switch tactic. The free seminar is the bait; the switch comes when the marketers urge investors to sell their investment portfolios and purchase other products—sold by representatives at the seminar, of course (Chu, 2005). Most certainly not all financial services promotions are fraudulent. Fundamentally, there's nothing more dishonest in the seminars than in advertising or direct mail. It becomes a problem when people are not well informed, have been drawn into a trusting relationship with the marketer (estate planner), and feel uncertain or overwhelmed by the complexity of the financial world.

Some fraudulent marketers contact older people to offer services in preparing living trusts, marketed as ways in which people can avoid probate court, the sometimes expensive process allocating assets following the person's death. They charge high fees, make phony pledges, and then go out of business or otherwise disappear with the money; no living trust is provided (Crenshaw, 1992).

Court-appointed guardians or conservators assigned to look after older persons are in a position to steal from them, and some do. To preserve an inheritance, children will sometimes force adult parents to live in inferior dwellings. Individual cases can become complicated, when the line between generosity and friendship and fraudulent taking advantage becomes blurred. In a California case, a savings and loan manager conservator as well as the corporation for which he works are being sued for $10 million on behalf of family members of a woman with Alzheimer's disease, whose friendship with the thrift manager clearly preceded her mental deterioration (Mintz, 1998). Some authorities have urged the government and courts to more closely audit estates so that abuse can be discovered. They also suggest a limit on the yearly fees that conservators can charge.

A study of guardianship revealed the following:

1. The elderly in guardianship courts are often afforded fewer rights than criminal defendants.

2. The overburdened court systems put elders' lives in the hands of others without enough evidence that such placement is necessary.

3. The more than 300,000 people age 65 and over who live under guardianship are "unpersons" in that they can no longer receive money or pay bills, marry, or choose where they will live or what medical treatment they receive.

Though most appointed guardians are dedicated, caring people, adequate safeguards against the minority who are corrupt and greedy must be maintained.

Medical and Health Care Fraud

Medical and health care fraud is a huge issue being paid growing attention. It is often easier to commit fraud against an older person who is managing multiple doctor visits, various prescriptions, and many kinds of medical care. In 1985 some private insurers and law enforcement personnel founded the National Health Care Anti-Fraud Association (NHCAA). This nonprofit organization "focuses on improving private and public sectors' ability to detect, investigate, prosecute and, ultimately, prevent fraud against our private and public health care systems" (NHCAA, 2009). They identified the most common types of health care fraud as:

- Billing for services that were never rendered
- Billing for more expensive services or procedures than were actually provided (billing for higher-priced treatment than was provided). . . commonly known as "upgrading"
- Performing medically unnecessary services for the purpose of generating insurance payments.
- Misrepresenting noncovered treatments as medically necessary to gain insurance payments (such as cosmetic procedures such as "nose jobs" billed as deviated-septum repairs)
- Falsifying a patient's diagnosis to justify tests or other procedures
- Unbundling—billing each step of a procedure as if it were a separate procedure
- Billing a patient more than the copay amount for services paid in full by the benefit plan
- Accepting kickbacks for patient referrals.
- Waiving patient copays or deductibles and overbilling the insurance carrier or benefit plan

Despite an elaborate system of safeguards, some experts say that the Medicare program is losing millions of dollars a year through fraud. It has long been seen as easy prey. Vague regulations, overworked investigators, swamped claims processors, and gullible consumers all play a part. It lures small-time swindlers such as the podiatrist who billed Medicare for $800,000 for phantom procedures, to large health care insurance companies, one of which paid $51 million to settle charges that one of its hospitals tried to collect on unnecessary cardiac surgeries (Vardi, 2005). Kickbacks come about in the form of "bonuses" given for each prescription written for wheelchairs for Medicare beneficiaries, to service providers who pay for doctor referrals.

The passage of Medicare Part D in 2003, part of the Medicare Prescription Drug and

Modernization Act, included broad prescription drugs coverage. It is expected to cover 29 million people age 65 and older, costing an approximate $720 billion over the first decade (Vardi, 2005). The concept is honorable; oversight of the program can be a problem. Multiple drug plans, each with its own pricing structure, are complicated and confusing and provide an opportunity for drug and insurance companies to take advantage of both government funding and seniors. Drug companies are forbidden to bill the government more than they charge private-sector insurers, but sometimes they do.

Oversight is crucial and oversight bodies need to be independent of the health care industry to assure objectivity in the process.

U.S. Health and Human Services has paid out nearly $2 million to recruit and train retired professionals to fight against fraud, waste, and abuse in Medicare and Medicaid. Twelve projects have been funded to develop demonstrations to train people in uncovering federal insurance fraud. The Administration on Aging estimates that as much as 14 percent of Medicare claims ($23.2 billion) were overpaid in 1996 because of improper fee-for-service billings (Bellandi, 1997). The program trains volunteers to work with people in their communities to help them understand their benefits and look for overbilling and unnecessary care. In California alone, more than 600 volunteers work in 30 health insurance counseling and advocacy programs (HICAP) to help older adults and Social Security offices navigate Medicare's explanation of benefits (Bellandi, 1997). The program is considered to be successful for both consumer education and the community involvement it generates in the issue of health care insurance.

Medical Quackery

Though **medical quackery**, the misrepresentation of either health or cosmetic benefits through devices or drugs that are presumably therapeutic, can target both young and old,

older people tend to be more prone to this kind of victimization. For one thing, older people often have more ailments than younger people; for another, the "youth culture" in the United States sometimes leads older people on a medical quest to look younger.

Many who offer medical goods and services are honest. Some are not, however, and older adults are often cheated. Americans spend about $27 billion a year on quack products or treatments. The top five health frauds in the United States, according to the FDA, are (1) ineffective arthritis products, (2) spurious cancer clinics (many of which are located in Mexico), (3) bogus AIDS cures (offered at underground clinics in the United States, the Caribbean, and Europe), (4) instant weight-loss schemes, and (5) fraudulent sexual aids (The Top Ten Health Frauds, 1990). We will cover four topics here: medical devices, youth restorers and **miracle drugs**, cancer cures, and arthritis cures.

Medical Devices

According to the Federal Trade Commission, many older Americans who need such medical devices as eyeglasses, hearing aids, and dentures are frequently victims of overpricing, misrepresentation, and high-pressure sales tactics. And because Medicare does not cover the cost of these items, many people cannot pay for the devices and must do without them. Some forms of device frauds are 200 to 300 percent variation in the cost of identical eyeglasses, dentures, or hearing aids and profits of 1000 percent or greater for life-saving devices such as heart pacemakers.

Youth Restorers and Miracle Drugs

Elders are likely targets for products that promise to restore the appearance of youth—cosmetics, skin treatments, hair restorers, male potency pills, wrinkle and "age spot" removers, and the like. In such a society as ours, which glamorizes youth, the desire to remain young is strong.

These women share recommendations for doctors, beauticians, face cream, child-rearing, and life satisfaction. The sharing of information can help older adults be less subject to misrepresentation or medical quackery.

Like the medicine shows that once traveled from town to town offering miracle tonics and multipurpose cures, those who today provide cosmetic surgery and breast implants are enjoying a booming business. Unfortunately, the cosmetic surgery field contains may quacks. Poorly trained "surgeons" have mutilated patient's faces and bodies and endangered lives.

Even ads for beauty products promise to restore wrinkled skin border on quackery. But though the Federal Drug Administration (FDA) believes that such claims as the following are misleading if not absolutely false, the interpretation of the law is fuzzy, and there's not much that can be done.

The skin-care market has grown into a huge moneymaking industry. Dermatologists say that despite the claims of all the antiaging creams and lotions, there is no substance that can alter the structure or function of your living skin. The only thing that any product can hope to do is add some moisture to the top layer of skin. Nevertheless, advertisers have a heyday because the FDA spends its time regulating the more harmful substances.

Even products that have been FDA-approved tend to be advertised with exaggerated claims. A product called Retin-A (generic name: tretinoin), originally used to treat acne, is now touted as a miraculous cure for wrinkles. Actually it is not a permanent cure. Further, use of the product, like the use of any chemical product, carries some risks and side effects. The FDA-approved antibaldness medication minoxidil, under the brand name Rogaine, may not produce a perfect head of hair and does not work for everybody.

Cancer and AIDS Cures

Cancer victims have been offered "cures" ranging from seawater at $10 per pint to irradiated grape juice to machines alleged to cure cancer. Scientific studies have not shown Laetrile, a substance extracted from apricot seeds, to have value in treating cancer although the case is still out on those studies. The FDA's refusal to approve Laetrile, however, caused many states to ban it; yet many cancer sufferers claim to have been helped by the drug.

Many factors account for the popularity of proven "cures" such as Laetrile. Because hundreds of thousands of new cancer cases are diagnosed each year and because two out of three victims will ultimately die of it, people are rightfully afraid of cancer. Conventional treatment is neither simple nor pleasant—surgery is often extensive, radiation can burn, chemotherapy can cause great physical discomfort. Laetrile, for example, offers a far more easy treatment. It comes in tablets to be taken with large doses of vitamin C on a low-sugar diet that avoids all foods containing additives.

Fraudulent cancer cures rob the victims not only of their money but also the most precious thing they have: time for proper treatment.

AIDS patients who can perceive no cure in the traditional medical establishment are especially willing to try unconventional treatments. AIDS patients are prime targets for medical quackery because of their desperate efforts to prolong their lives. Some have spent thousands of dollars on worthless cures. On the other hand, nutritional management of HIV/AIDS has shown to extend quality of life. What is fraudulent and what is simply not understood is a grey line.

Arthritis Cures

Arthritis is the most common chronic condition of elders. An inflammation that makes joints stiff and painful to move, arthritis appears in a hundred or so forms, of which the most severe and crippling is rheumatoid arthritis. Some forms of arthritis are painless; others cause severe pain.

Because no one knows exactly what causes arthritis, doctors can do little more than prescribe pain relievers. This lack of certainty leaves the field wide open for all kinds of fake cures. Copper bracelets have been sold to cure or ease arthritis pain. Magnets are sold as shoe inserts and bracelets. Wearing two kinds of metals in each shoe purportedly sets up chemical impulses that ward off pain. Various diets, cod-liver oil, brown vinegar with honey, "immune" milk, alfalfa tablets, mega vitamins, and snake or bee venom have all been sold to cure arthritis. Again the difference in truth and fiction is sometimes not absolutely clear.

Confidence Games

Con artists use various tactics. In a **confidence game**, the victim is tricked into giving up money voluntarily. Several games are common. Some examples are (1) the "block hustle," in which the swindler sells the victim a worthless item that he or she claims is both stolen and valuable; (2) the "pigeon drop," in which the victim is persuaded to put up money on the promise of making much more; and (3) the "lottery swindle," in which the victim pays cash for counterfeit lottery tickets. In one case a man was arrested for bilking older tenants by dunning them with fake water bills (James, 1994). Fraud can happen in big and small ways. It is apparently a lucrative business.

Drug Abuse

Older people, just like younger people, can be victims of drug abuse. Our society offers drugs—both legal and illegal—as a solution to a host of problems. The abuses discussed here are not those involving illegal drugs, such as heroin and cocaine, but those involving legal drugs such as prescription drugs, over-the-counter (OTC) drugs, and alcohol.

Older people are particularly vulnerable to the mental and physical effects of drugs because of changes in minds and bodies with age. Negative health effects can be seen in older adults consuming levels of alcohol or drugs considered to be light or moderate for younger people (Healthy Aging, 2009). Additionally, older people are more likely to be receiving multiple prescription drugs, taking nutritional supplements, and experimenting with a medicated cream or tonic, for example, to ease the discomfort of arthritis.

Prescription Drugs

Though people of all ages need prescription drugs for various health problems, elders need them in much greater proportions, because they are more likely to suffer from chronic illnesses or pain. Older adults, who comprise 14 percent of the population, consume 30 percent of all prescribed drugs (Blazer, 1990). More than 30 percent of all prescriptions for Seconal and Valium, two potent and potentially addictive sedatives, are written for people over age 60.

With multiple drug usage, a more significant chance exists for adverse drug reactions and drug abuse. Health professionals are responsible for some of the drug abuse suffered by the older population. Their errors can be of two types: (1) prescribing an incorrect drug or (2) prescribing fewer or more drugs than necessary. A study from Harvard University Medical School found that almost one-fourth of seniors are prescribed drugs that by themselves or in combination are dangerous or wrong (Stolberg, 1994). Digitized medical records, available to all of someone's authorized health care providers, will probably help coordinate drug prescription, though "paperless" medical records are fraught with potential problems, beginning with privacy issues.

Errors in medication can sometimes seem intentional. Health professionals in nursing homes, for example, may overprescribe drugs so that patients will be calm and easier to manage. When overused, drugs can stupefy, injure, and kill. Educational intervention is advised, not only for older drug consumers, but also for practicing physicians and pharmacists.

The nearly two million nursing home residents are perhaps the nation's most medicated people, often suffering depression and disorientation from receiving too high a dosage of drugs such as Valium, Calpa, and Elavil or a combination of such drugs. Elderly patients both in and out of nursing homes may get prescriptions from several doctors and have reactions from their combined ingestion. Friends may exchange prescription drugs without regard to side effects or combination (synergistic) reactions with other medications. They may also mix prescription and OTC drugs without realizing that the interactions can be harmful. Or they may take the wrong amounts of their own medications. Adverse drug reactions occur more frequently in old age, and multiple drug use should be closely monitored.

Stereotypes and the negative portrayal of older people have a considerable impact on a physician's prescribing habits. In ads in physicians' magazines, such as *Geriatrics,* elders tend to be pictured as inactive and described in negative terms: aimless, apathetic, disruptive, insecure, out of control, and temperamental. Drug advertising influences the medical professional to offer drugs as a first solution to emotionally disturbed elders; viewing old people in this manner may result in a physician giving increased prescriptions.

Over-the-Counter Drugs

Although OTC drugs may seem harmless, they are overused and consequently abused. One can blame either the drug companies, who push their products through advertising, or the poorly informed consumer. Aspirin is the most widely used OTC drug. Many old people use very high doses of aspirin for arthritis, even though it is a stomach irritant and can deplete the body of essential nutrients. (Some doctors use a milder drug, suprofen, rather than aspirin, to treat arthritis; but suprofen is not available OTC.) And, though laxatives have been advertised as "nature's way," nothing is further from the truth. Prolonged laxative use can impair normal bowel function. Nature's way is drinking plenty of water and including fiber in the diet; many people, however, get "addicted" to laxatives, thinking they are making the best choice. The same is true of those who take sleeping pills, which must be taken in escalating doses to be effective. Consequently, they can cause rather than cure insomnia (Kolata, 1992). Because people need less sleep as they get older, many are better off to accept this fact rather than to try to force themselves to sleep eight or more hours a night. Antacids such as Alka-Seltzer contain sodium bicarbonate, which may harm kidney function. Hemorrhoid medication ads exaggerate the effectiveness of their products, and mouthwash does little more than add to the complaints of dry mouth. These OTC drugs can be a waste of money, cause bodily harm, and delay proper treatment of potentially serious ailments.

Alcoholism

Alcoholism as a problem for elders has been largely ignored until recent years. If, after retirement, older people lead much less visible lives, alcoholism can remain well concealed, becoming a self-generating cycle of abuse: diminished contacts and a sense of loss may trigger drinking that goes unrecognized.

Alcoholism: ". . . a primary, chronic disease with genetic, psychosocial and environmental factors influencing its development and manifestations. The disease is often progressive and fatal. It is characterized by impaired control over drinking, preoccupation with the drug alcohol, use of alcohol

despite adverse consequences, and distortions in thinking, most notably denial."

NATIONAL COUNCIL ON ALCOHOLISM AND DRUG DEPENDENCE AND THE AMERICAN SOCIETY OF ADDICTION MEDICINE [REPORTED IN MORSE & FLAVIN, 1992].

Elders *do not* have higher rates of alcoholism than younger adults, however, and one review of the literature concluded that alcohol consumption goes down with age. In addition there are fewer alcohol-related problems among heavy drinkers who were elderly than among those who were not (Council on Scientific Affairs, 1996). Among those with problems, however, the strain on the body of alcoholism is profound. Coupled with smoking, the matter becomes worse and increases the risk of lung diseases and cancer.

Some older alcoholics began their drinking early in life and continue that pattern of behavior. Others do not drink heavily until old age. Between half and two-thirds of older alcoholics are estimated to have begun drinking early in life (Council on Scientific Affairs, 1996). The "early onset" alcoholics have had alcohol-related problems for years and are more identifiable because of earlier dysfunctional behavior that brought them to the attention of the medical or helping establishment. In contrast, the "late-life onset" alcoholics, who begin abusive drinking in their 50s or 60s, are often viewed as reactive drinkers, whose problem began after traumatic events such as the death of a spouse, retirement, or moving from an original home. They will have fewer chronic health problems and are likely to drink alone at home. In cases where alcoholism first occurs in old age, the alcoholic will often respond readily to intervention—help for depression or loneliness, for example, may reduce the need for alcohol. Treatment focuses on rebuilding a social support network and overcoming negative emotional states. Long-term alcoholics are more difficult, but not impossible, to help.

Psychiatric hospital and outpatient clinic records show that admission for alcoholism peaks in the 35-to-40 age group. Some researchers believe that alcoholism may be a self-limiting disease; that is, the decrease in alcoholism in the older age groups results not from treatment but from a spontaneous recovery with age due to such factors as a lowered social pressure to achieve. Others think that alcoholism kills many of its victims before they reach old age. Still, about 10 percent of elders manifest symptoms related to excessive use of alcohol. Older men are more at risk than older women. And two-thirds of older alcoholics are severe chronic alcoholics whose symptoms tend to be obvious and profound (Council on Scientific Affairs, 1996).

Risk factors for alcoholism in later life include: (a) family history of alcoholism; (b) personal history of excessive alcohol consumption; (c) discretionary time, money, and/or opportunity to drink; (d) age-related volume of alcohol distribution in the body; (e) increased central nervous system sensitivity to alcohol; (f) pain or insomnia secondary to chronic medical disorders; and (g) other psychiatric disorders, such as schizophrenia and depression (Goldstein et al., 1996).

Medical support is of great value. It begins with a thorough history, physical, and laboratory examination. Sudden withdrawal of alcohol for the older adult can be life-threatening and might require a "detoxification" process of weaning the body off the drug. Alcoholics Anonymous (AA) and other groups can be instrumental in stopping heavy drinking. Countless people have benefited from AA.

Rehabilitation for older adults is generally good. They are more likely than younger adults to stay in treatment and remain sober. One focus of rehabilitation among older adults is managing time. Keeping busy is important. Day programs and senior centers can be helpful. Sometimes supervised living arrangements (live-in companion, living with relatives, or living in assisted living or CCRC facilities) should be considered.

Promoting Consumer Education

Consumer education is vital in helping individuals to avoid fraud, medical quackery, and drug abuse. Adult education programs are one way of reaching the public. Videotape systems now provide advice on a variety of topics, such as how to buy a used car without being swindled, how to handle the estate of a deceased relative, or how to recognize and avoid con artists. Books and magazines offer consumer information aimed at the older population, covering topics such as prescription drugs and their side effects, as well as the use and abuse of OTC drugs and alcohol. Consumer action groups are available for people of all ages who want to support the creation of stricter consumer-protection laws and the enforcement of existing laws. Internet is an amazing source of information with sites sponsored by medical associations, the National Institutes of Health, pharmaceutical companies, blogs, and discussion rooms of people sharing information. As the boomer generation enters old age, they bring with them a far more sophisticated level of computer saavy, and many sites are specifically designed to provide older people with disease and health-based information.

Chapter Summary

Suicide continues to be a significant problem among those 65 and over. The suicide *attempt* rate for older women is much higher than that for older men, yet older men (i.e., white males) have the highest rate of *actual* suicides of any age group. Many factors may be involved, especially dependency, depression, physical illness, and social isolation.

The dimensions of the problem of crime against the elders are being studied by law enforcement and social agencies. National surveys show that elders are less likely to be victimized than younger adults, except in the area of personal larceny (e.g., purse and wallet snatchings), where attacks on elders are considerable. Over half of the violent victimizations occur in or near one's home. Fear of crime is widespread among elders and exists in a greater degree than the actual crime rate against them would suggest. It is the metropolitan elderly for whom fear of crime takes its greatest toll. Elders, themselves, can be the victimizers. Older criminals come from two groups: (1) those who started at a young age engaging in illegal activities and (2) those who committed offenses for the first time in their later years. The older prison population is growing, but adequate medical care is lacking in jails and prisons.

Elders are potential victims of hucksters and con artists. Land fraud, mail-order fraud, mail-order health insurance, and confidence games are used to deceive and cheat people out of money. Medical quackery often entices the sick and ailing. Medical devices, youth restorers, and cancer and arthritis cures have robbed many of their money. Elders are sometimes victims of drug abuse—both prescription and nonprescription drugs.

Education, self-help groups, and professional help should be available to assist elders with their special problems. Consumer education is useful in learning to detect frauds and gaining knowledge about OTC and prescription drugs.

Key Terms

attempted suicide
battered parent
   syndrome
boiler room
confidence games
credit card fraud
crib job
crimes against persons
double suicide
fear of crime
granny dumping

home-equity fraud
land fraud
mail-order fraud
medical quackery
miracle drugs
Neighborhood Watch
rational suicide
successful suicide
suicide rate
perceived vulnerability
Phishing

Questions for Discussion

1. List five types of elder abuse. What are some examples of each type?
2. Look through news items over the last month. What examples of fraud can you find that affected an elder?
3. Research prison populations in your state. What percentage of prisoners is over the age of 60? Are there special provisions for them?
4. Why might some elders be susceptible to con artists?
5. Imagine that you have just been cheated of $1,000 by a con artist who has promised that your money will be quadrupled in four days. What would you do? What *should* you do?

Fieldwork Suggestions

Clip articles from magazines and newspapers that advertise goods and health insurance that are probably fraudulent or questionable in their authenticity. What caught your attention with the ad? To whom is it targeted?

Internet Activities

1. Locate a grief forum on the Internet. What themes do you find? What ages to you guess are represented among the correspondents? Can you find a blog about depression, death, and suicide that reflects the perspectives of older adults?
2. Search a website that would provide useful and appropriate information for a family member concerned about alcohol consumption of an older person in their family.

Women and Ethnic Groups

13

Gertrude Baines, World's Oldest Person, Dies at 115

Los Angeles, California (CNN)—Gertrude Baines, the world's oldest person, has died in Los Angeles, California, at the age of 115, according to the home where she lived and Guinness World Records.

Gertrude Baines said she attributed her longevity to not drinking or smoking.

Gertrude Baines passed away at the Western Convalescent Hospital at 7:25 A.M. (10:25 A.M. ET) Friday, Guinness World Records said.

Born in 1894, Baines became the world's oldest person in January after the death of another 115-year-old, Maria de Jesus, from Portugal, Guinness World Records said.

At her 115th birthday party in April, Baines shook her head in disbelief when presented with the certificate saying she was now in the *Guinness Book of World Records* as the world's oldest person.

"She told me that she owes her longevity to the Lord, that she never did drink, never did smoke, and she never did fool around," her doctor, Dr. Charles Witt, said in April.

Baines, whose grandparents were slaves, worked as a maid in Ohio State University dormitories until her retirement, and lived at the Los Angeles convalescent home for more than 10 years.

Last November, she became the oldest African-American to vote for President Obama and received a letter from him on her 115th birthday, Guinness World Records said. Witt said Baines planned to vote for Obama again in 2012.

Asked why she voted for Obama, Baines said it was because "he's for the colored people," according to footage from *The Los Angeles Times.*

Gertrude Baines at 115 years of age.

She said she never thought a black man could become president.

"Everybody's glad for a colored man to be in there sometime," Baines said. "We all are the same on the skin. It's dark, and theirs is white."

Baines had few complaints, her doctor said. She fussed about the bacon not being crisp enough and the arthritis in her knees, Witt said.

The smooth skin on Baines' face belied her 115 years, but she didn't attribute that to any anti-wrinkle cream or miracle product, according to her best friend, Lucille Fayall. She said Baines simply washed her face in cold water.

Source: www.cnn.com/2009/US/09/12/oldest.person.dies/index.html?iref=hpmostpop

Multiculturalism is a central focus of the social sciences. This focus corrects a past tendency to ignore the diversity and richness that ethnic groups have added to American life and assume that Americans' experiences are relatively similar, despite their roots of origin and cultural traditions. Until recently the study of aging America was a study of the older white Americans who made up nearly 90 percent of America's elderly population, while the many ethnic groups were ignored. Generalizing from the larger population to all subcultures of elders is misleading, incorrect, and insensitive. Being old and a member of a minority group, or being an old white woman, is to experience the political economy—the context of aging—in ways strikingly different from that experienced by older white males in America.

By 2030, about 70 million people will be over 65 in the United States—more than two times their number in 1996. Of this number, 25 percent will be minority populations by 2030, and by 2050 minorities will constitute 33 percent of the older population (AoA, 2005). Thus, with every passing year the ethnic aged become a larger proportion of aging America. The bars in Figure 13.1 illustrate the proportional shift of elderly minorities from 1990 to 2030.

The non-Hispanic white elder population is projected to increase by 91 percent from 1990 to 2030. The proportion of African American elders will increase slowly from 8 percent in 1990 to 10 percent by 2030 (an increase of 159 percent). The Native American, Eskimos, and Aleut older populations will increase to 0.5 million by 2030 (a 249 percent increase). The Asian and Hispanic populations will represent the greatest proportional shift. Asian and Pacific Islander elders will increase to 8 percent of the elder population—an increase of 643 percent from 1990; and the elder

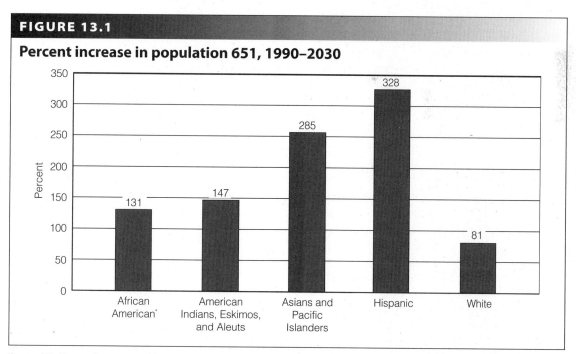

FIGURE 13.1

Percent increase in population 651, 1990–2030

Source: U.S. Census Bureau, population projections of the United States by age, sex, race, Hispanic origin, and nativity: 1999 to 2100; published January 2000.

Hispanic population will increase to 17 percent of the older population (570 percent increase) by 2030. Increases in all the groups are related to past fertility patterns, immigration patterns, and increasing life expectancy (National Institute on Aging, 2005). The categorization of minorities, however useful it might be for bureaucratic purposes, often flies in the face of group identification. Koreans, for example, identify strongly as Koreans and most certainly do not identify with, say, the Japanese. Both groups, however, are considered "Asians" under U.S. government standards.

Ethnicity involves (1) a culture and an internalized heritage not shared by outsiders, (2) social status, and (3) the composition and function of support systems (Hooyman & Kiyak, 2006). Discrimination is not necessarily associated with ethnicity: some German, Scandinavian, Eastern Bloc, and southern European immigrants have retained close cultural ties with their ethnic backgrounds, but do not necessarily hold minority status. **Minorities** in this context are ethnic elders identified by language, physical, or cultural characteristics who, based on those characteristics, have experienced unequal treatment in certain segments of society.

A particular challenge to current America is that its aged population is relatively homogeneous, but its younger population is quite heterogeneous, and this diverse population of younger adults will be responsible for developing national policy that affects a homogeneous older age group. It will not be until around 2020 that the older adult population will begin to reflect the diversity of the larger population. By 2050, elder diversity will be established: around 30 percent of the elderly will be African American, 36 percent will be elderly Hispanics, and 40 percent will be elderly whites (Hobbs & Damon, 1996).

In 2006, Dr. Christopher Murray of the Harvard University School of Public Health reported that based on his epidemiologic research, where you live and what your race and income is plays a huge role in how long you live (Associated Press, 2006). These gaps in lifespan have persisted for over 20 years, a time of relative national prosperity during which major leaps in health care and public health policy have taken place. One would hope that these positive developments in health care and policy would have an impact on aging for all Americans.

Murray found Asian American women living in Bergen County, New Jersey, are leading the nation in longevity. They typically reach their ninety-first birthdays. American Indian men in South Dakota, however, die at around 58, which is comparable to life expectancies in developing countries. Traditional considerations for longevity have been an issue of minorities and the poor, with a focus on good medical care. Murray's study shows a more complex issue: geography plays a role. Perhaps longevity is multiply related to health care options and availability, lifelong poverty and discrimination, and the culture in which one lives.

Murray's findings serve as a reminder that statistics are combined averages, and even when they are broken down by categories they remain averages. Three minority family incomes of $12,500 plus that of one family from the same ethnic minority of $159,000, for example, makes an average of four families with incomes of $49,125. The huge difference between living on $49,125 and living on $12,500 becomes lost in the statistics.

We must additionally acknowledge the limitations of the "ethnic" concept and the vagaries of group identification; however, in this chapter in seeking broad similarities and understanding the cost of the "average," we consider five groups: women of any ethnicity, African Americans, Hispanics, Asian Americans, and Native Americans. Women, despite not being an ethnic minority group, hold a lower status in our culture than men. Therefore, women have been judged a minority group not in terms of numbers, but in terms of status. Older women are considered first in this chapter, after which comes the section on ethnic minority elders.

Women

The minority status of women is based on the sexism that pervades U.S. society. For older women, sexism is compounded by ageism. Older women have trouble finding acceptance and equality in the work world, politics, and romance. Women are making progress in these areas; however, hundreds of years of established patterns cannot be changed overnight.

A well known talk radio host commented on an unflattering photograph of Senator Hillary Rodham Clinton, then a candidate for the American presidency: "Will this country want to actually watch a woman get older before their eyes on a daily basis?" www.mediamatters.org/research/ 200712190002

One advantage for women is their willingness to reach out and get help: from one another, books and seminars, re-entry programs at colleges, and various counseling services. Models of positive aging are Meryl Streep and Susan Sarandon in their early 60s; Jane Fonda, in her 70s; and Gloria Steinem, born in 1934, in her mid-70s. Each is looked to for advice and inspiration. Dr. Ruth Jacobs (1993), a university professor and author of books for older women, also offers support to aging women. She recommends some of her and others' books:

Be an Outrageous Older Woman (Jacobs, 1993)

I Am Becoming the Woman I've Wanted (Martz, 1994)

Flying Solo: Single Women in Midlife (Anderson and Stewart, 1994)

Moving Beyond Words (Steinem, 1994)

Going Strong (York, 1991)

Look Me in the Eye: Old Women, Aging and Ageism (Macdonald and Rich, 1991)

Old and Smart: Women and Aging (Nickerson, 1991)

To this list we might add:

At 82: A Journal (Sarton, 1995)

And God Created Wrinkles (Mall, 1990)

Handbook on Women and Aging (Coyle, 1997)

Revisioning Aging: Empowerment of Older Women (Onyx, Leonard, and Reed, 1999)

Aging Artfully: Profiles of Visual and Performing Women Artists 85–105 (Gorman, 2000)

The range of popular books suggests that the present generation of older women is receiving more attention and validation than previous generations. Roles for older women, which historically have been narrowly constructed in the United States, are broadening. Even the topic of menopause, previously considered a process that was experienced individually and silently, is out of the closet as women begin to view it not as a marker to the end, but as a bridge to a new stage of adulthood. Increasingly, older women have more options for experiencing a rewarding and fulfilling later life. Two areas of struggle for older women are discussed in the following sections: finances and the double standard of aging.

Adequate health care and income are related: affluent individuals receive better health care. The lifetime histories of lower wages among minorities create a double jeopardy in later life.

Financial Status

Poverty among older people is unevenly distributed, with women being among the most poor in the United States. Poverty among married women is fairly low whereas that among unmarried women is three to four times higher than that of their married counterparts, reflecting their work and marital histories (Cooke, 2007; Branch, 2009).

More than 50 percent of women 65 and older would live in poverty if it were not for Social security. Women receive more than 48 percent of retired worker benefits and almost 46 percent of disabled worker benefits, but over 12 percent of all older women are poor, despite receiving Social Security benefits. Among unmarried older women, Social Security makes up 48 percent of their total income, whereas it makes up only 37 percent of unmarried elderly

men's income, and only 30 percent of couples' income (SSA, 2007).

Some of the major ways in which women's lives are compromised in later life include:

- *Lower wage histories*: Women on average earn only 76 percent of what men earn, which is unlikely to change dramatically over the next few decades. As a result, women have less money to invest than do men (Shaker, 2006).

- *Low retirement income*: Women are more likely than men to leave jobs before becoming vested in retirement plans; low wages mean many women cannot save for retirement.

- *Vulnerability of divorced women*: Divorced women are generally better educated than divorced men; however, their retirement incomes are lower, and the likelihood of remarriage is less for women than for men (Cooke, 2007).

- *Lack of financial planning skills:* Among most older adults, managing money was traditionally left to husbands; without money management skills, women jeopardize their long-range security and independence.

- *Longevity of women:* Women live longer than men, which makes them especially vulnerable during retirement. They outlive their husbands by an average of four to six years, and once alone they often have less money to pay the same bills.

- *Need for public benefits:* Women are almost twice as likely to be poor than men. This means they are far more likely to need public services to survive in later life.

- *Caregiving responsibilities:* Women provide three-quarters of family caregiving; for 12 percent of working caregivers, the demands are so intense they leave paid employment.

- *Inadequate housing:* Because of longer life spans, women are more likely to live alone in old age than are men. Added to the lower incomes of women, affordable housing is a challenge to living independently for many older women.

Middle-Aged Boomer Women

Financial problems for women frequently originate in middle age or even earlier for some of the reasons mentioned earlier: sporadic lifetime work histories at lower wages than men's, greater chance of work in a business not providing a pension, and less information about investment and retirement planning.

For both men and women, being single in later life is associated with greater poverty. For the boomer generation, divorce has been very common. Because women are less likely than are men to remarry, a cohort is formed of divorced women entering later life. This results in women being more likely than men to spend old age in poverty—in part because many have spent their lives at an economic disadvantage. Hundreds of thousands of women in their 60s are part of the surge of divorces that began in the 1990s. They find themselves forced to stay in the workforce because they lack sufficient resources to retire. Wages in effect are becoming their pensions.

The middle-aged woman who can save some money—or, at a minimum, pay into Social Security—improves her chances for a fulfilling old age. Though middle-aged women are now in the workforce in large numbers, the persisting pay gap predicts greater poverty in their later life.

Single, Widowed, and Divorced Older Women

Single women

More than 25 percent of women age 65 and older who live alone or with nonrelatives live below the poverty level (U.S. Bureau of the Census, 2005). This percentage would almost double if the Federal Poverty Index were updated as experts recommend. In contrast, 8 percent of those living with their husbands are poor. Those married and living with their husbands have the benefits of another income.

Ever-single women (having never married) tend to have greater retirement incomes than do divorced or widowed single women. This success is attributable to a large increase in their share of private pensions as a main source of income. Never-married women exhibit similar commitments to the labor force and household responsibilities as do men, earning 96 cents for every dollar earned by their male counterparts whereas married and divorced women earned about 78 cents on every dollar in 1998 (Cooke, 2007). Those with lower work history incomes are less likely to have private pension assets.

Lifetime income impacts the quality of life in many aspects, from health care to living arrangements to opportunities for socialization.

Widows

Among widows, the scenario is a bit different because they benefit from the wealth of their husbands. However, many older widowed women depended entirely on their husbands' incomes, and, when retired, on their husbands' private pension plans or Social Security. More often than not, private pension plans fall sharply when a retired spouse dies. Additionally, the death of a spouse lowers the amount of Social Security benefits. In this case, widows' low incomes expose them to greater social and economic risks than other segments of the older population.

Data from a national sample of widows of all ages found that widowhood decreased living standards by 18 percent and pushed into poverty 10 percent of women whose prewidowhood incomes were above the poverty line (National Center on Women and Aging, 2002). The problem is so great that some laws have been passed to protect women during times of marital transition, including the Consolidated Omnibus Budget Reconciliation Act (COBRA), allowing women to keep their health insurance coverage under their husbands' plans, at their own expense, for up to three years after divorce, separation, or widowhood.

The opportunity for older widows to remarry is limited, due primarily to the relatively small number of eligible males in their age group. Older females who are eligible for marriage outnumber eligible males by a ratio of more than three to one. In addition, males who marry after age 65 tend to marry women from younger age groups.

Divorced women

The socioeconomic well-being of divorced women is significantly lower than that of married, single, or widowed women. Given current statistics and expected trends in marriage, divorce, and widowhood, the numbers of married and widowed older women will decline, but the proportion of divorced older women will dramatically increase. It is projected that each year, at least 50,000 people older than 60 dissolve marriages of 30 to 40 years' duration (Uchitelle, 2001).

If current remarriage rates persist, few women who enter midlife divorced, or who divorce after midlife, will ever remarry. Remarriage rates have fallen by half since 1970 for women between the ages of 45 and 64 (Uchitelle, 2001). According to rough projections, by 2025 no more than 37 percent of women between the ages of 65 and 69 will be in their first marriage. Half will not be in any marriage; this figure could be considerably higher if the divorce rate after age 40 continues to increase and the remarriage rate continues to decline (Uhlenberg, 1996). There is good reason for public concern over these statistics.

Upgrading the Financial Status of Older Women

In our society, in spite of positive steps toward equality, women remain disadvantaged in the workplace. Inequalities in income for older women will not totally disappear until women achieve equality in the workplace from the beginning of their careers.

If women are homemakers or caretakers of children or older parents during their working years, they suffer financially in old age. Being removed from the paid labor market reduces their Social Security benefits (Devlin & Arye, 1997). These important cultural activities could be compensated in a number of ways, however. The "motherhood penalty" could be eliminated by granting caregiver credits for up to five years spent out of the full-time workforce to care for children. Granting credits could eliminate five additional "zero" years from the Social Security benefits calculation. Additionally, four "drop out" years could be allowed to include care for family members other than children (Women and Social Security, 2008). Additional options might include expanding the special minimum

benefits to compensate long-term low-wage workers and improving benefits for divorced women by increasing benefits for lower-earning spouses from 50 percent of the former spouse's benefit to 75 percent.

Younger adult women have observed the divorce rates of the generation preceding them, and many are more active retirement planners than were their mothers. Attention to what percent of a salary can be put in a 401(k) plan and be matched by the employer is not uncommon information requested by younger women in employment research or interviews. Employees at Google, for example, can contribute up to 60 percent of their salary to their retirement fund to be matched by the employer. Ford Motor Company offered several attractive employee packages to encourage employment termination because of the company's financial difficulties. Employees under the Special Termination of Employment Program could receive a lump sum payment of $100,000 after one year of service. Or they could select the Family Scholarship Program for $100,000 to be used to fund the college educations of their children, spouse, and grandchildren (Branch, 2009).

These are only two examples of corporate policies or programs already on the radar for younger boomers and the following cohort. This awareness will vastly improve future retirement income for those appropriately educated and skilled workers, including hundreds of thousands of women.

Double Standard of Aging

The term **double standard of aging** has been used to suggest the standard of aging for a woman that progressively destroys her sense of beauty and self-worth, whereas the standard of aging for a man is much less wounding. More recent literature questions the belief that changes in body self-concept are more difficult for women than for men, but opinion is unified around the issue of intense social standards for women's appearance.

> *Like wolves, women are sometimes dis-*
> *cussed as though only a certain temperament, only a certain restrained appetite, is*
> *acceptable. And too often added to that is*
> *an attribution of moral goodness or badness*
> *according whether a woman's size, height,*
> *gait, and shape conform to a singular or ex-*
> *clusionary ideal. When women are relegated*
> *to moods, mannerisms, and contours that*
> *conform to a single ideal of beauty and be-*
> *havior, they are captured in both body and*
> *soul, and are no longer free.*
>
> PINKOLA ESTES, 1992

Women from an early age learn to care about their physical beauty. As a result, women spend a tremendous amount of time and money on their better appearance. Women are concerned about being "fat" or "ugly"—indeed, dieting and eating disorders are far more a woman's issue than a man's (Abbott et al., 1998). Cosmetic and plastic surgery and face-lifts are performed more often on women than on men. Many women's exercise programs emphasize appearance rather than strength or endurance: in the self-help section at any local video store, note the number of books and DVDs that promise shapelier breasts, thighs, or buttocks. Or attend an aerobics class at a local health club and note the female clients' concerns about their exercise clothing.

The youth culture in our society exerts an intense social pressure for women to remain young. In a personal account of her own aging Ruth Thone, an activist from Lincoln, Nebraska, gave her reason for writing her book: the "subtle, deep, pervasive, unspoken distaste and derision" for old Americans in general and old women in particular (Thone, 1993). She wrote of her own "internalized aging," in which she is filled with self-loathing and anxiety about aging. She is also furious at being sexually invisible to men and being patronized by younger people. She is sensitive about any jokes putting down older women. She gave this account.

> *My husband found a joke in a magazine that*
> *he added to his repertoire, about two women*
> *in a nursing home who decided to streak*
> *their fellow residents. Two startled old men*
> *looked up and one asked, "What was that?"*
> *"I don't know," the other replied, "but what-*
> *ever they were wearing sure needs ironing."*
> *My husband did not understand how hurt I*
> *was by that joke, by that ridicule of women's*
> *aging skin and by the double standard that*
> *does not make a mockery of men's aging*
> *skin. He insisted it was my feminism, not any*
> *ageism in him that kept me from knowing*
> *the joke was harmless. (p. 54)*

She wondered, "Am I an object of scorn as my body ages?" She dealt with self-criticism, self-rejection, and self-hate, coming to terms with aging by writing in her journal, meditating, and becoming more spiritual. In a chapter titled "The Grief of Aging," she described working through her sadness at the loss of her youth.

Not all women can confront their aging so directly and honestly. They buy into the idea that they must stay young and beautiful in appearance forever, and when physiological and health changes occur to alter that image, healthy adjustment to aging becomes difficult. Self-neglect in old age is thought to come about because of the negative self-concept that results in part from comparing one's physical condition and personal competence to a standard that is unattainable (Karpinski, 1997; Rathbone-McCuan, 1996).

The successful exploitation of women's fears of growing older has been called **age terrorism**. Pearlman (1993) speaks of **late midlife astonishment**—a developmental crisis in which women aged 50 to 60 work through society's devaluation of their physical appearance. Women suffer a loss of self-esteem, depression, and feelings of shame and self-consciousness. Feminist therapists believe that body image disturbances are not limited to eating-disordered clients and that they occur to women of all ages. One example is Helen Gurley Brown, former editor of *Cosmopolitan*, the "Cosmo Girl," who at age 71 referred to age as the great destroyer. She discusses having had silicone injections, cosmetic surgery, shrink sessions, endless dieting, and a 90-minute daily "killer" exercise regime, all with the intent of delaying the appearance of age. She says:

> *I'm afraid of losing my sexuality. I'm desperately afraid of retirement. I fear that with age, I'll cease being a woman, that I'll be neuter. I fear losing my looks and ending up looking like . . . like an old crumb. (quoted in an interview with Marian Christy, 1993)*

One dangerous outcome of negative shifts in self-image can be self-neglect. Women's fear of age-related physical changes can take on a pathological flavor, since personal changes can cause a woman to doubt her social capability, which leads to low self-assessment, leading to low self-esteem, leading to self-neglect (Watson, 1997; Rathbone-McCuan, 1996). Self-concept itself, however, does not seem to undergo age-related changes. The key to successful aging appears to be that women develop adequate strategies to maintain their self-identities even as key aspects of their bodies, their minds, and their health change (Matthews, 1979; Kaufman, 1986; Barusch, 1997).

Our nation is moving toward two societies, one Black, one White—separate and unequal; discrimination and segregation have long permeated much of American life; they now threaten the future of every American.

U.S. Riot Commission report, 1967.

African Americans

The African American older population is the largest minority group among older Americans. The older population was over 3 million in 2007 and is projected to grow to just under 10 million by 2050. In 2007, African American people constituted 8.3 percent of the older population; by 2050 that percentage is projected to account for 11 percent of the older population (AoA, 2009).

Income and Housing

Economic security of people 65 and older has improved over the last 30-plus years. Poverty rates for this age group fell from 28.5 percent in 1966 to 10.1 percent in 2005. For most older African Americans, Social Security benefits provide the foundation for their retirement income security (AARP, 2007).

African Americans as a group differ widely in socioeconomic factors. As a result of the civil-rights movement, African Americans have gained a large middle class, a class having grown to roughly equal the same size group as the African American poor or marginally poor. As comparatively well-off African Americans move to better neighborhoods, they leave behind an African American "underclass"—chronic welfare recipients, the unemployed, high school dropouts, and struggling single-parent families. Although many African Americans moved into the upper middle class in the 1990s, one-third remained locked in deprivation. As a result, a deep class division now exists among African Americans. Elders in inner cities are often left to cope with deteriorating neighborhoods, high crime rates, and the threat of violence; or they are left in rural areas to struggle with poverty and a lack of medical and social services, and there are often more limited social support networks.

African Americans are a diverse group, yet overall pictures do emerge. Described as a group, households with families headed by black people aged 65 or older reported a median income of $32,025 in 2007. The comparable figure for all older households was $41,851. The median incomes for African

American men and women were $16,074 and $11,578, respectively. Comparable figures for all elderly were $24,323 for men and $14,031 for women (AoA, 2009).

The poverty rate in 2007 for older African Americans was 23 percent—more than twice the rate for all older adults (9.7 percent). Nevertheless, this represents a significant decline from 65 percent in 1965 in the poverty rate for black elderly (AoA, 2009). This reality is a social reality: it can add to the complexity of intergenerational relations between older adults having experienced a lifetime of comparative disadvantage and younger adults for whom the struggle for social equality is more assumed and for whom a lifetime of aggressive discrimination has been less pervasive.

In terms of living arrangements, a majority of older African American men (57 percent) lived with their spouses, 10 percent lived with other relatives (most often a daughter), 4 percent lived with nonrelatives, and 29 percent lived alone. Among older black women, 25 percent lived with their spouses, 32 percent lived with other relatives, 2 percent lived with nonrelatives, and 40 percent lived alone. These differences imply much different later life experience for older men and women.

Compared with Caucasian elders, older African Americans have less adequate income and poorer quality housing. This constitutes a serious social problem because older people cannot easily move elsewhere because of those limited incomes and the likelihood or the fear of housing and/or neighborhood discrimination.

Unemployment rates for African Americans of all ages are far higher than those for whites, and have been for many years. Employment is increasingly associated with level of education in our technologically advanced world, and increased educational opportunities for people of color over the past 30 years have made a substantive difference in employment histories of aging African Americans. In 2007, over 57 percent of the black population aged 65 and older had completed high school. This compares with only 9 percent in 1970. Also in 2007, over 10 percent of older African Americans had a bachelor's degree or higher (AoA, 2009).

A small percentage of retired African Americans are from the upper class, having owned large businesses or real estate, headed large corporations, or worked in the highest levels of industry. Other retired African Americans are from the middle class, having been schoolteachers, owners of small businesses, or government employees. Still others are retired from manual labor or domestic service jobs. Overall, however, older black Americans suffer higher levels of poverty, having not paid as much into Social Security as white Americans and having worked predominantly in jobs with no pensions.

Older African Americans are slightly more likely than older white Americans to be looking for work after age 65. Older African American men have had lifetimes of higher unemployment rates than older white men. Older people in lower socioeconomic groups regardless of race have a different understanding of retirement. Health permitting, they often work at lower-paying jobs well beyond retirement age in order to meet basic expenses for food, medical care, and housing. Work becomes the retirement pension equivalent. This work is usually part-time and temporary; however, those under age 62 are not eligible for Social Security or other pension programs and, therefore, without work, are more financially needy than those eligible for pensions. In the near future, following the steep economic decline and accompanying job losses of 2008, the availability of even low-wage part-time jobs for older people is quite uncertain.

African American elders are more likely than older whites to reside within decaying central cities and live in substandard housing. They are also more likely than whites to live in public housing. Those who work with minority elders must be able to advise them accurately on low-cost public housing, low-interest housing loans, and other forms of available property relief.

Older African Americans are admitted to nursing homes at between one-half and three-quarters of the rate of whites. Although the physical health of black elders has at least the same proportion of chronic physical illnesses as that of white elders, low use of long-term care cannot be explained only with the statement that African Americans prefer to care for elders within their families. Having the economic option to make long-term care a choice is not available to many black American families.

Health Care and Life Expectancy

Low-income African American elders have health care problems. Many poor elders have lacked the resources for adequate health care throughout their lives, resulting in a life expectancy rate that is much lower than that for whites. African American life expectancy is 70.2 years, compared to an average of 76.5 years for all population groups. The difference is even more striking among African American men, whose life expectancy of 66.1 years compares to the national average of 73.6 years for all men (About.Com, 2008). The age at which maximum Social Security benefits begin is scheduled to gradually increase.

Options for appropriate housing, health care, and overall quality of life in later life are income-related.

As such, many boomer-generation African American men, whose benefits begin when they reach age 66, will not live to be recipients.

The differences in life expectancy between African and white Americans fade away if a number of social variables are held constant, including marital status, income, education, and family size. Marriage, high income and education, and small family size are correlated with longevity.

Between 1986 and 1991, life expectancy for African Americans actually *dropped*. Though African Americans had been sharing in life expectancy increases over the decades, the trend reversed in the late 1980s. One major reason was the large number of African American babies who died in their first year. But this was only one factor. Thousands of African Americans die in the prime of life from illnesses that could be cured or treated by routine medical care, for example, appendicitis, pneumonia, hypertension, cervical cancer, tuberculosis, and influenza. Given early detection and good treatment, all these illnesses are curable; few people should be dying of curable diseases and those headed off by early detection. The lifelong experience of inadequate health care beginning in childhood has become a national-level issue. At the time of publication, it was still uncertain whether the American public will see fit to support health care for all Americans, much as do Europeans and other industrialized nations. The debate is passionate, and it highlights the reality and experience of group disadvantage for the world to see.

Assuming that low social status generates repressed negative emotions and inner tensions, some medical and social scientists believe that the high incidence of hypertension among black Americans reflects the pressures of low social status. Both physical and mental illness can result from the chronic stress of prejudicial attitudes and behavior of others. Combined with the possibility of genetic predisposition for hypertension, the health of older African Americans is precarious for tragically unjust reasons.

Whatever the cause, the incidence of high blood pressure among African Americans is nearly two-and-a-half times that of whites, and the mortality rate from high blood pressure is higher for African Americans than for whites. The three most frequently occurring conditions among African American elderly in 2004 to 2005 were hypertension (58 percent), arthritis (46 percent), and all types of heart disease (22 percent). Comparative figures for all older persons were hypertension (48 percent), arthritis (47 percent), and all types of heart disease (29 percent) (AoA, 2009).

The story is complex, however, representing diversity among African American elders. In the years 2004 to 2006 about 63 percent of African American older men and about 59 percent of older black women reported good or excellent health. Among all the 65 and older population, these figures were 74 percent for men and 74 percent for women. This has been attributed to cultural differences in self-acceptance, partly in terms of who different cultures anticipate themselves to be in old age. One interesting study, however, found that older Caucasian and African American adults may interpret the question "How's your health?" quite differently. Self-rated health was a better predictor of four-year mortality among Caucasians than among African Americans. Fewer than 5 percent of older white adults who rated their health as excellent died within the four-year period; however, roughly 8 percent of African Americans rating their health as excellent died within that period. Researchers in this study concluded that education accounted for much of the difference in the association of self-rated health and four-year mortality between black and non-black Americans (Weller, 2006). It is unclear whether their analysis included the cultural differences of religiosity, belief systems, and social support.

The provision of health care through Medicare has radically changed the proportion of older African Americans who have a usual source of care. In 2007, 32 percent of black elderly had both Medicare and supplementary private health insurance, while 53 percent of all elderly had both Medicare and supplementary private health insurance (U.S. Census Bureau, 2008).

Though yoga, aerobics, and biofeedback programs to reduce blood pressure have typically been attended by white, middle-class people, African Americans are now joining programs focused on prevention of disability. This is probably related to greater educational differences between the oldest generation of African Americans and the emerging generation of the young-old.

Family and Social Relationships

More than 40 percent of African American women age 65 and over live in poverty. Some groups urge African American women to become politically active to lobby for equality on behalf of themselves and their mothers, aunts, and sisters. Organizations such as the National Association for the Advancement of Colored People (NAACP) and the National Black Women's Political Caucus speak and work on behalf of issues that are most salient to black Americans.

Traditional African American culture incorporates a resolve to persevere. The solid family ties, which are one source of this strength, are indicated in the concept of **familism**—a notion of family extending beyond the immediate household. Family network roles are flexible and interchangeable. A young mother, an aunt, a grandfather, or an older couple may head a family. Grandmothers often help raise children while their parents work. The high divorce rate has encouraged reliance on older relatives. Families tend to value their older members because they have survived in the face of hardship and because they play important roles within the family (Choi, 1996; Mullen, 1996).

One well-designed study of 154 African American and white grandparents rearing their children's children as a consequence of biological parents' neglect and drug addiction showed that

Traditional cultural values of family support across the life course are critical for sustaining quality of life in later life.

grandparents bore a large burden (Kelley et al., 1997). It was emotionally rewarding but exacted many costs, both psychological and financial. Similar findings appear in studies of grandmothers raising grandchildren in the crack cocaine epidemic (Minkler & Roe, 1993; Burton, 1992).

Religion has long been a resource of support in the African American family (Nye, 1993; Walls, 1992). The black American church has been a source of strength for coping with racial oppression and has played a vital role in the survival and advancement of African Americans. The church has provided a place of importance, belonging, and values. Within the church, older people are valued as members, choir members, deacons, and treasurers. A study of a Pennsylvania church, which revealed that church membership contributed to feelings of well-being among older African Americans, recommended that these churches act as a link between families and aging agencies. The church is a likely information and referral institution because so many older individuals are active participants (Walls & Zarit, 1991).

A spirit of survival has seen older African Americans through hard times. Thankful to have survived, they are more likely to appreciate aging; thus, they accept hardship more easily than those who have not experienced such hardship.

Future Outlook

Studies of fear of crime and victimization show much higher rates of fear for African Americans than for whites, and, in fact, their rates of victimization are significantly higher than those of white Americans. The major reason is thought to be geographic: more black elders than white elders live in high-crime areas such as the inner city and in or near public housing. Because feelings of alienation and mistrust of police may have existed from their youth, older African Americans are less likely than their white counterparts to reach out for help. It is in this regard

The election of Barack Obama, a U.S. President of mixed ethnicity, has altered perceptions and stereotypes about ethnicity throughout the world.

that family ties and community strengths are most critical. Fortunately, the values of family (by a nonexclusive definition) and church community are integral parts African American culture.

The election in 2008 of an African American president—a man of multiracial background—has shifted current perceptions of black Americans by much of the American and world population. He is seen as being a brilliant, articulate, principled, well-educated, charismatic, person of color. This does not mean he or his policies are universally accepted. But beyond question, he and his family have become a symbol of progress and possibility for American people of color and a symbol to the world of America's ability to embrace cultural and racial differences. The 2008 recession demonstrated the extent to which we are globally integrated by commerce and industry. America's symbolic acceptance of differences from a "standard" established by the majority may turn out to be a key tool for encouraging global cooperation.

Hispanic Americans

Older Hispanics, the *ancianos*, are not a homogeneous group. Researchers are often unprepared for the cultural and socioeconomic diversity of the Hispanic community. One of the obstacles preventing **Hispanic elders** from being understood and served is the lack of a clear-cut definition of who they are. Census counting often uses two inclusive terms that exacerbate the problem: Spanish heritage (having Spanish blood or antecedents) and Spanish origin (having been born in a Spanish-speaking country or having antecedents who were). Theoretically, then, a person could be of Spanish origin but not of Spanish heritage, or vice versa. The term *Hispanic* will be used here to mean Spanish people in a broad sense including either term.

Demographics

All Americans are living longer and the same is true for the Hispanic population. The older population was 2.5 million in 2007 and is projected to grow to over 17 million by 2050. In 2007, Hispanic people made up 6.6 percent of the older population. By 2050, older Hispanic people are projected to account for almost 20 percent of the older population (AoA, 2009).

The 1970 and 1980 censuses identified only four categories of Hispanics: Mexican, Puerto Rican, Cuban, and other. The 1990 census, reflecting the large influx from still other countries, identified 14 categories.

Mexico
Cuba
Puerto Rico
Dominican Republic
Central America
El Salvador
Guatemala
Nicaragua
Ecuador
Honduras
Panama
Other Central Americans
South America
Peru
Colombia
Other South Americans

The 2000 census included the option of selecting more than one category of racial/ethnic identity, and just as some of those selections became, for example, Asian/African American, so too some of those selections are Mexican/Guatemalan or some other combination of Hispanic background.

In 2007, 65 percent of Hispanic older men lived with their spouses whereas 17 percent

The Hispanic model of la familia, the primacy of family, continues to be seen generationally.

other relatives is almost twice that of the total older population (AoA, 2009). That finding is an outcome of the strongly valued *la familia*: family is foremost. This cultural value explains some of the lower utilization of social services and community resources for older adults among Hispanics.

The Hispanic population, one of the fastest growing ethnic groups, became the largest U.S. minority group in the 2000 census. The older Hispanic population more than doubled from 1990 to 2010 and is projected to be 11 times greater by 2050.

Minority Status

Several social factors indicate the minority status of Hispanic elders: (1) high percentage living below the poverty level, (2) inadequate health care brought about by poverty and cultural contingencies, (3) high illiteracy rates (second only to rates of Native American elders), and (4) low occupational levels such as operatives, artisans, unskilled laborers, and farm workers. Many traditional job categories of Hispanics have had few benefits, including retirement and pension. Some older Mexican Americans are unlikely to seek social services support because they entered the country illegally or are uncertain about their legal status, and they fear detection and expulsion.

Households of families headed by Hispanic people age 65 and older reported a median income of $31,544 (compared with $43,654 for non-Hispanic whites) in 2007. Among these households, 16 percent had an income of less than $15,000 (compared with 5.4 percent for non-Hispanic whites) and 45 percent had an income of $35,000 or more (compared with 62 percent for non-Hispanic whites) (AoA, 2009).

The poverty rate in 2007 for Hispanic older people was over 17 percent, which was more than twice the percent for non-Hispanic whites (7.4 percent) (AoA, 2009).

lived with other relatives. Three percent lived with nonrelatives, and 15 percent lived alone. For older women, 39 percent lived with their spouses, 33 percent lived with other relatives (mostly adult daughters), 2 percent lived with nonrelatives, and 26 percent lived alone. Older women are generally more likely to live alone than are older men, but the percent of Hispanic elderly men and women living alone is lower than that of the general population, and the percent of Hispanic older people living with

Migration Patterns

The two largest Hispanic subgroups are Mexican and Puerto Rican, whose portion totals 76 percent of all Hispanics. The majority of older Mexican Americans were born in the United States; other Hispanic elders are more likely to be foreign born. Foreign-born people of any ethnicity are not as acculturated as native-born citizens: They need more help to understand and utilize services. Furthermore, social service programs and medical services must acknowledge the traditions and values of their clients in order to be utilized and be effective. This has not always been the norm.

Differing patterns of migration have brought Hispanics to various locations in the United States. Hispanic American immigrants tend to live in urban areas. In 2007, 70 percent of Hispanic people age 65 and older lived in four states: California (27 percent), Texas (19 percent), Florida (16 percent), and New York (9 percent) (AoA, 2007). States with the largest proportion of Hispanic elders are the most likely to have culture-sensitive social service programs and community resources. Most other states, however, do not.

Immigrants from six of the twelve Hispanic nations identified in the 1990 census have more than 80 percent of their populations in the nation's 20 largest cities (e.g., New York; Los Angeles; Chicago; Miami; Washington, D.C.; Boston; Philadelphia; and San Diego), and three others have between 70 percent and 79 percent in the 20 largest cities. Because so many Mexican Americans are born in the United States, they are not as extensively urbanized as other Hispanics. Their occupational histories have also been more agricultural and less industrial in nature.

A certain proportion of the American Hispanic population became citizens when the rural Mexican territory in the Southwest was incorporated into the United States. This area remains rural in nature and is highly populated with Mexican Americans who have retained the values of family as more than just each person, an inclusion of parental values in parenting, father as head of the household, and religious traditions of the Catholic Church (Ethnic Awareness, 2009).

Most older Hispanic people who immigrate, however, tend to locate in urban areas. In Texas, for example, the older urban immigrants outnumber their rural counterparts by a ratio of five to one. San Antonio, Houston, and Dallas have large numbers of foreign-born Mexican Americans. Puerto Rican and Cuban elders are almost exclusively city dwellers. The Hispanic populations of Florida and the Northeast are more urbanized than those in the Southwest.

Hispanic elders today are an ethnic composite that has suffered with major linguistic and cultural barriers to assimilation. Forty-five percent of Hispanic elders are not proficient in English (Seelye, 1997). This has contributed to a history of jobs with no pension plan in America, much as are the employment histories of older American women of any color. Generally, older Hispanic worker pension coverage rates are lower than those for whites or African Americans regardless of gender.

Utilization of Services

Hispanic elders are generally somewhat suspicious of governmental institutions and of service workers and researchers not of their culture. Their suspicion, along with a lack of education and money, results in isolation and a low utilization of available services. This underutilization tends to conceal a very real need. Further, because the census undercounts minorities, underutilization is even greater than is generally recognized. Social program providers need to have and train staff to develop greater sensitivity to and communication skills with Hispanic seniors.

In 2007, almost 7 percent of Hispanic older people reported they had no usual source of medical care; 6.5 percent reported delays in obtaining health care due to cost, and in 2001,

20.7 percent reported they were not satisfied with the quality of the health care they received. Comparable figures for the total population reported that 5 percent reported no usual source of medical care, 4.8 percent reported delays in obtaining health care due to cost, and in 2001, 15.6 percent reported they were not satisfied with the quality of the health care they received (AoA, 2009). These differences reflect the result of poverty and also the perception and experience of systemic bias toward Hispanic elders in the medical system.

Popular beliefs characterize Mexican Americans as living in extended families, and, in fact, Hispanic families do tend to be larger than white families in the United States. But that does not mean that the extended family system runs smoothly. Generally speaking, the more acculturated the family members, the less extensive the family interaction, and the more complete the breakdown of extended family. Traditionally, adult children provide a great deal of support to Hispanic parents in terms of chores such as laundry, housework, transportation, and shopping. Both cultural values (the needs of the larger family go above one's individual needs) and economic need dictate these close family ties. The patterns of intergenerational assistance are strong compared to white Americans, for example; however, older parents have expressed dismay at the shift in filial responsibility of their adult children.

Studying Ethnic Variations

For all Hispanic groups, researchers study the types of ethnic communities and institutions that develop in given localities. This inquiry includes the impact of these organizations on the lives of older people. Mexican Americans, for example, participate heavily in senior citizen clubs. Actually, the culture of being an older person is strong in various representations of Hispanic community (Torres-Gill, 1988). Though minority elders tend to underutilize government and health services, high

ethnic population density seems to correlate with higher rates of utilization. The status of the elders in the various Hispanic groups is better in large, fully developed ethnic communities than in small or scattered ones. Here are some statistics about high ethnic population density (Winsberg, 1994):

- Eighteen percent of all Hispanics live in Los Angeles.
- Twelve percent of all Hispanics live in New York.
- New York's Puerto Rican population is double that of San Juan, Puerto Rico.
- Seventy-seven percent of immigrants from the Dominican Republic live in urbanized New York.
- Laredo, Texas, constitutes 94 percent Hispanics, and the proportion of Hispanics is nearly that high in several other Texas border towns.
- Cubans are clustered in Miami and other cities of southeastern Florida.
- Panamanians are the most geographically diverse.
- Other groups cluster in enclaves of large cities according to national origin.

Another area of research considers the unique cultural traditions that are maintained in each Hispanic group. Cuban elders, for example, tend to be much more politically active than are Mexican Americans. A professor from Miami described some unique aspects of Cuban culture as they affect elders (Hernandez, 1992): Cuban culture is a blend of whites from the Iberian Peninsula and blacks from Africa. The Afro-Cuban culture emphasizes respect for elders, stemming from their folk healing beliefs and practices. **Familism** (adherence to strong family values) is strong in Cuban culture and evokes guilt when members do not fulfill expected roles. For example, adult children typically feel very guilty if they do not care for

Inuit Alaskan Native American grandfather.

their aged parents. Cuban women tend to marry older men who care for them financially. The implicit exchange is that eventually the younger wife cares for the aged husband. This tradition is a reflection of a care-provision culture for women as well as men. Complete dependence on family members, such as adult children or spouses, is welcomed and encouraged by the culture and not interpreted as pathology.

Neglected areas of research are social stratification within ethnic groups, rate of return migration, and degree of cultural adaptation. Hispanic elders who lived most of their lives in their native countries and do not immigrate until late adulthood no doubt experience more culture shock and isolation, but this phenomena needs to be explored.

Asian Americans

The term *Asian American* refers in the broadest sense to people of Chinese, Korean, Japanese, Filipino, East Indian, Thai, Vietnamese, Burmese, Indonesian, Laotian, Malayan, and Cambodian descent who live in the United States. Most older **Asian Americans** are concentrated in California, Hawaii, New York, Illinois, Washington, and Massachusetts. The total U.S. population of Asians and Pacific Islanders was 14.4 million in 2005, comprising about 5 percent of the total population (U.S. Census Bureau, 2007). Of this number, over 6 percent were 65 and older.

A single description cannot encompass the Asian communities in the United States. Differences of culture, language, and religion make each group unique. Chinese Americans are the largest Asian group, followed by Filipinos (2.8 million), Asian Indians (2.5 million), Vietnamese (1.5 million), Koreans (1.4 million), and Japanese (1.2 million) (U.S. Census Bureau, 2007).

American laws historically discriminated against Asians. The Chinese Exclusion Act of 1882, passed during high unemployment by fearful whites, made all Chinese immigration illegal. It was amended twice to allow immigration under certain circumstances, and then in 1924 it became totally exclusionary again. It was not repealed until 1943, when China became a U.S. ally in World War II. The legislative history is not an honorable one and has important social implications, as outlined by Hooyman and Kiyak (1996):

Laws discriminating against Asians are numerous, including the Japanese Alien Land Law of 1913, the Executive Order of 1942 for the internment of 110,000 people of Japanese ancestry, denial of citizenship to first-generation Asians in 1922, the anti-miscegenation statue of 1935, and more recently Public Law 95-507 excluding Asians as a protected minority under the definition of "socially and economically disadvantaged" (U.S. Commission on Civil Rights, 1979). Such legislation, combined with a history of prejudice and discrimination, has contributed to feelings of mistrust, injustice, powerlessness, and fear of government among many Pacific Asian elders, and thus a reluctance to utilize services. (p. 455)

The current generation of Asian Americans, conditioned to traditional American social and cultural folkways and mores, may be just as likely to regard their elders as an unwelcome burden as do some middle-class white Americans. According to traditional culture in China, Japan, and Korea, the eldest son assumes responsibility of his elder parents. **Filial piety** is a custom demanding that family members respect and care for elders. Japanese American families have preserved to some extent the norm of *enryo*, a pattern of deference and modesty, as well as *amaeru*, the value of dependency, which encourages special dependent relationships (Johnson, 1996). These patterns are carried into respectful attitudes toward elders; however, they contrast markedly with dominant American values. As children and grandchildren become more enculturated, intergenerational tensions result. This becomes a moral generation gap between young Asian Americans and their elders. Older family members hold on to traditions, especially those concerning moral propriety, while the young move away from them.

Because health and welfare agencies have few bilingual staff members, and because they therefore have difficulty publicizing their available services to the Asian community, outreach programs to Asian American seniors have been limited in their success. These deficits, in addition to their socially conditioned reluctance to seek aid from their adopted land, result in neglect of Asian American elders.

Japanese Americans

Most of the Japanese who first came to the United States were single men, often times younger sons who did not inherit any family wealth. The bulk of the Japanese immigration, which took place between 1870 and 1924, constituted men who wanted to have traditional families. Many waited until they could afford a wife and then paid for one to come from Japan. This pattern reinforced traditional values of high status for men and elders. The survivors of that earliest immigration period, called *issei*, are now mostly women because the men, being much older, have died.

The first generation worked primarily on farms or as unskilled laborers or service workers. However, within 25 years of entering the United States, they showed great economic mobility. Though their internment during World War II was economically as well as morally devastating, Japanese Americans as a group rebounded remarkably. First-generation Japanese, the *issei*, learned to live socially segregated from American culture. The children of the *issei*, the *nisei*, generally born between 1910 and 1940, are more likely to be integrated into the American mainstream. Most *nisei* are now over 65 and doing well economically and socially.

Older Japanese Americans, on the whole, have adequate savings or family support in their retirement years. Through the normative values of *amaeru*, *filial piety*, and other Confucian moral principles, Japanese Americans have largely replicated the traditional pattern of family care for elders. Nearly half of Japanese elders live with an adult child in addition to or in lieu of a spouse. In traditional Japanese society, upon retirement the retiree joined the ranks of elders and assumed religious duties in the community.

Chinese Americans

Because past restrictive immigration laws denied entry to wives or children, a disproportionate share of Chinese American and Filipino American elders are men. Male immigrants have outnumbered females by at least three to one in census counts. These Asian men were valuable as cheap labor in U.S. mines, canneries, farms, and railroads, but their wives and children were neither needed nor wanted. Though we cannot fully assess the damage to the family life of the elderly Chinese American, such damage has no doubt been extensive, traumatic, and demoralizing.

The immigration law of 1924, which halted Asian immigration, forbade males of Chinese descent from bringing their foreign-born wives to the United States. As a result, many Chinese men in the United States could not marry. Although the men who originally came in the early 1900s are a rapidly vanishing group, a few are still around, typically living in poverty and without close family ties.

Chinese Americans retain a tradition of respect for elders based on Confucian ethics. Traditionally, the Chinese family was embedded in a larger system of extended family and clans than was the Japanese family. Older family members held wealth and power, not only in their immediate family but all the way up the family hierarchy to the encompassing clan. Though this traditional structure has never been reproduced in the United States, respect for elders persists. The Chinese American pattern is for adult children to bring a widowed parent into their household.

Increasing proportions of Chinese elders are second generation. This generally means that they are more educated and acculturated and have a more comfortable financial situation.

A vast difference exists between the lifestyles of those who are foreign born and who have never learned English and those who were born in the United States. The second generation is retiring with pensions and savings, reaping the harvest of their hard work in this country. Despite discrimination, Chinese Americans have achieved a high rate of occupational mobility; many have gone from restaurant and laundry businesses to educating children who have entered professional and technical occupations.

Southeast Asian Americans

The settlement of Vietnamese, Cambodians, and Laotians in the United States has included a small percentage of elders, who enter a world alien in all facets of life, from language, dress, and eating habits to religious beliefs. Family ties are strong for most of these people. Traditionally, extended families are standard. Southeast Asians have a special respect for their elders, especially for fathers and grandfathers.

Though immigration policy initially places Southeast Asians throughout the entire United States, once on their own, many have gradually migrated to areas where the weather is similar to their native countries. California has by far the largest number of Southeast Asians. Studies of Southeast Asian refugees show that, like other Asian groups, they adjust fairly well to U.S. culture. Older first-generation immigrants who have suffered, living near or below the poverty line, take pride in children or grandchildren who have achieved financial and other successes.

In Los Angeles, interviews of 19 older Hmong refugees and their families clearly demonstrated the pain and culture shock of displacement (Johnson, 1996). In Laos, they lived in extended families with households containing as many as 35 members. Order and authority were maintained through respect for age. The oldest male of each clan sat on a governing council that handled all problems. Both male and female elders experienced a high status within the village.

Grandparents and great-grandparents are the source of cultural tradition for minority groups whose historical values are not necessarily represented by the major culture.

possibly on a reservation. It was a major blow to be scattered in cities. They were shocked to learn about American housing standards; they were unable to understand, for example, why a family of 10 could not live in a two-bedroom apartment. One elder Hmong recalled, "When I found out that some of my children would not live with me, my life stopped." The older Hmong had had no formal education in Laos, and many could not face the rigors of learning English in an American classroom.

This cultural tradition clearly has not been maintained in American society and has resulted in generational conflict that creates greater assimilation by the younger generation of the cultures of *their* birthcountry—America. The role of elders in the Hmong family has therefore changed dramatically. In Laos, the elders acted as counselors for adult children experiencing marital difficulties. In the United States they are more out of touch with young couples' marital problems, and they rarely act as advisers and mediators. The older women try to help with child care but older men do not have much to do. Many would like a farm and animals to tend. Elderly Hmong have experienced loss of function, mobility, religious customs, and status. Some are very depressed. Studies of Southeast Asian refugees find that, because of the language barrier and lack of communication with the larger society, displacement is more difficult for the elderly than for the young.

Cultural values among South East Asians remain, to some extent, that family is not just one person's needs; mothers are able to work because other family members care for younger children; older people are highly valued; children are expected to be well behaved and to obey; and the value of intelligence is very strong (Ethnic Awareness, 2009).

The Hmong believed that they would be resettled in the United States in a large group,

Native Americans

Measured by numbers alone, **Native American elders** constitute a very small percentage of American society. By any social or economic indicator of living conditions, however, Native American elders are possibly the *most* deprived of all ethnic groups in the United States.

Many of the problems of Native American elders are due to minority status rather than to age. American Indians on reservations and in rural areas experience extremely high unemployment rates. Few jobs exist on the reservations, and those who leave in search of work pay the high price of losing touch with family, lifestyle, and culture. Many houses on reservations are substandard, and, despite substantial improvements in health, disparities still exist, especially in sanitation and nutrition. More than 140 years of federal programs have done little to improve the lives of Native Americans.

Cultural Uniformity and Diversity

Though Native Americans are a diverse group, they do share some values that set them apart generally from the larger society. Their lifestyle and spirituality dictate a deep reverence for the land, animals, and nature; and, generally, they believe in attaining harmony between human beings and nature.

Family structure, values, and norms among Native American tribes are diverse. Generally speaking, Native Americans have close family ties. Though many family structures are patriarchal, a wide variety of descent systems exist. The largest tribe, the Navajo, follows a matrilineal structure. The position of older people varies from tribe to tribe, as does emphasis on peace versus war and many other values. And although some tribes are rivals, an extensive pan-Indian network exists today, which promotes intertribal networking, visiting, and cooperation.

The United States has approximately 278 federally recognized reservations, more than 300 recognized tribes, and around 100 non-recognized tribes (John, 1991a). This includes Eskimo and Aleut populations.

Population Data

The 1990 census reported an increase from the 1980 census in people identifying themselves as American Indians or Eskimos. The largest Native American groups in descending order are Cherokee, Navajo, Chippewa, and Lakota. In 1990, 54 percent of Native Americans lived on

This grandfather teaches his granddaughter through example: respect for many cultural traditions.

reservations or tribal lands, whereas the others lived in rural and urban areas (U.S. Census Bureau, 1996).

For decades, despite rapid growth in the population of the country as a whole, the Native American population declined. At the time of the first European settlement in what is now the United States, the number of Indians is estimated to have been between 1 and 10 million.

By 1800, the native population had declined to approximately 600,000; by 1850, it had shrunk to 250,000. This mortifying decrease, the result of malnutrition, disease, and an all-out military assault on Native Americans, was a unique occurrence in our national history. However, the population eventually stabilized and is now increasing (U.S. Census Bureau, 1996).

Most older Native Americans live on reservations, staying behind while the young seek work in the city. Young or old, those who go to the city often expect to return to rural reservations to retire. Although the American population as a whole is more urban than rural, the reverse is largely true for the Native American population. Native Americans are the most rural of any ethnic group in the country (Hoyer & Roodin, 2009).

Life expectancy among Native Americans is substantially lower than that for whites. The Native American population is largely young; in 1995 nearly 50 percent were under the age of 25 (Hobbs & Damon, 1996). Elders constitute slightly more than 6 percent of the total Native American population; by comparison, they represent 13 percent of the total U.S. population. As in the white population, Native American women live longer than Native American men.

Native Americans who leave the reservations are usually scattered throughout urban areas rather than forming ethnic enclaves. Over 48,000 Native Americans, for example, live throughout the Los Angeles area. San Francisco, Tulsa, Denver, New York City, Seattle, Minneapolis, Chicago, and Phoenix all have sizable Native American populations; Minneapolis has a Native American enclave. Although a multitude of tribes are scattered throughout the United States, 44 percent of all Native Americans reside in California, Oklahoma, Arizona, and New Mexico (Kitano, 1991). Findings for one tribe cannot be generalized to include all others; values and behavior vary greatly. Specific patterns of aging need to be examined in each tribe. Some generalizations can be observed, however.

Education, Employment, and Income

The educational attainment of Native Americans today is behind that of whites. A sizable percentage of unemployed older Navajo adults, for example, cannot speak English; nor can they read or write it. Among Native American youths, the school dropout rate is twice the national average. Because education has been a traditional means of social advancement in the United States, these data suggest that future generations of Native American elders may continue to suffer from functional illiteracy. A high percentage of Native American elders have graduated neither from elementary nor from high school.

Historically, in the late nineteenth century the federal government created the reservation system where tribes were isolated and had little prospect for development or economic growth. Indian education was the responsibility of the federal government, and day schools were operated in the nineteenth century with the intent of acculturating Indian children to the dominant society (Nagel, 1996). Most of the schools were poorly equipped and poorly run. As a result of this tradition, American Indians are the most poorly educated of all minority groups.

Following World War II many American Indians left reservations and did not return. Federal relocation programs established in the 1950s and 1960s were designed to end reservation life as the only option and to train and relocate people to urban areas. Between 1952 and 1972, more than 100,000 American Indians relocated to urban areas, remaining there to raise their own families. As a result, many Indian elderly were left isolated on the reservations.

Unemployment is a continuing problem in the Native American community. An extremely high percentage of the Native American workforce is unemployed, and most of those who work hold menial jobs with low pay and few if any fringe benefits. Because they have often paid only small amounts into Social Security, retirement for Native Americans is extremely limited. White Americans often associate major difficulties of growing old with retirement from the workforce. Native Americans usually have no work from which to retire. For most over 65, old age is a continuation of poverty and joblessness that has lasted a lifetime.

Health Characteristics

Native American elders are more likely to suffer from chronic illnesses and disabilities than any other ethnic aged group, and they have the lowest life expectancy. However, though the mortality rates for younger Navajos are relatively high, mortality rates among elder Navajos are lower than those for non-Indians of the same age. The maintenance of strongly held cultural beliefs and practices is believed to result in reduced stress levels among older, traditional tribal people.

Native Americans are more likely than the general population to die from diabetes, alcoholism, influenza, pneumonia, suicide, and homicide (Hobbs & Damon, 1996). Lack of income leads to poor nutrition and health care and is associated with lower education levels, which compound other variables. The Native American accident rate, for example, is high. Native Americans are more likely than the general population to be killed in motor-vehicle accidents; and death from other types of accidents is also more likely.

Native Americans suffer disproportionately from alcoholism and alcohol-related diseases. American Indians are generally disposed genetically to have a low alcohol tolerance; however, the abuse rate of alcohol in the Native American population is probably related to the stressors of minority status and poverty. Health problems are compounded by the alien nature of the dominant health-care system to the traditional culture of Native Americans.

Older Native Americans remain an enormously needy group. Today, Native Americans suffer both from dependency on the federal government and from the impact of conflicting federal policies. They are sometimes denied assistance from various government agencies under the excuse that the Bureau of Indian Affairs (BIA) is responsible for providing the denied service. According to the *United States Government Manual 1989/90*, the principal objectives of the BIA are to "actively encourage and train Indian and Alaska Native people to manage their own affairs under the trust relationship to the Federal Government; to facilitate, with maximum involvement of Indian and Alaska Native people, full development of their human and natural resource potential; to mobilize all public and private aids to the advancement of Indian and Alaska Native people for use by them; and to utilize the skill and capabilities of Indian and Alaska Native people in the direction and management of programs

Minority women will represent a far larger proportion of elders in the 21st century than in the past. This shift to greater diversity in later life will impact social policy.

for their benefit." The BIA has twelve area offices in the United States, so that the distance among the service provider, the work site, and the reservation (place of residence) compounds the difficulties of eligibility and availability of assistance.

Improving the Status of Ethnic Elders

According to Census Bureau reports, the United States can be divided into the following eight demographic groups based on disparities in life expectancy (Murray et al., 2006).

1. 10.4 million upper-income Asians with a life expectancy of 85.
2. 3.6 million rural whites in the mid-north of the country with a life expectancy of 79.
3. 214 million middle-income people living to approximately 78.
4. 16.6 million whites in Appalachia with a life expectancy of 75.
5. 1 million reservation-living Native Americans in the Western Mountain and plains areas with an average life expectancy of 73.
6. 23.4 million blacks, living to an average of 73 years.
7. 5.8 million blacks in southern parts of the United States with a life expectancy of 71.
8. 7.5 million blacks living in big cities with homicide rates in the 95th percentile has a life expectancy of 71.

Differences in life expectancy are due in large part to risks associated with lifestyle among different groups and geographic regions. This means that, with the exception of high-crime mortality, educational interventions to lower cancer and heart disease could have profound impact on minority group longevity.

A large number of minority elders spend the last years of their lives with inadequate income and housing, poor medical care, and few necessary services. Aging accentuates the factors that have contributed to a lifetime of social, economic, and psychological struggle. Rather than achieving comfort and respect with age, minority people may get pushed further aside.

Upgrading the status of ethnic elders in the short term requires, first, recognition of the various factors that prevent them from utilizing services and, second, outreach programs designed to overcome those factors. The object should be to expand present programs, develop more self-help programs, increase the **bilingual** abilities of social service staffs, and recruit staff members from the ethnic groups they serve.

With these goals in mind, the Administration of Aging (AoA) has funded four national organizations to improve the well-being of minority elders.

1. The National Caucus and Center on Black Aged (NCCBA) in Washington, D.C.
2. The National Indian Council on Aging (NICOA) in Albuquerque, New Mexico
3. Asociación Nacional Pro Personas Mayores (National Association for Hispanic Elderly) in Los Angeles, California
4. National Pacific/Asian Resource Center on Aging (NP/ARCA) in Seattle, Washington

These centers educate the general public and advocate for their groups.

Ultimately, though, spending a bit of extra money and adding a few services will not solve the real problem. The disadvantages tend to derive from economic marginality that has lasted a lifetime. One's work history is a central factor in how one fares in later life. A work history that allows for a good pension or maximum Social Security benefits is a big step toward economic security in old age. Also, those who have made good salaries have had a chance to save money for their retirement years—a safety net not possible for low wage earners. Until members of minority groups are from birth accorded full participation in the goods and services our society offers, they will continue to suffer throughout their lives. The critical perspective in sociology calls for addressing these basic inequalities and taking major steps toward making all citizens equal.

Chapter Summary

Women, including older ones, in the United States, are in a minority status. They do not participate equally in the political and economic structures. Older women are victims of the double standard of aging—a standard that judges them more harshly as they age. A large percentage of elders in poverty are females.

Ethnic elders suffer from inequality in the United States. Older blacks are poorer than older whites and have a lower life expectancy, poorer health, more inferior housing, and less material comforts. Although some older African Americans are well-to-do, others are impoverished. Family ties, religion, and a resolve to persevere are special strengths. Hispanic elders are not a homogeneous group. About half of all Hispanic elders are Mexican Americans; others come from Cuba, Puerto Rico, and various Central and South American countries. Hispanic elders have suffered major linguistic and cultural barriers to assimilation, and their socioeconomic status is low. Asian American elders also come from many countries. They, too, have encountered racial hatred, language barriers, and discrimination. Native American elders are a small group. But by any social or economic indicator, they are possibly the most deprived group in the United States. More efforts need to be directed at correcting inequities for ethnic minorities and women.

Key Terms

age terrorism
amaeru
Asian American

bilingual
double standard of aging
enryo

ethnic group
Ethnicity
familism
Filial piety

Hispanic elders
late midlife astonishment
minority
minority elders

Questions for Discussion

1. What is the attitude about aging in your family? Has it changed down the generations? What other cultures have similar attitudes?

2. Compare the financial options of a 70-year-old man who has worked all of his life and collected Social Security benefits and a 70-year-old single woman who worked privately doing housework and childcare all of her life and received no Social Security benefits. What are his options? What are her options?

3. Divide up into pairs in the classroom and interview your partner about his or her cultural and ethnic background. What are the traditions in each of your families that affect the elders in the family?

Fieldwork Suggestions

Interview an older person from a racial group other than your own. Note carefully the person's lifestyle and outlook on life. In what ways has this person's ethnicity shaped his or her perspective on life? Is there anything unique about being an older person for your interviewee?

Internet Activities 🌐

1. Locate the AARP home page, and search for information on retired women's issues. Now search for information on minority issues. What levels of analysis do you find? Are the issues of retired women and minorities well addressed, or are they addressed as being "other" to the dominant (white male) population of retired people?

2. Go to the U.S. Government Bureau of the Census home page and search out the number of children living with grandparents in the country. Can you locate this information by race and ethnicity? Does your home state have parallel information? Where else might you search for data on ethnicity and family household structure?

3. Seek out Internet information on minority elder status, other than U.S. Census data. Maintain a log of your Internet "travels." How easy or hard was it for you to locate the information? Who is the intended audience of the information you have?

Death and Dying

14

Time to Address the Need to Improve End-of-Life Care

Susan Keller

For a few decades now, I've been steeped in work dedicated to improving care for people living with serious life-limiting illness that ends in death. This work grew out of concern for the impacts of high-tech medicine on elderly people who really needed "high touch" care instead.

I'm delighted to see health care reforms in progress aimed at providing more appropriate care and assistance for older adults, the frail elderly and dying.

My end-of-life care work with the Community Network Journey Project is rooted in personal experiences with old folks close to me, including neighbors, family and friends. In this way, I experienced what it was like to be old and frail, traveling the journey through life's end.

Routinely, I helped people learn about choices and find their way despite tremendous pressure to pursue aggressive medical interventions they didn't want. That's what helped to alleviate needless trauma and suffering most of all. What made the difference was just plain common sense. It was simply taking the time to learn of the elder's hopes and fears, their values and wishes. It was about being a good listener and offering help in the smallest of ways, nothing out of the ordinary, really.

They all knew they were dying and chose to live out their remaining life on their own terms. After all, that was how they had found their way through lives spanning most of the 20th century.

They weren't willing to risk dying in an Intensive Care Unit plugged into machines and they accepted dying in old age as a natural part of living.

In the health care arena, things have improved somewhat these past few years, but there is much more to be done. Important changes concerning end-of-life care are now under review in Congress. Changes proposed are essential given the fact that nowadays most elders die as a result of multiple life-limiting illnesses spanning several years. It is rare for an elder to benefit from aggressive treatment when already dependent upon others to get through activities of daily life.

New benefits proposed for seniors include the right (not the obligation) to have timely, meaningful conversations with health providers about advance health care planning.

Palliative care, a life-affirming supportive way of caring for people living with serious illness and frailty, would be covered as any other medical program. Models of care—the bedrock for reforms proposed—are time-tested and well proven. Health outcomes are better, patient/family satisfaction is top notch, burnout is less for family caregivers and health care workers, and cost-savings are remarkable.

Leaders at our local, regional, state and national levels clearly understand how this all works. They know how critically important proposed reforms are to community health and well-being.

Here in Sonoma County, a coalition of health and human service leaders is working with the Community Network to nurture development of palliative (comfort) care and values-based advance health care planning across the health care system.

This work complements [other] effort[s] dedicated to improving health and health care for all residents. Changes now under way and on the horizon are badly needed and essential for community health and well-being.

Source: www.pressdemocrat.com/article/20090919/
ARTICLES/909199951/1307/OPINION?Title=Time-
to-adddress-the-need-to-improve-end-of-life-care#
Downloaded September 20, 2009

Death is one of the few certainties of life. This statement is neither pessimistic nor morbid. Despite our wildest fantasies about immortality, no one has yet escaped death permanently. Despite its universality, however, death is not easily discussed in American society. People tend to be sensitive and shy about discussing the topic openly.

Facing death, dealing with the fact of death in a rational way, and exerting control over the manner of one's dying are all difficult situations in a society that denies death. Just because one is old does not mean that one does not fear death or that one welcomes dying. In this chapter, we will examine the ability of older individuals to experience a "good" death to the extent our society permits it.

Life itself is a sexually-transmitted, terminal condition.

ANONYMOUS

A Death-Denying Society?

Our words, attitudes, and practices suggest that ours is a **death-denying society**. Have you ever used the word *died* and had the uncomfortable feeling that those with whom you were talking considered the word too direct and in bad taste? "Passed away" or "passed on" might be the preferred phrase—but not "dead." Dozens of euphemisms are used for the process of dying. Funerals, presenting an embalmed and painted body, try to project an illusion of life. Even a sort of prescribed script exists that mourners follow: "She looks just like she did yesterday when I was talking to her" or "He looks so natural." Those selling cemetery plots advise us not to buy a grave but to invest in a "pre-need memorial estate." The funeral industry in the United States is enormous, now offering services, in addition to burial, from funeral planning to bereavement counseling.

Watching death on television or killing after violent killing in the movies is not facing death. Violent deaths portrayed in cartoons and action films do not evoke the human emotions—the deep sorrow, anger, and guilt—that the actual death of a loved one would. "Sensitive" television dramas that depict the death of a family member tend to be just as unrealistic.

At the end of the nineteenth century, dying and death were household events that encompassed not only elders but also the young and middle aged. At the beginning of the twenty-first century, death is characterized by life-extending medical procedures, hospital and insurance bureaucracies, and secularism. As chronic diseases and increased technology keep people in a prolonged state of dying, bureaucracy makes the setting impersonal, and a secular society robs a person of the religious significance of dying. These three trends promote the impersonalization of the process through which the dying individual must go. Small wonder that death is perceived as such a fearful, lonely experience.

> *Death frights us. Death is a perpetual torment, for which there is no sort of consolation. There is no way by which it may not reach us. We may continually turn our heads this way and that, as if in a suspected country, but we can't forget death.*
>
> MONTAIGNE, *ESSAYS*

Death avoidance is not unique to modern times, however. Dealing with the unknowable, whatever its outcome, produces anxiety for humans, whose driving desire is toward *understanding*. French philosopher Michel de Montaigne (1533–1592) argued in the sixteenth century that to become free of death's grip we must immerse ourselves in all the aspects and nuances of death (Montaigne, quoted in Marrone, 1997):

> *We come and we go and we trot and we dance, . . . and never a word about death. All well and good. Yet when death does come— to them, their wives, their children, their friends—catching them unawares and unprepared, then what storms of passion overwhelm them, what cries, what fury, what despair! . . . To begin depriving death of its greatest advantage over us, let us adopt a way clean contrary to that common one; let us deprive death of its strangeness, let us frequent it, let us get used to it; let us have nothing more often in mind than death. . . We do not know where death awaits us: so let us wait for it everywhere. To practice death is to practice freedom. [One] who has learned how to die has unlearned how to be a slave.* (p. 328)

Denial of death is possible on a grand scale now not because we choose to deny it more than our ancestors did, but because we have greater institutional resources for helping us to avoid its reality.

About 85 percent of American deaths occur in hospital rooms and convalescent hospitals (Marrone, 1997). The final moments of life are seldom observed, even by family. Health care personnel are the predominant care providers for the dying, but health professionals are seldom trained in the process of death. They are trained to extend life, and death is a failure of the promise of that training. Death is often as hard a reality to health professionals as it is to everyone else.

Despite social customs that decrease the interaction of the living with death and dying, interest in the topic has escalated in the last decade or so. College courses on death and dying, seminars instructing health professionals and clergy in understanding the dying person, and many books on the subject have become available.

Fear of Death

Fear of death is a normal human condition—we struggle for our life, and for the lives of others, and we desperately desire to avoid unknowns. But are people more fearful, or less fearful, at different ages?

Old people are commonly stereotyped as waiting fearlessly for death. And, though any given individual may be an exception to the rule, older people as a group appear to be less fearful than younger persons. Thanatologist Richard Kalish (1987) attributed this difference to three things: (1) many older people feel that they have completed the most important tasks of life; (2) they are more likely to be in pain or to be suffering from chronic diseases and view death as an escape from that pain; and (3) they have lost many friends and relatives, and so those losses have made death more a reality and sometimes a welcome relief.

Some people who work with elders, however, believe that old people may fear death *more* because they are closer to it. There is less time available to complete important life tasks, make amends, and say goodbye. Other observers say that fear is greater in late middle age, when midlife brings a heightened awareness of aging and death. Once that transitory phase passes, fears decline. Evidence for which age group is most fearful is conflicting, likely because understanding and quantifying such an emotion as fear is nearly impossible to manage. Research does appear to converge on findings that the education and emotional/spiritual support patients receive from hospice team care appears to decrease fear of death well into the dying process. At some point, most people acknowledge the reality that they will not continue to live much longer. The uncertainty of impending death is no longer present.

Most observers and philosophers agree that fear of death is innate in all individuals, regardless of age, and that it provides direction for life's activities. Understanding that we will die may be a precondition for a fuller and deeper understanding of life.

There are two ways of not thinking about death: The way of our technological civilization, which denies death and refuses to talk about it; and the way of traditional civilization, which is not a denial but a recognition of the impossibility of thinking about it directly or for very long because death is too close and too much a part of daily life.

PHILIPPE ARIES

Given painful experiences, and the power of storytelling and imagination, people have many reasons to fear death itself or have fears about the process of dying, such as the imagined pain, helplessness, and dependence. People fear a long, painful death; illness, such as cancer; senility; the unknown; judgment in the afterlife; and the fate of one's body. They fear dying alone or, conversely, of being watched; dying in a hospital or nursing home; and losing bodily control.

Fear of the *condition* of death—of being dead, and the *process* of dying are important to understand if we are to help people who are dying and to help ourselves in our own deaths.

Fear of death can sometimes be healthy. The fear can prompt us to be more careful and take appropriate precautions such as wearing a helmet and seatbelts. A healthy fear of death can also serve as a reminder for us to make the very most of our time here and especially not to take our relationships for granted. It can push us to work hard and leave a legacy to those we love. George Bernard Shaw was quoted as saying "I want to be thoroughly used up when I die, for the harder I work the more I live" (Morrow, 2009).

Some fears seem normal, that is, justified and within reason. Others seem to be exaggerated. The term **fear of death** is generally used when such apprehension has a specific, identifiable source. In comparison, **death anxiety** describes feelings of apprehension and discomfort that lack an identifiable source. **Death competency,** on the other hand, is a capability and skill in dealing with death. But how fearful, anxious, or competent are people? Does an individual's fear of death or overriding anxiety about death impact their process of grieving or shape their own deaths? If we knew more about which aspects of death were most problematic for a given person, could that person be better assisted through a "good" death, or with grief and mourning over the death of someone beloved?

Several measures of death fear have been developed in an attempt to understand more about the fears and anxieties surrounding death for groups of people and provide a better understanding about a particular person. The questions generally cover a range of fears, such as fear of dependency, fear of pain, fear of indignity, fear of isolation or loneliness, fears related to an afterlife, fear of leaving people we love, and fear of the fate of the body (Lemming & Dickinson, 1994; Neiman, 1994). Findings from more general research provide a road map for understanding individual issues with death and dying.

Traditional religious beliefs including heaven (as a sacred, perfect place) and survival of the body after death have been found to be related to a decrease in fears of death, as are beliefs in the paranormal, such as sightings of angels or other personal guides (Alvardo et al., 1995; Lange & Houran, 1997). People with lower death depression scores seem to have greater strength of conviction and greater belief in an after life, and they are less likely to say that *the* most important aspect of religion is that it offers the possibility of life after death (Alvarado et al., 1995).

Research consistently finds that the strength of belief—whatever that belief might be—is more strongly related to lower death anxiety scores than the knowledge of dogma, attitudes toward mortality, or frequency of church attendance (Thorson & Powell, 1996; Conn et al., 1996). This reinforces the suggestion that fear of death is an internal, psychological relationship with the state of dying, based on an individual's beliefs and values. We die as we live; we are the same person as we "go into that dark night" as we were before the specter of death was upon us. This might give us pause to consider just how we choose to live—our coping skills, choices, and personality patterns are what we will have to help us cope in our own deaths.

The fear of death is understood in terms of specific fears that dying people express. Some people have many, if not all, of these fears. For some, one particular fear becomes haunting, dominating their dying process.

- *Fear of pain and suffering*: This fear is common and often seen in people who are indeed in pain.
- *Fear of the unknown*: Human nature is to want to understand and make sense of the world around us, but death is the ultimate unknown, which cannot be understood.

- *Fear of nonexistence*: Many people fear they will cease to exist after death. This is confined not only to nonreligious people but also to people of faith who worry that their belief might not actually be true.
- *Fear of eternal punishment*: This fear is shared among nonreligious as well as religious people. People, while they are still alive, fear for their eternal punishment for their actions or inactions.
- *Fear of loss of control*: Human nature is to want to control life situations. Death is completely out of the realm of control, and it becomes very frightening for many people.
- *Fear of what will become of loved ones*: This might be the most common fear of death among new parents, single parents, and caregivers. What will happen to those who depend on me?

Though the experience of death will be different for every person, in general, our fears about the suffering of death may be at least somewhat unfounded. Most people have relatively painless deaths. In most cases, an individual's health does not deteriorate until very close to death. Most people are in good or excellent health a year before they die, and many are in good health the week of their death. It is the haunting stories of lingering, painful, lonely death that many people carry in their heads when they think of death that feeds our innate fears. In reality, the more we know and understand about death, the less fearful it will seem.

Older people's fears and concerns about dying must be dealt with as openly as possible, difficult as that might be. Because of their own anxieties about death, adult children will often not allow parents or grandparents to fully express their views on the subject. Older people often attempt a discussion about their own death, and/or what the family will do after (or just before) the elder dies, only to be told in various ways, "Oh, you don't have to think about this yet! You have a lot of time for that." Yet this "death talk" can be realistic, practical, and therapeutic. Family members would be wise to listen.

As for man, his days are as grass: as a flower of the field, so he flourishes. But the wind passé over, and soon all disappears; and his place will no more exist.

PSALMS, THE BIBLE

Living Fully Until Death

The marginal status of elders in our society affects their ability to live fully until death. Richard Kalish (1985) discovered that although the death of an elder is less disturbing than that of a younger person in American culture, this attitude is not universal. He told this story:

> I once asked a group of about two dozen Cambodian students in their mid-twenties whether, given the necessity for choice, they would save the life of their mother, their wife or their daughter. All responded immediately that they would save their mother, and their tone implied that only an immoral or ignorant person would even ask such a question. (p. 116)

Although all of us hope to live fully from the moment of birth until death, for some elders, living fully until death may not be possible. Some spend their final years confined to nursing homes or other care-taking institutions. Oddly, researchers have generally found a more positive attitude toward death among institutionalized elderly than among those living in their own homes. Upon closer examination, this positive attitude may not represent an acceptance of dying as much as it represents a desire not to continue living. Elders living in nursing homes are generally frailer and less physically competent than those

living at home. Death for some elders, rather than being a point of acceptance, is one simply of resignation in a life that is no longer meaningful.

During her pioneering work in hospitals with death and the dying, Elisabeth Kübler-Ross invited dying patients to share their wants, anxieties, and fears with professionals involved in their care. Though talking about death was off-limits for doctors, it was definitely not so for the dying patients:

> With few exceptions, the patients were surprised, amazed, and grateful. Some were plain curious and others expressed their disbelief that a... doctor would sit with a dying old woman and really care to know what it is like. In the majority of cases the initial outcome was [like] opening floodgates... The patients responded with great relief to sharing some of their last concerns, expressing their feelings without fear of repercussions. (Kübler-Ross, 1969, pp. 157–158)

Stages of Grief

Kübler-Ross's work provided the foundation for modern thanatology—she moved dying out of the medical closet, in which it had lived throughout the twentieth century.

The following section summarizes Kübler-Ross's theory of the process of dying. Although the process is articulated as stages, Kübler-Ross observed that a person might be in any stage or in several stages simultaneously, and the "stages" are not necessarily sequential. People might move back and forth, in and out of the categorizations, or skip some altogether. The Kübler-Ross model provided the most fundamental groundwork for developing a greater understanding of the process of dying, and, importantly, understanding and talking about the reality that one is actually dying. It also provides a good framework to understand the process of grief and bereavement, which is, indeed, what the dying person experiences.

Family support is crucial to the dying person. It is likewise important to family members who will survive their loved one's death.

Stage 1: Denial and shock

Of the more than 300 dying patients interviewed by Kübler-Ross, most reported first reacting to the awareness of a terminal illness with statements such as, "No, not me, it can't be true." The reactions of denial and shock are necessary to give us time to come to terms with the reality of the impending death.

Stage 2: Anger

Following the initial shock, a new reaction sets in: anger or resentment of those whose lives will continue. "Why me? I have [young children]. . . [finally found love in life]. . . [responsibilities to family]. . . [not yet done something important]. . . " or a myriad of other realities that give our lives meaning. This anger often becomes displaced to the doctors, who are no good; the nurses, who do everything wrong; the family, which is not sympathetic; and the world, which is a mess. I'm not dead yet—so pay attention to me! Do not abandon me with my fears. The anger serves to give emotion to feelings of helplessness, despair, and frustration—it can be a very important process to help the dying person move vague anxieties to a psychological point at which he or she can deal more openly with the experience of dying. Understanding that anger has a function for the dying person can help the people around him or her to cope with those expressions of dissatisfaction or rage.

Stage 3: Bargaining

Bargaining is an attempt to postpone the inevitable. If life can be extended only until they can do the one thing they have always wanted to do, or until they can make amends for something they have regretted, or until they can see someone again, they promise to accept death. Some pledge their life to God or to service in the church in exchange for additional time. Others might promise to give their bodies to science, if only life could be extended. Bargaining is a compromise stage that follows the realization that death cannot be denied or escaped and functions to help the individual cope with the reality of the approaching death.

Stage 4: Depression

When terminally ill patients can no longer deny an illness because of its advancing symptoms, a profound sense of loss is experienced. The loss may be of physical parts of the body removed by surgery, money that is being spent on treatment, functions that can no longer be performed, or relationships with family and friends. A deep sorrow generally accompanies the recognition that death is undeniable. The depression might be reactive (the reaction to loss) or preparatory (preparation for one's impending loss of life; such preparation facilitates a state of acceptance), but it comes with the profound recognition that nothing can be done about the approaching death.

Stage 5: Acceptance

Most people ultimately move into a weak, quiet state of submission to that which fate has to offer. Kübler-Ross referred to this state as acceptance, but it is apparently a time devoid of emotion—it is a twilight for the dying person, who is often sleepy or half-conscious, but no longer depressed or angry. "The struggle for survival is over," and the dying person seems to be taking "a final rest before the long journey," said one of Kübler-Ross's patients (Kübler-Ross, 1969). The individual's vital capacity has ebbed and interaction seems unimportant. At this time, the dying person often wants to be with only one particular, special person. Despite the disinclination to engage at this time, Kübler-Ross emphasized the profound importance for the dying person to be touched and spoken to, as a loving reminder of the context of their life.

Facing death is a difficult task upon which age, status, and experience have little bearing, if any. Young children, old people, and those in between seem to experience a similar process. We die as we have lived, the saying goes. If we are angry, or energetic fighters, or philosophers—those will be the characteristics we bring with us to the process of our death. People do have choice in the way they die, however. Lynne Lofland (1979) proposed that four dimensions of choice shape the role of a dying person: (1) how much "life space" to devote to the dying role—the degree of activity to give over to dying, (2) whether to surround oneself with others who are dying, (3) whether to share information about the facts and feelings of one's death with others, and (4) the sort of personal philosophy one wishes to express in the dying role.

Making decisions about the way in which one will die might be the final act of self-determination the dying person can make. Understanding the process makes the possibility of addressing it and making choices about it far more possible.

The Life Review

The study of geropsychology attempts to provide insight into the factors most likely to affect the life satisfaction of older people. In early works, Erik Erikson (1982) and Robert Butler (1981) described a process of **reminiscence** as a source of life satisfaction for older adults. This process of recollecting memories from the past helps to integrate past experiences (who I was then) with the present (who I am now) and make reasonable projections about the future (who will I be. . . what will happen to me?).

During reminiscence, a person remembers more past experiences and reexamines and reintegrates unresolved conflicts with the present. This process can bring new significance and meaning to life, reduce anxiety, and prepare one for death. It may well occur in all individuals in the final days of their lives. Though a life review can occur at a younger age, the drive to put one's life in order seems strongest in old age—and it seems to be a necessary process in order to make sense of the life as lived. Chapter 3 discusses the theory of gerotranscendence, which describes an internally driven process of integrating life experiences during the last chapter of life. It suggests that gerotranscendence is not a process all people engage in, but it is a critical part of continued growth and development in late life. Activities involving reminiscence appear to trigger the process of integrating one's life experience into a more cohesive whole—gerotranscencence. This process allows people to grapple with their vulnerability and mortality as they reassess the meaning of their lives.

The life review resulting from this process may be told to other people, or it may be preserved as private reflection. Finding the process of expression therapeutic, some older people will tell their life history to anyone who will listen; others share their thoughts with no one. Those who cannot resolve the issues their life review uncovers may become anxious or depressed or even enter a state of terror or panic. Those who cannot face or accept the resolution of their life conflicts may become deeply depressed or even commit suicide. Others gain a sense of satisfaction, a sense of tranquility, a capacity to enjoy to the utmost the remainder of their life. Counselors are advised to be attentive and ready to question and listen with patients about life and dying. By creating the opportunity for a dying person to talk about his or her life in terms of the integration of past, present, and future, the listener helps the dying person to deal more effectively with stress (Kennard, 2006).

These ideas have generated much research. Though the terms *life review* and *reminiscence* have often been used interchangeably in the literature, some researchers have separated the concepts by considering the life review to be one of many forms of reminiscence. Reminiscence may provide materials for life review, or it may be just storytelling for fun or social activity. By comparison, **life review** is the form of reminiscence in which the reviewer actively evaluates the past and attempts to resolve conflicts. It is not always comfortable; the reviewer is not glossing lightly over a topic, creating a good story for the benefit of the listener. In the life review processes, the listener is merely the excuse for the elder's process of working with past material. Indeed, some older people proceed with the process of life review through their writing—of journals, stores, of letters. Butler, on introducing the concept in 1963, said:

> *life review is a naturally occurring, universal mental process characterized by the progressive return to consciousness of past experiences, and particularly, the resurgence of unresolved conflicts; simultaneously, and normally, these revived experiences and conflicts can be surveyed and reintegrated. Presumably this process is prompted by the realization of approaching dissolution and death, and the inability to maintain one's sense of personal invulnerability. It is further shaped by contemporaneous experiences and its nature and outcome are affected by the lifelong unfolding of character. (p. 66)*

It is possible that people more commonly engage in life review in their later years, though it is not limited to that period of the life course. Butler contends that thoughts of death initiate the life review, which is, developmentally, more expected as one nears the end of life.

> *We must heal our misfortunes by the grateful recollections of what has been and by the recognition that it is impossible to make undone that what has been done.*
>
> EPICURUS, THE EXTANT REMAINS

Although it may seem that older people, whether close to death or not, are wasting time by talking of the past and dwelling on details that have little meaning to younger listeners, these elders may be engaging in life review. Bernice Neugarten (1987) coined the term *interiority* to refer to a process common in mid- and later life of becoming more focused on internal messages and stories and less focused on outside, social issues. This disengagement from the social world may accurately describe what happens just prior to death: people turn inward, mentally evaluate their lives, and gradually reduce connections with the larger social world. In contrast, seeking to remain actively, externally connected may potentially damage the self-concept of the dying person by discouraging this reflection, reevaluation, and resolution.

Older residents in nursing homes who belong to reminiscence groups seem to have higher rates of life satisfaction than do residents who do not engage in reminiscing (Cook, 1997). In the early 1980s, a sense of closeness to death was examined among nursing home residents. Researchers named this sense "awareness of finitude" and found that this awareness was a better predictor of disengagement than chronological age. The residents who gave a shorter estimate had altered their behavior by constricting their life space and had already become more introverted (Still, 1980). This process is thought to assist the individual in focusing on the process he or she is engaged in—that is, preparing for death.

Systematically, we insist on the occasional nature of death—accidents, illnesses, infections, advanced age—revealing in this way our deep desire to deprive from death all its necessary element, thus making it become just an accidental event.

SIGMUND FREUD

The process of reminiscence can be elicited visually, through photographs, for example. Familiar music is also often associated with particular memories or times in one's life, so also are smell and taste. Touching objects, feeling textures, and painting or engaging in creative activities can also stimulate memories—a particularly important source of expression for people with dementia or aphasia.

Care of the Dying in Hospitals

Medical anthropologist Sharon Kaufman identified a conflict between the **medicalization paradigm** (belief in scientific answers to health) and **individualism and autonomy** paradigm (belief in noninterference in personal choice).

She suggested that this conflict is the source of the "problem" of frailty in old age in American society (Kaufman, 1994), and by extension, the "problem" of dying in a hospital.

In their pioneer study of patient-staff interactions in a number of hospitals, sociologists Glazer and Strauss (1966) created a typology of **awareness contexts of dying** for terminally ill patients: (1) closed awareness, (2) suspicion awareness, (3) the ritual drama of mutual pretense, (4) open awareness, and (5) disconnection.

Closed Awareness

The patient does not yet know that he or she is going to die. The physician in charge decides to keep the patient from knowing or even suspecting the actual diagnosis, and the rest of the staff do all they can to maintain the patient's lack of knowledge. The physician gains the patient's trust and at the same time avoids revealing the fact of his or her impending death.

Suspicion Awareness

The patient suspects the truth, but no one will confirm his or her suspicions. Because the patient is afraid to ask outright questions, he or she therefore receives no clear answers. Still, the patient, who wants evidence, tries to interpret what the staff says and does. Peeking at medical charts, eavesdropping on medical conversations, and watching listeners' reactions after declaring, "I think I'm dying," are typical behaviors.

Ritual Drama of Mutual Pretense

This is the third type of awareness. Both patient and staff now know that death is impending, but both choose to act as if it is not. The patient tries to project a healthy, well-groomed appearance and behaves as if he or she will be leaving the hospital soon; the staff makes comments such as

"You're looking well today"; and both follow the script as if acting out a drama. Blatant events that expose the mutual pretense are ignored.

Open Awareness

This may or may not extend to the time of death and mode of dying, but both patient and staff openly acknowledge that the patient is dying. Game playing is eliminated.

Disconnection

Doctors and nurses for whom daily encounters with dying patients produce painful and bewildering emotions can be distant and tense with dying people. To face a dying person is a reminder, in the health care context, of the "failure" at keeping that person alive and vigorous as well as a reminder of one's own mortality and the finality of death. Some hospital staff members react to the dying person by withdrawing emotionally at a time when the patient most needs their support.

Fortunately, in no small part because of hospice and chaplaincy services, the health care system is showing an increased sensitivity to issues about death and dying. If health professionals can face their own mortality, they can more easily face mortality in others and are less likely to transmit their fear, shock, or horror to the dying person. Generally, the goal for hospital staff is to maintain an open awareness context, as appropriate to the patient. Sometimes other contexts are called for; closed awareness, for example, may be appropriate if the patient indicates a strong desire not to know any details about his or her condition and clearly denies his or her own illness.

Care of the dying in a hospital involves more than the decision to tell or not to tell a patient of his or her fate. A number of other problems are evident.

One problem occurs when doctors and nurses stereotype or label a terminally ill person as "the dying patient." The word *dying* fades all other facets of the individual's personality and colors others' behavior toward him or her. Forgetting that the person is still alive, aware, and unique, staff members may talk as if the patient were insensate or absent. Paradoxically, the full richness of the human being is often not acknowledged, at the very time such validation is central for the dying person to reassess with a shifting identity.

How much hope to offer the dying patient is another task health care professionals must face. Some degree of hope can be helpful to the terminally ill. A glimpse of hope can maintain a person through tests, surgery, and suffering. Patients show the greatest confidence in doctors who offer hope without lying, who share with them the hope that some unforeseen development will change the course of events. Hope does not particularly mean that people who understand they are dying believe they will live. It means they have something to live for a little while longer.

The Dying Person's Bill of Rights, drawn up by those who nurse cancer patients, includes this statement about hope: "I have the right to be cared for by those who can maintain a sense of hopefulness, however changing this might be."

The National Cancer Institute provides a seminar to physicians planning to specialize in the care of cancer patients. The physicians meet to come to grips with their own anxieties about death and better understand their interactions with dying patients and their families. Doctors and nurses often feel an involuntary anger at the dying patient, who comes to represent their own sense of failure and helplessness. "When you've exhausted everything you can do for a patient medically, it becomes difficult to walk into the room every day and talk to that patient." The seminar seems to relieve stress on the cancer ward's nurses, who used to regularly ask for transfers to other wards. It also teaches the staff to be more comfortable in discussing death and other potentially sensitive topics with the patient.

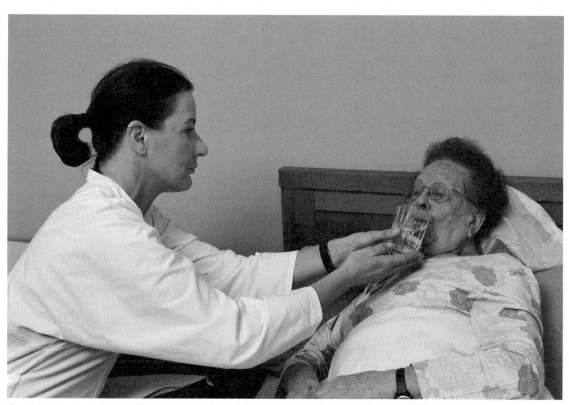

Consistent, appropriate, and loving care of the dying person can help make this final life stage meaningful.

Courses in death and dying are now available at Harvard, Tufts, Stanford, and other medical schools. Nine of ten medical schools in the country have at least one or two lectures for students on the topic, but few schools offer an entire course on death and dying. In only half the schools with such a course is it required; in the other half it is an elective. Personal involvement of students with dying patients was minimal in almost every course at every medical school. One exception is the Yale School of Medicine, which developed a course in which medical students interact with very sick patients. Course objectives include that students develop and demonstrate their compassion, without being afraid.

Older people, as we might expect, have a higher death rate than any other age group.

Morbidity and mortality statistics indicate that health declines and the death rate increases at the upper age levels. Care of the dying, then, has special relevance to older persons, especially the very old, who stand to benefit the most from improved care for the dying.

A Matter of Choice

Individuals who face the prospect of imminent death increasingly seek to expand their rights in determining the manner of their dying. Deciding to die at home and not to prolong life by artificial means are becoming two areas of choice for terminally ill patients. In addition, suicide and funeral plans, which also involve the matter of choice, have special implications for elders.

> *Everything hangs on one's thinking. [. . .]*
> *A man is as unhappy as he has convinced*
> *himself he is.*
>
> SENECA, *LETTERS TO LUCILIUS*

Hospice

Hospice was virtually unheard of in the United States until the mid-1970s in response to the work of Cicely Saunders in the United Kingdom. The United States has an extensive system of volunteers, emphasizing the psychological well-being of patients and their families in dealing with dying and the death of a loved family member.

Hospice received a tremendous boost in 1982 when Medicare payments were authorized to cover care for terminally ill patients. In 1996, the Hospice Association of America reported that over 2,100 Medicare-certified hospice programs were in operation across the country, compared with 158 in 1985. Medicare claims for hospice coverage are suitable if the following criteria are met:

- The patient is terminally ill and has elected Medicare hospice coverage.

- The attending physician and medical director of the hospice team must certify in writing that the patient was terminally ill.

- The patient or representative has signed and filed a hospice election form with the hospice of choice.

- The hospice provider is Medicare certified.

The hospice model, which stresses effective pain relief for terminal cancer patients, is designed to care for the dying person—the whole person, not the disease. Although many organizations are associated with hospitals, a hospice can be any organization that provides support services to dying patients and their loved ones in the place where that care is given—the home,

a hospital, or a nursing home (Worsnop, 1997). Hospice goals include that the patient be as comfortable and pain-free as possible, that they be independent for as long as possible, that they receive care from family and friends, that they receive support through the **stages of dying**, and that the person die with dignity.

Palliative care is the medical specialty focused on relief of pain, stress, and other debilitating symptoms of illness. Palliative care is not the same as hospice care. Palliative care may be provided at any time during a person's illness, and it may be given at the same time as curative treatment. Relief of pain is the core of hospice practice. An individual's death is neither hastened nor prolonged, under with hospice care no effort to "cure" is made, because the patient is dying. It provides, in the words of Christine Cassel (1996), "aggressive comfort care."

Tremendous advances in the management of pain have occurred in the past 25 years: drugs can be delivered through skin patches, through topical creams, and by intravenous pumps. Some new drugs and delivery systems have helped to minimize the side effects of many painkillers, such as grogginess, nausea, and reduced mental clarity (Horgan, 1997).

Social attitudes that have limited prescriptions of analgesics—opiates such as morphine, particularly—have shifted somewhat in recent years. The Pain and Policy Studies Group at the University Wisconsin has conducted some outstanding research on attitudes of the public and physicians about pain killers and found changing perceptions on the use of "addicting" drugs for pain control (Horgan, 1997). It is striking to note that medical research in the field of pain control has not been directed by humanitarian philosophy, but by the market potential brought about by longevity, the pharmaceutical industry, and governmental funding through Medicare for palliative care.

The first hospice program in the United States, Hospice of Connecticut, in New Haven, offers a model of care for the dying. The New Haven hospice began as a home-care program.

Outreach workers visited and counseled patients, and they also provided home nursing to the dying. Family members could receive individual or group counseling throughout the dying process to get support from others experiencing the same shock and grief. The program later developed a residential hospice in which families and friends could be with the dying person, away from the home.

Hospice works with an interdisciplinary team of workers, including nurses, health aides, counselors, physical therapists, nutritionists, chaplains, volunteer visitors, doctors, and long-term care administrators. Volunteers provide care so that family members can get out of the house or feel comfortable leaving the bedside. Volunteers also might do light housekeeping, laundry, meal preparations, and run errands for the family or the ill person.

Some hospitals run hospice programs, and programs are becoming available at nursing homes as well. These programs offer an integrated approach to death in which medical and nursing staffs, chaplains and visiting clergy, and social service staff members work together to meet the physical, spiritual, and psychological needs of the dying.

Awareness of hospitals as potentially grim and depersonalized places to die has encouraged a trend toward dying at home. Some health-care organizations enable patients to go home to die by providing doctors and nurses who will counsel the family, monitor the patient's drug intake, and provide emotional support.

A dying person often takes a renewed interest in life when he or she finds people who are sensitive to both medical needs and emotions. The hospice or its wing of a hospital often does not appear to be institutional. In the visitors' area, the radio may play upbeat music while rambunctious children play in the hallways. Visitors are welcome over a wide range of hours—indeed, the model encourages open interactions designed around the needs of real people, rather than requiring that those people fit their needs into an institutional format.

Hospice care in the 1990s focused a great deal of attention on AIDS patients. Nursing homes that, until recently, have cared mostly for elders now provide facilities for young adult AIDS patients. Hospice care, though stretching to meet the needs of all clients, is, nevertheless, overburdened in areas where the numbers of AIDS patients are high. Additionally, they are becoming far more bureaucratized and regulated, especially where Medicaid and Medicare reimbursements apply (Magno, 1990). The epidemic of HIV/AIDS has underscored the need for the hospice orientation, an approach that emphasizes care when there is no cure. Hospice care promotes the philosophy that comfort and quality of life are in and of themselves, worthwhile and essential goals to the end of one's life.

Although most people, when asked, express a desire to die at home, only a small percentage actually do. Not everyone should die at home. For a patient who has an unhappy or unstable family, or who needs complicated nursing care, a hospital or nursing home may be a better option. Not all homes are safe places in which to die. Parents and grandparents sometimes choose to die in hospitals to avoid traumatizing young children. However, some psychiatrists believe that the very fact of a death occurring at home can ward off psychological damage to the family. The child who is involved throughout the dying process does not have to face the sudden, unexplained disappearance of a parent or grandparent. In contrast, children who are not involved may believe, for example, that they are somehow responsible for the death. Keeping a dying family member at home can both make it easier for relatives to accept the death and prevent the patient from being alone in the dying process.

Reframing the Craft of Dying

Lynne Lofland (1979) refers to the hospice movement and a number of other groups that work toward improving attitudes and conditions surrounding death as the "happy death movement."

This movement opposes the conventional view that the dying person be stoic, strong, and silent. It began, she says, in academia and filtered out to ordinary citizens who are embracing the "happy death" concept (Hoefler, 1994). Lofland notes that although the new movement has helped some people to freely express their fears, others still refuse to express their concerns.

The new movement has advocated the view that dying can be a learning, growing, and positive experience. Lofland observes that this emphasis, in itself, may pressure dying people to assume a certain way of thinking. Those who will not or cannot share their thoughts and feelings, or who do not develop positive attitudes, may feel like failures. She believes that alternative ideologies are critical to provide dying people the greatest degree of choice in their mental preparation for death.

In a similar vein, some who work in the field of death and dying have criticized the phrase "death with dignity" as placing too much emphasis on being proper or accepting (Nuland, 1994). "Dignity" has a Latin root, *dignitas*, meaning "worthiness" (Meyer, 1995). A potential exists for "death with dignity" that might place a burden of particular behavior on the dying individual, rather than creating an environment in which the dying person is helped to be aware of his or her worthiness.

The anguish of death hangs over and leads the human spirit to wonder about the mysteries of existence, man's destiny, life, the world.

E. MORIN, *METHOD V*

Right to Die

Why is choice in dying so important? Enjoying personal freedoms is a way of life in democratic countries. We have the freedom to choose how to celebrate our religious beliefs, to participate in democratic elections, and to speak out minds without fear. We generally do not, however, have freedom of choice at the end of our lives to hasten death. As the boomer generation enters the aging population and advances in health care prolong both life and dying process, there is a huge need to face the issues, learn about them, and discuss them openly (Dying with Dignity, no date).

We have grappled with the **right-to-die issue** for many decades. The first bill to legalize doctor-assisted suicide was introduced in 1906 in the state of Ohio, which proposed that a doctor could ask a terminally ill person in the presence of three witnesses whether he or she wished to die. If the answer was "yes," and three other physicians agreed that the patient's condition was hopeless, they could assist in his suicide. Opponents feared a lack of safeguards, and the measure was defeated.

The issue slowly grew in public acceptance, fermenting until the liberation of the Nazi death camps in 1945 at the end of World War II. The full horror of medically assisted death—systematic termination of life—suddenly became so abhorrent a concept that assisted suicide ceased to be a public topic (Worsner, 1996).

In the 1970s, the parents of Karen Quinlan, a 21-year-old woman in a permanent vegetative state, initiated a court battle in the state of New Jersey for the right to disconnect the life-support machines to which Karen was attached. Though a trial judge initially ruled against them, in 1976 after taking the case to the New Jersey Supreme Court, the Quinlans were allowed to have the machines removed. Karen lived in a comatose state another nine years.

A second case that galvanized the public's awareness was that of *Cruzan v. Missouri Department of Health*. Following an auto accident rendering her in a vegetative state, Nancy Cruzan had been kept alive for seven years by a feeding tube. In a suit filed in 1988, her parents, acting on her behalf, requested that her food and fluids be stopped. A lengthy and extraordinarily expensive process ensued for the Cruzans. In a summary of the case in the *Congressional Quarterly*, Richard Worsnop reported (1996):

> *The Missouri Department of Health appealed the decision to the State Supreme Court, which reversed it. As a result, the Cruzans appealed to the US Supreme Court, which held on June 25, 1990, that a person whose wishes were clearly known had a constitutional right to refuse life-sustaining medical treatment. At the same time, however, it ruled that states could require that comatose patients be kept alive unless there was "clear and convincing evidence" that they would not want to live under such conditions. Such evidence, said the court, was lacking in the* Cruzan *case.*

> *Instead of giving up, the Cruzans returned to Jasper County Circuit Court. Three of Nancy's former co-workers now testified that she had told them years earlier she would not want to live out her days in a coma. . . . [The court ruled that] this disclosure amounted to clear and convincing evidence. . . . The Cruzans had the right to order the removal of their daughter's feeding tube. (p. 780)*

Nancy died 12 days following removal of the feeding tube. Her death came two months after a U.S. law requiring hospitals and other health-care facilities receiving Medicare or Medicaid payments to inform patients of their right to executive living wills or other advance directives (Worsnop, 1996).

In 2005, 41-year-old Terri Schiavo died; she was a woman in an advanced vegetative state with no written advance directive. Although she was medically determined to have whole-brain death, her death became the centerpiece of a national battle over the right to die. For seven years Schiavo's husband and legal guardian requested the removal of her life-sustaining feeding tube. Her parents strongly disagreed with the request, and a bitter battle in the courts became a larger, public battle. The case culminated in an unprecedented Congressional intervention in the judicial process, although ultimately the judicial system supported Michael Schiavo's decision to remove the feeding tube. The court supported the husband's contention that she would not have recovered and that her suffering was being prolonged by keeping her on life support for 15 years. The deeply personal and grieving decision had become national media headlines.

Families are fighting similar court battles today. When someone you love is in what is determined to be a persistent vegetative state, do you withdraw the feeding tube, as Michael Schiavo wanted to do over the objections of her parents, the state legislature, and the government?

It is a moral issue facing and dividing many families. Who decides for the person who has never considered the possibility of being unable to make a decision regarding extraordinary measures of life support? How many of us have considered that possibility for ourselves? Yet, as in the cases outlined earlier, it happens. Sometimes, someone else must be our voice, must make that decision for us—hopefully with our best good in mind.

Doctors traditionally have felt they must prolong life as long as possible without questioning the circumstances, an idea that constitutes an intrinsic part of the Hippocratic oath. However, values and laws traditionally lag behind technological advancements, and the value system doctors have followed for centuries is now in a state of flux. The right to die is but one of many issues confronting **medical ethics**.

In the following sections, we will discuss this issue and several others. The issues are complex: at what point should the decision not to prolong life be made, who should make the decision, and does an individual have the right to choose death when life could be extended in some fashion? Some of the right-to-die issues can be stated as specific questions:

- What is the difference between killing and allowing a person to die?
- What is the difference between stopping treatment and not beginning it?
- Are there reasonable and unreasonable treatments?

The answers to these questions, which cannot be easily answered, vary on both moral and legal grounds.

In an odd way, the right-to-die movement has improved the quality of care of the dying in America. After Measure 16 passed [in Oregon] in 1994, the health profession started sponsoring seminars and lectures on pain control and set up task forces on end-of-life care. Now people reckon Oregon is the finest place to die in America. Doctors and nurses here will do their very best for you so that you don't ask for assisted dying.

DEREK HUMPHRY, HEMLOCK SOCIETY FOUNDER, 1997

Theoretically, there are a number of ways that medical personnel could hasten the end for patients who wish to die. The terminally ill patient could be killed by injection. The decision could be made not to begin an intravenous drip or a respirator or to stop one that was being used. A decision could be made to avoid antibiotic use with a patient with pneumonia. A fatally high dose of a narcotic or barbiturate could be administered. The legal consequences of these alternatives vary, and doctors vary, as well, in their attitudes toward assisted death. Some use code words on charts for "hopeless case" patients. Such a patient

may be labeled "Code 90 DNR" (Do Not Resuscitate), for example, or "CMO" (Comfort Measures Only). Both indicate that extraordinary lifesaving measures should not be applied (Kastenbaum, 2004).

Advance Directives

Margaret Jones's 88-year-old mother was in terminal stages with brain cancer. When Margaret arrived to visit, she found her mother unconscious on the bathroom floor. Margaret telephoned 911; when the paramedics arrived, they resuscitated the elder Jones, as required by law. The painful intubation procedure and a portable ventilator kept Jones alive for the remainder of the month, when she finally died. She had no advance directives. In trying to be sensitive to her mother's sense of hope, Margaret probably did her a great disservice. Margaret's choice also spared her the terrible decision of requesting that her mother be removed from life support.

Though nearly 90 percent of all Americans will have a "managed death" in a hospital or skilled nursing facility, which can lengthen life for up to several years through medical and nursing interventions, only about 30 percent of patients have advance directives (AoA, 209).

An **advance directive** instructs the doctor regarding what kind of care a person wants to have if he or she becomes unable to make medical decisions (e.g., in a coma). A good advance directive will describe the treatment you would want depending on how debilitated you are. You might (or might not) make, for example, a distinction between an illness you are unlikely to recover from with what decisions you want to make if you are permanently unconscious.

A **living will** is one type of advance directive. It is a written document, legal in most states, that describes the kind of life-sustaining treatments you would want if you were terminally ill. A living will does not allow you to select someone to make decisions for you.

The durable power of attorney (DPA) for health care is another advance directive, stating whom you want to make decisions for you. It becomes active when you become unconscious or unable to make medical decisions (e.g., as with dementia). A requirement is having a person you trust to make these decisions for you. A DPA is a legal document in most states, and even if not legal in a state, it provides guidance for medical care.

Do not resuscitate (DNR) is another kind of advance directive. It requests that in the case of your heart stopping or you are not breathing, you do not want cardiopulmonary resuscitation (CPR). In emergency cases, first responders do not have access or time to check your medical records: many people on life-support systems had DNR in their medical records.

Two terms are important to a discussion of the right to die: (1) **passive euthanasia,** or the process of allowing people to die without using extraordinary means to save their lives; and (2) **assisted suicide,** or performing a deliberate act to end a person's life, such as administering a fatal injection. There are acts between active and passive euthanasia that are difficult to categorize. Not treating a person for pneumonia is one example. Is the intentional lack of action in itself "active?"

The Patient Self-Determination Act

The **Patient Self-Determination Act (PSDA)** of 1991 requires all health care facilities receiving Medicare and Medicaid funding to recognize the living will and DPA for health care (health care proxy) as advance directives. It further requires "all hospitals, nursing homes, health maintenance organizations, hospices, and health care companies that participate in Medicare or Medicaid to provide patients with written information on their rights concerning advance directives under state and federal law" (Duffield & Podzamsky, 1996). The PSDA created no new rights for patients, but it reaffirmed

the common-law right of self-determination as guaranteed in the Fourteenth Amendment (Haynor, 1998).

The decision to commit suicide with the assistance of another may be just as personal and profound as the decision to refuse unwanted medical treatment, but it has never enjoyed similar legal protection. Indeed, the two acts are widely and reasonably regarded as quite distinct. . . . First, when a patient refuses life-sustaining medical treatment, he dies from an underlying fatal disease or pathology; but if a patient ingests lethal medication prescribed by a physician, he is killed by that medication.

Supreme Court Chief Justice William Rehnquist, 1997

Suicide and Assisted Suicide

Adults of all ages support the right to die by refusing treatment, as reflected in living will legislation in all 50 states. When it comes to actively hastening death, or physician-assisted suicide, people are less certain. Between 40 percent and 50 percent of physicians believe that physician-assisted suicide may be an ethically permissible practice, and around 66 percent of the general public favors its legalization (Koenig et al., 1996). Polls, however, have included only a small proportion of older adults (a proportionally representative sample).

In November, 1997, Oregon voters overwhelmingly reaffirmed their support for the nation's only law allowing physician-assisted suicide. The legislation was originally passed in 1994, immediately challenged in court, and prevented from going into effect. The Oregon legislature determined that there was a lack of public information about the law and brought to a second vote in 1997, with a public referendum.

The law includes a 15-day waiting period after a terminally ill person has requested physician assistance with suicide. At least two doctors must agree that the patient's condition is terminal, and the patient must request the drugs in writing and administer the drugs to himself or herself.

In 2006, the U.S. Supreme Court ruled that it was up to states to regulate medical practice, including assisted suicide. This was controversial decision, opposed by the Republican Party and the administration. However, 60 percent of the state of Washington created a voter-approved measure allowing assisted suicide.

> *They tell us that suicide is the greatest piece of cowardice . . . that suicide is wrong; when it is quite obvious that there is nothing in the world to which every man has a more unassailable title than to his own life and person.*
>
> ARTHUR SCHOPENHAUER

When older patients are asked about their attitudes toward physician-assisted suicide and the types of assistance that should be provided, interesting attitude differences emerge. In two separate studies, older patients were found to be less likely than their relatives to support physician-assisted suicide. Those patients who oppose the practice represent a particularly vulnerable portion of society: women, African Americans, the poor, and people with lower educational achievement (Koenig et al., 1996; Duffield & Podzamsky, 1996). This implies that special protection might be necessary for some groups of individuals who by tradition have less "voice" in the system.

Whether anyone has the right to take his or her own life is a complicated issue with moral implications. Older persons may first contemplate, then commit, suicide because they feel life no longer holds meaning or because they are in pain that feels unbearable. Some deaths of older people lie between accident and suicide. Older adults who are disappointed and frustrated with life may refuse to eat, keep to themselves, and refuse care, with fatal consequences. Such deaths may not be regarded as suicides.

> *The thought of suicide is a great source of comfort; with it a calm passage is to be made across many a bad night.*
>
> FRIEDRICH NIETZSCHE

Such suicides pose this question: Do people have a right to choose death by suicide? Several groups, such as the Hemlock Society, advocate control over one's dying which includes committing suicide, if necessary, to escape profound pain or great bodily deterioration. AIDS advocates, for example, often argue for the right to commit suicide, if that is the choice of the suffering person. But no one has received as much publicity in terms of assisted suicide as Dr. Jack Kevorkian—"Doctor Death."

Pathologist Jack Kevorkian has become the most well-known advocate in the United States for physician-assisted suicide. He has sparked national debate on the ethical issues involved in assisted suicide following the design of a machine that started an intravenous drip into the arm of the person wishing to die. The chemical would put the person into a deep sleep and then a coma. After one minute, the timer would send a lethal dose of potassium chloride, stopping the heart in minutes. The patient would die of a heart attack while in a deep sleep.

By 1998, Kevorkian had helped over 100 commit suicide. All of his suicide patients were in terminal stages of disease, generally accompanied by excruciating pain. He has referred to assisted suicide as patholysis (Greek for freedom from suffering) and dedicated his life and career to bringing to public awareness the need, in his perspective, for the legalization of assisted suicide. He was imprisoned in 1999 and initially charged with violating the state statute prohibiting assisted suicide. That charge was dropped, however, and he was convicted of second-degree murder for delivering a controlled substance without a license (PBS, 2004).

Patients of Kevorkian who requested assistance with their suicide were interviewed at length by him and his associates, usually for several weeks before actually meeting with him. His agreement to provide assistance was based on his assessment of the mental competence and the physical condition of the person, among other factors. If he agreed to provide assistance, the patient met with Kevorkian and his attorney for personal interviews and reviewing the case. Documents assuring the patient's competence and desire to commit suicide were signed. Those willing to do so also make a video-taped personal statement, often in interview format with Kevorkian and/or his attorney.

Examples of people whose deaths were assisted by Kevorkian are a 72-year-old woman, a double-amputee in severe pain from rheumatoid arthritis and advanced osteoporosis. Another person with multiple sclerosis, who was unable to sit or lie down because of excruciating pain, was assisted. People who completed their suicide with Kevorkian's assistance left a note directing family and/or police to the location of their body; generally the note included a personal statement articulating the person's choice in the matter. A 78-year old Canadian emigrant from India with severe Parkinson's disease left a note including the statement, "I am grateful to the merciful hands of Dr. Kevorkian . . . to bring about a happy deliverance—Maha Samadhi—a final embrace with the divine" (U.S. News Bulletins, 1997).

Jack Kevorkian and his associates, including an attorney and assistant, attempted to challenge the legal system into making a definitive ruling on the legal status of physician-assisted suicide. His objectives were to make it an accepted medical practice. "It's time to clear the air," said Kevorkian in a news interview. "I admit I assisted in [the deaths of 100 people]. If there

was a crime committed, charge me. If there isn't, don't bother me" (U.S. News Bulletins, 1998).

Kevorkian served five and one-half years of a 10- to 25-year prison sentence, during which time his health deteriorated. He was released on parole in 2007. Although personally he is a controversial man, his activism has increased public awareness of some of the most difficult ethical issues surrounding death and dying.

Assisted Suicide

Law in the United States did not uphold assisted suicide—although the assistant is a physician, friend, or family member—until November 1997, when Oregon became the first state in the country to pass a law allowing doctors to hasten death for the terminally ill. The law provides that a doctor may prescribe but not administer a lethal dose of medication to a patient who has less than six months to live. Two doctors must agree that the patient is mentally competent and that the decision is voluntary. The Oregon State Health Division reviewed the first year of the law's implementation and concluded that the law was working well and had not been abused.

In 2008, 60 percent of voters in the state of Washington passed legislation allowing a doctor to prescribe lethal drugs to patients given less than six months to live. It was an extremely controversial passage, opposed by the Catholic Church, conservative Christians, right-to-life groups, and advocates for the disabled.

In other states, the assistant to a suicide might be prosecuted under a general charge of manslaughter; in still others, the law is unclear and not vigorously pursued. The issue has created intense public debate and split the health care community.

A review by John Horgan (1997) summarized:

The American Medical Association (AMA), the American Nursing Association, the National Hospice Organization and dozens

Having a choice of how to die once the dying process has begun and there is no other alternative can be tremendously comforting to the dying person, regardless of his or her decision.

of other groups have filed briefs with the Supreme Court opposing physician-assisted suicide. Supporters of legalization include the American Medical Student Association, the Coalition of Hospice Professionals, and Marcia Angell, editor of the New England Journal of Medicine.

Nonetheless, the dispute masks a deep consensus among health care experts that much can and should be done to improve the care of the dying. (p. 100)

The arguments both in support of and against assisted suicide touch upon powerful issues of ethics and American values. On the one hand is fear that laws allowing physician-assisted suicide will start us down a "slippery slope" toward involuntary euthanasia. Ethicists worry that the power to decide who should live and who should die will get into the hands of the wrong person. The ability to make a decision as powerful as this might move to the "next step" of making the determination for a "medically defective" person, or someone who is "too old" or "too poor" to warrant expensive medical care.

On the other hand is the belief that self-determination by competent individuals is an inviolable right and essential to the maintenance of human dignity. Validation for this perspective comes in part from teaching hospitals, which report that 40 percent of terminally ill patients experience severe pain "most of the time," (Horgan, 1997). The perspective also emerges from a greater sensitivity to cultural diversity. The Japanese, for example, have a less restrictive attitude toward suicide than do European Americans. It is seen as a "sad but not morally ambiguous response to human relational struggles" (Matzo, 1997).

Those in the center of the argument believe the stage is set for an assisted suicide if doctor and patient have thoroughly explored all options, and if the patient is not depressed. If the patient is terminally ill and near death and cannot face either the pain of the disease or the agonizing efforts of more useless treatments, then physician-assisted suicide is acceptable from this position.

Even in the very shadow of death, one's living experience can give rise to accomplishment. Dying is just a time of living. It's a particularly difficult time of living that has a propensity for problems and suffering . . . People are dignified even in their dying, even in their physical dependency, if they're treated in a dignified manner.

IRA BYOCK, *DYING WELL*, 1996

Another approach to end-of-life comfort and control is to sidestep the issues of ethics and morality and address the need for pain control. The capability exists to control pain far better than is often done. Pain control is simply not made available in all cases (Matzo, 1997). The seriously ill may become trapped in a cycle of intensifying distress in which chronic pain becomes more severe over time, leading to psychological distress, making the pain more difficult to endure. Even with the advances in pain medications and their delivery, pain is not adequately addressed by health care services (Beresford, 1997). Too little, too late creates too much suffering.

Informed Consent to Treatment

Norms have radically changed in acceptable medical practice in the past 30 years. Issues of autonomy, self-determination, and quality of life now make a huge difference in how medicine is delivered. Somewhere around 85 percent of deaths in the United States occur in institutions, and 70 percent of those involve elective withholding of life-sustaining treatment (Matzo, 1997). The role of the patient as an autonomous individual responsible for making medical and life decisions has preempted the role of patient passivity.

Previously, doctors felt comfortable making all informed decisions for their patients, certain that their judgment, though imperfect, was the best judgment available. Nurses acquiesced without question in the doctor's judgments—they were considered technicians whose job it was to support the physician's bidding. This is no longer the case. Patients have the right—and are given the responsibility—to accept or reject any treatment or prescription that their physician may offer them, and they have the right to be informed of their choices. Those suffering a serious or life-threatening illness for which several treatment plans are possible or optional are now in the difficult position of needing to select what they believe to be the best plan, under the informed guidance of their physician. This requires a communication process between physician and patient, which is both an ethical obligation and a legal requirement in all 50 states.

This means that patients have an obligation to inform themselves of the options as well as the risks involved with each option and then base their consent to treatment on three principles:

1. They are competent to give general consent.

2. They give consent freely (e.g., not coerced by economic situation or relatives).

3. They have a full understanding of the meaning of the options (AMA, 2009).

The physician is required to inform the patient of alternative medical approaches (sometimes called complementary) only of "medically indicated" or "medically viable" options. This is a gray area because it is not directly addressed in case law, and what is meant by "indicated" and "viable" has not been clarified. The approach taken by physicians is that alternative or complementary medical approaches are mentioned as options when the nonmainstream approach is endorsed by a "respectable minority" of physicians.

Informed consent means that patients share in their health care decision making by becoming informed and by basing their choice, acceptance, or rejection of treatment on the information available to them. It is a big responsibility for patients and physicians alike, and complete information comes at risk of provoking or increasing patient anxiety and confusion.

More Ethics in Aging

Other ethical issues have been arisen now that medical technology has the capacity to prolong life almost indefinitely, and disease control keeps more people alive into advanced age. Here are some of those issues:

1. What is the appropriate level of direction family members and the health care system should provide when dealing with a person who is losing competency? For example, is the person with early Alzheimer's allowed self-determination in deciding whether to drive a car? With more advanced Alzheimer's, how is a person who wanders restrained and who makes the decision?

2. With regard to the ethics of nursing home placement and ethical dilemmas in the nursing home, how much self-determination and autonomy is allowed the older person to make decisions and act for himself or herself?

3. What obligations do adult children have for the financial, physical, and emotional support of frail elderly parents?

Hospital ethics committees have helped make strides toward better doctor/patient communication, particularly regarding end of life decisions.

Funerals

Dying in institutions is depersonalized for the dying individual and dissociates society from the process of dying. The person dies, usually alone. The body is sent to a funeral parlor, where it is prepared for burial by strangers, who never knew the living person. Preparation includes washing the body, injecting preservatives, dressing it for burial in whatever garment the family or care taker chooses, applying flesh-colored makeup to whatever skin is exposed after dressing, arranging the hair, and arranging the facial expression to give the impression of peaceful slumber. It is often *after* this point that family members spend time with the body to say goodbye to the person who has died.

The majority of Americans who die are buried, but cremations are more frequently chosen now than in the past. The option has roughly doubled in acceptance since 1990 to 53 percent of Californians choosing cremation over plot burial (Hasemyer, 2009). The economic crash of 2008 left some families unable to provide funerals and burials for relatives even when their culture or religion called for more elaborate rituals. For them, cremation is an alternative. The cost of a cremation runs between $700 and $1,200 whereas a "high end" funeral with elaborate casket, limousines, and a burial plot in the cemetery can exceed $10,000 (Hasemyer, 2009).

Both burial and cremation, however, allow for a funeral or memorial service at which the grieving friends and family can commemorate the life of the person they loved. These rituals serve as a way that families or groups can share their identity and reestablish their collective presence in the absence of the person who has died.

The cost of a funeral, however, can be high, and in the United States mortuaries are a necessary business. Charges for limousines, burial clothes, use of a viewing room, sometimes rented gathering space, and a headstone are not necessary but are culturally appropriate in most areas of the country. The ritual ceremony of funeral or memorial service has evolved as a way to assist people through a process of grieving, which is a highly personal experience. As with informed consent to treatment, however, individuals need to be aware that they do have choices.

Because of complaints about the funeral industry, the Federal Trade Commission (FTC) ruled in 1984 that funeral directors must provide itemized lists of goods and services and their prices. This may keep prices from varying depending on a client's vulnerability, and it shows clearly that a price range exists. The FTC also ruled that misrepresentation of state laws concerning embalming and cremation is against the law. In addition, funeral parlors must itemize on a funeral bill every charge—the cost of

Funerals and memorial services are powerful reminders of the relationship of culture to grief and the process of bereavement.

flowers, death notices in newspapers, the provision of hearses and limousines, guest registers, and other related services (Aiken, 1994).

Consumer groups are available to educate the public about the laws governing the funeral industry and inform individuals about their rights and choices. A family member needs to know about the itemized price list, for example, to ensure that the service is received. The greatest protection a person can have, however, is to be well informed far before a death occurs. That means thinking about and discussing funeral preferences while people are in good health and making burial arrangements before a death. This can be a difficult topic to broach for family, friends, and a seriously ill person. Some people are more accepting of death than are others. Some elderly people see their lives drawing to an end and accept that ending, whereas others cling to life. The relationship to dying held by the dying person can make a huge difference in family survivors' coping.

Facing and Preparing for Death

Realizing that death is the ultimate destiny for each of us, we need to understand, regardless of our age, the importance of preparing for it. Facing death, or at least easing personal anxiety about it, can improve the quality of an individual's life. Whether you have thought about or have already answered any of the following questions tells something about your degree of preparation for dying.

- Do you have a will?
- Do you have life insurance?
- Did you prepare advance directives?
- Are you willing to have an autopsy done on your body?
- Have you chosen to donate the organs of your body for use after you die?

- Do you know how you want your body to be disposed of (burial, cremation, donated to a medical school)?
- What kind of last rites do you want (funeral, memorial services, a party)?

Families can ease fears and concerns of the dying person by assuring him or her it is OK to let go. A dying person commonly tries to hang on to life even though it might mean prolonged discomfort, in order to be assured that those left behind will be all right. A family's ability to reassure and release the dying person of this concern is the greatest gift of love they can give at this time. Saying goodbye in personal ways is the closure allowing for final release.

When we reduce our own fears of death and dying, we can offer others, including the terminally ill, better care and help them cope. If thoughts of death elicit thoughts of satisfaction with a life well-lived, we have reached what psychologist Erik Erikson means by integrity, his final stage of psychosocial development. In the later years of life, those with a sense of integrity are able to reflect on their past with satisfaction; those filled with despair dwell on missed opportunities and missed directions. With luck and understanding, the path each person takes will prepare them for death, which Kübler-Ross calls "the final stage of growth."

Bereavement

The loss of a loved one can be the most tragic event a person experiences in a lifetime. It is an experience that happens some time or another in nearly everyone's life. In old age, these events occur with increasing frequency, as friends and family members die. Psychologists believe that their goal, after empathizing with another person's suffering, is to step aside from the maze of emotion and sensation and to make sense of it—what is the function of suffering to the human condition? A folk wisdom states "Pain is inevitable. Suffering is not." Is that so? If we

Mourning the loss of someone important in our life is part of life's process. It is shared, differently, by many people, and in that sharing, grief becomes manageable.

can understand more about bereavement, we can understand more about supporting those who are bereaved.

Death leaves a heartache no one can heal; love leaves a memory no one can steal.

HEADSTONE IN IRISH CEMETERY

Bereavement is the experienced loss of someone important in our lives and the adjustment to that loss. Grief is the emotional response experienced by bereavement, and it takes many forms depending on the individual and the culture in which they live. Mourning is not the feelings, but our behavioral responses to the grief. Customs shape our responses to grief. In Japan, religious rituals are designed to help survivors maintain connection with the deceased. An altar in the home, dedicated to ancestors, is tended to and treated with honor and respect. In some areas of Africa the dead are considered to be part of the community. A person's family includes living aunts, uncles, and so forth, as well as the beloved grandmother buried in the local cemetery. There is no one, best way, to cope with sadness and loss.

Bereavement theory and research has expanded into such areas as why some bereaved people themselves die soon after their loved one. Though there are vast individual differences and bereaved persons tend to recover within a two-year period, a smaller high-risk group of people simply cannot make a recovery.

Complicated mourning is a cognitive disorganization and emotional chaos that can lead some people into profound depressions that do not move toward resolution (Family Caregiver Alliance, 2007). Types of complicated mourning reactions include chronic grief reactions, masked grief reactions, exaggerated grief reactions, and chronic depression.

Chronic grief reactions involve reactions to death that are of long duration and do not lead to resolution. The survivor feels sorrow for years after the loss. Chronic grief is often associated with unresolved relationship issues between the bereaved and the dead person.

Masked grief reactions are the complete absence of grief reactions accompanied with substitute psychosomatic complaints (headaches, insomnia, pain). The broken heart syndrome, referring to findings that a widow or widower is more likely to die within the first two years following the death of his or her spouse, may be connected to this type of grief reaction (Stroebe & Stroebe, 1987). It is related to the inability to consciously express the depth of sadness and loss.

Exaggerated grief reactions are repressed reactions to former grief-causing situations that erupt in response to a current loss. Old losses that were covered up erupt as phobias, psychosomatic symptoms, psychiatric disorders, and so forth (Marrone, 1996).

Clinical depression is similar to the profound sadness and depression experienced in mourning; however, its effects are more powerful and it lasts longer. Clinical depression is considered to be a mood disorder that includes sadness, helplessness, guilt, and a pervasive sense of suffering. The inability to concentrate and general apathy accompany clinical depression.

Complicated grief reactions occur based on personality characteristics of the bereaved individual, the degree of support available to him or her in the grief process, and the relationship of the deceased person with the bereaved individual. Feelings of ambivalence toward the deceased person can bring about a complicated grief reaction, as can the circumstances of the death.

Mourning is a profoundly individual matter. Bereavement is a state of loss—to bereave, in fact, means "to take away from, to rob, to dispossess" (Kalish, 1985). The ways in which we respond to that loss will be as different as are our personalities and our life experiences.

Chapter Summary

Death is one of the few certainties of life, yet denial of death is common. We have euphemisms for the word "death"; we protect children from hearing about it; and death on TV is not real. Instead, death is disassociated with everyday events by typically occurring in a hospital setting. Elders, just as any other age group, may have fears about dying. The phases that a terminally-ill person goes through include: (1) denial and isolation, (2) anger, (3) bargaining, (4) depression, and (5) acceptance. Efforts must be made so that terminally ill patients are given care, concern, and support. Elders are being given more choices in dying, but suicide is still not socially accepted. Yet changes are slowly taking place to broaden choices for the dying.

Physician-assisted suicide has become a topic of wide debate and has been legalized in Oregon. The hospice movement has supported people in their choice to die at home and to refuse life extending measures that bring pain and discomfort. Medical ethics are challenging the assumption that lives should be extended at any cost. Patients are being informed of their rights, especially the right to refuse treatment. The "right-to-die" movement has gained momentum. Social scientists are now seriously exploring the issue of near-death experiences. It is important for elders to face death and prepare for it.

Key Terms

active euthanasia
advance directive
assisted suicide
awareness contexts
   of dying
bereavement
death anxiety
death competency
death-denying
   society
fears about dying
fear of death
happy death
hospice
individualism and
   autonomy

individualism and
   autonomy paradigm
informed consent
life review
living will
medical ethics
medicalization
   paradigm
near-death experiences
passive euthanasia
Palliative care
Patient Self-
   Determination Act
reminiscence
right-to-die issue
stages of dying

Questions for Discussion

1. Imagine being told that you have six months to live. What would be your reaction? Explain in depth. What would you think? Where would you go? What would you do?

2. List 10 things and/or people that are most important in your life. Now remove each one, one at a time, until there are none left. Does that exercise bring up some emotions for you? How did you feel as you "lost" each of the important things in your life?

3. Do you think that euthanasia is ethical? Why or why not? Develop an argument either for or against it.

Fieldwork Suggestions

1. Go to a cemetery, walk around, and examine your own feelings about death and dying.

2. Talk with three people about their attitudes toward physician-assisted suicide. What do they think about Kevorkian? Did you find anyone who is not familiar with his name?

3. Call a local funeral home and ask for a tour. Record notes of your experience and the feelings that it brought up for you. Share your experience with the class.

4. Find a local hospice center and learn about what they do. Share your findings with the class.

Internet Activities

1. Locate the website of a company producing caskets. What products and services are offered? To whom is the marketing addressed? What is your assessment of the website? Can you think of ways in which it might be improved or otherwise changed?

2. Find information on cremation and other on-line resources for burial and body disposal. How difficult was it to find this information? Who do you think uses this site, and for what purpose?

3. Find information on estate and funeral planning, living wills, legal assistance, and directories concerning other advance directives. How difficult was it to locate this information? To whom is it directed? If you were in the terminal stages of a disease, would the information you have found be useful for you in developing a living will or other advance directive?

15 Politics, Policies, and Programs

New President Encounters Old Politics

Pete Golis

"Ask not what you can do for your country, but ask what your country can do for you."

—President John F. Kennedy, Inaugural (adjusted for real time)

During a Sonoma State University panel reviewing election results, a young man asked a subversive question. I was surprised the trouble-maker wasn't booed and removed from the hall. Now that the election is over, he asked, what can citizens do to help the new president be successful?

Yes, it was outrageous. Where was this young man's sense of ownership? Why hadn't someone taught him that politics is about claiming what's in it for me and sticking it to the other guy? Obviously, this young man's professors had failed him.

Meanwhile, in the town where I live, liberals were proclaiming that Democratic victories heralded the arrival of a new liberal nirvana in which "just about anything is possible."

Here, you recognize the old politics in which one faction, imagining itself the winner, rushes to claim the spoils (while people on the other side are quietly reaching for the long knives). Congratulations, Mr. President-elect, and good luck.

Here's something the new president can't say, but I can.

Note to liberals: You did not win this election, and left to your own devices, you would not have won this election. Barack Obama won this election. You can pretend all you want, but Ohio, Pennsylvania and Virginia are not like Sonoma County, and if a new Democratic president is to succeed, he needs the support of people in those states, too. [. . .] By the way, the country is broke. Maybe you didn't get the memo.

Uh-oh. Now I've done it. I've suggested activists keep one foot in the real world. This will never do.

This is Barack Obama's problem, isn't it? How does he deal with the ideological wars that have obsessed the baby boomers for all these years? He knows he will be hammered from the right, and if he gets beat up by the left as well, his presidency has no chance to succeed. [. . .]

Being very smart about his elders, Obama knows the risks. Note the word, immaturity, in this section from Tuesday night's victory speech: "Let us resist the temptation to fall back on the same partisanship and pettiness and immaturity that has poisoned our politics for so long."

Calling all adults.

Here's a question: Last week while you were crying for joy and congratulating yourself on the election of Barack Obama, did you think about the ways you could help him reunite your country, or did you think about how happy you were that the people on the other side were suffering? If you want to refuse to answer on the grounds of self-incrimination, go ahead. We all know the answer. [. . .]

OK, enough. After a couple of days, we need to be grown-ups.

The question posed by the young man at Sonoma State University goes to the heart of the matter. Obama, a president from a new generation, cannot fix the economy and put the nation back on the right track unless a majority of Americans agrees to move beyond the selfishness and recriminations that have poisoned politics over these last decades. [. . .] Obama's success depends on Americans' capacity to come together.

Older politicians, Hillary Clinton and later John McCain, called Obama naïve. In their world view, Americans are not capable of being more tolerant of each other. Politics will always be bare knuckled. Is the president-elect naïve? Over the next several months, we'll test the proposition. Political insiders will be eager to destroy his presidency. By disposition and by profession, many of them thrive on conflict.

It's a familiar pattern. Winning parties can't handle success because factions within the party want it all. It happened to the Republicans in 1992, and it happened to the Democrats in 2000, and now it's happening to the Republicans again. All I know is that if liberal Democrats can't agree to put aside the old politics, the country will continue to lose ground, Obama's presidency will fail, and in 2012 or 2016, we will elect some other ineffective president.

Source: www.pressdemocrat.com/article/20081109/OPINION03/811099971

To judge strength by size alone, the older population of the United States is stronger than ever before. Numbers command political power at the voting booth and in lobbying efforts. Yet because of fiscally conservative federal policies in the 1980s and harsh economic turns in the 1990s, the United States has a general political environment increasingly hostile to older adults or, for that matter, to any group that makes a request of local, state, or federal budgets. Though older Americans have made significant strides over the last 50 years in gaining the wherewithal to maintain healthy and meaningful lives, their progress, like that of our society in general, depends primarily on the condition of the U.S. economic system. The deep recession of 2008 brought about the need for the reassessment of all government spending as well as review of all governmental regulatory agencies. The American population remains deeply divided on the responsibility taxpayers have for fundamental national ideals: public education, health and well-being of citizens; and the protection of children, the disabled, and older people.

Proactive social policy is necessary if we are to adequately meet the needs of our citizens but it takes specific, targeted, cause-based organizational dedication to make that policy happen. Focused advocacy often comes at the expense of a broader vision—one that includes ancillary social issues or that has a long range perspective. Appropriate and adequate social policy addressing the emerging aging boomer generation, for example, needed to begin in the 1980s but did not.

The threat of cutbacks in government programs exists in the United States and most Western countries. It will grow larger in the next 20 years as the global boomers reach age 65. The Netherlands provides a clear example of threatened cutbacks: their social welfare policies provide substantial benefits to older people which citizens have come to take for granted. A conservative political environment has challenged these policies. In 1994 the Dutch spent $100 billion a year on health and Social Security costs, about 10 times what they paid for their military defense (Drozdiak, 1994). Older people fear a decrease in their standard of living as cost-cutting measures become inevitable, and they support political candidates who do not advocate the reduction of their entitlements.

We begin this chapter by briefly reviewing periods during which political activism flourished among older adults in the United States. Today's political climate can then be examined in light of these earlier decades of struggle and progress.

Early Rumblings

Political movements are the combined result of social, economic, and historical events. Circumstances develop that make political action imperative. Those who are affected respond by joining forces and asserting their need to improve their situation. This happened in the early 1900s, when a large proportion of older people were living in poverty. Although retirement had become increasingly mandatory, pensions were not yet generally available, which often left the elderly with neither work nor money. In 1921, a limited pension system was established for federal employees, but the vast majority of elders still had no pensions at all. In 1928, 65 percent of those over 65 were receiving assistance of some sort from their children. The Depression had wiped out the pensions and savings of millions, leaving elders in the most desperate poverty of any age group. Described in *Endangered Dreams* by Kevin Starr (1996):

> *With 12 million people out of work by 1933, voluntary assistance to parents plummeted along with the rest of the economy. To whom could the elderly now turn? Not to government. American culture had no discernible tradition of old-age pensions from government. Only 6 American states had any form of old-age assistance program. By 1934 a mere 180,000 elderly, out of a total population of 15 million—plus senior citizens, were receiving any form of legally mandated assistance. Yet fully 50 percent of*

> *the elderly in America were in need of some form of outside aid if they were to make it through the slump. (p. 134)*

Their plight created the social environment for the nation's first elderly uprising, which took political form as the **Townsend Plan**. Named after the retired physician, Francis E. Townsend, who headed the program, the Townsend plan was organized through local clubs. By 1935 a half million Americans had joined Townsend Clubs, and by the end of 1936, about 2.2 million people belonged (Starr, 1996). The Townsend plan proposed placing $200 of government funds per month in the hands of people aged 60 and older, with the requirements that the recipients be retired and spend the money within 30 days. The plan was to end the economic problems of elders, solve unemployment, plus introduce new spending to stimulate the economy. The plan would not increase the deficit, because it was funded by a multiple sales tax.

The plan was resoundingly criticized: the political Left argued that the transactions tax was regressive; the political Right feared the tax would cut profits and the spending would undermine the incentive to work (Amenta et al., 1992). Scholars and policy experts felt the tax implied higher levies on finished goods, and thought it would not work to register farmers for the tax. Ultimately, following internal conflict among Townsend plan leaders over charges of graft, plus vigorous opposition from the Roosevelt administration, less than 4 percent of the national electorate approved the Townsend plan (Starr, 1996). The plan was not a utopian scheme, however. It provided for the aged to have a large national pension, whose size would be determined by an earmarked tax (Amenta et al., 1992). It provided the model for old age assistance to come.

The Social Security Act, part of the New Deal legislation, adopted one of Townsend's main interests—a pension for elders. However, the Social Security Act was a more conservative measure. Rather than "giving" pensions to the retired, it paid pensions out of sums collected from workers; the amount of each pension was determined by each individual's employment record. Those who did not work were not eligible for benefits. In the decades that followed, amendments to the Social Security Act added coverage for more kinds of workers and their dependents.

Few Californians could be found in the inner circle of President Roosevelt or at the helm of New Deal agencies. While eager for a myth of individualism and self-improvement, Californians seemed incapable, the majority of them, of moving to an expression of such ambitions beyond the instant solution of an old-age pension plan. The affluence and altered attitudes of the 1950s and 1960s would reverse this orientation and make of California, briefly, the very model of the social democratic experiment. (p. 222)

KEVIN STARR, 1996

Other political movements, many of which originated in California, attempted to help the elderly during this era. Even before the Townsend plan, there had been EPIC (End Poverty in California). The EPIC movement was based on campaign promises Upton Sinclair made when he ran for governor of California in 1934. His platform of sweeping social reform was energized by utopianism, which had popular support in California at the same time. The movement's slogan was decidedly socialist in orientation: production for use, not profit.

Also from California, the **Ham and Eggs Movement** proposed that the government issue a large amount of special script, the value of which would expire at the end of a year; therefore, the elderly pensioners who received it would spend it promptly and stimulate the economy. In the late 1940s, George McLain, a leader of the Ham and Eggs movement, lobbied the California legislature and advocated his reforms through a daily radio program. He believed that all elderly were entitled to pension increases, low-cost government housing, medical aid, and elimination of a financial means test. Before any of his proposals were widely adopted, the McLain movement, which drew a membership of about 7 percent of those aged 65 and over in California, ended in 1965 with McLain's death.

A significant characteristic of the various movements in the 1930s and 1940s, including McLain's, was that they extracted great sums of money from the elderly themselves. Indeed, although the movements set a precedent for political organization, they also demonstrated how vulnerable elders could be to corrupt leaders who preyed on their trust:

> *Certainly old-age power was helped to grow because of them; however, the secretiveness, greed, and paranoia with which the advocates held power stymied grassroot growth. (Kleyman, 1974, 70).*

Although the civil rights movement began in the 1950s, this decade is often described as a time of political inactivity and apathy, an era of general economic prosperity chiefly characterized by inattention to needy groups. The early 1960s brought more fervor to the civil rights movement and President Lyndon B. Johnson declared the War on Poverty. The high incidence of poverty among some groups—racial minorities, the rural, and elders—began to attract public attention. With the advent of the antiwar movement and the hippie lifestyle, the 1960s also saw increasing political activism among students and other adults. Gay liberation and the women's movement evolved in the 1970s as civil libertarian ideals gathered support. The women's movement in particular dealt with the issue of aging by attacking job discrimination against middle-aged and older women. The National Organization for Women (NOW) established an energetic task force of older women to identify issues and organize political action.

I have come to the conclusion that politics are too serious a matter to be left to the politicians.

CHARLES DE GAULE, FRENCH GENERAL AND POLITICIAN.

The 1970s also ushered in the advocacy of rights for the disabled. These humanistic movements created an environment in which one movement could combine forces with another. In this sense, all of the social movements of the 1960s and 1970s contributed to the social movement of **Senior Power.** A number of prosenior interest groups and programs materialized in the 1960s; their chief accomplishments were the passage of the Older Americans Act [OAA] (1965) and Medicare (1966).

The early 1970s brought forth a more visible and vital seniors movement. For the first time since the Great Depression, older people, with cries of "gray power" and slogans such as "Don't Agonize—Organize," showed evidence of considerable political activism. Demonstrations, sit-ins, and sleep-ins by old people wearing "Senior Power" buttons made the headlines. The new militancy of old-age groups such as the Gray Panthers presented a surprising and exciting image for those who had once been stereotyped as powerless, dependent, slow, and unenthusiastic. The name Gray Panthers captured the new consciousness of the elderly. It suggested strength, power, and radical—if not revolutionary—political and social behavior.

Political action begins with voting. A larger proportion of older adults vote than do younger adults.

Senior Power Today

Voting

When enough older people turn out to vote on a given issue and vote as a bloc, they wield considerable power. Approximately 90 percent of Americans over age 50 are registered to vote, compared with an overall national figure of less than 75 percent. In all recent elections; older people are more likely to vote than younger ones (Hooyman & Kiyak, 2005).

Four in ten 18- to 29-year olds are not even registered to vote, which is double the proportion of 30- to 49-year olds, and nearly three times greater than those aged 50 or older (Pew Research Center, 2009). Older voter turnout (those 65 and over) is generally double the voter turnout of young adults under 25 years of age. Thus, the voting power of older people is greater than their actual numbers in the population would indicate, even though disability and lack of transportation keep a few older people from the polls.

Those who voted in their younger years continue to vote for as long as they are able to do so, so that pattern remains similar. The patterns do show, however, a steady or growing rate of voter turnout among elders. People over 65 years of age have steadily voted in increasing proportion since the 1970s, whereas turnout rates of other age groups have remained stagnant or declined. This might be because registration requirements and suffrage laws became more lenient since the 1950s and because the educational levels of the American electorate overall increased (Turner et al., 2001).

Despite large voter turnout among older people, there is no old-age voting bloc—not even on "old-age issues." In national politics, people of similar social classes are more likely to vote alike than are people of similar age groups. The wealthy older person, for example, might not support welfare measures for older people living in poverty; neither would the wealthy younger voter. Political organization and mobilization would be required to encourage elders to use their potential power by voting as a bloc. So far, however, there is no indication of intergenerational warfare, wherein younger people might fear a loss of equity brought about by "greedy geezers" (Turner et al., 2001).

Politics is the art of the possible.

OTTO VON BISMARCK, 1867

Older people are neither more nor less conservative than younger people, unlike the common stereotype that people grow more conservative with age. The older people's politics are balanced fairly equally between the Republican and Democratic parties, and they have good voting records. The 75-plus age group has been more likely to vote in the last five presidential elections than the voters under 35. When older voter turnout is low, factors such as gender, ethnic minority, and education are more important than age: only 51.5 percent of African American elders voted in the 2000 presidential election, and less than 50 percent of the older Hispanic voters regularly vote (Hooyman & Kiyak, 2005). These voting proportions have probably shifted with data on the 2008 election of Barack Obama.

Holding Office

Another measure of political power is the ability to be elected or appointed to public office. Only two U.S. presidents have entered office after the age of 65: William Henry Harrison (age 68) and Ronald Reagan (age 69). Many more have turned 65 while in office. Because life expectancy has increased greatly since George Washington's time, one might guess that presidents have entered office at increasingly older ages; however, this has not happened. Though the first four presidents were 57 or older, the nation has seen some very young presidents in the following years. The youngest was Theodore Roosevelt (age 42), followed by John F. Kennedy, who was 43 years old at his inauguration in 1961. Bill Clinton was aged 47 at his inauguration in 1993, as was Barack Obama on his inauguration in 2009.

Children, believe me: Things were not better in the old days. . . . I remember black people riding in the back of the bus. I remember summer being a season of fear because summer was when polio stalked the neighborhood, crippling and killing. I remember cars being cranked by hand and seeing men with arms broken from cranking carelessly. I remember headlines about gangsters spraying crowded streets with Tommy guns, remember hoboes at kitchen doors begging for sandwiches. . . . Just like right now, back then was chock full of meanness.

RUSSELL BAKER, 1996

Members of Congress, because they are elected officials, are not subject to mandatory retirement laws. Some members have served into their 80s and, occasionally, into their 90s. Since 1974, the average age has risen to over 50; in 1996, many congressional members were over age 65. The seniority system in Congress allows older members who have served many years to wield the most power, often as heads of important committees.

Cabinet members and ambassadors tend to be older, perhaps because they are appointed by the president rather than elected. These positions, which imply years of experience, are often filled by men and women who, in any other area of work, might have been forced by age alone to retire. Supreme Court justices also tend to be older. There is no mandatory retirement for them, and they generally serve as long as their health allows. In 1991, the controversial appointment of Clarence Thomas at age

43 made him the youngest justice to serve on the Court. In 2009 the ages of the justices ranged from John Paul Stevens at age 89 to John G. Roberts at age 54. The newest appointee, Sonia Sotomayor, was 55.

Despite the representation of older people in elected and appointed positions, we do not seem to be headed toward a **gerontocracy**, or rule by the old. Both young and old are involved in the political arena. Although it has long been acceptable for members of Congress to be middle aged or older, a political youthful image has been favorable since the time of Teddy Roosevelt. In addition, even among older politicians age does not appear to be a major variable affecting their position on political issues. A presidential candidate or Congress member does not support "old age" measures simply because he or she is near age 65. The support that members of Congress give to issues reflects not only their political ideology but also generally the wishes of their constituents and party affiliations. Supreme Court justices make decisions based on philosophy and ideology; impartiality dictates that age not be an influencing variable. Older politicians and other older people involved in the political process provide good role models and typically work hard at their jobs.

Political Associations

Interest groups representing older Americans have increased in number and political effectiveness over the past several decades.

One type of specialized interest group is the trade association. Trade associations represent specific concerns and lobby to achieve their purposes and goals. Some examples are the American Association of Homes for the Aging, the American Nursing Home Association, the National Council of Health Care Services, and the National Association of State Units on the Aging. There are also professional associations, such as the Gerontological Society, the Association for Gerontology in Higher Education, and the American Society of Aging, composed primarily of academicians in the field of aging. States and regions also have professional associations, such as the California Coalition for Gerontology and Geriatrics. Interest groups, such as the National Association of Retired Federal Employees (NARFE) and the National Retired Teachers Association (NRTA), are composed of retired persons.

A few organizations representing older persons have enjoyed special growth, success, and media attention. Here we consider six of them: the American Association of Retired Persons, the National Committee to Preserve Social Security and Medicare, the National Council of Senior Citizens, the National Council on the Aging, the National Caucus and Center on the Black Aged, and the Gray Panthers.

American Association of Retired Persons

Currently the premier lobbyist for causes primarily impacting older people in the United States is the American Association of Retired Persons (AARP), with 37 million members over 50, and two-thirds of them are over 65 (American Association of Retired Persons [AARP], 2006). This nonprofit, nonpartisan organization is dedicated to helping people 50+ to have "independence, choice and control in ways that are beneficial and affordable to them and society as a whole (AARP, 2006)." The AARP Mission Statement states "we seek through education, advocacy, and service to enhance the quality of life for all by promoting independence, dignity, and purpose" (AARP, 2002).

The group's founder, Ethel Percy Andrus, was an educator—the first woman in California to become a high school principal. She founded the NRTA in 1947 when she was forced to retire from her position after reaching the age of 65. She was 76 in 1958 when she created AARP as a parallel organization to bring benefits of the NRTA to older people who had not been teachers. The organization's cofounder was Leonard Davis, a young insurance agent who developed the group health insurance plan for NRTA members. It has been suggested that Davis financed the creation of AARP to expand his own mail-order health insurance market. Until being forced out in the early 1980s, Davis was in control of AARP, "operating as a sales network to hawk very high-priced insurance and a host of other ... products to old people" (Morris, 1996, p. 10). He left the organization a very wealthy man.

Membership makes the AARP one of the largest voluntary organizations represented in Washington and therefore one of the nation's largest lobbying groups. The group, which is nonpartisan, has directed its political efforts at improving pensions, opposing mandatory retirement, and improving Social Security benefits—widely shared goals that unite the association's members. Their congressional lobbyists support causes for poor elders, and their magazine *Modern Maturity* stresses an image of well-to-do, healthy aging.

The following are included in recent contributions for its membership:

- Successful leadership in efforts to preserve social security for future generations.

- Educational campaign regarding the Medicare Part D benefit, ensuring that millions benefit.

- Publication of Watchdog reports to keep the spotlight on rising drug prices.

- Lobbying efforts for state and federal laws to allow safe and legal prescription drugs to be imported from Canada.

- Through the AARP Foundation, leadership in a funding effort to provide disaster relief to older victims of the 2005 Gulf Coast hurricanes; more than 41 grants totaling almost $1.6 million were issued to local organizations in the affected areas to provide food, shelter, health services, housing, and legal assistance.

- AARP Services, Inc., an affiliated entity offers a range of products and services such as Medicare supplement, long-term care, and automobile/homeowners and life insurance. Additionally manages services providing member discounts and savings on prescription drugs, eye-health services, and eyewear products.

The National Committee to Preserve Social Security and Medicare

Founded by James Roosevelt, the son of President Franklin Roosevelt, in 1983, the National Committee to Preserve Social Security and Medicare (NCPSSM) is concerned about the solvency of the Social Security Trust Funds. Its mission is "to protect, preserve, promote, and ensure the financial security, health, and the well-being of current and future generations of maturing Americans (NCPSSM, 2009). Its membership is currently more than six million. Its staff of professional lobbyists works to protect and improve Social Security and Medicare benefits.

Services provided by the organization include:

- Lobbying on behalf of older adult benefits
- Email news and updates: Legislative Alerts about developments in Washington
- Monthly online newsletter featuring news, information, and updates on issues related to Social Security and Medicare

Both the NCPSSM and AARP have been instrumental in educating their members on the status of health care reform legislation. Both organizations have a large lobbying presence in Washington.

National Council of Senior Citizens

The National Council of Senior Citizens (NCSC) originated in 1961 as a pressure group to support the enactment of Medicare legislation. Its leadership has always come from labor unions, mainly the AFL-CIO and other industrial unions. From a small, highly specialized group, the NCSC has grown to include a general membership of about one million. It is an organization of autonomous senior citizens clubs, associations, councils, and other groups. Their aims are to support Medicare, increases in Social Security, reductions in health costs, and increased social programs for seniors.

The organization's stated goals are broad and straightforward: "We work with, persuade, push, convince, testify, petition, and urge Congress, the Administration, and government agencies to get things done on behalf of the aged." The organization, located in Washington, D.C., lobbies for Social Security revision, a national health insurance program, higher health and safety standards in nursing homes, and adequate housing and jobs for elders. Their Mission Statement is: "To enhance the lives of older adults through education, recreation, nutrition and social services in welcoming community settings (NCSC, 2009)."

National Council on the Aging

Founded in 1950, the National Council on the Aging, a confederation of social welfare agencies and professionals in the field of aging, has been at the forefront of advocacy, policy, and program development on all issues affecting the quality of life for older Americans. The organization, which has its headquarters in Washington, D.C., has regional offices in New York and California. The council publishes books, articles, and journals on aging; sponsors seminars and conferences; and funds eight special-interest groups.

National Association of Older Worker Employment Services (NAOWES)

National Center on Rural Aging (NCRA)

National Institute on Adult Day Care (NIAD)

Older adults have experience with civic issues and are more likely to have the time to commit to civic and political organizations than are younger adults. Activism at all ages, however, is important.

National Institute on Community-Based Long-Term Care (NICLC)

National Institute of Senior Centers (NISC)

National Institute of Senior Housing (NISH)

National Voluntary Organizations for Independent Living for the Aging (NVOILA)

The Health Promotion Institute (HPI)

The National Caucus and Center on Black Aged

Formed in 1980 when two groups—the National Caucus on Black Aged and the National Center on Black Aged—joined, and the National Caucus and Center on Black Aged (NCCBA) attempts to improve the quality and length of life for senior African Americans.

The NCCBA believes that problems can be resolved through effective and concentrated political action on behalf of and by older African Americans. It was initially formed as an ad hoc group of African American and white professionals who shared a concern for older African Americans. NCCBA activities have been largely national, focusing on employment, health, and housing. Their mission is to continue to address equality and access issues in these areas.

The five groups discussed so far are neither revolutionary nor radical. Rather than advocating a redistribution of wealth, they work to improve the status of the elders by ensuring that more federal money is channeled to them and by being a voice in Washington for their members. The organizations have, for example, successfully lobbied the Department of Health and Human Services, the Department of Education, the Office of Economic Opportunity, the Administration on Aging, and other government agencies to increase resources for the older population. All currently focus on health care reform, with a particular focus on health care for older people.

The Gray Panthers

The Gray Panthers is a political group that identifies itself as radical nonviolent organization. It is a loosely organized intergenerational, multiissue organization "working to create a society that puts the needs of people over profit, responsibility over power and democracy over institutions" (Gray Panthers, 2009). Its e-newsletter informs people of health care reform issues, threatened shifts in Social Security policy, and other issues concerning social and economic justice. Through it gained an accredited nongovernmental organization (NGO) status with the United Nations, it communicates with other activists throughout the world with shared interests in democracy, peace, and social justice.

The late Maggie Kuhn, a dynamic woman who was unwillingly retired at age 65, founded the movement in her search for new ways to constructively use her energies. She and the group's other members describe themselves as a consciousness-raising activist group, drawn together by deeply felt concerns for human liberation and social change. Their goals are to bring dignity to old age, eliminate poverty and mandatory retirement, reform pension systems, and develop a new public consciousness of the aging person's potential.

Describing themselves as a movement rather than an organization, the Gray Panthers have no formal membership requirements and collect no membership dues. They have opposed the national budget deficit, the nuclear arms race, and cuts in Social Security. Currently, the group advocates a national health insurance program.

Gray Panthers also defend the rights of gays, lesbians, and people with HIV/AIDS. They are opposed to violence and abuse of any kind and draw special attention to the presence of hate crimes in our society. They also sponsor jointly with other organizations an international "People's Summit for Peace" and work for world peace in any way they can.

Researchers offer divergent views on the political impact of the "gray lobby." In some areas they are organized and strong. The AARP, along with the other organizations, for example, has been effective in staving off large cuts in Medicare. Reductions in Social Security benefits proposed by the Clinton administration were abandoned as politically dangerous because of the "gray lobby" groups (Heyser, 1996). And because of senior power, higher taxes on Social Security benefits are also viewed as a politically risky move by politicians. On the other hand, the state of the economy and the growing number of elders, along with a shrinking workforce, are not factors that the "gray lobby" can control.

Assessing the impact of the senior movement is difficult because it is so complex—representing, as it does, the vast diversity of that collection of Americans who happen to be 65 years old or older.

The Older Americans Act and Other Programs

The **Older Americans Act (OAA)** is designed to alter state and local priorities to ensure that older people's needs are represented in social services allocations. It was created to form a national network for comprehensive planning and delivery of aging services. Funding for the OAA comes up for Congressional renewal every four years; therefore, the U.S. political and economic climate is a major determinant of OAA's level of funding.

The OAA was adopted by Congress in 1965, a period very favorable to programs for the aging. (Medicare was enacted during the same general time period.) Whereas the 1960s and 1970s saw budget increases and expanded programs under the OAA, the 1990s brought decreases in funding. Since participation rates in many OAA services have been highest among middle-income older persons, cuts in funding alongside proposals for cost sharing have been introduced in Congress. The major portion of OAA funds is devoted to congregate meals, senior centers, and transportation programs.

The OAA contains six major sections, or titles, which outline the intentions and objectives of the act. Title II establishes the Administration on Aging (AOA) in Washington, D.C., the organization that administers the programs and services that the act mandates. Title III provides for the distribution of money to establish state and community agencies—Area Agencies on Aging (AAAs). These agencies disseminate information about available social services and are responsible for planning and coordinating such services for the elderly in local settings. A local community that has a council or center on aging is probably receiving federal funds through Title III. Title IV, its funding slashed in the 1980s, and not increased in the 1990s, provides funds for training people to work in the field of aging as well as research and education on aging. Other titles provide funds for multipurpose senior centers to serve as community focal points for the development and delivery of social services.

The specific titles of the act are intended to achieve the following objectives for older citizens, as spelled out in Title I: (1) adequate income; (2) good physical and mental health; (3) suitable housing at reasonable cost; (4) full restorative services for those who require institutional care; (5) equal employment opportunities; (6) retirement with dignity; (7) a meaningful existence; (8) efficient, coordinated community services; (9) benefits of knowledge from research; and (10) freedom, independence, and individual initiative in the planning and management of one's life.

The act's major programs are developed to address these goals at local levels, with guidance from state units and the federal AOA. Funding comes partially from the federal government and partially from state and local governments and charity organizations. Some of the programs funded and administered through this system are discussed next.

Meals and Nutrition

The success of pilot projects in the 1960s resulted in federal funding for home-delivered meals and, in 1972, funding for congregate, or group, dining projects. Under the OAA, funds are available to all states to deliver meals to the homebound elderly or to serve meals in congregate sites. Meals on Wheels, for example, delivers meals to the homebound at a minimal charge. Congregate dining is aimed at getting the elderly out of their homes and into a friendly environment where they can socialize with other people as well as enjoy a hot meal.

Meals and activities take place in churches, schools, restaurants, or senior centers. Besides serving hot meals, some agencies provide transportation to and from the dining site, information and referral services, nutrition education, health and welfare counseling, shopping assistance, and recreational activities. The act establishing the congregate dining program specifies that participants not be asked their income; they may make a contribution if they wish, but there is no set charge.

Friendly Visiting and Telephone Reassurance

The OAA provides limited funding for volunteer community services such as friendly visitor and telephone reassurance programs. Because of decreasing funds, however, local staffs have had to become increasingly adept at fund-raising to keep programs afloat.

The Friendly Visitor Program has improved the quality of life for many people. Visitors are volunteers of all ages who have an interest in developing a friendly relationship with older persons. They are matched with elderly persons in their community on the basis of such things as common interests or location. In some cases, the Friendly Visitor has proved to be a lifeline. The program has a small paid staff that organizes and instructs volunteers.

Telephone reassurance programs have staff, either paid or volunteer, to check on older persons daily by phone at a given time. For the homebound, this call may be an uplifting and meaningful part of the day. If the phone is not answered, a staff member immediately goes in person to check on the elder's welfare.

Employment

The Title V employment program is small but is one of the most politically popular programs. The major program, the Senior Community Service Employment Program (SCSEP), which provides community service work opportunities for unemployed low-income persons aged 55 and over, employs 62,500 older applicants in part-time minimum-wage community service jobs throughout the United States.

A related program is Green Thumb. Men and women aged 55 and older who live in rural areas may work up to 20 hours per week at minimum wage in public projects that include landscaping, horticulture, and highway maintenance.

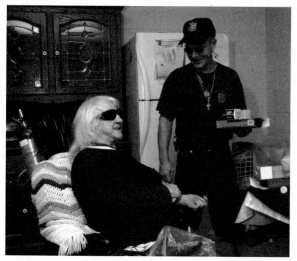

The delivery of Meals on Wheels to this woman makes a huge difference in her quality of life. It likewise provides opportunity for the volunteer (also a retiree) to engage in meaningful work.

Adult Day Centers

The OAA partially funds multipurpose day programs to provide services on an outpatient basis to the many older people who need services on a daily or weekly basis but do not need 24-hour institutional care. Older people may come to such centers for health care, meals, social activities, physical exercise, or rehabilitation. Practitioners from the centers may, on occasion, deliver home services. These centers allow people to enjoy the advantages of both institutional care and home living.

Other Services

Local agencies on aging may offer legal services to low-income elders for civil matters such as landlord disputes; other agencies may offer home health services as well. Still others may offer housing services such as counseling on housing options, home equity conversion information, and assistance in locating affordable housing. Information and referral is a hallmark of agencies on aging.

Considering the budget constraints within which the AoA must work, fulfilling all its goals is largely impossible. Though the limited funds cannot reach far enough to include those above and below the poverty line, many elders living in modest if not meager circumstances do meet eligibility requirements.

Two Programs Not Funded by the OAA

Certainly not all successful programs for older persons have been funded by the OAA. Some have sprung up locally and are funded by local tax dollars. Others are funded by private donations or charity organizations or sponsored by churches or various federal or state government agencies separate from the AOA.

Exploritis (Formerly Elderhostel)

Exploritis is the new name for the programs of Elderhostel. It is a not-for-profit organization founded in 1975 to provide lifelong learning opportunities for adults.

It's such fun to watch a group of strangers meet, share experiences, bond, and become good friends.

IMOGENE K

Elderhostel was founded by Marty Knowlton—a social activist, traveler, and former educator—and David Bianco—an organized university administrator. In the early 1970s Knowlton returned from a four-year walking tour of Europe, having stayed in youth hostels and experiencing inexpensive, safe lodgings and opportunities to meet other people. He had also experienced the Scandinavian institutions called folk schools, in which older adults instructed younger generations on traditional music and crafts. Why, he questioned, should older Americans retreat quietly into their "golden age" rather than continuing to challenge their minds with affordable travel and education?

In 1975 five colleges and universities in New Hampshire sponsored the first Elderhostel programs to 220 participants. By 2000, more than 30,000 people participated in programs in all 50 states and throughout Canada (Elderhostel, 2009).

In 2004 the program launched Road Scholar, a series of adventurous, experiential learning programs, and in 2006 it created a diversity outreach program to attract African Americans to its lifelong learning experiences. In 2008 Elderhostel acquired a travel partner, Lyon Travel, now called Exploritas Travel Services. This allowed the organization to offer a more full range of services and a more seamless customer/student experience.

Elderhostel was renamed Exploritas in 2009 to focus on lifelong-learning goals to "explore" the world. From its Boston headquarters, the program directs a network of several thousand high schools, universities, national parks, environmental education centers, and many other educational institutions throughout the United States and Canada and in more than 100 overseas locations. Noncredit courses with no exams, no grades, and no required homework are offered in multitudinous subjects; no-cost extracurricular activities are also provided. Scholarships are available, and students are charged a reasonable flat fee for class, room, and board (Elderhostel, 2009).

Foster Grandparent Program

This program pairs low-income older adults with troubled, disabled, or hospitalized children. Foster Grandparents volunteer from 15 to 40 hours every week, providing kindness and attention at hospitals, homeless shelters, and special care facilities. They serve as a role model, mentor, and friend to disadvantaged children (Senior Corps, 2009). Funded through participating states, ACTION (the Federal Domestic Volunteer Agency), and the Office of National and Community Service, the Foster Grandparent Program (FGP) allows low-income citizens of age 60 or older to use their time and talents to provide love, care, and attention to disadvantaged youngsters.

Begun in 1965, the FGP was the first federally sponsored program to offer challenging activity to retirees. It provides opportunities for close relationships to two different groups: low-income elders who want to feel needed and who want to participate in and contribute to the lives of institutionalized children and/or children with special or exceptional needs.

Volunteer grandparents receive 40 hours of orientation in the specific area of their choice. In most cases, they work four hours a day, five days a week, assisting youngsters in physical or speech therapy or with their homework. Their main task is to provide for the emotional, mental, and physical well-being of children by affording them intimate and continuing relations through uncomplicated activities such as going for a walk or reading stories, or within family intervention structures. Foster grandparents receive a nontaxable stipend of an amount per hour that is approximately minimum wage, transportation if needed, a meal each day that they work, accident and liability insurance coverage, and an annual physical exam. Nonstipended volunteers may enroll in the program under certain conditions. Many older people work in this program for years both because they need the extra income and, more important, because their "grandchildren" love them.

Activism and Advocacy

Activism connotes more than simply voting or joining an organization. The term *political activist* brings to mind an individual who is politically committed to a cause and who uses a variety of means to push for social change.

Older adults and people interested in their causes participate in community activities by serving on boards of trustees, advisory councils, and committees in all kinds of social service agencies. Interested people may get involved in police/community relations, join voluntary organizations, and enter local politics as either campaign workers or as candidates. They may lobby to maintain or increase funding for federal programs, including those of the OAA.

Grandparenting, whether of biological grandchildren or "foster" grandchildren, is enormously satisfying for many older adults.

Legal activity is another area for activism. Class-action lawsuits can be brought against organizations and institutions that discriminate against or do not properly serve elders. Voter registration drives are important. They ensure that older persons can support those who support Social Security, Medicaid, home-delivered meals, public housing, employment programs, improved transportation, health benefits, and dozens of other programs for the elderly. Grievance procedures can be initiated to aid complaints against Social Security or welfare departments.

Collective activity is a good tactic for the activist who joins with others in membership drives, marches, and demonstrations. That activism can be high profile, such as a Gray Panthers-sponsored march, or it can be on a far less media-oriented scale. In either case, it is taking the responsibility to put values into action, and the result can be profoundly validating.

Resistance activity can be an effective form of social activity. It includes boycotts, rent strikes, wheelchair sit-ins, and voting against bond issues. Designated individuals may act as watchdogs on agencies that serve the elderly. In any type of organized activity, the media can provide an important tool in airing the interests and concerns of elders.

Advocacy is a term widely used in the field of aging, perhaps because of numerous individuals and organizations doing just that. Advocacy involves using one's resources and power for the benefit of a special-interest group, such as the elderly. Stereotypes of the 1950s and 1960s portrayed elders as dependent and passive, in need of someone to advocate for them. The 1980s and 1990s changed that idea. Gray Panthers were in the forefront of creating a stronger, more forceful image for elders. It is quite clear that older adults are capable of advocating for themselves; nonetheless, it takes a village, as the saying tells us. Advocacy combines the efforts of individuals, regardless of age, who desire to promote a more satisfying later life for all Americans.

The Equity Issue

Generational Equity is a controversial issue that was first popularized by the media in the mid-1980s. Critics of programs for older adults, fearing that the older generation is advancing at the expense of the younger one, have described older people as demanding programs targeting only their age group, at the expense of children and children's programs. An example of the argument in action is the criticism of a politician who supports Meals-on-Wheels programs for older adults, but votes to cut federal nutrition programs for school lunches. The reason? "The elderly vote, and children don't" goes the usual reply. Perhaps so, perhaps not. The issue of social programs for special-needs groups is far too complex to be dealt with simplistically. From a national policy perspective, decisions on programmatic support must be made in the context of the total net of services available, as well as the need for those services. Sloganism brought to the public by a media searching for the 20-second sound bite does not provide reasoning adequate to make decisions that will affect millions of American citizens, whatever their ages.

Those who advocate generational equity believe that younger generations will suffer because of the older population's unprecedented size and affluence. Around the country suburbs have a growing number of older residents. Older people have no direct stake in the public school systems, and the constituency for supporting schools and other youth services is shrinking proportionally as the nation ages. On the other hand, it is the grandchildren of those older adults who are being educated by a public school system, supported by tax dollars. It is complex and noncategorical.

The idea that increased affluence for older people leads to decreased resources for children is unsound. Social Security coverage that increases with cost of living has dramatically improved the financial status of older people. It has made huge steps to reducing poverty rates among older people, and the essentially small cost of living increases are not the reason for poverty among children and other younger generations. Nor is the increased cost of medical care for elders. If, as some suggest, Medicare premiums tripled in cost for the well-to-do and expensive surgeries were rationed for the very old, women and children would still be poor in the United States. The growing number of older people is hardly the major reason for spiraling health care costs. It is one contributing factor in a complex economic system. The basic reason for the poverty of children is their parents' unemployment or underemployment and the increasing number of female-headed families with low incomes. To protect children, our economic house must be in order.

Economic prosperity and public support for education are strongly associated. American industry must have workers capable of making informed decisions, learning new tasks, and following complex instructions. A system of public education in the United States has provided such competent workers in the past and must be strong enough to do so for the future. Blaming either Medicare and Social Security, or blaming poor parents, does not address issues necessary for national-level, workable, effective policy for caring for America's citizens.

In 1950 Social Security was amended to establish a program of aid to permanently and totally disabled persons and to broaden aid to dependent children to include a relative with whom the child lives. Social Security actually provides benefits to various age groups, and every worker can rest easier knowing that, despite recent doubts about the program's continued solvency, he or she will likely be covered in retirement. A working father or mother who has paid into Social Security can know that, should he or she die, the children will be provided for to some extent by Social Security. Social Security is a retirement *and disability* program—in reality, an intergenerational safety net to reduce poverty.

Public policy based on the competition to "get" limited resources will ultimately be unworkable because it will not meet the needs of those who require assistance, and it flies in the face of the deeply held value of interdependence across generations. Rather than competing for human services dollars—dividing the piece of pie into yet smaller sections—politicians and citizens need to be reformulating the way in which the pie is sliced.

The costs of providing services to an aging nation are mighty, however. The retirement and disability programs today are financed by a payroll tax of approximately 12.4 percent of covered wages and salaries (van Blaricom, 2009). This is a bigger tax burden on most households than personal income tax. Additionally, Medicare plus the means-tested Medicaid benefits for the aged costs as much as the Social Security pensions. Together these programs cost about 10 percent of the Gross Domestic Product. What must a nation do to address the dramatic shifts in the age of its population? With advances in medical science and improvements in life styles, it is possible that people will live even longer than current projections predict. The "contract between generations" is the relationship of expectations between older and younger age groups in a context of changing resources and values. The relationship goes beyond parents and children to that of generations in a nation: who owes what to whom and how, and when is it transferred. The exchanges, or **intergenerational transfers,** are based on the question, "To what extent do we hold older generations accountable for the effect of current policies on future generations?" At the beginning of the twenty-first century, we might take a lesson from ancient Native American philosophy: the consequences of all actions must be considered through the seventh generation. In what ways do my actions (my political perspective) affect those generations surrounding me as well as future generations?

The "new" old age is different than it once was. Later life is now a time of vigor and exploration, regardless of where that exploration takes place—within the family or in the larger world.

Epilogue: A New Generation of Older Adults

What does the future hold for older individuals? In some respects, the future looks bright. Elders are living longer and healthier lives, and they are becoming more visible not only in numbers but also in political activism. They have required, and to some degree have received, a bigger slice of that pie than did their parents, who were not as long-lived. As a result of dollars put into health, education, nutrition, and socialization for elders, the quality of life for the average older person has dramatically improved over that of several decades past.

On the other hand, the future is uncertain. The line between paying for programs for the well-being of older people and seeing thousands of elders live in poverty, as many once did, is a fine one. The nation is faced with a stark no-fault reality: people are not growing old because they are mean-hearted or self-centered. Longevity has increased so suddenly in terms of human history that those who are currently living into very old age are at the cutting edge. They have few role models, and they did not "plan" to live to be 80, 90, and older. The questions are of the nation itself: Where do national values lie in relation to responsibility to provide a safety net for the country's older adults now and in the future? What services should be provided; who should be targeted to receive those services; and what cost, if any, should participants pay for those services?

The 2006 reauthorization of the OAA provided an opportunity (and requirement) for advocates and policymakers to debate the role and future of OAA—essentially, to develop a plan for the future of services to elders (OAA, 2008). The amendments within the reauthorization broadened the role of the Agency on Aging (AoA) related to elder justice, elder abuse, and mental health issues. They update and strengthen the role and functions of AoA and the Assistant Secretary for Aging, address coordination of Medicare and Medicaid programs with federal benefits, and coordinate volunteer and civic engagement activities for all ages in supportive services and community capacity-building initiatives (OAA, 2008).

No public service program is without criticism. The image of a large administrative overhead looms for many people. The choices are not easy, the task is enormous, and the outcomes are monumental for the lives of many older adults.

The most exciting part of the future, however, is that there now is a new generation of older people. First, their parent generation will have given them the experience of extended longevity. They will know so much more about living to be 90, and more of them will *expect* to do so, than did their parents. They will have made plans, even if those plans are psychological and not practical. The gift of the previous generation is an understanding that the *possible* is likely to be the *probable*. The new generation of elders will have a better understanding of the relationship between lifestyle and health, and they will have lived in greater affluence than did their parents and have had greater opportunity to make lifestyle choices. They will bring with them the ability to redefine the timing of life course transactions. They just might choose to work (or introduce the *possibility*) until they are 55, then work for five years, take two years off, and repeat the cycle until they become completely unable to do so. They will bring with them different attitudes toward marriage, responsibility to the community, and responsibility for personal growth.

The future is a very exciting place. The present responsibility is to understand the ways in which an aging nation and globe will shape that future, economics will shape that future, and both will combine to shape the cultural environment of the nation.

Chapter Summary

Judging by numbers, the political power of elders is stronger than ever before. However, numbers do not tell the whole story. The political environment for elders in the 1990s was rather harsh because of the recession and budget deficits.

In earlier times, there have been examples of social movements by and for elders. Two examples are the Townsend plan and Ham and Eggs movement in the 1930s. The activism of the 1970s symbolized by the Gray Panthers is a third example.

Today there are a number of indicators that older people have reasonable political clout. They vote in larger numbers than do younger persons. We have recently seen two presidents (Reagan and Bush) over 65, and the seniority system in Congress insures power to elder members. A number of political associations are strong—the leading one being the American Association of Retired Persons. The Older Americans Act of 2006 authorizes increased taxpayer spending on older adults. To some extent this compensates for funding that did *not* increase in the 1990s and the first six years of 2000. The major funding presently goes to congregate meals, senior centers, and transportation programs. Only if there is effective advocacy and activism can these programs remain in place to meet the needs of an increasing number of older people in need.

The dramatic recession of 2008 has as yet an uncertain outcome for middle-aged and younger people's later life well-being. All that is certain at this point is that many life plans are changing in the face of high unemployment and economic uncertainty. The future will definitely be interesting: it will differ from the past as it always does; it will catch some people unprepared for change, as it always does; and with adequate interest, concern, and public discussion of options, it can be made more effective to support later life well-being in the years to come.

Key Terms

advocacy
aging network
American Association of Retired Persons
Elderhostel
Foster Grandparents Program
generational equity
Generations United
Gray Panthers
Ham and Eggs movement
intergenerational transfers

National Caucus and Center on Black Aged
National Committee to Preserve Social Security and Medicare
National Council of Aging
National Council of Senior Citizens
Older Americans Act
senior power
Townsend plan

Questions for Discussion

1. If you were born 100 years ago, what services would have been available to your parents when they turned 65? How would they have supported themselves if they needed to work at age 60, but no one would hire an "old" person?

2. Where could a childless elderly couple turn to for medical help in the 1930s if they could not afford a doctor? What has changed today?

3. What do you predict will be the future social movements for seniors? Why? How politically active do you expect these seniors to be?

4. Do you expect to be politically active as an elder? Why or why not?

Fieldwork Suggestions

1. Contact three different political action groups that lobby for seniors. Find out what they do and what legislation they are currently working to pass.

2. Interview the head of the AARP in your area. How politically active is this person? Do you think that the AARP will represent your best interests when you are eligible to join? If you are eligible to join, why or why not are you a member? How well informed are the people you know about the AARP?

3. Interview an old person who considers himself or herself to be an activist or radical or to have been one at an earlier point in their life. What were the issues that brought them into activism? What are the issues they have addressed since?

Internet Activities

1. Multidisciplinary resource centers of gerontology are funded by the Administration on Aging. They include the Institute of Gerontology, University of Michigan; Wayne State University; Andrus Gerontology at University of Southern California; and the Center for Aging and Human Development at Duke University. Updates on the status of the centers are posted on the Community Resources for Older Adults web page. Locate this page and develop a summary of the current status of resource centers supported by AOA funding. (http://www.hhs.unco.edu/geron.htm)

2. Locate the American Association of Retired Persons (AARP) home page and create a list of resources available from that website.

3. Using the resource list at the back of the text, locate an appropriate agency or interest group and write an analysis of (1) the description and content of the site, and (2) your evaluation of its intent and purpose.

INTERNET INFORMATION RESOURCES

Administration on Aging
http://www.aoa.dhhs.gov/
Extensive links to other national organizations and programs; statistics on aging population.

Administration on Aging
Eldercare Search
http://www.eldercare.gov/Eldercare/Public/Home.asp
The Eldercare Locator connects older Americans and their caregivers with sources of information on senior services.

Administration on Aging
Elder Rights
http://www.aoa.gov/eldfam /Elder_Rights/Elder_Rights.asp
Elder rights and resources.

Administration on Aging
Family Caregiver Options
http://www.aoa.gov/eldfam /For_Caregivers/For_Caregivers.asp
Good place for information about respite and caregiver support services. Updated frequently.

Aging Network Services
http://www.agingnets.com/
A guide for children who are coping with their aging parents.

Aging in the Media
http://caregiver.org/caregiver/jsp/publications

Alliance of Information and Referral Systems (AIRS)
http://www.airs.org/
More than 1,000 information and service providers.

Alzheimer's Association
http://www.alz.org/
Provides details on Alzheimer's research and support for patients and their families.

Alzheimer's Association
Treatment Options for Alzheimer's Disease
http://www.alz.org/AboutAD/Treatments.asp
Latest research on Alzheimer's Disease.

Alzheimer's Association
What Is Alzheimer's Disease?
http://www.alz.org/AboutAD/WhatIsAD.asp
Information on Alzheimer's Disease, with fact sheets and resource lists.

American Association of Retired Persons (AARP)
http://www.aarp.org/internetresources/
Guide to Internet Resources; includes alphabetical list of sites.

American Association of Suicidology
www.cyberpsych.org
Professional and caregiver resources, death.

Association for Death Education and Counseling
www.adec. org
Professional and caregiver resources, death.

BenefitsLink
http://benefitslink.com/index.html
Links to benefit-plan information.

Bereavement resources
http://www.funeral.net/info/brvres
http://www.rivendell.org
http://www.cmhc.com
Assorted resources for Information on bereavement.

Canada Pension Plan (CPP)
http://www.sdc.gc.ca/en/isp/cpp/cpptoc.shtml
Compare Canada's retirement pension program with the U.S. Social Security program.

Canadian Institute for Health Information
http://secure.cihi.ca/cihiweb/products
Caring for Nursing Home Residents with Behavioral Symptoms. March 2008.

Caregiver's Handbook
http://www.sdc.gc.ca/en/isp/cpp/cpptoc.shtml
Complete caregiver's handbook; also guide to choosing a residential facility.

East Bay (California) Elder Abuse Prevention
http://www.elderabuseprevention-eastbay.org/
Definition of elder abuse, how to recognize it, who must report it, and where to report it.

ElderWeb
http://www.elderweb.com/home/
Information for caregivers and older adults.

Family Caregiver Alliance
http://www.caregiver.org/caregiver/jsp/home.jsp
Assistance to caregivers of persons with Alzheimer's,
Parkinson's, and other brain disorders.

FedStats
http://www.fedstats.gov/
Charts and graphs of population projections.

FindLaw
Living Wills
http://estate.findlaw.com /estate-planning/living-wills.html
Information on living wills, legal assistance, and directories
and associations on advance directives.

Florida Health Stat
http://www.floridahealthstat.com/fit_nh.shtml
Tips on finding a nursing home; good resource even for
non-Floridians.

Funeral and memorial societies of America
www.ubiweb.c.champlain.edu/famsa.directory
www.cremation.org
www.interlog.com/cemetery
www.rights.orga/deathnet/open.html

Gay and Lesbian Retirement Communities
http://www.ourtownvillages.com/ or
http://www.gaylesbianretiring.org/
Information regarding retirement communities for
gay adults.

Gerontological Society of America
http://www.geron.org
Organization maintains a national/international calendar.

The Gerontologist
http://www.cpr.maxwell.syr.edu/gerontologist/ira/
Internet resources on aging. Includes a guide to other
resources.

GriefNet
http://www.griefnet.org/
Resources related to death, dying, bereavement, and major
emotional and physical losses.

Hospice Foundation of America
http://www.hospicefoundation.org/
A resource for locating a hospice organization, with
related links.

Institute of Gerontology
http://www.iog.wayne.edu/
Wayne State University resource for researchers, educators,
and practitioners.

Kairos Support for HIV/Aids caregivers
http://www.the-park.com/kairos
Professional and caregiver resources, death.

Last Acts
www.lastacts.org
Professional and caregiver resources on death.

Mental Health Net Directory
http://www.mentalhealth.net/
For locating a mental health professional in your
community.

Mental Help Net
http://mentalhelp.net/
Help for individuals on a variety of mental health topics.

Mid-America Congress on Aging
http://www.lgrossman.com/macareso.htm
Internet resources for the aging; many links to associations
and support services.

National Association for Home Care and Hospice
http://www.nahc.org/
Information about state organizations; consumer and
legislative information.

National Center on Elder Abuse
http://www.elderabusecenter.org/
Information and statistics about elder abuse; links and
publications listing.

National Coalition for the Homeless (NCH)
http://www.nationalhomeless.org/
A national advocacy network of homeless persons,
activists, and service providers.

National Coalition for Homeless Veterans
http://www.nchv.org/
Resources to governmental, legislative, advocacy, and
political issues. With guides and brochures for veterans.

National Institute on Aging
http://www.nia.nih.gov
www.agingcenters.org
Information on aging research, clinical trials, grants, and
scientific resources; demography.

National Senior Citizens Law Center
http://www.nsclc.org/
Information on Social Security, Supplemental Security
Income, Medicare, Medicaid, age discrimination and
mandatory retirement, and the Older Americans Act.

Older Americans publication
http://www.agingstats.gov
Latest data on 38 key indicators to portray lives of older Americans.

Profile of Older Americans
www.aoa.gov/prof/Statistics/profile
Statistics on the aging population.

Project on Death in America
www.sworos.org/death.html
Professional and caregiver resources, death.

Resources for Aging
http://www.ageinfo.com/
Seeks to empower the growing number of informal caregivers to be more effective and efficient in managing the care process.

Resources for Caregivers
http://www.stls.org/caregiver/DevelopmentalDisabilities/CGLinks_1.htm
Central listing on personal and professional caregiving.

SAMHSA's Mental Health Information Resources
http://www.mentalhealth.samhsa.gov/
Substance Abuse and Mental Health Services Administration provides mental illness facts, services, and tips.

SeniorLaw Home Page
http://www.seniorlaw.com/
Specialists in elder law.

Seniorlink
http://www.seniorlink.com /content/view/12/59/
Elder care resource, with referral and consultation services for older adults, families, and care providers.

Senior Sites
http://www.seniorsites.com/
Information on long-term care and housing for older adults.

Seniors-Site.com
http://www.seniors-site.com/
Concerns about aging; financial information.

Social Security Online
http://www.ssa.gov/
Comprehensive site. Visitors can request earnings record, review statistical information about benefits and beneficiaries, and review information on history and legislation on Social Security.

Stress Management and Emotional Wellness Links
http://www.imt.net /~randolfi/StressLinks.html
Links to emotional wellness sites.

U.S. Census Bureau
http://www.census.gov/
U.S. Census Bureau links, with data tools, population estimates, and a list of publications.

U.S. Department of Housing and Urban Development
http://www.hud.gov/
Information about housing policy and programs.

U.S. Department of Labor
http://www.dol.gov/
Labor resources and links to pension information.

Yahoo!'s Death Page
www.yahoo.com/Society_and_Culture/Death
Resources on death and dying.

REFERENCES

AARP Bulletin Today (2008). *Why are more older Americans sleeping in their cars?* http://bulletin.aarp.org/yourmoney/personalfinance. Retrieved August 25, 2009.

Abel, S. A. (1998). Social Security retirement benefits: The last insult of a sexist society. *Family and Conciliation Courts Review, 36* (1), 54–64.

About.com (2008). Senior Health: AOA information for African American elders. http://seniorhealth.about.com. Retrieved September 18, 2009.

Achenbaum, W. A. (1996). Historical perspectives on aging. *In* R. H. Binstock and L. K. George (Eds.), *Handbook of aging and the social sciences* (4th ed). San Diego: Academic Press.

Adams, R. G., & Blieszner, R. (1995). Aging well with family and friends. *American Behavioral Scientist, 38* (2), 209–224.

Adams, T. (1996). Informal family caregiving in older people with dementia: research priorities for community psychiatric nursing. *Journal of Advanced Nursing, 24*, 703–710.

Aday, R. H., Sims, C. R., McDuffie, W., & Evans, E. (1996). Changing children's attitudes toward the elderly: The longitudinal effects of an intergenerational partners program. *Journal of Research in Childhood Education, 10* (2), 143–151.

Adelman, R. (1987). A community-oriented geriatric rehabilitation unit in a nursing home. *Gerontologist, 27*, 143–146.

Adler, J. M., Kissel, E., & McAdams, D. P. (2006). Emerging from the CAVE: Attributional style and the narrative study of identity in midlife adults. *Cognitive Therapy and Research, 30*, 39–51.

Administration on Aging [AoA] (2008b). *Challenges of Global Aging.* Health and Human Services, Administration on Aging. U.S. Government printing office: Washington, DC. Retrieved August 14, 2009.

Administration on Aging [AoA] (2007). *Profile of Older Americans 2007.* Health and Human Services, Administration on Aging. U.S. Government printing office: Washington, DC. Retrieved July 30, 2009.

Administration on Aging [AoA] (2008a). *A Profile of Older Americans 2008.* Health and Human Services, Administration on Aging. U.S. Government printing office: Washington, DC. Retrieved July 30, 2009.

Administration on Aging [AoA] (2009). *Aging into the 21st century: Growth of the elderly population.* Health and Human Services, Administration on Aging. U.S. Government printing office: Washington, DC. Retrieved August 2, 2009.

Administration on Aging [AoA] (1999). *Challenges of Global Aging.* Health and Human Services, Administration on Aging. U.S. Government printing office: Washington, DC.

Aiken, L. (1994). *Dying, death, and bereavement* (3rd ed.). Boston: Allyn & Bacon.

Albert, C. M. (2007a). *Stress, coping, and development* (2nd ed.), NY: Guilford.

Albert, M. S. (2007). Projecting neurologic disease burden: Difficult but critical. *Neurology, 68*, 322–333.

Albert, M. S., & Killiany R. J. (2001). Age-related cognitive changes and brain-behavior relationships. *In* J. E. Birren, & K. W. Schaie (Eds.), *Handbook of the psychology of aging* (5th ed.). San Diego: Academic Press.

Albert, M. S., Dickerson, B. C., Fenstermacher, E., Salat, D. H., Wolk, D. A., Maguire, R. P., Desikan, R., Pacheco, J., Quinn, B. T., Van der Kouwe, A., Greve, D. N., Blacker, D., Killiany, R. J., & Fischl, B. (2008). Detection of cortical thickness correlates of cognitive performance: Reliability across MRI scan sessions, scanners, and field strengths. *NeuroImage, 38* (1), 10–18.

Alexy, B., & Belcher, J. (1997). Rural elderly present need for nursing continuity. *Nursing Economics, 15* (3), 146–150.

Allaire, J. C., & Willis, S. L. (2006). Competence in everyday activities as a predictor of cognitive risk and mortality. *Aging, Neuropsychology, and Cognition, 13*, 207–248.

Allan, M. A. (1998). Elder abuse: A challenge for home care nurses. *Home Healthcare Nurse, 16* (2), 103–110.

Almberg, B. E., & Grafstrom, W. (2000). Caregivers of relatives with dementia: Experiences encompassing social support and bereavement, *Aging and Mental Health, 4* (1), 82–89.

Amenta, E., Carruthers, B. G., & Zylan, Y. (1992). A hero for the aged? The Townsend movement, the political mediation model, and U.S. old-age policy, 1934–1950. *American Journal of Sociology, 98*(2), 308–339.

American Association of Retired Persons [AARP] (1997). *Profile of older Americans: 1997.* Administration on Aging, U.S. Department of Health and Human Services, Washington, DC.

American Association of Retired Persons [AARP] (2002). Introducing AARP. http://www.aarp.org/aarp/articles/introducingaarp.html Retrieved September 18, 2009.

American Association of Retired Persons [AARP] (2005). *Using the experience of a lifetime: 21st Century update.* Washington, DC: American Association of Retired Persons.

American Association of Retired Persons [AARP] (2006). AARP announces 37 millionth member. http://www. aarp.org/presscenter/pressrelease. Retrieved September 18, 2009.

American Association of Retired Persons [AARP]. (2000). *Fixing to stay.* http://www.aarp.org/publications. Retrieved August 18, 2009.

American Association of Retired Persons [AARP]. (2003). *These four walls: Americans 45+ talk about home and community.* http://www.aarp.org/publications. Retrieved August 18, 2009.

American Association of Retired Persons [AARP]. (2004). *Beyond 50.05: Civic involvement in America.* http:// www.aarp.org/publications. Retrieved August 13, 2009.

American Association of Retired Persons [AARP]. (2005). *Sexuality at midlife and beyond.* Washington, D.C.

American Association of Retired Persons [AARP]. (2007). Closing in on Alzheimer's. *AARP Bulletin Today.* http:// bulletin.aarp.org/yourhealth/diseases/articles/alzheimer Retrieved June 7, 2009.

American Cancer Society (2008). *Cancer Facts & Figures 2008.* Atlanta GA: American Cancer Society.

American Heart Association (2008). *Heart Disease and Stroke Statistics—2009 Update.* Statistics Committee and Stroke Statistics Subcommittee. Dallas: AHA.

American Housing Survey (2007). *American Housing Survey for the United States: 2007, Current Housing Reports,* H150/07. Administration on Aging. Washington DC.

American Medical Association [AMA] (2009). Informed consent. www.ama-assn.org/ama/pub/physician-resources. Retrieved September 20, 2009.

American Psychiatric Association [APA]. (1994). *Diagnostic and statistical manual for mental disorders (DSM-IV).* 4th edition. Washington, DC: American Psychiatric Association.

American Psychiatric Association [APA]. (1997). Practice guidelines for the treatment of patients with schizophrenia. *American Journal of Psychiatry, 15,* (supplement), 1–63.

Amir, Y. (1969). Contact hypothesis in ethnic relations. *Psychological bulletin, 71,* 319–342.

Anonucci, T. C., & Akiyama, H. (1995). *Convoys of social relations: Family and of aging and the family.* Westport, CT: Greenwood.

Anstey, K., Lord, S. R., & Philippa, W. (1997). Strength in the lower limbs, visual contrast sensitivity, and simple reaction time predict cognition in older women. *Psychology and aging, 12* (1), 137–144.

Antonucci, T. C., Fuhrer, R., & Dartigues, J. F. (1997). Social relations and depressive symptomatology in a sample of community-dwelling French older adults. *Psychology and aging 12* (1), 189–195.

Apple, D. D. (1956). The social structure of grandparent-hood." *American anthropologist, 58,* 656–663.

Aquino, J. A., Russell, D. W., Cutrona, C. E., & Altmaier, E. M. (1996). Employment status, social support, and life satisfaction among the elderly. *Journal of Counseling Psychology, 43* (6), 480–489.

Archbold, P. (1983). Impact of parent caring on women. *Family Relations, 32,* 39–45.

Arias, E. (2002). *United States life tables, 2000.* National vital statistics report, 51(3). National Center for Health Statistics.

Aries, E. J., & Johnson, F. L. (1983). Close friendship in adulthood: Conversational content between same-sex friends. *Sex Roles: A Journal of Research, 9,* 1183–1196.

Aristotle (2007). *Rhetoric.* Published http://www.public. iastate.edu. Retrieved April 29, 2009.

Arnold, S. E., & Trojanowski, J. Q. (1996). Cognitive impairment in elderly schizophrenia: A dementia (still) lacking distinctive histopathology. *Schizophrenia Bulletin, 22* (1), 5–9.

Associated Press (2006). Study: Race, location affects longevity. http://www.diverseeducation.com/artman/ Retrieved September 12, 2009.

Atchley, R. (1976). *The sociology of retirement.* Cambridge, MA: Schenkman.

Attar-Schwartz S., Tan J.-P., Buchanan A., Flouri E., & Griggs J. (2009). Grandparenting and adolescent adjustment in two-parent biological, lone-parent, and step-families. *Journal of family psychology, 23* (1) 67–75.

Attar-Schwartz, S., Tan, J.-P., & Buchanan, A. (2009). Adolescents' perspectives on relationships with grand-parents: The contribution of adolescent grandparent, and parent-grandparent relationship variables. *Children and Youth Services Review, 31* (9), 1057–1066.

Bailey, T. A. (2009) Effect of message type and source in advocacy communication: Investigating message strategies to combat ageism. *Dissertation Abstracts International [0419-4209] Section A: Humanities and Social Sciences, 69* (7-A).

Baldassare, M. (2006). PPIC Statewide Survey: Californians and the Future, November 2006. http://www.ppic.org.

Baltes, P. B., & Horgas, A. L. (1996). Long-term care institutions and the maintenance of competence. Willis, S. L., Schaie, K. W., & Hayward, M. (Eds.), *Social mechanisms for maintaining competence in old age.* New York: Springer.

Baltes, P. B. (1993). The aging mind: Potential and limits. *Gerontologist, 33* (5), 580–594.

Bank, S. P. (1995). Before the last leaves fall: Sibling connections among the elderly. *Journal of Geriatric Psychology, 28* (2), 183–195.

Barbato, C. A., & Freezel, J. D. (1987). The language of aging in different age groups. *Gerontologist 27,* 527–531.

Bargh, J. A., Chen, M., & Burrows, L. (1996). Automaticity of social behaviors: Direct effects of trait construct and stereotype activation on action. *Journal of Personality and Social Psychology, 71,* 230–244.

Barnes, P. M., & Schoenborn, C. A. (2003). *Physical activity among adults: United States, 2000.* Advance data from Vital and Health Statistics, 333, National Center for Health Statistics.

Bartels, S. J., Muester, K. T., & Miles, K. M. (1997). Functional impairments in elderly patients with schizophrenia and major affective illness in the community: Social skills, living skills, and behavior problems. *Behavior Therapy, 28,* 43–63.

Bastien, S. (2006). 12 benefits of hiring older workers. http:www/entrepreneur.com, 20 September. Retrieved July 30, 2009.

Bauer, J. J., McAdams, D. P., & Sakaeda, A. R. (2005). Interpreting the good life: Growth memories in the lives of mature, happy people. *Journal of personality and social psychology, 88,* 203–217.

Bauer, J. J., & McAdams, D. P. (2004a). Growth goals, maturity, and well-being. *Developmental Psychology 40,* 114–127.

Bauer, J. J., & McAdams, D. P. (2004b). Personal growth in adults' stories of life transitions. *Journal of personality, 72,* 573–602.

Baumeister, R. F., & Tice, D. M. (2001). *The social dimension of sex.* Toronto: Allyn and Bacon.

Beard, B. B. (1991). Centenarians: The new generation. Westport, CT: Greenwood Press.

Bechara A., Damasio H., Tranel D., & Anderson, S. W. (1998). Dissociation of working memory from decision making within the human prefrontal cortex. *Journal of Neuroscience, 18*(1), 428–437.

Beck, J. G., & Stanleyk, M. A. (1997). Anxiety disorders in the elderly: The emerging role of behavior therapy. *Behavior therapy, 28,* 83–100.

Beck, M. (1990). Trading places. *Newsweek,* July 16, 48–54.

Beckman, L. J. (1981). Effects of social interaction and children's relative inputs on older women's psychological well-being. *Journal of Personality and Social Psychology, 4,* 1075–1086.

Beckman, M. (2004). Low-cal connections. *Science of Aging Knowledge & Environment, 25,* 60.

Bellandi, D. (1997). Seniors on patrol. *Modern Healthcare,* 44–45, August 25.

Belliveau, F., & Richter, L. (1970). *Understanding human sexual inadequacy.* New York: Bantam Books.

Benbow, S. M., & Jagus, C. E. (2002). Sexuality in older women with mental health problems. *Sexual and relationship therapy, 17*(3), 261–270.

Bengtson, V. L., Giarrusso, R., Silverstein, M., & Wang, H. (2000). Families and intergenerational relationships in aging societies. *Hallym International Journal of Aging,* 3–10.

Bengtson, V. L., & Achenbaum, W. A. (1993). *The changing contract across generations.* New York: Aldine de Gruyter.

Bennett, K. M. (2005). Psychological wellbeing in later life: The longitudinal effects of marriage, widowhood and marital status change. *International Journal of Geriatric Psychiatry, 20*(3), 280–284.

Beresford, L. (1997). The good death. *Hospitals and Health Networks, 71*(12), 60–62.

Bergman I., Blomberg, M., & Almkvist, O. (2007). The importance of impaired physical health and age in normal cognitive aging. *Scandanavian Journal of Psychology, 48,* 115–125.

Berzlanovich, M., Leil, W., Waldhoer, T., Sim, E., Fasching, P., & Fazeny-Dorner, B. (2005). Do Centenarians die healthy? An autopsy study. *The Journals of Gerontology Series A: Biological Sciences and Medical Sciences, 60,* 862–865.

Binstock, R. (1983). The aged as scapegoats. *Gerontologist 23,* 136–143.

Bisconti, T. L., Bergeman, C. S., & Boker, S. M. (2006). Social support as a predictor of variability: An examination of the adjustment trajectories of recent widows. *Psychology and aging, 21,* 217–239.

Blagov, P. S., & Singer, J. A. (2004). Four dimensions of self-defining memories (specificity, meaning, content, and affect) and their relationships to self-restraint, distress, and repressive defensiveness. *Journal of Personality, 72,* 481–511.

Blazer, D. (1990). *Emotional problems in later life.* New York: Springer.

Blieszner, R. (2006). A lifetime of caring: Close relationships in old age. *Personal relationships, 13,* 1–18.

Bliwise, N. (1987). The epidemiology of mental illness in late life. *In* E. Lurie, and J. Swan (Eds.), *Serving the Mentally Ill Elderly.* Lexington, MA: D. C. Health and Company.

Blouin, A. S., & Brent, N. J. (1996). Downsizing and potential discrimination based on age. *Journal of nursing administrators, 26* (11), 3–5.

Bluck, S., & Gluck, J. (2004). Making things better and learning a lesson: Experiencing wisdom across the lifespan. *Journal of personality, 72,* 543–572.

Bock J., & Johnson, S. E. (2008). Grandmothers' productivity and the HIV/AIDS pandemic in sub-Saharan Africa. *Journal of Cross-Cultural Gerontology, 23,* 131–145.

Bolland, M. J., Barber, P. A., Doughty, R. N., *et al.* (2008). Vascular events in healthy older women receiving calcium supplementation: randomized controlled trial. *British Medical Journal, 336,* 262.

Boomsma, D., Cacioppo, J., Muthen, B., Asparouhov, T., & Clark, S. (2007). Longitudinal genetic analysis for loneliness in Dutch twins. *Twin Research and Human Genetics, 10,* 267–273.

Botkin, D. R., Weeks, M. O., & Morris, J. E. (2000). Changing marriage role expectations: 1961–1996. *Sex roles, 42* (9), 933–942.

Boudreau, J. W., Boswell, W. R., & Judge, T. A. (2001). Effects of personality on executive career success in the United States and Europe. *Journal of Vocational Behavior, 58,* 53–81.

Bouman, W. P., & Arcelus, J. (2001). Are psychiatrists guilty of 'ageism' when it comes to taking a sexual history? *International journal of geriatric psychiatry, 16*(1), 27–31.

Bouman, W. P., Arcelus, J., & Benbow, S. M. (2006). Nottingham study of sexuality and ageing (NoSSA I). Attitudes regarding sexuality and older people: a review of the literature. *Sexual and Relationship Therapy, 21*(2), 149–161.

Bowen, J., Teri, L., Kukull, S., McCormick, W., McCurry, S. M., & Larson, E. B. (1997). Progression to dementia in patients with isolated memory loss. *Lancet, 349,* 1763–1765.

Boxer, A. M. (1997). Gay, lesbian, and bisexual aging into the twenty-first century: An overview and introduction. *Journal of Gay, Lesbian, and Bisexual Identity, 2* (3/4), 187–193.

Boynton, S. (1993). Aging boomers! The graying of a generation. *Santa Rosa Press Democrat,* March 22, p. B1.

Brackley, M. H. (1994). The plight of American family caregivers: Implications for nursing. *Perspectives in Psychiatric Care, 30*(4), 14–20.

Brandon, E. (2007). A list of job sites for older workers. http://www.usnews.com. Retrieved August 2, 2009.

Brecht, S. B. (1996). Trends in the retirement housing industry. *Urban Land, 55*(11), 32–39.

Bretschneider, J. G., & McCoy, N. L. (1988). Sexual interest and behavior in healthy 80–102 year olds. *Archives of Sexual Behavior, 17*(2), 109–129.

Brignolo, D. (1998, March 10). Investment fiasco ends in sentencing. *San Jose Mercury News,* 1B.

Brody, E. (1984). Caregivers, daughters and their local siblings: Perceptions, strains, and interactions. *Gerontologist, 29*(4), 529–538.

Brown, A. S., & Nix, L. A. (1996). Age-related changes in tip-of-the-tongue experience. *American Journal of Psychology, 109* (1), 79–91.

Bryant, J., & Fernald, L. (1997). Nursing knowledge and use of restraint alternatives: Acute and chronic care. *Geriatric Nursing, 18*(2), 57–60.

Buckwalter, J. A. (1997). Decreased mobility in the elderly: The exercise antidote. *Physician and Sportsmedicine, 25*(9), 127–133.

Buckwalter, K., Smith, M., & Martin, M. (1993). Attitude problem. *Nursing times, 89* (5), 54–57.

Buettner, D. (2008). *The Blue Zone: Lessons for Living Longer From the People Who've Lived the Longest.* National Geographic Books.

Buhrich, N., & Teeson, M. (1996). Impact of a psychiatric outreach service for homeless persons with schizophrenia. *Psychiatric services, 47*(6), 644–646.

Bureau of Justice Statistics [BJS] (2005). Intimate partner violence. Bureau of Justice Statistics Special Report, May.

Bureau of Labor Statistics [BLS] (1996). *Suicide among older persons: United States, 1980–1992.* U.S. Department of Health and Human Services *MMWR, 45 (1),* 3–6.

Bureau of Labor Statistics [BLS] (2004). Occupational Injuries and Illnesses: Counts, Rates, and Characteristics, 2004. U.S. Department of Labor. Washington DC.

Bureau of Labor Statistics [BLS] (2004). *The employment situation: March 2004.* Washington DC United States Department of Labor.

Burger, J. M., Horita, M., Kinoshita, L., Roberts, K., & Vera, C. (1997). Effects of time on the norm of reciprocity. *Basic and Applied Social Psychology 19* (1), 91–100.

Burke, J. L. (1981). Young children's attitudes and perceptions of older adults. *International Journal of Aging and Human Development, 14,* 205–221.

Burkhauser, R. V. (1996). Editorial. Touching the third rail: Time to return the retirement age for early social security benefits to 65. *The Gerontologist, 36* (6), 726–727.

Burkhauser, R. V., Couch, K. A., & Phillips, J. W. (1996). Who takes early Social Security benefits? The economic and health characteristics of early beneficiaries. *The gerontologist, 36*(6), 789–799.

Burnette, D. (1998). Grandparents rearing grandchildren: A school-based small group intervention. *Research on Social Work Practice, 8* (1), 10–27.

Burtless, G., & Quinn, J. F. (2002). *Is working longer the answer for an aging workforce? An issue in brief.* Center for retirement Research, 11.

Busan, T. (1991). *Use your perfect memory.* New York: Plume Division, Penguin Books.

Butler, R. N., & Lewis, M. I. (1983). The second language of sex. *In* S. Sprecher & K. McKinney (Eds.), *Sexuality.* Toronto: Little, Brown.

Butler, R. N., & Lewis, M. R. (2002). *The new love and sex after 60.* New York: Ballantine.

Butler, R. N. (1963). The life review: An interpretation of reminiscence in the aged. *Psychiatry, 26,* 65–76.

Butler, R. N. (1975). *Why survive?* New York: Harper & Row.

Butler, S. S., & DePoy, E. (1996). Rural elderly women's attitudes toward professional and governmental assistance. *Affilia, 11*(1), 76–94.

Butrica, B. A., Iams, H. M., & Smith, K. E. (2003/2004). The changing impact of Social Security on retirement in the United States. *Social security bulletin, 65*(3), 1–13.

Bytheway, B. (2005). Ageism and age categorization. *Journal of Social Issues, 61*(2), 361–374.

Cabeza, R. (2001). Cognitive neuroscience of aging: Contributions of functional neuroimaging. *Scandinavian Journal of Neuroscience, 42,* 277–286.

Cacioppo, J. T., Hawkley, L. C., Rickett, E. M., & Masi, C. M. (2005). Sociality, spirituality, and meaning making: Chicago Health, Aging and Social Relations Study. *Review of General Psychology, 9,* 143–155.

Cahill, S., South, K., & Spade, J. (2001). *Outing age: Public policy issues affecting gay, lesbian, bisexual and transgender elders.* Policy Institute of the National Gay and Lesbian Task Force. New York.

Camp, C., & Camp, G. (1996). *The corrections yearbook.* South Salem, NYU: Criminal Justice Institute.

Campbell, L. D. (2003). Gendered nature of men's filial care. *Journals of gerontology: Series B: Psychological sciences and social sciences, 58B*(6), 287–298.

Campbell, T. C., Lyman, H., & Robbins, J. (2006). *The China Study.* TX: BenBella Publishing.

Camplin, J. C. (2009). Volunteers leading volunteers. *Professional Safety, 54*(5), 36–32. *Caring for the dying: Identification and promotion of physician competency.* Washington, DC: AMA.

Carney, M. (2009). Clinical leadership effectiveness. *Journal of Nursing Management, 17*(4), 435–445.

Carr, M. (1997). Not at your age! *Accountancy—International Edition, 41,* not paginated.

Carstensen, L. L., Mickels, J. A., & Mather M. (2006). Aging and the intersection of cognition, motivation, and emotion. *In* J. E. Birren & K. W. Schaie (Eds.), *Handbook of the psychology of aging.* San Diego: Elsevier.

Carstensen, L. L. (2006). The influence of a sense of time on human development. *Science, 312,* 1913–1915.

Carstensen, L., Mason, S. E., & Caldwell, E. C. (1982). Children's attitudes toward the elderly: An intergenerational technique for change. *Educational Gerontology, 8,* 291–301.

Cassel, C. K. (1996). Overview on attitudes of physicians toward caring for the dying patient. *In* American Board of Internal Medicine (Ed.),

Cassel, C. K. (1996). *Medicare matters: What geriatric medicine can teach American health care.* California: Milbank Books on Health and the Public, 14.

Cauley, J. A., Hochbreg, M. C., Lui, L. Y., *et al.* (2007). Long-term risk of incident vertebral fractures. *Journal of the American Medical Association, 298,* 2761–2767.

Cavanaugh, J., & Blanchard-Fields, F. (2006). *Adult development and aging,* 5e. Belmont, CA: Wadsworth/Thomson Learning.

Center on Aging Society (2003). *Obesity Among Older Americans. Data Profile.* Center on Aging: Georgetown University.

Centers for Disease Control and Prevention [CDC] (1996). *Suicide among older persons: United States, 1980–1992.* U.S. Department of Health and Human Services *MMWR, 45* (1), 3–6.

Centers for Disease Control and Prevention [CDC] (2007). *Deaths: Leading causes for 2004.* National Vital Statistics Reports, 56(5).

Centers for Disease Control and Prevention [CDC] (2007). *Health, United States, 2007 with Chartbook on Trends in the health of Americans.* http://www.cdc.gov/nchs/data. Retrieved June 2, 2009.

Centers for Disease Control and Prevention [CDC] (2008a). *Older Americans 2008.* Washington DC: National Center for Health Statistics.

Centers for Disease Control and Prevention [CDC] (2008b). *Strength training for older adults.* Online Source for Credible Health Information. http://www.cdc.gov/physicalactivity. Retrieved June 22, 2009.

Centers for Disease Control and Prevention [CDC] (2009). *Deaths and Mortality: FastStats.* http://www.cdc.gov/nchs/FASTSTATS/deaths. Retrieved June 24, 2009.

Centers for Medicare & Medicaid Services [CMS] (2009). *Medicare & you: 2009.* U.S. Government Printing Office Washington DC.

Cerella, J., Onyper, S. V., & Hoyer, W. J. (2006). The associative-memory basis of cognitive skill learning: Adult age differences. *Psychology and Aging, 21,* 483–498.

Ceridian, Control Data Systems settle age bias lawsuits. (1997). *Wall Street Journal,* 6 March, 13.

Chappell, N. L., & Dujeka, C. (2009). Caregivers—Who copes how? *International Journal of Aging and Human Development,* 69(3), 221–224.

Cheek, C., & Piercy, K. W. (2008). Quilting as a tool in resolving Erikson's adult stage of human development. *Journal of Adult Development, 15,* 13–34.

Cherlin, A. J., & Furstenberg, F. F., Jr. (1994). Stepfamilies in the United States: A reconsideration. *Annual Review of Sociology, 20,* 359–381.

Chew, S. C. (1962). *The pilgrimage of life.* New Haven, CT: Yale University Press.

Choi, N. G. (1994). Changes in labor force activities and income of the elderly before and after retirement: A longitudinal analysis. *Journal of Sociology and Social Welfare, 21* (2), 5–26.

Choi, N. G. (1996). Changes in the living arrangements, work patterns, and economic status of middle-aged single women, 1971–1991. *Affilia,* 11(2), 164–178.

Chou, S-C., Boldy, D. P., & Lee, A. H. (2003). Factors influencing residents' satisfaction in residential aged care. *The gerontologist, 43,* 459–472.

Chu, K. (2006). Financial scams expected to boom as boomers age. *USA Today.* http://www.usatoday.com. Retrieved September 10, 2009.

Cicirelli, V. G. (1982). Sibling influence throughout the lifespan. *In* M. E. Lamb & B. Sutton-Smith (Eds.), *sibling relationships: Their nature and significance across the lifespan.* Hillsdale NJ: Erlbaum.

Cicirelli, V. G. (1997). Relationship of psychosocial and background variables to older adults' end-of-life decisions. *Psychology and Aging, 12* (1), 137–151.

Clark, D. E., Knapp, T. A., & White, N. E. (1996). Personal and location-specific characteristics and elderly interstate migration. *Growth and Change, 27,* 327–351.

Clarke, L. H. (2006). Older women and sexuality: Experiences in marital relationships across the life course. *Canadian Journal on Aging, 25* (2), 129–140.

Cloninger, S. C. (2003). *Theories of personality: Understanding persons,* 2nd ed. NJ: Prentice-Hall.

Cohen, C. I. (1992). Project rescue: Serving the homeless and marginally housed elderly." *Gerontologist, 32* (4), 467–471.

Cohen, C. I., & Sokolovsky, J. (1989). *Old men of the Bowery.* New York: Guilford Press.

Cohen, E., & Kruschwitz, A. (1990). Old age in America represented in ninteenth and twentieth century popular sheet music. *Gerontologist 30* (3), 354–354.

Colburn, D. (1987). Facing the certainty of death. *Washington Post, 112,* 14 April, p. 16.

Colcombe, S., & Kramer, A. F. (2003). Fitness effects on the cognitive function of older adults: A meta-analysis study. *Psychological Science (14)* 2, 125–131.

Cole, T. (1992). *The journey of life: A cultural history of aging in America.* New York: Cambridge University Press.

Collins, C., & Frantz, D. (1994). Let us prey: Cults and the elderly. *Modern Maturity,* June, 22–26.

Colsher, P. L., Wallace, R. B., Loeffelholz, P. L., & Sales, M. (1992, June). Health status of older male prisoners. *American Journal of Public Health, 82* (6), 881–884.

Comfort, A. (1976). Age prejudice in America. *Sociology Policy, 17*, 3–8.

Commonwealth Fund (1993). *The untapped resource: Americans over 55 at work*. New York: Commonwealth Fund, November.

Congressional Budget Office Study. (1993). *Displaced workers: Trends in the 1980s and implications for the future*. Washington, DC: Congress of the United States.

Connidis, I. A., & Campbell, L. D. (1995). Closeness, confiding, and contact among siblings in middle and late adulthood. *Journal of family issues, 16* (6), 722–745.

Connidis, I. A. (2001). *Family ties and aging*. Thousand Oaks CA: Sage Publications.

Continuing Care Retirement Community [CCRC] (2009). Commission on Accreditation of Rehabilitation Facilities. http://www.carf.org/. Retrieved August 22, 2009.

Conwell, Y. (2001). Suicide in later life: a review and recommendations for prevention. *Suicide and life threatening behavior, 31*, 32–47.

Cook, E. A. (1997). The effects of reminiscence on psychological measures of ego integrity in elderly nursing home residents. *Archives of Psychiatric Nursing, 5* (5), 292–298.

Cooke, T., & Marchant, S. (2006). The changing intra-metropolitan location of high-poverty neighbourhoods in the U.S. 1990–2000. *Urban Studies, 43*(11), 1971–1989.

Cooley, C. H. (1902). *Human nature and the social order*. New York: Charles Scribner's Sons.

Corbin, D. E., Metal-Corbin, J., & Barg, C. (1989). Teaching about aging in the elementary school: A one-year followup. *Educational Gerontology, 15*, 103–110.

Corder, E. H., Saundres, A. M., & Strittmatter, W. J. (1993). Gene dose of apolipoprotein E type 4 allele and the risk of Alzheimer's disease in late onset families. *Science, 261*, 921–923.

Costa, P. T., Jr., Terracciano, A., & McCrae, R. R. (2001). Gender differences in personality traits across cultures: Robust and surprising findings. *Journal of Personality and Social Psychology, 81*(2), 322–331.

Council on Scientific Affairs (1996). American Medical Association Council report: Alcoholism in the elderly. *Journal of the American Medical Association, 275* (10), 797–801.

Coupland, N., Coupland, J., & Giles, H. (1991). *Language, society and the elderly*. Oxford: Basil Blackwell.

Courtenay, B. C., Poon, L. W., Martin P., Clayton G. M., & Johnson, M. A. (1992). Religiosity and adaptation in the oldest-old. *International Journal of Aging and Human Development, 34* (1), 47–56.

Covey, H. C. (1991). Old age and historical examples of the miser. *The gerontologist, 31* (5), 573–678.

Cowan, C. A., & Hartman, M. B. (2005). Financing health care: Businesses, households, and governments, 1987–2003. *Health Care Financing Review, 1*(2), 1–26.

Coward, R. T., Cutler, S. J., Lee, G. R., Danigelis, N. L., & Netzer, J. (1996). Racial differences in the household composition of elders by age, gender, and area of residence. *International Journal of Aging and Human Development, 42* (3), 205–227.

Cowgill, D., & Holmes, L. D. (1972). *Aging and modernization*. New York: Appleton-century-Crofts.

Cowgill, D. (1974). Aging and modernization: A revision of the theory. *In* J. F. Gubrium (Ed.,), *Late life communities and environmental policy*. Springfield, Il: Charles C. Thomas.

Cowgill, D. (1986). *Aging around the world*. Belmont, CA: Wadsworth.

Crane, M. (1994). Elderly homeless people: Elusive subjects and slippery concepts. *Ageing and Society, 14*, 631–640.

Crenshaw, A. (1992, September 27). Living trusts lure some people not to be trusted. *Washington Post,* H3.

Crockett, W., & Hummert, M. L. (1987). Perceptions of aging and the elderly, p. 217–41. *In* Maher, B. A. (Ed.), *Progress in experimental personality research*, vol. 2. New York: Academic Press.

Cronk, M. (1998). Thief poses as Cupertino City worker, steals cash. *San Jose Mercury News,* 2B, 21 March.

Crowley, M. (1980). *Gerontology: a behavioral approach*. Reading, MA: Addison-wesley.

Crystal, S., & Beck, P. (1992). A room of one's own: The SRO and the single elderly. *Gerontologist, 32* (5), 684–692.

Crystal, S., Sambamoorthi, U., Walkup, J. T., & Akincigil, A. (2008). Diagnosis and treatment of depression in the elderly Medicare population: predictors, disparities, and trends. Published in *Journal of the American Geriatric Society*. NIH-PA author manuscript. National Institutes of Health. Retrieved November 16, 2008.

Cumming, E., & Henry, W. E. (1961). *Growing Old: The Process of Disengagement*. New York: Basic Books.

Cunningham, C. M., Callahan, C. M., Plucker, J. A., Roberson, S. C., & Rapkin, A. (1998). Identifying Hispanic students of outstanding talent: psychometric integrity of a peer nomination form. *Exceptional children, 645* (2), 197–209.

Cutler, N. E., & Devlin, S. J. (1996). A framework for understanding financial responsibilities among generations. *Generations, 20* (1), 24–28.

Cyr, D. (1996). Lost and found: Retired employees. *Personnel Journal, 75* (11), 40–46.

Czopp, A. M., & Monteith, M. J. (2003). Confronting prejudice (literally): Reactions to confrontations of racial and gender bias. *Personality and Social Psychology Bulletin, 29*, 50–52.

Daniels, W. (1997). "Derelicts," recurring misfortunate, economic hard times and lifestyle choices: Judicial images of homeless litigants and implications for legal advocates. *Buffalo Law Review, 45*(3), 687–736.

Daniluk, J. C. (2003). *Women's sexuality across the life span: Challenging myths, creating meanings*. New York: Guilford Press.

Dannefer, D. (2003). Cumulative advantage/disadvantage in the life course: Cross-fertilizing age and social science theory. *Journal of Gerontology, 58B*, S327–S337.

Davidson, M., Harvey, P. D., Powchick, P., Parrella, M., White, L., Knobler, H. Y., Losonczy, M. F., Keefe, R. S., Katz, S., & Frecska, E. (1995). Severity of symptoms in chronically institutionalized geriatric schizophrenic patients. *American Journal of Psychiatry, 152,* 197–207.

Dawson, K. (1982). *Serving the older community.* Sex Education and Information Council of the United States, Report No. SIECUS Report.

De Augelli, A. R., & Garnets, L. D. (1995). Lesbian, gay, and bisexual communities. *In* A. R. de Augelli & C. J. Patterson (Eds.), *Lesbian, gay and bisexual identities over the lifespan: Psychological perspectives.* New York: Oxford University Press.

De Beauvoir, S. (1973). *The Coming Of Age.* Warner Communications: NY

De Epiro, N. W. (1996). Treating Alzheimer's disease: Today and tomorrow. *Patient care,* November, 62–83.

De Grey, A. D. (2005). The unfortunate influence of the weather on the rate of ageing: Why human caloric restriction or its emulation may only extend life expectancy by 2–3 years. *Gerontology 51*(2), 73–82.

De Magalhaes, J. P., & Sandberg, A. (2005). Cognitive aging as an extension of brain development: A model linking learning, brain plasticity, and neurodegeneration. *Mechanisms of Ageing and Development, 126*(10), 1026–1033.

De Parle, J. (1994). Build single room occupancy hotels. *The Washington Monthly,* March, 52–54.

De Poy, E., & Butler, S. S. (1996). Health: Elderly rural women's conceptions. *Affilia,* 11 (2), 207–220.

De Rijk, M. C., Breteler, M. M.B., Graveland, G. A., Ott, A., Grobbee, D. E., van der Meche, F. G.A., & Hofman, A. (1995). Prevalence of Parkinson's disease in the elderly: The Rotterdam study. *Neurology, 45,* 2143–2146.

De Spelder, L. A., & Strickland, A. L. (2009). *The last dance: Encountering death and dying, 8e.* NY: McGraw Hill.

De Viney, S. (1995). Life course, private pension, and financial well being. *American behavioral scientist, 39* (2), 172–185.

Deary, I. J. & Der, G. (2005). Reaction time explains IQ's association with death. *Psychological Science, 16,* 64–69.

Deary, I. J., (2006). Intelligence, destiny and education: The ideological roots of intelligence testing. *Intelligence, 34,* 5621–622.

Deiner E., Lucas R. E., & Scollon, C. (2006). Beyond the hedonic treadmill: Revising the adaptation theory of well-being. *American Psychologist, 61,* 305–314.

Dench, G., Ogg, J., & Thompson, K. (1999)The role of grandparents. *In* R. Jowell (Ed.), *British social attitudes: the 16th report* p. 136–156). Ashgate, United Kingdom: National Centre for Social Research.

Denis, M. (1996). Implications of offering early retirement benefits in exchange for a release from employment claims. *Employment Relations Today, 23* (3), 65–69.

Denney, N. W., & Palmer, A. M. (1987). Adult age differences in traditional and practical problem solving measures. *Psychology and Aging, 4,* 438–442.

Devlin, S. J., & Ayre, L. (1997). The Social Security debate: A financial crisis or a new retirement paradigm? *Generations, 21*(2), 27–33.

Diener, E., Lucas, R. E., & Scollon, C. (2006). Beyond the hedonic treadmill: Revising the adaptation theory of well-being. *American Psychologist, 61,* 305–314.

Dierckx, E., Engelborghs S., deRaedt, R., van Buggenhout M., deDeyn, P. P., Verleye, G., Verte, D., & Ponjaert-Kristoffersen, I. (2008). Differentiation between dementia and depression among older persons: Can the difference between actual and premorbid intelligence be useful? *Journal of Geriatric Psychiatry Neurology, 21*(4), 242–249.

Dion, K. K., & Dion, K. L. (1996) Cultural perspectives on romantic love. *Personal Relationships, 3*(1), 5–7.

Downe-Wamboldt, B. L., & Melanson, P. M. (1990). Attitudes of baccalaureate student nurses toward aging and the aged: Results of a longitudinal study. *Educational Gerontology, 16,* 49–57.

Drew, L. A., & Smith, P. K. (1999)The impact of parental separation/divorce on grandparent-grandchild relationships. *International Journal of Aging and Human Development, 48,* 191–216.

Drozdiak, W. (1994, May 3). Elderly Dutch reach for political power in "granny revolution." *Washington Post,* A16.

Duff, C. (1994). These big clocks actually are made by grandfathers." *Wall Street Journal* (6 April 1994): A1(W), A1(E).

Duffield, P., & Podzamsky, J. E. (1996). The completion of advance directives in primary care. *Journal of Family Practice, 42* (4), 378–383.

Duncan, C. (2009). Five decades of crisis. *National Review, 61*(15), 35–37. August 24.

Dunkel, C. S., & Sefcek, J. A. (2009). Eriksonian lifespan theory and life history theory: An integration using the example of identity formation. *Review of General Psychology, 13*(1), 13–23.

Dunn, J. (2002). The adjustment of children in stepfamilies: Lessons from community studies. *Child and Adolescent Mental Health, 7,* 154–161.

Dying with Dignity (no date). Why is choice so important? http://www.dyingwithdignity.ca. Downloaded June 29, 2009.

Eisenhandler, S. (1994). A social milieu for spirituality in the lives of older adults. *In* L. Thomas and S. Eisenhandler (Eds.), *Aging and the Religious Dimension,* 140–168, Westport, CT: Auburn House.

Ekstrom, M. (1994). Elderly people's experiences of housing renewal and forced relocation: Social theories and contextual analysis in explanations of emotional experiences. *Housing Studies, 9*(3), 369–391.

Elder Fraud in Denver. (2006). http://www.ojp.usdoj.gov/ocv/publications. Retrieved September 2, 2009.

Elder, G. H., & O'Rand, A.M. (1994). Adult lives in a changing society. *In* K. Cook, G. fine, & J. S. Hosue [Eds.], *Sociological perspectives on social psychology* [pp. 102–121]. New York: Basic.

Elder, G. H. (1979). Historical patterns and personality. P. B. Baltes and O. G. Brim, Jr. (Eds.) *Life-span development and behavior* (Vol. 2). New York: Academic Press.

Elderhostel (2009). The history of Elderhostel, Inc. and Exploritas. www.exploritas.org/about. Retrieved September 21, 2009.

Elder Law Answers (2008). To combat fraud, elderly may need to be a bit less trusting. http://www.elderlawanswers.com. Retrieved September 6, 2009.

Erikson, E. (1966). Eight ages of man. *International Journal of Psychiatry, 2,* 281–297.

Erikson, E. (1982). *The life cycle completed.* New York: W. W. Norton.

Estrada, L. (1995). The dawn of assisted living centers. *Washington Post, WB11.*

Evans, M. (1998). Caregiving daughters suffer stress. *Santa Rosa Press Democrat,* March 27, A13.

Families and Work Institute (1998). *Business work-life study.* Families and Work Institute. http://familiesand-work.org. Retrieved June 2, 2009.

Family Caregiver Alliance (2007). *Work and elder care.* National Center on Caregiving.

FamilyDoctor.org (no date). Dementia: Also called senility. http://familydoctor.org/online/famdocent/home/seniors/mental-health. Retrieved July 9, 2009.

Federal Interagency Forum on Aging-Related Statistics [FIFAS] (2009). *Older Americans 2008: Key Indicators of Well-Being.* http://www.agingstats.gov/Agingstatsdotnet. Retrieved August 9 2009.

Fehring, R. J., Miller, J. F., & Shaw, C. (1997). Spiritual well-being, religiosity, hope, depression, and other mood states in elderly people coping with cancer. *Oncology Nursing Forum, 24* (4), 663–671.

Feldstein, I. (1970). *Sex in later life.* Baltimore, MD: Penguin Books.

Feldstein, M. (1997). The case for privatization. *Foreign Affairs, 76* (4), 24 –38.

FIFAS [Federal Interagency Forum on Aging-Related Statistics] (2009). *Older Americans 2008: Key Indicators of Well-Being.* http://www.agingstats.gov/Agingstatsdotnet. Retrieved August 9, 2009.

Finch, C. E., & Pike, M. C. (1996). Maximum life span predictions from the Compartz mortality model. *Journal of Gerontology: Biological Sciences, 51B,* B183–B194.

Fincham, F. D. (2003). Marital conflict, correlates, structure, and context. *Current Directions in Psychological Science, 12,* 23–27.

Fischer, D. H. (1977). *Growing old in America.* New York: Oxford University Press.

Fisk, J. E., & Warr, P. (1996). Age and working memory: The role of perceptual speed, the central executive, and the phonological loop. *Psychology and Aging, 11* (2), 316–323.

Fitch, C. A., & Ruggles, S. (2003). Building the national historical geographic information system. *Historical Methods, 36*(1), 41–50.

Flanagan, C. A., Syvertsen, A. K., Gill, S., Gallay, L. S., & Cumsille, P. (2009). Ethnic awareness, prejudice, and civic commitments in four ethnic groups of American adolescents. *Journal of Youth Adolescence, 38* (4), 500–518.

Fleeson, W., & Heckhausen, J. (1997). More or less "me" in past, present, and future: Perceived lifetime personality during adulthood. *Psychology and Aging, 12* (1), 125–136.

Foley, K. T., & Mitchell, S. J. (1997). The elderly driver: What physicians need to know. *Cleveland Clinic Journal of Medicine, 64* (8) 423–428.

Foundation on Aging (2006). Identifying the issue and challenges women face as they age. *Proceedings of the February 2, 2006 Roundtable on Women and Aging.* University of California, Riverside Extension and Osher Lifelong Learning Institute.

Froeschle, N. (2007). Older people warned of health-care fraud. *Tulsa World.* http://www.tulsaworld.com/site. Retrieved September 5, 2009.

Frolik, L. A. (1996). The special housing needs of older persons: An essay. *Stetson Law Review, 26*(2), 647–666.

Frost, J. C. (1997). Group psychotherapy with the aging gay male: Treatment of choice. *Group, 21* (3), 267–285.

Fry, P. S. (2003). Perceived self-efficacy domains as predictors of fear of the unknown and fear of dying among older adults. *Psychology and Aging, 18*(3), 0882–7974.

Fullmer, E. J. (1995). Challenging biases against families of older gays and lesbians. *In* G. C. Smith, S. S. Tobin, E. A. Robertson-Tchabo, & P. W. Power (Eds.) *Strengthening aging families: Diversity in practice and policy.* Thousand Oaks CA: Sage.

Fullmer, H. T. (1984). Children's descriptions of an attitude toward the elderly. *Educational Gerontology, 10,* 22–107.

Furst, A. (1988). Unforgettable…That's what he is. *50 plus,* June, 72.

Gall, T. L., Evans, D. R., & Howard, J. (1997). The retirement adjustment process: Changes in the well-being of male retirees across time. *The Journal of Gerontology: Psychological Sciences, 52B* (3), P110–P117.

Gamliel, T. (2001). Social version of gerotranscendence: Case study. *Journal of Aging and Identity, 6*(2), 105–114.

Garcia, E. (2009). Raising leadership criticality in MBAs. *Higher Education, 58*(1), 113–130.

Garrard, J., Cooper, S. L., & Goertz, C. (1997). Drug use management in board and care facilities. *Gerontologist, 37*(6), 748–756.

Generation Birth Years. http://www.lifecourse.com/mi/insight/turnings. Retrieved May 9, 2009.

Geyelin, M. (1994). Age bias verdict. *Wall Street Journal,* 17 May, B7(W), B5(E).

Giles, H., & Reid, S. A. (2005). Ageism across the lifespan: Towards a self-categorization model of ageing. *Journal of Social Issues, 61(2),* 389–404.

Giles, H., Noels, K. A., Williams, A., Ota, H., Lim, T-S., Ng, S. H., *et al.* (2003). International communication across cultures: Young people's perception of conversations with family elders, non-family elders, and same-age peers. *Journal of Cross-Cultural Gerontology, 18,* 1–30.

Glascock, A. P. (1997). When is killing acceptable: The moral dilemma surrounding assisted suicide in America and other societies. *In* J. Sokolovsky (Ed.), *The cultural context of aging* (3rd ed.). Westport, CT: Bergin and Garvey.

Glazer, B., & Strauss, A. (1996). *Awareness of dying.* Chicago: Aldine.

Glenn, N. D., & Weaver, C. N. (1983–1984). Enjoyment of work by full-time workers in the U.S.: 1955–1980. *Public Opinion Quarterly, 46,* 459–470.

Goering, P., Wasylenki, D., Lindsay, C., Lemire, D., & Rhodes, A. (1997). Process and outcome in a hostel outreach program for homeless clients with severe mental illness. *American Journal of Orthopsychiatry, 67(4),* 607–617.

Golant, S. M., & LaGreca, A. J., (1994). Differences in the housing quality of white, black, and Hispanic U.S. elderly households. *Journal of Applied Gerontology, 13* (4), 413–437.

Goldie, S. J., Gaffikin, L., Goldhaber-Fiebert, J. D., *et al.* (2005). Cost-effectiveness of cervical cancer screening in five developing countries. *New England Journal of Medicine, 353,* 2158–2168.

Goldman, N., Korenmaan, S., & Weinstein, R. (1995). Marital status and health among the elderly. *Social Science and Medicine, 40* (12), 1717–1730.

Goldman, K. (1993). Seniors get little respect on Madison Avenue. *Wall Street Journal,* September 20, B8.

Goldman, R. J., & Goldman, J. D. (1981). How children view old people and aging: A developmental study of children in four countries. *Australian Journal of Psychology, 33* (3), 405–418.

Goldner, E. M., Hsu L., Waraich P., & Somers, J. M. (2002). Prevalence and incidence studies of schizophrenic disorders: A systematic review of the literature. *Canadian Journal of Psychiatry, 47*(9), 833–843.

Goldstein, K. (1980). *Language and language disturbances.* New York: Grune & Stratton.

Goldstein, M. Z., Pataki, A., & Webb, M. T. (1996). Alcoholism among elderly persons. *Psychiatric Services, 47* (9), 941–943.

Gordon, A. (2007). The Jewish view of death: Guidelines for mourning. Reprinted from E. Kubler-Ross (Ed.) (1975) *Death: The final stage of growth.* Englewood Cliffs, NJ: Prentice Hall.

Gordon, H. R. (2006). Allies within and without: How adolescent activists conceptualize ageism and navigate adult power in youth social movements. *Journal of Contemporary Ethnography, 36*(6), 631–668.

Goronzy, J. (2004). Prognostic markers of radiographic progression in early rheumatoid arthritis. *Arthritis and Rheumatism, 50* (1), 43–54.

Gott, M., & Hinchliff, S. (2003). Sex and ageing: A gendered issue? *In* S. Arber, K. Davidson, & J. Gin (Eds.), *Gender and ageing: new directions.* Buckingham: Open University Press.

Gott, M. (2005). *Sexuality, sexual health and ageing.* Maidenhead, Berkshire: Open University Press.

Gotthardt, M. (2005). Across the divide. *AARP: The magazine.* March/April 2005; p. 106.

Gouldner, A. (1960). The norm of reciprocity: A preliminary statement. *American Sociological Review, 25,* 161–178.

Graham, J. E., Rockwood, K. Beattie, B. L., Eastwood R., Gauthier S., Tukko, H., & McDowell I., (1997). Prevalence and severity of cognitive impairment with and without dementia in an elderly population. *Lancet, 349,*1793–1796.

Graham, J. E., Rockwood, K., Beattie, B. L., Eastwood, R., Gauthier, S., Tukko, H., & McDowell, I. (1997). Prevalence and severity of cognitive impairment with and without dementia in an elderly population. *Lancet, 349,* 1793–96.

Gray Panthers (2009). Gray Panthers Home Page. http://www.graypanthers.org. Retrieved September 18, 2009.

Greco, A. J., & Swayne, L. E. (1999). Sales response of elderly consumers to point-of-purchase advertising. *Journal of Advertising Research, September/October,* 43–53.

Green, F. (2009). Growing old behind bars. *Richmond times-dispatch.* January 4. Retrieved November 20, 2009.

Greenberg, J. S., Seltzer, M. M., & Greenley, J. R. (1993). Aging parents of adults with disabilities: The gratifications and frustrations of later-life caregiving. *Gerontologist, 33,* 542–550.

Greenhalgh, T. (1997). I told you I was ill. *Accountancy, 119* (1241), 16.

Griffin, K. (2005). You're wiser now: A new look at the surprising resilience and growth potential of the human brain. AARP [American Association of Retired Persons], http://www.aarpmagazine.org/health. Retrieved June 30, 2090.

Griffin, R. From sacred to secular. *In Aging and the Religious Dimension,* edited by L. Thomas and S. Eisenhandler, 90–112, Westport, CT: Auburn House, 1994.

Guilford, J. P. (1967). *The nature of human intelligence.* NY: McGraw-Hill.

Guilford, J. P. (1966). Intelligence: 1965 model. *American Psychologist, 21,* 20–26.

Guillemard, A-M., & Rein, M. (1993). Comparative patterns of retirement: Recent trends in developed societies. *Annual Review of Sociology, 19,* 469–503.

Gutmann, D. L. (1992). Toward a dynamic geropsychology. *In* J. Birren, M. Eagle, and D. Welitzky (Eds.), *Interface of psychoanalysis and psychology,* p. 284–295. Washington, DC: American Psychological Association.

Gutmann, D. L. (1987) *Reclaimed powers: Toward a new psychology of men and women in later life*. New York: Basic Books.

Haass, C., & Selkoe, D. J. (1998). A technical KO of amayloid-B peptide. *Nature, 391 (2),* 339–340.

Hagenaars, A., & de Vos, K. (1988). The definition and measurement of poverty. *Journal of human resources, 23,* 211–221.

Hagestad, G. O., & Uhlenberg, P. (2995). The social separation of old and young: A root of ageism. *Journal of Social Issues, 61(2),* 343–360.

Haight, B. K. (1996). Suicide risk in frail elderly people relocated to nursing homes. *Geriatric Nursing, 16 (5),* 104–107.

Hajjar, R. R., & Kamel, H. K. (2003). Sexuality in the nursing home, part I: Attitudes and barriers to sexual expression. *Journal of the American Medical Directors Association, 4*(3), 152–156.

Harman, D. (1995). Free radical theory of aging: Alzheimer's disease pathogenesis. *Age, 18* (3), 97–119.

Harris, M. B. (1994). Growing old gracefully: Age concealment and gender. *Journal of Gerontology: Psychological Sciences, 49,* 149–158.

Hartup, W. W., & Stevens, N. (1999). Friendships and adaptation across the life span. *Current Directions in Psychological Science, 8,* 76–79.

Harwood, J., & Giles, H. (2001). Older adults' trait ratings of three age-groups around the Pacific rim. *Fournal of cross-cultural gerontology, 16*(2), 157–202.

Harwood, J., & Giles, H. (1993). Creating intergenerational distance: Language, communication and middle-age. *Language Sciences, 15,* 1–24.

Hasemyer, D. (2009). More cremations seen amid tough economy. *San Diego Sign On.* http://signonsandiego.com. Retrieved September 20, 2009.

Havighurst, R. J. (1972). *Developmental tasks and education* (3rd ed.). New York: McKay.

Haynor, P. (1998). Meeting the challenge of advance directives. *American Journal of Nursing, 98* (3), 26–32.

HCFA [Health Care Financing Administration] (2000). *Program statistics: Medicare and Medicaid data book.* Publication No. 0333154. Washington, DC: Health Care Financing Administration.

He, W., Sengupta, M., Velkoff, V. A., & deBarros, K. A. (2005). *65+ in the United States: 2005.* U.S. Census Bureau, Current Population Reports, P23–209. Washington DC.

He, W. (2000). *The older foreign-born population in the United States.* U.S. Census Bureau, Current Population Reports. U.S. Government Printing Office, Washington, DC.

Health Fraud (no date). Health Fraud. http://www.nlm.nih.gov/medlineplus Retrieved September 9, 2009.

Healthy Aging (2009). Aging in the know: Substance abuse. http://www.healthinaging.org. Retrieved September 11, 2009.

Heckhausen, J., Dixon, R. A., & Baltes, P. B. (1989). Losses in development throughout adulthood as perceived by different adult age groups. *Developmental Psychology, 25,* 109–21.

Heidrich, S. L., & Ryff, C. D. (1993). The role of social comparisons processes in the psychological adaptation of elderly adults. *Journal of gerontology: Psychological sciences, P48,* 127–136.

Henderson, C. E. (2003). Grandparent-grandchild attachment as a predictor of psychological adjustment among youth from divorced families. *Dissertation abstracts international: Section B: The sciences and engineering, 63*(9-B), 4371.

Henry, N. J. M., Berg, C. A., Smith, T. W., & Florsheim, P. (2007). Positive and negative characteristics of marital interaction and their association with marital satisfaction in middle-aged and older couples. *Psychology and Aging, 22,* 428–441.

Herdt, G., Beeler, J., & Rawls, T. (1997). Understanding the identities of older lesbians and gay men: A study in Chicago. *Journal of gay, lesbian, and bisexual identity, 2* (3/4), 231–245.

Hernandez, G. (1992). The family and its aged members: The Cuban experience. *In* T. Brink (Ed.), *Hispanic aged mental health.* Binghamton, NY: Haworth.

Hertzog, D., Lindenberger, U., Ghisletta, P., & Oertzen, T. (2006). On the power of multivariate latent growth curve models to detect individual differences in change. *Psychological Methods, 11,* 244–252.

Hess, T. M. (2005). Memory and aging in context. *Psychological Bulletin, 131*(3), 383–406.

Hevesi, D. (1994). Anger wins out over fear of a gang. *New York Times, 31L,* March 27.

Hickey, N. (1990). Its audience is aging…So why is TV still chasing the kids? *TV Guide,* 20 October, 22–24.

Hicks, L. L., Rantz, M. J., Petroski, G. F., Madsen, R. W., Conn, V. S., Mehr, D. R., & Porter, R. (1997). Assessing contributions to cost of care in nursing homes. *Nursing Economics, 15* (4), 205–212.

Hillier, S. (2008). Classroom exercise: Basic gerontology journey through adulthood. Sonoma State University, California. Unpublished.

Hillman, J., & Stricker, G. (1996). Predictors of college students' knowledge of attitudes toward elderly sexuality: The relevance of grandparental contact. *Educational Gerontology, 22,* 539–555.

Hobbs, F. B., & Damon, B. L. (1996). *65+ in the United States.* http://www.census.gov. Retrieved May 21, 2006.

Hoch, C., & Slayton, R. (1989). *New homeless and old.* Philadelphia, PA: Temple University.

Hoefler, J. (1994). *Deathright: Culture, medicine, politics and the right to die.* Boulder, CO: Westview Press

Holden, K. C., & Kuo, H.-H. D. (1996). Complex marital histories and economic well-being: The continuing legacy of divorce and widowhood as the HRS cohort approaches retirement. *The Gerontologist, 36* (3), 383–390.

Hooker, K., & Kaus, C. R. (1994). Health-related possible selves in young and middle adulthood. *Psychology and Aging, 9,* 126–133.

Hooyman, N. R., & Kiyak, H. A. (2002). *Social gerontology: A multidisciplinary perspective* (6th ed.). Boston, MA: Allyn & Bacon. (First published 1996.)

Hooyman, N. R., & Kiyak, H. A. (2006). *Social gerontology: A multidisciplinary perspective* (7th ed.). Boston, MA: Allyn & Bacon.

Horgan, J. (1997). Seeking a better way to die. *Scientific American, 100* –105.

Horn, J. L., & Noll, J. (1997). Human cognitive abilities: Gf-Ge theory. *In* D. P. Flanagan & J. L. Genshaft (Eds.), *Contemporary intellectual assessment,* (53–91). NY: Guilford Press.

Horton, S., & Deakin, J. M. (2007). Role models for seniors and society: Seniors' perceptions of aging successfully. *Journal of Sport & Exercise Psychology, 29,* S14–S15.

Hoyer, W. J., & Roodin, P. A. (2009). *Adult Development and Aging.* SF: McGraw Hill Publishing.

Hoyert, D. L., & Rosenberg, H. M. (1997). Alzheimer's disease as a cause of death in the United States. *Public health reports, 112,* 497–505.

HRSA (2008). Homeless and elderly: understanding the special health care needs of elderly persons who are homeless. http://bphc.hrsa.gov. Retrieved August 20, 2009.

http://people-press.org/reprt/97 (1998). Young, old differ on using surplus to fix Social Security. Retrieved August 12, 2009.

http://www.ag.ohio-state.edu/~seniors/. (no date) Older adults. Retrieved August 6, 2009.

http://www.bmj.com/cgi/content. (2009). Plight of elderly made homeless in hospital. Retrieved August 25, 2009.

http://www.buildingonline.com/addurl (2009). NAHB Survey Shows Credit Woes Threaten Housing Recovery. BuildingOnline Inc. Dana Point, California. Retrieved December 2, 2009.

http://www.dyingwithdignity.ca. Retrieved September 15, 2009.

http://www.ich.gov/library (2008). Letter from Philip F. Mangano, Executive Director, United States Interagency Council on Homelessness to the Congressional Budget Justification committee. Retrieved August 21–2009.

http://www.ich.gov/library (2008). Letter from Philip F. Mangano, Executive Director, United States Interagency Council on Homelessness to the Congressional Budget Justification committee. Retrieved August 21, 2009.

http://www.iwpr.org. (2008). Women and Social Security Institute for Women's Policy Research. Retrieved June 2, 2009.

http://www.myhealthcaremanager.com/Files/Update_8_Senior%20Fraud.pdf

http://www.ninds.nih.gov/disorders/cjd/. Retrieved July 15, 2009.

http://www.un.org/esa/policy/wess/wess2007files/

Hueston, W. J., Mainous III, A. G., & Schilling, R. (1996). Patients with personality disorders: Functional status, health care utilization, and satisfaction with care. *The Journal of Family Practice, 42* (1), 54–60.

Hummert, M. L. (1995). Judgments about stereotypes of the elderly: Attitudes, age associations, and typicality ratings of young, middle-aged, and elderly adults. *Research on Aging, 17* (2), 168–189.

Hummert, M. L. (1999). A social cognitive perspective on age stereotypes. *In* T. M. Hess & F. Blanchard-Fields (Eds.), *Social cognition and aging* (pp. 175–196). San Diego: Academic Press.

Hummert, M. L., Garetka, T. A., Shaner, J. L., & Strahm, S. (1994). Stereotypes of the elderly held by young, middle-aged and elderly adults. *Journal of Gerontology: Psychological Sciences, P49,* P240–49.

Humphry, D. (2005). Farewell to hemlock: Killed by its name: Modern history of the U.S. right-to-die movement. http://www.assistedsuicide.org/farewell 2005. Retrieved September 9, 2009.Humphry…1985, 1991.

Huyck, M. H. (2001). Romantic relationships in later life. *Generations, 25*(2), 9–17.

Hyse, K., & Tornstam, L. (2009). Recognizing aspects of oneself in the theory of gerotranscendence. Online publication from the Social Gerontology Group, Uppsala. Department of Sociology, Uppsala University. http://www.soc.uu.se/research/gerontology. Retrieved April 30 2009.

Identifying and Preventing Frauds Against Older People. http://www.ci.worcester.ma.us/wpd. (no date). Retrieved Retrieved September 2, 2009.

ILC [Internatonal Longevity Center-USA] (2009). *Alzheimer's: The disease of the century.* ILC Occasional Paper. Mount Sinai School of Medicine.

Intrieri, R. C., Kelly, J. A., Brown, M. M., & Castilla, C. (1993). Improving medical students' attitudes toward and skills with the elderly. *The Gerontologist, 33,* 373–378.

Italian Multicentric Study on Centenarians [IMSC]. (1997). Epidemiological and socioeconomic aspects of Italian centenarians. *Archives of gerontology and geriatrics, 25,* 149–157.

Jacob, M., & Guarnaccia, V. (1997). Motivational and behavioral correlates of life satisfaction in an elderly sample. *Psychological Reports, 80,* 811–818.

Jacobs, R. (1993). *Be an outrageous older woman: ARASP.* Manchester, CT: Knowledge, Ideas and Trends Publishing.

Jail doctor wins age-bias settlement. (1998). *New York Times, 5* January, A13.

James, G. (1994). Man is held in swindles of elderly. *New York Times, 29,* B3 (L)

Jellison, F. M. (2006). *Managing the dynamics of change: The fastest path to creating an engaged and productive workforce.* NY: McGraw-Hill.

John, R. (1991). Family support networks among elders in a Native American community: Contact with children and siblings among the Prairie Band Potawatomi. *Journal of Aging Studies, 5*(1), 45–59.

Johns Hopkins (1999). Alzheimer's disease retrospective. *Medical Letter: Health After 50, 11*(8), 7–8.

Johnson, C. L. (1996). Cultural diversity in the latelife family. *In* R. Blieszner & V. H. Bedford (Eds.), *Aging and the family: Theory and research,* 218–223. Westport, CT: Praeger.

Joseph, J. (1997). Fear of crime among black elderly. *Journal of Black Studies, 27*(5), 698–717.

Jung, C. (1916/1956). *Two essays on analytical psychology.* New York: Meridian.

Jung, C. *Modern man in search of a soul.* San Diego, CA: Harcourt-Brace, 1955.

Jung, C. G. (1918/1959). *Aion: Researches into the phenomenology of the self,* (2nd ed.) (R.F.C. Hull, Trans.). Princeton, NJ: Princeton University Press.

Kalet, A. L., Fletcher, K. E., Ferdman, D. J., & Bickell, N. A. (2006). Defining, navigating, and negotiating success: The experiences of mid-career Robert Wood Johnson clinical scholar women. *Journal of Internal Medicine, 21,* 920–925.

Kalish, R. A. (1985). *Death, grief, and caring relationships.* Belmont, CA: Brooks/Cole.

Kalish, R. A. (1987). Death and dying. *In* G. Busse (Ed.), *Elderly as pioneers* (pp. 360 –385). Bloomington: Indiana University Press.

Kalymun, M. (1992). Board and care vs assisted living. *Adult residential care journal, 6*(1), 35–44.

Kant, G. L., d'Zurilla, T. J., & Maydeu-Olivares, A. (1997). Social problem solving as a mediator of stress-related depression and anxiety in middle-aged and elderly community residents. *Cognitive Therapy and Research, 21*(1), 73–96.

Karpinski, J. (1997). Engaging and treating the self-neglecting elder. *Journal of Geriatric Psychiatry, 42,* 133–151.

Kastenbaum, R. J. (2004). Death, society, and human experience. Pearson Press: New York.

Kaufman, A. S., & Horn, J. L. (1996). Age changes on tests of fluid and crystallized ability for women and men on the Kaufman adolescent and adult intelligence test (KAIT) at ages 17–94 years. *Archives of Clinical Neuropsychology, 11* (2), 97–121.

Kaufman, S. (1986). *The ageless self: Sources of meaning in later life.* Madison: University of Wisconsin Press.
Kaufman, S. (1993). Reflections on the ageless self. *Generations, 17,* 13–16.

Kaufman, S. (1987). *The ageless self: Sources of meaning in later life.* Madison: University of Wisconsin Press.

Kausler, D. H., & Kausler, B. C. (1996). *The graying of America: An encyclopedia of aging, health, mind, and behavior.* Chicago: University of Illinois Press.

Kelley, S. J., Yorker, B. C., & Whitley, D. (1997). To grandmother's house we go…and stay: Children raised in intergenerational families. *Journal of Gerontological Nursing, 23*(1), 12–20.

Kemper, S. (1994). "Elderspeak": Speech accommodation to older adults. *Aging and Cognition, 1,* 17–28.

Kennard, C. (2006). Reminiscence therapy and activities for people with dementia. http://dying.about.com/od/thedyingprocess. Retrieved September 9, 2009.

Kimmel, D., Rose, R. T., Orel, N., & Green, B. (2006). Historical context for research on lesbian, gay, bisexual, and transgender aging. *In* D. Kimmel, T. Rose, & S. David (Eds.), *Lesbian, gay, bisexual, and transgender aging: Research and clinical perspectives* (p. 2–19). New York: Columbia University Press.

King, V., & Elder, G. H. (1997). The legacy of grandparenting: Childhood experiences with grandparents and current involvement with grandchildren. *Journal of Marriage and the Family, 59,* 848–859.

King, V., Elder, G. H., & Conger, R. D. (2000). Wisdom of the ages. *In* G. H. Elder, Jr., & R. D. Conger, *Children of the land: Adversity and success in rural America.* Chicago: University of Chicago Press.

Kinsella, K., & Gist, Y. J. (1998). *International brief: Gender and aging.* U.S. Department of Commerce, Bureau of the Census: Washington, DC.

Kinsella, K., & He, W. (2009). *An aging world 2008.* U.S. Census Bureau, International Population Reports, P95-09-1. U.S. Government Printing Office, Washington, DC.

Kinsella, K., & Velkoff, V. A. (2001). Gender stereotypes: Data needs for ageing research. *Ageing International, 24*(4), 18–31.

Kinsey, A. (1948). *Sexual behavior in the human male.* Philadelphia, PA: W. B. Saunders.

Kinsey, A. (1953). *Sexual behavior in the human female.* Philadelphia, PA: W. B. Saunders.

Kirkland, R. I. (1994). Why we will live longer—and what it will mean. *Fortune, February 21,* 66–77.

Kitano, H. (1991). *Race relations.* Englewood Cliffs, NJ: Prentice Hall.

Kleemeier, R. W. (1962). Intellectual change in the senium. *Proceedings of the social statistics section of the American statistical association, 1,* 290–295.

Klemmack, D., & Roff, L. L. (1984). Fear of personal aging and subjective well-being in later life. *Journal of Gerontology, 39,* 756–758.

Klohnen, E. C., & Vandewater, E. A. (1996). Negotiating the middle years: Ego-resilience and successful midlife adjustment in women. *Psychology and Aging, 11* (3), 431–442.

Knox, R. (1989). Crisis predicted in state's elder care. *Boston Globe,* 14 March, A2.

Kobrin, S. (2005). They've gone the distance. *Los Angeles times.* 17 January.

Koenig, H. G., Wildman-Hanlon, D., & Schmader, K. (1996). Attitudes of elderly patients and their families toward physician-assisted suicide. *Archives of Internal Medicine, 156* (9), 2240 –2248.

Koh, A. (1999). Non-judgmental care as a professional obligation. *Nursing Standard, 13* (37), 38–41.

Kolata, G. (1992). Elderly become addicts to drug-induced sleep. *New York Times, E4,* February 2.

Kramer, A. F., & Madden, D. J. (2008). Attention. *In* F. I. M. Craik & T. A. Salthouse (Eds.), *Handbook of aging and cognition.* NJ: Erlbaum.

Kratch, P., deVaney, S., deTurk, C., & Zink, M. H. (1996). Functional status of the oldest-old in a home setting. *Journal of Advanced Nursing, 25*, 456–464.

Krause, N. (2006a). Religious doubt and psychological well-being: A longitudinal investigation. *Review of Religious Research, 47* (3), 287–302.

Krause, N. (2006b). Church-based social support and morality. *Journals of Gerontology Series B—Psychological Sciences and Social Sciences, 61*(3), S35–S43.

Krause, N. (2008). *Aging in the church: How social relationships affect health.* PA: Tempelton Foundation Press.

Krause, N. (2009). Meaning in life and mortality. *Journals of Gerontology Series B: Psychological Sciences and Social Sciences, 65B*(4), 517–427.

Kruger, A. (1994). The midlife transition: Crisis or chimera? *Psychological Reports, 75*, 1299–1305.

Kübler-Ross, E. (1969). *On death and dying.* New York: Macmillan.

Kung, H. C., Hoyert, D. L., Xu, J., & Murphy, S. L. (2008). Deaths: final data for 2005. National Vital Statistics Reports. 56(10).

Kurdek, L. A. (2005). What do we know about gay and lesbian couples? *Current Directions in Psychological Science, 14*, 251–254.

Kurdek, L. A. (2006). Differences between partners from heterosexual, gay, and lesbian cohabiting couples. *Journal of Homosexuality, 12*, 85–99.

Kutty, N. (1998). The scope for poverty alleviation among elderly home-owners in the United States through reverse mortgages. *Urban Studies, 35* (1), 113–129.

Kvale, J. N. (2003). International aging—India. *Journal of the American Geriatrics Society, 51*(1), 137–138.

Lachs, M. S., Williams, C., O'Brien, S., Hurst, L., & Horwitz, R. (1997). Risk factors for reported elder abuse and neglect: A nine-year observational cohort study. *Gerontologist, 37* (4), 469–474.

Laird, J. (1996). Invisible ties: Lesbians and their families of origin. *In* J. Lairs & R.-J. Green (Eds.), *Lesbians and gays in couples and families: A handbook for therapists.* San Francisco: Jossey-Bass.

Lamb, H., Christie, J., Singelton, A. B., Leake, A., Perry, R. H., Ince, P. G., McKeith, I. G., Melton, L. M., Edwardson, J. A., & Morris, C. M. (1998). Apolipoprotein E and alpha-1 antichymotrypsin polymorphism genotyping in Alzheimer's disease and in dementia with Lewy bodies. *Neurology, 50*, 388–391.

Lambert, W., & W., J. (1994). CBS settlement. *Wall Street Journal,* 29 April, B5(W), B4(E).

Lang, F. R., & Carstensen, L. L. (1994). Close emotional relationships in late life: Further support for proactive aging in the social domain. *Psychology & Aging, 9* (2),

Lang, F. R., Baltes, P. B., & Wagner, G. G. (2007). Desired lifetime and end-of-life desires across adulthood from 20 to 90: A dual-source information model. *Journal of Gerontology: Psychological Sciences, 62B*, P268–P276.

Lanier, D. N., & Dietz, T. L. (2009). Elder criminal victimization: Its relative rate compared to non-elders, 1992–2005. *Social Science Journal, 46*, 442–458.

Lanza, M. L. (1996). Divorce experienced as an older woman. *Geriatric Nursing, 17*, 166–170.

Larson, C. (2008). Keeping your brain fit. *U.S. News and World Report Online.* http://health.usnews.com/articles/health. Retrieved July 7, 2009.

Laumann, E., Michaels, M. R., & Gagnon, J. (1994). *The social organization of sexuality.* Chicago: University of Chicago Press.

Lee, W. M. L., Blando, J. A., Mizelle, N. D., & Orozco, G. L. (2006). Introduction to multicultural counseling for helping professionals. *Psychological Bulletin.*

Lemming, M. R., & Dickinson, G. E. (1994). *Understanding dying, death, and bereavement* (3rd ed). Orlando, FL: Harcourt Brace.

Lenze, D. G., & von Kerczek M., (2009). Personal Income: First Quarter 2009. http://www.bea.gov/newsreleases/rels.htm. Retrieved August 10, 2009.

Lester, D., & Savlid, A. C. (1997). Social psychological indicators associated with the suicide rate: A comment. *Psychological Reports, 80*, 1065–1066.

Levinson, D. (1977). *The seasons of a man's life.* New York: Knopf.

Levinson, D. (1996). *The seasons of a woman's life.* New York: Knopf.

Levitt, K. R., Lazenby, H. C., Sivarajan, L., Stewart, M. W., Braden, B. R., Cowan, C. A., Donham, C. S., Long, A. M., McDonnell, P. A., Sensenig, A. L., Stiller, J. M., & Won, D. K. (1996). Health care expenditures, 1994. *Health Care Financing Review, 17* (3), 205–224.

Levy, B. R., & Banaji, M. R. (2002). Implicit ageism. *In* T. Nelson (Ed.), *Ageism: Stereotypes and prejudice against older persons* (pp. 49–75). Cambridge, MA: MIT Press.

Levy, B. R. (2003). Mind matters: Cognitive and physical effects of aging self-stereotypes. *Journals of Gerontology: Series B: Psychological Sciences and Social Sciences, 58B* (4), 203–211.

Levy, B. R. (2008). Rigidity as predictor of older persons' aging stereotypes and aging self-perceptions. *Social Behavior and Personality, 36*(4), 559–570.

Levy, B. R., & Langer, E. (1994). Aging free from negative stereotypes: Successful memory in China and among the American Deaf. *Journal of Personality and Social Psychology, 66* (6), 989–997.

Lewinsohn, P. M., Seeley, J. R., Roberts, R. E., & Allen, N. B. (1997). Center for eipdemiologic studies depression scale (CES-D) as a screening instrument for depression among community-residing older adults. *Psychology and Aging, 12* (2), 277–287.

Liberto, J. G., Oslin, D. W., & Ruskin, P. E. (1992). Alcoholism in older persons: A review of the literature. *Hospital Community Psychiatry, 43*, 975–983.

Lieberman, M. S., & Tobin, S. (1983*). The experience of old age: Stress, coping, and survival.* New York: Basic Books.

Lillard, L. A., & Panis, C. W. A. (1996). Marital status and mortality: The role of health. *Demography, 33*(3), 313–327.

Lindau, S. T., *et al.* (2007). A national study of sexuality and health among older adults in the U.S. *New England Journal of Medicine, 357,* 762–774.

Linewater, T. T., Berger, A. K., & Hertzog, C. (2009). Expectations about memory change across the life span are impacted by aging stereotypes. *Psychology and Aging, 24 (1)*, 169–176.

Linville, P. W. (1982). The complexity-extremity effect and age-based stereotyping. *Journal of personality and social psychology, 42,* 183–211.

Liu, H., & Umberson, D. (2008). The times they are a changin': Marital status and health differentials from 1972 to 2003. *ScienceDaily.* http://www.sciencedaily.com. Retrieved July 24, 2009.

Lockenhoff, C. E., Terracciano, A., Bienvenu, O. J., Patriciu, N. S., Nestady, G., McCrae, R. R., Eaton, W. W., & Costa, P. T. (2008). Ethnicity, education, and the temporal stability of personality traits in the East Baltimore Epidemiologic Catchment Area study. *Journal of Research in Personality, 42*(3), 577–598.

Lodi-Smith, J., Geise, A. C., Roberts, B. W., & Robins, R. W. (2009). Narrating personality change. *Journal of Personality and Social Psychology, 96*(3), 679–689.

Loevinger, J. (1976). *Ego development: Conceptions and theories.* San Francisco: Jossey-Bass.

Lofland, L. H. (1979). *The craft of dying: The modern face of death.* Beverly Hills, CA: Sage.

Lopata, H. Z. (1988). Support systems of American urban widowhood. *Journal of Social Issues, 44 (3),* 113–128.

Lu, P.-C. (2007). Sibling relationships in adulthood and old age. *Current Sociology, 55 (4),* 621–637.

Lucas, R. E. (2007). Long-term disability is associated with lasting changes in subjective well-being: Evidence from two nationally representative longitudinal studies. *Journal of Personality and Social Psychology, 92,* 717–730.

Lucas, R. E., Clark, A. E., Georgellis, Y., & Diener, E. (2003). Reexamining adaptation and the set point model of happiness: Reactions to changes in marital status. *Journal of Personality and Social Psychology, 84,* 527–539.

Lussier, G., Deater-Deckard, K., Dunn, J., & Davies, L. (2002). Support across two generations: Children's closeness to grandparents following parental divorce and remarriage. *Journal of Family Psychology, 16*(3), 363–367.

Lustig, C., May, C. P., & Hasher, L. (2001). Working memory span and the role of proactive interference. *Journal of Experimental Psychology: General, 130,* 199–207.

Luszcz, M. A., Bryan, J., & Kent, P. (1997). Predicting episodic memory performance of very old men and women: Contributions from age, depression, activity, cognitive ability, and speed. *Psychology and Aging, 12 (2),* 340–351.

Lyles, K. W., Colon-Emeric, C. S., Magaziner, J. S., *et al.* (2007). Zoledronic acid and clinical fractures and mortality after hip fracture. *New England Journal of Medicine, 357,* published online 2007-09-17. Retrieved April 2, 2009.

Madden, M. M. (1997). Strengthening protection of employees at home and abroad: The extraterritorial application of Title VII of the civil rights act of 1964 and the age discrimination in employment act. *Hamline Law Review, 20 (3),* 739–768.

Magno, J. (1990). The hospice concept of care: Facing the 1990s. *Death Studies, 14,* 109–119.

Mahay, J., & Lewin, A. C. (2007). Age and the desire to marry. *Journal of Family Issues, 28*(5), 706–723.

Mangen, D., & Westbrook, G. (1988). Measuring intergenerational norms. *In* D. Mangen (Ed.), *Measurement of Intergenerational Relationships.* Newbury Park, CA: Sage.

Manson, M., & Gutfeld, G. (1994). Losing the final five. *Prevention Magazine 46 (5),* 22–24.

Marcellini, F., Sensoli, C., Barbini, N., & Fioravanti, P. (1997). Preparation for retirement: Problems and suggestions of retirees. *Educational Gerontology, 23,* 337–388.

Markson, E. W. (2003) *Social gerontology today: An introduction.* Los Angeles, CA: Roxbury Publishing.

Markus, H. R., & Herzog, A. R. (1991). The role of the self-concept in aging. *In* K. W. Schaie & M. P. Lawton (Eds.), *Annual Review of Gerontology and Geriatrics* (Vol. 11, pp. 110–143). New York: Springer.

Marrone, R. (1997). *Death, mourning, and caring.* Pacific Grove, CA: Books/Cole.

Marsiske, M., & Margrett, J. A. (2006). Everyday problem solving and decision making. *In* J. E. Birren, & K. W. Schaie (Eds.), *Handbook of the psychology of aging* (6th ed.), 315–342. San Diego: Elsevier.

Masters, W. M., & Johnson, V. E. (1966). *Human sexual response.* Boston: Little Brown.

Mather, M. (2006). A review of decision-making processes: Weighing the risks and benefits of aging. *In* L. L. Carstensen & C. R. Hartel (Eds.), *When I'm 64.* Washington, DC: National Academics Press.

Mattson, M. P., Cutler, R. G., & Camandola, S. (2007). Energy intake and amyotrophic lateral sclerosis. *Neuromolecular Medicine, 9 (1),* 17–20.

Matzo, M. L. (1997). The search to end suffering: A historical perspective. *Journal of Gerontological Nursing, 23 (3),* 11–17.

Mayeux, R., Saunders, A., Shea, S., Mirra, S., Evans, D., Roses, A., Hyman, B., Crain, B. Tang, M.-X., & Phelps, C. (1998). Utility of the apoliproprotein E genotype in the diagnosis of Alzheimer's disease. *The New England Journal of Medicine, 338,* 506–511.

Mayo Clinic (2007). *Metabolic Syndrome.* http://www.mayoclinic.com/health. Retrieved May 22, 2009.

Mays, V. M., Cochran, S. D., & Barnes, N. W. (2007). Race, race-based discrimination, and health outcomes among African Americans. *Annual Review of Psychology, 58,* 24.1–24.25.

McAdams, D. P. (1995). What do we know when we know a person? *Journal of Personality, 63* (3), 365–375.

McAdams, D. P. (2005). Tracing religions belief, practice, and change. *In* M. Dillon and P. Wink (Eds.), *In the course of a lifetime.* California: UC Press.

McAdams, D. P. (2006). *The Person: A New Introduction to Personality Psychology,* (4th Ed.). New York: Wiley.

McAdams, D. P. (2007). *The Redemptive Self: Stories Americans Live By.* New York: Oxford University Press.

McAdams, D. P., & Bowman, P. J. (2001). Narrating life's turning points: Redemption and contamination. *In* D. P. McAdams, R. Josselson, & A. Lieblich (Eds.), *Turns in the road: Narrative studies of lives in transition.* Washington, DC: American Psychological Association.

McAdams, D. P., & de St. Aubin, E. (1992). A theory of generativity and its assessment through self report, behavioral acts, and narrative themes in autobiography. *Journal of Personality and Social Psychology, 62* (6), 1003–1015.

McAdams, D. P., Diamond, A., de St. Aubin, E., & Mansfield, E. (1997). Stories of commitment: The psychosocial construction of generative lives. *Journal of personality and social psychology, 72* (3), 678–694.

McAdams, D. P., & Pals, J. L. (2006). A new big five: Fundamental principles for an integrative science of personality. *American Psychologist, 61,* 204–217.

McCoy, H. V., Wooldredge, J. D., Cullen, F. T., Dubeck, P. J. & Browning, S. L. (1996). Lifestyles of the old and not so fearful: Life situation and older persons' fear of crime. *Journal of Criminal Justice, 24*(3), 191–205.

McGarry, K. (1996). Factors determining participation of the elderly in supplemental security income. *The Journal of Human Resources, 31* (2), 331–358.

McIntosh, J. (2008). How to help a suicidal older men and women. http://www.healthyplace.com. Retrieved September 2, 2009.

McMorris, F. A. (1997). Age-bias suits may become harder to prove. *The Wall Street Journal,* 20 February, B1.

Meyer, D. R., & Bartolomei-Hill, S. (1994). The adequacy of supplemental security income benefits for aged individuals and couples. *The Gerontologist, 34* (2), 161–172.

Meyer, M. J. (1995). Dignity, death and modern virtue. *American Philosophical Quarterly, 32* (1), 45–55.

Michaels, S. (1996). The prevalence of homosexuality in the United States. *In* R. P. Cabaj & T. S. Stein (Eds.), *Textbook of homosexuality and mental health.* Washington, DC and London: American Psychiatric Press.

Miller, B., & Kaufman, J. E. (1996). Beyond gender stereotypes: spouse caregivers of persons with dementia. *Journal of Aging Studies, 10*(3), 189–204.

Mills, C. W. (1959). *The sociological imagination.* Oxford University Press: NY

Minda, G. (1997). Aging workers n the postindustrial era. *Stetson law review, 26* (2), 561–597.

Minkler, M., & Roe, K. (1993). *Grandmothers as caregivers: Raising children of the crack cocaine epidemic.* Newbury Park, CA: Sage.

Mintz, H. (1998). Trial opens for thrift accused of elder fraud. *San Jose Mercury News,* 1B, 17 January.

Mittelman, M. S., Ferris, S. H., Shulman, E., Steinberg, G., Ambinder, A., Mackell, J. A., & Cohen, J. (1995). A comprehensive support program: Effect on depression in spouse-caregivers of AD patients. *Gerontologist, 35*(6), 792–802.

Moffitt, T. E., Caspi, A., & Rutter, M. (2006). Strategy for investigating interactions between measured genes and measured environments. *Archives of General Psychiatry, 62,* 473–481.

Mokni, M., Elkahoui, S., Limam, F., Amri, M., & Aouani, E. (2007). Effect of resveratrol on antioxidant enzyme activities in the brain of healthy rat. *Neurochem Research, 32* (6), 981–987.

Montaigne, quoted in Marrone, 1997.

Montepare, J. M., & Zebrowitz, L. A. (1998). Person perception comes of age: The salience and significance of age in social judgments. *In* M. P. Zanna (Ed.), *Advances in Experimental Social Psychology, 30* (pp. 93–161). San Diego, CA: Academic Press.

Moody, H. R. (200). *Aging: Concepts and controversies.* Thousand Oaks, CA: Pine Forge Press.

Moorman, S. M., Booth, A., & Fingerman, K. L. (2006). Women's romantic relationships after widowhood. *Journal of Family Issues, 27,* 1281–1304.

Morgan, D. L., Neal, M. B., & Carder, P. C. (1997). Both what and when: The effects of positive and negative aspects of relationships on depression during the first 3 years of widowhood. *Journal of Clinical Geropsychology, 3* (1), 73–91.

Morin, R. (2009). Most middle-aged adults are rethinking retirement plans: The Threshold Generation. Pew Research Center, http://pewsocialtrends.org/pubs/735. Retrieved July 23. 2009.

Morin, C., & Cohn, A. (2008). The natural history of insomnia. *Archives of internal medicine, 169*(5), 447–453.

Morin, R. (2009). Most middle-aged adults are rethinking retirement plans. The Threshold Generation. Pew Research Center, http://www.pewsocialtrends.org/pubs. Retrieved August 3, 2009.

Morley, J. E., & Kaiser, F. E. (1989). Sexual function with advancing age. *Medical Clinics of North America, 73* (6), 1483–1495.

Morris, J. N. (1996). Exercise in the prevention of coronary heart disease: Today's best buy in public health. *Medicine and Science in Sports and Exercise, 26,*

Morris, J. N., Heady, J. A., & Raffle, P. A. B. (1953). Coronary heart disease and physical activity at work. *Lancet, 265,* 1053–1105.

Morris, P. L., Robinson, R. G., & Samuels, J. (1993). Depression, introversion and mortality following stroke. *Australian and New Zealand Journal of Psychiatry, 23,* 443–449.

Morrow, A. (2009). Scared to death—of death. The fear of death and dying. http://dying.about.com/od/thedyingprocess. Retrieved September 12, 2009.

Mortmier, J. A., Borenstein, A. R., Gosche, K. M., & Snowdon, D. A. (2005). Very early detection of Alzheimer neuropathology and the role of brain reserve in modifying its clinical expression. *Journal of Geriatric Psychiatry and Neurology, 18*, 218–223.

Mroczek, D., & Little, T. (2006). *Handbook of Personality Development*. Mahwah, NJ: Erlbaum.

Mrozcek, D. K., Spiro, A., III, & Griffin, P. W. (2006). Personality and aging. *In* J. E. Birren and K. W. Schaie (Eds.), *Handbook of the psychology of aging, 6e,* 363–377.

Mudrack, P. E. (1997). Protestant work-ethic dimensions and work orientations. *Personality and Individual Differences, 23* (2), 217–225.

Mueller, M. M., Wilhelm, B., & Elder, G. H., Jr., (2002). Variations in grandparenting. *Research on Aging, 24*(3), 360–388.

Mui, T. Y., Leng, L. T., & Traphagan, J. W. (2005). Introduction: Aging in Asia—Perennial concerns on support and caring for the old. *Journal of Cross Cultural Gerontology, 20*, 257–267.

Mulroy, T. M. (1996). Divorcing the elderly: Special issues. *American Journal of Family Law, 10*, 65–70.

Murer, M. J. (1997). Assisted living: The regulatory outlook. *Nursing homes: Long Term Care Management, 46*(7), 24–29.

Mutter, S. A., & Goedert, K. M. (1997). Frequency discrimination vs. frequency estimation: Adult age differences and the effect of divided attention. *Journal of Gerontology: Psychological Sciences, 52B*(6), P319–P328.

My Health Care Manager (2008). Update: Senior fraud. http://www.myhealthcaremanager.com. Retrieved September 12, 2009.

NAHB Research Center (2005). *National Older Adults Housing Survey.* http://www.nahbrc.org. Retrieved August 22, 2009.

Nakazawa, D. J. (2006). Living longer: Diet. *AARP (American Association of Retired Persons) Magazine,* September & October.

National Center for Health Statistics (1999). *Vital statistics of the United States, 2* (A). Washington: U.S. Public Health Service.

National Center for Health Statistics [NCHS] (1999). Homelessness among elderly persons. National Coalition for the Homeless Fact Sheet #15. June.

National Center for Health Statistics [NCHS] (1999). *National Vital Statistics.* Government Printing Office: Washington, DC.

National Center for Health Statistics [NCHS] (2004a). *Older Americans 2004: Key Indicators of Well-Being.* Federal Interagency Forum on Aging Related Statistics, http://www.agingstats.gov. February 2005.

National Center on Women and Aging (2002). Executive summary: 2002 national poll; women 50+. Heller School for Social Policy and Management, Brandeis University.

National Committee to Preserve Social Security and Medicare (2009). Mission statement of the National Committee. http://: www.ncpssm.org/mission. Retrieved September 20, 2009.

National Consumers League (1999). They can't hang up: Help for elderly people targeted by fraud. http://www.fraud.org/elderfraud. Retrieved September 7, 2009.

National Council for Senior Citizens [NCSC] (2009). Council for Senior Citizen home page. http://www.councilseniorcitizens.org. Retrieved September 20, 2009.

National Institute of Mental Health [NIMH] (2009). Suicide in the U.S.: statistics and prevention. http://www.nimh.nih.gov. Retrieved August 29, 2009.

National Institute of Mental Health [NIMH] (2002). Mental Health: A Report of the Surgeon General: Older Adults and Mental Health. http://www.surgeongeneral.gov/library/mental health/chapter5/. Retrieved July 2, 2009.

National Institute of Mental Health [NIMH] (2003). *Older adults: Depression and suicide facts.* NIH Publication No. 03-4593. Department of Health and Human Services.

National Institute of Mental Health [NIMH] (2009). Older Adults' Mental Health Facts (Fact Sheet). http://www.himh.hih.gov. Retrieved June 7, 2009.

National Institute of Neurological Disorders [NINDS] (2008). Parkinson's Disease: Hope Through Research. http://www.ninds.nih.gov. Retrieved July 2, 2009.

National Institute of Neurological Disorders [NINDS] (2009). Creutzfeld-Jackob fact sheet. http://www.ninds.nih.gov. Retrieved July 13, 2009.

National Library of Medicine & National Institutes of Health (2009). Medline Plus. http://www.nlm.nih.gov/medlineplus. Retrieved May 4, 2009.

National Public Radio [NPR] (2009). *Hunger in America.* July 3.

Nelson, L., Brown, R., Gold, M., Ciemnecki, A., & Docteur, E. (1997). Access to care in Medicare HMOs, 1996. *Health Affairs, 16* (2), 148–156.

Neugarten, B. L. (1977). Personality and aging. *In* J. E. Birren and K. W. Schaie (Eds.), *Handbook of the psychology of aging.* New York: Academic Press.

Neugarten, B. L. (1987). Interpretative social sicences and research on aging. *In* A. Rossi (Ed.), *Gender and the life course.* New York: Aldine.

Newbold, D. C. (1996). Determinants of elderly interstate migration in the United States, 1985–1990. *Research on Aging, 18*(4), 451–476.

Newman, E. (1984). *Elderly criminals.* Cambridge, MA: Oelgeschlager, Gunn, & Hain.

NHCAA [National Health Care Antifraud Association] (no date). The problem of health care fraud. http://www.nhcaa.org/eweb. Retrieved September 5, 2009.

Niederehe, G. (1997). Future directions for clinical research in mental health and aging. *Behavior Therapy, 28*, 101–108.

Nielsen, L., Knutson, B., & Carstensen, L. L. (2008). Affect dynamics, affective forecasting, and aging. *Emotion, 8*(3), 318–330.

Nolan, D. C. (1997). Assisted living: Moving its availability down the income scale. *Nursing Homes: Long Term Care Management, 46*(7), 29–31.

Nuland, S. (1994). *How we die: Reflections on life's final chapter.* New York: Alfred A. Knopf, 1994.

Nye, W. (1993). Amazing grace: Religion and identity among elderly black individuals. *International Journal of Aging and Human Development, 36*(2), 103–105.

O'Connell, L. J. (1994). The role of religion in health-related decision making. *Generations, 18* (4), 27–30.

Office of Personnel Management (2002). *Elder care responsibilities of federal employees and agency programs.* http://www.opm.gov. Retrieved June 19 2009.

Ogg, J., & Bennett, G. (1992). Elder abuse in Britain. *British Medical Journal, 305,* 988–989.

Older Americans Act [OAA] (2008). http://www.medicareadvocacy.org.

Olson, S. (1998). Senior housing: A quiet revolution. *Architectural Record, 186*(1), 103–106.

Online Fraud (2007). Online fraud and the 50-plus. http://www.20plus30.com/blog/2007. Retrieved September 3, 2009.

Optarny, S. (1991). Women who stay vital past seventy. *San Francisco Examiner,* 3 March, p. A1–A3.

Paffenbarger, R. S., Jr., & Lee, I.-M. (1996). Physical activity and fitness for health and longevity. *Research quarterly for exercise and sport, 67* (3), 11–28.

Palmore, E. (2001). The ageism survey: First findings. *The Gerontologist, 41,* 572–575.

Palmore, E. B., Nowlin, J. B., & Wang, H. S. (1985). Predictors of function among the old-old: A ten-year followup. *Journal of Gerontology, 40*(2), 244–250.

Palmore, E. B. (1990). *Ageism, negative and positive.* New York: Springer.

Parker, R. (1995). Reminiscence: A continuity theory framework. *The gerontologist, 35* (4), 515–525.

Parkin, D. M., & Bray, F. (2006). International patterns of cancer incidence and mortality. *In* D. Scottenfeld & F. J. Fraumeni, Jr. (Eds.), *Cancer Epidemiology and Prevention,* p. 101–138. New York: Oxford University Press.

Pascucci, M. A., & Loving, G. L. (1997). Ingredients of an old and healthy life: A centenarian perspective. *Journal of Holistic Nursing, 15*(2), 199–213.

Payne, L., Mowen, A., & Montoro-Rodriguez, J. (2006). The role of leisure style maintaining the health of older adults with arthritis. *Journal of Leisure Research, 28,* 20–45.

Pearlin, L. I., & LeBlanc, A. J. (1997). Bereavement and the loss of mattering among Alzheimer's caregivers. *In* J. T. Mullan (Chair), *Bereavement and the life course: Some views from AIDS and Alzheimer's caregivers.* Symposium presented at the 50th Annual Scientific Meeting of the Gerontological Society of America, Cincinnati.

Peate, I. (1999). Need to address sexuality in older people. *British journal of community nursing, 4*(4), 174–180.

Peracchi, F., & Welch, F. (1994). Trends in labor force transitions of older men and women. *Journal of labor economics, 12* (2), 2210–243.

Perlman, M. (1993). Late mid-life astonishment: Disruptions to identity and self-esteem. *In* N. David (Ed.), *Faces of women and aging.* Binghamton, NU: Haworth.

Peters, E., Hess, T. M., Vastfjall, D., & Auman, C. (2007). Adult age differences in dual information processes: Implications for the role of affective and deliberative processes in older adults' decision making. *Perspectives on Psychological Science, 2,* 1–23.

Peters, S. H. (1994). Book reviews: Social Sciences. *Library Journal, 119*(21), 102.

Pillemer, K., & Moore, D. (1989, June). Abuse of patients in nursing homes: Findings from a survey of staff. *Gerontologist, 29* (3), 314 –320.

Pinquart, M., & Sorensen, S. (2005). Ethnic differences in stressors, resources, and psychological outcomes of family caregiving: A meta-analysis. *Gerontologist, 45,* 90–106.

Pitts, M. J., & Nussbaum, J. F. (2006). Integrating the past and paving the future. *Journal of Language and Social Psychology, 25*(3), 197–202.

Plath, D. (2009). International policy perspectives on independence in old age. *Journal of Aging and Social Policy, 21*(2), 209–223.

Plumb, J. D. (1997). Homelessness: Care, prevention, and public policy. *Annals of Internal Medicine, 126*(12), 973–975.

Podkieks, E. (1992). National survey on abuse of the elderly in Canada. *Journal of Elder Abuse and Neglect, 4,* 5–58.

Pollan, M. (2008). *In Defense of Food.* NY: Penguin Press.

Pollet, T. V., Nettle, D., & Nelissen, M. (2006). Contact frequencies between grandparents and grandchildren in modern society: Estimates of the impact of paternity uncertainty. *Journal of Cultural Evolutionary Psychology, 4,* 203–213.

Poon, L. W. (1992). *The Georgia centenarian study.* Amityville, NH: Baywood Publishing.

Posner, R. A. (1995). *Aging and old age.* Chicago: University of Chicago Press.

Pratt, M. W., Norris J. E., Cressman, K., Lawford, H., & Hebblethwaits, S. (2008). Parents' stories of grandparenting concerns in the three-generational family: Generativity, optimism, and forgiveness. *Journal of Personality, 76,* 313–337.

Preidt, R. (2009). Dementia increasing among the 'oldest old.' Medline Plus, http://www.nim.nih.gov/medlineplus. Retrieved July 13, 2009.

Price, C. A. (2000). Facts about retirement. Ohio State University Extension Senior Series.

Prinzinger, R. (2005). Programmed ageing: the theory of maximal metabolic scope. How does the biological clock tick? *EMBO Special Report, 6,* S14–19.

Prohaska, T., Belansky, E., Belza, B., Buchner, D., Marshall, V., McTigue, K. Satariano, W., & Wilcox, S. (2006). Physical activity, public health and aging: Critical issues and research priorities. *Journals of Gerontology: Series B: Psychological Sciences and Social Sciences, 61B,* S267–273.

Quadagno, J. (2005). *Aging and the life course: An introduction to social gerontology.* NY: McGraw Hill.

Quinn, K. M., Laidlaw, K., & Murray, L. K. (2009). Older peoples' attitudes to mental illness. *Clinical Psychological Psychotherapy, 16* (1), 33–45.

Quinn, E. (2006). Seniors make quick improvements when they start to exercise. *About.com: Health and Disease.* http://sportsmedicine.about.com. Retrieved June 24, 2009.

Quinn, J. B., & Ehrenfeld, T. (2008). There goes the 401(k). *Newsweek, 152* (25), 57–57.

Rathbone-McCuan, E. (1996). Self-neglecting in the elderly: Knowing when and how to intervene. *Aging,* 44–49.

Rattan, S. I. S., & Singh, R. (2009). Progress and prospects: Gene therapy in aging. *Gene Therapy, 16*(1), 3–9.

Raz, N., Lindenberger, U., Rodrigue, K. M., Kennedy, K. M., Head, D., Williamson, A., Dahle, C. Gerstorf, D., & Acker, J. D. (2005). Regional brain changes in aging healthy adults: General trends, individual differences, and modifiers. *Cerebral Cortex, 15,* 1676–1689.

Reday-Mulvey, G. (1996). Why working lives must be extended. *People Magazine,* 16 May, 24–29.

Redfoot, D., & Gaberlavage, G. (1991). Housing for older Americans: Sustaining the dream. *Generations,* 35–38.

Revenson, T. A., Danoff-Burg, S., Trudeau, K. J., & Paget, S. A. (2004). Unmitigated communion, Social Constraints, and Psychological Distress Among Women With Rheumatoid Arthritis. *Journal of Personality, 72*(1), 29–46.

Richardson, C. A., & Hammond, S. M. (1996). A psychometric analysis of a short device for assessing depression in elderly people. *British Journal of Clinical Psychology, 35*(4), 543–551.

Richardson, J. P., & Lazur, A. (1995). Sexuality in the nursing home patient. *American Family Physician, 51,* (1), 10–14.

Richman, J. (1995). From despair to integrity: an Eriksonian approach to psychotherapy for the terminally ill. *Psychotherapy, 32* (2), 317–322.

Roan, C. L., & Raley, R. K. (1996). Intergenerational coresidence and contact: A longitudinal analysis of adult children's response to their mother's widowhood. *Journal of Marriage and the Family, 58,* 708–717.

Roberto, K. A. (1998). Qualities of older women's friendships: Stable or volatile? *International Journal of Aging and Human Development, 44* (1), 1–14.

Roberto, K., & Scott, J. (1996). Friendship patterns among older women. *International Journal of Aging and Human Development* 19 (1984/85): 1–9.

Roberts, B. W., & Wood, D. (2006). Personality development in the context of the n-socioanalytic model of personality. *In* D. Mroczek & T. Little (Eds.), *Handbook of Personality Development.* Mahwah, NJ: Erlbaum.

Roberts, W. (2008). *The No-Nonsense Guide to World Food.* Oxford UK: New Internationalist Publications, Ltd.

Robine, J.-M., Crimmins, E. M., Horiuchi, S., & Zeng, Y. (2007). *Human Longevity, Individual Life Duration & the Growth of the Oldest-Old Population.* Netherlands: Springer Publ.

Robine, J.-M., & Michel, J.-P. (2004). Looking forward to a general theory on population aging. *Journals of Gerontology Series A: Biological Sciences and Medical Sciences, 59A*(6), 590–597.

Robinson, B. E. (1997). Guideline for initial evaluation of the patient with memory loss. *Geriatrics, 52* (12), 30–39.

Robinson, K. M., Roberts, K. T., Topp, R., Newman, J., Smith, F., & Stewart, C. (2008). Community Perceptions of Mental Health Needs in an Underserved Minority Neighborhood. *Journal of Community Health Nursing, 25*(4), 203–217.

Robinson, R. G., Gorge, R. E., Moser, D. J., Acion, L., Solodkin, A., Small, S. L., Fonzetti, P., Hegel, M., & Arndt, S. (2008). Escitalopram and problem-solving therapy for prevention of poststroke depression: a randomized controlled trial. *Journal of the American Medical Society, 299*(20), 2391–2400.

Rogers, C. C. (2000). *Changes in the Older Population and Implications for Rural Areas.* U.S. Department of Agriculture, Economic Research Service. (http://www.ers.usda.gov/publications/rdrr90/)

Roha, R. R. (1998). Medigap: One size doesn't fit all. *Kiplinger's Personal Finance,* January, 107–111.

Rowe, J. W. (1997). The new gerontology. *Science, 278* (5337), 367–369.

Rowe, J. W., & Kahn, R. L. (1997). Successful aging. *The Gerontologist, 37,* 433–440.

Rowland, V. T., & Shoemake, A. (1995). How experiences in a nursing home affect nursing students' perceptions of the elderly. *Educational Gerontology, 21,* 735–748.

Rubenstein, R. (1987). A cross-cultural comparison of children's drawings of same- and mixed-sex peer interaction. *Journal of Cross-Cultural Psychology, 18,* 234–250.

Rubenstein, R. (1994). Generativity as pragmatic spirituality." *In* L. Thomas and S. Eisenhandler (Eds.), *Aging and the Religious Dimension.* Westport, CT: Auburn House.

Ruiz, S. A., & Silverstein, M. (2007). Relationships with grandparents and the emotional well-being of late adolescence and young adult grandchildren *Journal of social issues, 63,* 793–808.

Ryan, E. B., Hummert, M. L., & Boich, L. H. (1995). Communication predicaments of aging: Patronizing behavior toward older adults. *Journal of Language and Social Psychology, 14* (1–2), 144–166.

Ryan, M. (1993). Undercover among the elderly. *Parade Magazine, 8,* July.

Ryff, C. D., Love, G. D., Urry, H., Muller, D., Rosenkranz M. A., Friedman E. M., *et al.* (2004). Psychological well-being and ill-being: Do they have distinct or mirrored biological correlates? *Psychotherapy and Psychosomatics, 75,* 85–95.

Sadovsky, R., *et al.* (2006). Sexual problems among a specific population of minority women aged 40-80 years attending a primary care practice. *The journal of sexual medicine, 3*(5), 795–803.

Salthouse, T. A. (2006). Mental exercise and mental aging: Evaluating the validity of the "Use it or lose it" hypothesis. *Perspectives on Psychological Science, 1,* 68–87.

Samuels, S. C. (1997). Midlife crisis: Helping patients cope with stress, anxiety, and depression. *Geriatrics, 52,* 55–63.

Sandecki, R. (1993). Ex-AT&T manager wins verdict. *Los Angeles Times, 112,* 5 July, C18.

Sandell, S. H., & Iams, H. M. (1997). Reducing women's poverty by shifting Social Security benefits from retired couples to widows. *Journal of Policy Analysis and Management, 16* (2), 279–297.

Sanders, G. F., Montgomery, J. E., Pittman, J. F., Jr., & Blackwell, C. (1984). Youth attitudes toward the elderly. *Journal of applied gerontology, 3,* 59–70.

Sassler, S., & Schoen, R. (1999). The effect of attitudes and economic activity on marriage. *Journal of Marriage and Family, 61,* 147–159.

Savage, R. D., Britton P., Bolton, H., & Hall, E. H. (1973). *Intellectual functioning in the aged.* London: Methuen.

Sawyer, C. H. (1996). Reverse mortgages: An innovative tool for elder law attorneys. *Stetson Law Review, 26* (2), 617–646.

Schaie, K. W. (1996). *Intellectual development in adulthood: The Seattle longitudinal study.* New York: Cambridge University Press.

Schaie, K. W. (2005a). What can we learn from longitudinal studies of adult intellectual development? *Research in Human Development, 2,* 133–158.

Schaie, K. W. (2005b). *Developmental influences on adult intelligence:* The Seattle Longitudinal Study. New York: Oxford University Press.

Schaie, K. W., Krause, N., & Booth, A. (2004). *Religious influences on health and well-being in the elderly.* NY: Spring Publications.

Schaie, K. W., & Willis, S. L. (1996). *Adult development and aging* (4th ed.). New York: HarperCollins.

Scharlach, A. E. (1987). Relieving feelings of strain among women with elderly mothers. *Psychology and Aging, 2*(1), 9–13.

Scheibe, S., Freund, A. M., & Baltes, P. B. (2007). Towards a developmental psychology of *Sehnucht* (life longings): The optimal (Utopian) life. *Developmental psychology, 43*(3), 778–795.

Scheibe, S., Kunzmann, U., & Baltes, P. B. (2008). New territories of positive lifespan development: Wisdom and life longings. *In* C. R. Snyder & S. J. Lopez (Eds.), *Handbook of Positive Psychology.* New York: Oxford University Press.

Schieman, S., Pearlin, L. I., & Nguyen, K. B. (2005). Status inequality and occupational regrets in late life. *Research on Aging, 27,* 692–724.

Schlesinger, B. (1995). The sexless years of sex rediscovered. *Journal of Gerontological Social Work, 26*(1), 117–131.

Schmitt, E., New York Times (2001). For 7 million people in census, one race category isn't enough. *New York Times,* March 13, 2001.

Schooler, C. (2007). Use it—and keep it, longer, probably: A reply to Salthouse (2006). *Perspectives on Psychological Science, 2,* 24–29.

Schooler, C., Caplan, L. J., Revell, A. J., Salazar, A. M., & Grafman, J. (2008) Brain lesion and memory functioning: Short-term memory deficit is independent of lesion location. *Psychonomic Bulletin and Review, 15* (3), 521–527.

Seelye, K. Q. (1997). U.S. of future: Grayer and more Hispanic. *New York Times, 9,* March 27.

Segal, D. L., Coolidge, F. L., & Rosowsky, E. (2006). *Personality disorders and older adults: Diagnosis, assessment and treatment.* Hoboken NH: Wiley.

Segell, M. How to live forever, part I. *Esquire Magazine* (September 1993): 125–132.

Sejnowski, T. (1995). Sleep and memory. *Current Biology, 5* (8), 832–834.

Senior Corps (2009). Foster Grandparent Program. National and Community Service. http://www.seniorcorps. gov. Retrieved September 17, 2009.

Seppa, N. (2006). Looking ahead: Tests might predict Alzheimer's risk. *Science News, 169*(7), 102.

Seshadri, S., Wold, P. A., Beiser, A., Au, R., McNulty, K., White, R., & d'Agostino, R. B. (1997). Lifetime risk of dementia and Alzheimer's disease. The impact of mortality on risk estimates in the Framingham study. *Neurology, 49,* 1498–1504.

Sethi, N. K. (2008). Tip of the tongue and "senior moments": The truths behind dementia. Comprehensive Epilepsy Center, Department of Neurology, Cornell Medical Center. http://www.braindiseases.wordpress. com. Retrieved July 5, 2009.

Severance offers change to teach. (1991). *USA Today,* 26 November, 4B.

Shaker, L., Scott, J. A., & Reid, M. (2006). Parental attitudes toward breastfeeding: their association with feeding outcome at hospital discharge. *Issues in Perinatal Care, 31*(2), 125–131.

Shapses, S. A., & Riedt, C. S. (2006). Bone, body weight, and weight reduction: what are the concerns? *Journal of Nutrition, 136*(6), 1453–1456.

Shea, D. G., Miles, T., & Hayward, M. (1996). The health-wealth connection: Racial differences. *The Gerontologist, 36*(3), 342–349.

Sheeder, J., Lezottte, D., & Stevens-Simon, C. (2006). Maternal age and the size of white, black, Hispanic, and mixed infants. *Journal of Pediatric & Adolescent Gynecology, 19* (6), 385–389.

Sheehy, G. (1976). *Passages.* New York: Dutton.

Sheehy, G. (1995). *New passages: Mapping your life across time.* New York: Ballantine Books.

Shenk, J. W. (2009). What makes us happy? *Atlantic Magazine, June,* 24–33.

Shepard, M. (1997). Site-based services for residents of single-room occupancy hotels. *Social work, 42* (6), 585–592.

Shephard, R. J., Rhind, S., & Shek, P. N. (1995). The impact of exercise on the immune sytem: NK cells, interleukins 1 and 2, and related responses. *In* J. O. Holloszy (Eds.), *Exercise and sports sciences reviews, 23,* 214–241.

Shepherd, M. D., & Erwin, G. (1983). An examination of students' attitudes toward the elderly. *American Journal of Pharmaceutical Education, 47,* 35–38.

Shoemake, A. F., & Rowland, V. T. (1993b). An examination of students' attitudes toward the elderly. *American journal of pharmaceutical education, 47,* 35–38.

Shoemake, A. F., & Rowland, V. T. (1993a). Do laboratory experiences change college students' attitudes toward the elderly? *Educational Gerontology, 19,* 295–309.

Should Hollywood keep aging actresses? (2009). http://www.cinematical.com. Retrieved May 2, 2009.

Siegel, S. R. (1993). Relationships between current performance and likelihood of promotion for old versus young workers. *Human Resources Development Quarterly, 4* (1), 39–50.

Silberman, P. R., Weiner, A., & El Ad, N. (1995). Parent-child communication in bereaved Israeli families. *Omega, 31* (4), 293–306.

Silver, M. (1990). Retire early at Hewlett-Packard. *Santa Rosa Press Democrat,* 30 January, B1.

Silverman, P. (1987). Community settings. *In* P. Silverman (Ed.), *The elderly as modern pioneers,* 185–210. Bloomington: Indiana University Press.

Silverstein, M., The Aging Dilemma (2009). http://womenandhollywood.com . Retrieved May 30, 2009.

Simmons, L. W. (1945). *The role of the aged in primitive society.* New Haven, CT: Yale University Press.

Simmons, L. W. (1960). Aging in preindustrial societies. *In* Clark Tibbitts (Ed.), *Handbook of social gerontology.* University of Chicago Press: Chicago.

Simonen, R. L., Videmaan, T., Battie, M. C., & Gibbons, L. (1997). Differences in hand and foot psychomotor speed among 18 pairs of monozygotic twins discordant for lifelong vehicular driving. *Archives of Occupational Environmental Health 70,* (4), 227–281.

Sinclair, D. A., & Guarente, L. (2006). Unlocking the secrets of longevity genes. *Scientific American.* http://www.supercentenarian.com. Retrieved July 1, 2009.

Singh, R., Kolvraa, S., & Rattan, S. E. (2007). Genetics of human longevity with emphasis on the relevance of HSP70 as candidate genes. *Frontiers in Bioscience, 12,* 4504–4513.

Skultety, K. M. (2007). Addressing issues of sexuality with older couples. *Generations, 3,* 31–37.

Slaven, A. (2008). The art of leadership. *Community Care, 1762,* 6–7.

Sloan, J. D., & Graves, J. (1995). Age discrimination: A trial lawyer's guide for bringing suit. *Trial, 31* (3), 48–53.

Slotterback, C. S., & Saarnio, D. A. (1996). Attitudes toward older adults reported by young adults: Variation based on attitudinal task and attribute categories. *Psychology and aging, 11* (4), 563–571.

Small, S. A., Tsai, W.Yl., deLaPaz, R., Mayeux, R., & Stern, Y. (2002). Imaging hippocampal functional acrtoss the human life span. Is memory decline normal or not? *Annals of Neurology, 51,* 290–295.

Small, B. J., Rosnick, C. B., Fratiglioni, L., & Backman, L. (2004). Apoliproprotein E and cognitive performance: A meta-analysis. *Psychology and Aging, 19,* 592–600.

Small, J. P. (1995). Recent scientific advances in the understanding of memory. *Helios, 22* (2), 156–162.

Smith, G. (2005). *How DNA is Transforming the Way We Live and Who We Are.* NY: AMACOM Books, a division of American Management Association.

Smyer, T., Gragert, M. D., & laMere, S. (1997). Stay safe! Stay healthy! Surviving old age in prison. *Journal of Psychosocial Nursing, 35* (9), 10 –17.

Social Security Administration [OASI]. (2009). The 2009 annual report of the Board of Trustees of the Federal Old-Age and Survivors Insurance and Federal Disability Insurance Trust Funds. May 12, 2009 Referred to the Committee on Ways and Means. Washington DC: U.S. Government Printing Office.

Social Security Administration [SSA] (2004). *Fast facts and figures about Social Security, 2004.* Social Security Administration. Washington DC.

Social Security Administration [SSA] (2006). http://www.ssa.gov/policy. Office of Policy publications.

Social Security Administration [SSA] (2009). Full retirement age. http://www.ssa.gov/pubs. Retrieved August 7, 2009.

Sokolovsky, J. (Ed.) (1997). *The cultural context of aging* (3rd ed.) Westport, CT: Bergin and Garvey.

South Salem, NY: Criminal Justice Institute.

Spindler, S. R. (2005). Rapid and reversible induction of the longevity, anticancer and genomic effects of caloric restriction. *Mechanisms of Aging Development, 127*(9), 960–966.

Squire, L. R. (2004). Memory systems of the brain: A brief history and current perspective. *Neurobiology of Learning and Memory, 82,* 171–177.

Stafford, R. J., & Cyr, P. L. (1997). The impact of cancer on the physical function of the elderly and their utilization of health care. *Cancer, 80*(10), 1973–1980.

Starr, K. (1996). *Endangered dreams: The Great Depression in California.* New York: Oxford University Press.

State Health Facts (2006). Distribution of Medicaid spending for dual eligibles by service, 2005. http://www.statehealthfacts.org/comparemaptable. Retrieved October 20, 2009.

Stein, M. B., Chartier, M. J., Hazen, A. L., Kozak, M. V., Tancer, M. E., Lander, S., Furer, P., Chubaty, D., & Walker, J. R. (1998). A direct-interview family study of generalized social phobia. *American Journal of Psychiatry, 155* (1), 90–97.

Stern, M. R. (1987). "At 93, Blessings and Memories." *Washington Post* (April): 23.

Sternberg, R. J. (1990). *Wisdom: Its nature, origins, and development*. Cambridge University Press: New York.

Sterns, H. L., & Kaplan, J. (2003). Self-management of career and retirement. *In* G. A. Adams & T. A. Beehr (Eds.), *Retirement: Reasons, processes, and results* (pp. 188–213). New York: Springer.

Sterns, H. L., Begovis, A., & Sotnak, D. L. (2003). Incorporating aging into industrial/organizational psychology courses. *In* S. E. Whitbourne & J. C. Cavanaugh (Eds.), *Integrating aging topics into psychology: A practical guide for teaching* (pp. 185–199). Washington, DC: American Psychological Association.

Stevens, N. (1995). Gender and adaptation to widowhood in later life. *Aging and Society, 15*, 37–58.

Still, J. S. (1980). Disengagement reconsidered: Awareness of finitude. *Gerontologist, 20*, 457– 462.

Stokes, G. (1992). On being old: *The psychology of later life*. London: The Falmer Press.

Stolberg, S. (1994, July 27). Many elderly too medicated, study finds. *Los Angeles Times,* A1.

Stolley, J. M., & Hoenig, H. (1997). Religion/spirituality and health among elderly African Americans and Hispanics. *Journal of Psychosocial Nursing, 35* (11), 32–38.

Strawbridge, W. J., Wallhagen, M. I., Shema, S. J., & Kaplan, G. A. (1997). New burdens or moe of the same? Comparing grandparent, spouse, and adult-child caregivers. *Gerontologist, 37*(4), 505–510.

Stroebe, M., Gergen, M. M., Gergen, K. J., & Stroebe, W. (1992). Broken hearts of broken bonds: Love and death in historical perspective. *American Psychologist, 47*(10), 1205–1212.

Strom, R. D., Buki, L. P., & Strom, S. K. (1997). Intergenerational perceptions of English speaking and Spanish speaking Mexican American grandparents. *International Journal of Aging and Human Development, 45*(1), 1–21.

Szinovacz, M., DeViney, S., & Atkinson, M. (1999). Effects of surrogate parenting on grandparents' well-being. *Journal of Gerontology, 54B,*(6), S376–389.

Takamura, J. C. (2001). Towards a new era in aging and social work. *Journal of Gerontological Social Work, 36*(3/4), 1–11.

Talamantes, M. A., Cornell, J., Espino, D. V., Lichenstein, M. J., & Hazuda, H. P. (1996). SES and ethnic differences in perceived caregiver availability among young-old Mexican Americans and non-Hispanic whites. *Gerontologist, 36*(1), 88–99.

Tandemar Research. (1988). *Quality of life among seniors*. Toronto, Ontario: Tandemar Research.

Tanenbaum, L. (1997). Changing the images makes a big difference. *Extra!* March/April, 22–23.

Tang, B. M., Eslick, G. D., Nowson, C., Smith, C., & Bensoussan, A. (2007). Use of calcium or calcium in combination with vitamin D supplementation to prevent fractures and bone loss in people aged 50 years and older: a meta-analysis. *Lancet, 370*(9588), 657–666.

Taubes, G. (2002). What if it's all been a big fat lie? *New York Times,* July 7.

Taylor, S. E., & Brown, J. D. (1988). Illusion and well-being: A social psychological perspective on mental health. *Psychological bulletin, 103*, 193–210.

Temple-Smith, M. J., Mulvey, G., & Keogh, L. (1999). Attitudes to taking a sexual history in general practice in Victoria, Australia. *Sexual transmitted infections, 75*, 41–44.

Terracciano, A., McCrae, R. R., Brant, L. J., & Costa, P. T. (2006). Personality plasticity after age 30. *Personality and Social Psychology Bulletin, 32*, 999–1009.

Thank, L. L. (2001). *Generations in touch: Linking the old and young in a Tokyo neighborhood*. Ithaca, NY: Cornell University Press.

The Week (2009). Talking Points: News, *9*(415), 18.

They Can't Hang Up (1999). Help for Elderly People Targeted by Fraud. http://www.fraud.org/elderfraud. Retrieved 9/7/09.

Thomas, J. (1992). *Adult and aging*. Boston, MA: Allyn and Bacon.

Thomas, L. (1994). *Introduction, aging and the religious dimension. In* L. Thomas and S. Eisenhandler (Eds.). Westport, CT: Auburn House.

Thomas, W. I. (1923). *The unadjusted girl*. Boston: Little, Brown.

Thone, R. (1993). *Women and aging: Celebrating ourselves*. Binghamton, NY: Haworth.

Tiberti, C., Sabe, L., Kuzis, G., Cuerva, A. Garcia; Leiguarda, R., & Starkstein, S. E. (1998). Prevalence and correlates of the catastrophic reaction in Alzheimer's disease. *Neurology, 50*, 546–548.

Times of Our Lives. http://www.trinity.edu. May 2, 2009. No author given.

Tischler, L. (2005). Extreme jobs (and the people who love them). *FastCompany, 93* (April), 54. http://www.fastcompany.com/magazine/93. Retrieved May 22 2009.

Tornstam, L. (2005). *Gerotranscendence: A Developmental Theory of Positive Aging*. NY: Springer Publishing.

Tornstam, L. (2006). The complexity of Ageism: A proposed typology. *International Journal of Ageing and Later Life, 1*(1), 43–68.

Torres-Gil, F. (1988). Interest group politics: Empowerment of the *ancianos. In* S. Applewhite (Ed.), *Hispanic elderly in transition*. New York: Greenwood.

Touron, D. R., Hoyer W. J., & Cerella J. (2004). Cognitive and skill learning: Age-related differences in strategy shifts and speed of component operations. *Psychology and Aging, 20*, 565–580.

Towers, P. (2007). *Riding the wave of growth and restructuring: Optimizing the deal for today's workforce*. Global Workforce Study.

Townsend, A. L., & Franks M. M. (1997). Quality of the relationship between elderly spouses: Influence on spouse caregivers' subjective effectiveness. *Family relations, 46*, 33–39.

Tracy, K. R. (1996). Diversifying into adult day care: A learning experience. *Nursing Homes: Long Term Care Management, 45*(10), 39–41.

Trifunovic, A., Hansson, A., Wredenberg, A., Rovio, A. T., Dufour, E., Khvorostov, I., Spelbrink, J. N., Wibom, R., Jacobs, H. T., & Larsson, N. G. (2005). Somatic mtDNA mutations cause aging phenotypes without affecting reactive oxygen species production. *Proceedings of the National Academy of Sciences, 102*(50), 17993–17998.

Tuckman, J., & Lorge, I. (1953). Attitudes toward old people. *Journal of Gerontology 32*, 227–232.

Tulvig, E. (1993). Varieties of consciousness and levels of awareness in memory. *In* A. Baddeley & L. Weiskrantz (Eds.), *Attention: Selection, awareness, and control. A tribute to Donald Broadbent* (283–299). London: Oxford University Press.

Turjanski, H., Lees, A. J., & Brooks, D. J. (1997). In vivo studies on striatal dopamine D-1 and D-2 site binding in L-dopa-treated Parkinson's disease patients with and without dyskinesias. *Neurology, 49*, 717–723.

Turner, L. (2001). Time out with half-time: Job sharing in the nineties. *Canadian Journal of Counseling, 30* (2), 104–113.

Turvey, C. L., Schultz, S., Arndt, S., Wallace, R. B., & Herzog, A.R. (2000). Memory complaint in a community sample aged 70 and older. *Journal of the American Geriatrics Society, 48*, 1435–1441.

U.S. Census Bureau (1991). *Statistical abstract of the United States, 1990.* Washington, DC: Department of Commerce, Bureau of the Census.

U.S. Census Bureau (1996). *Profiles of general demographic characteristics, 1990 census.* Census of Population and Housing. Washington, DC: U.S. Department of Commerce.

U.S. Census Bureau (2002). *We the people: Aging in the United States.* http://www.census.gov, retrieved January 2005.

U.S. Census Bureau (2004). *Annual estimates of the population by sex and five-year age groups for the United States: April 1, 2003 to July 1, 2003.* Retrieved February 14, 2009 at http://www.census.gov.

U.S. Census Bureau (2005). *Marital Status and Living Arrangement.* Current Population Reports (Series P20, No. 612). Washington, DC, Government Printing Office.

U.S. Census Bureau (2007a). Older Americans 2008: Key indicators of well-being. http://www.agingstats.cov/agingstats. Retrieved July 12, 2009.

U.S. Census Bureau (2007b). *Statistical abstract of the United States.* Washington DC http://www.census.gov, retrieved June 7, 2009.

U.S. Census Bureau (2008a). International database. Washington, DC http://www.census.gov/ipc. Retrieved June 2, 2009.

U.S. Census Bureau (2008b). *Older and more diverse nation by midcentury.* U.S. Census Bureau News: Washington, DC.

U.S. Census Bureau (2009). *Older Americans month: May 2009.* U.S. Census Bureau News: Washington, DC.

U.S. Commission on Civil Rights (1979). Reports on Asian Pacific Americans. http://ssrn.com/abstract=589431. Retrieved May 22, 2009.

U.S. Department of Labor, Bureau of Labor Statistics. (1989). *Handbook of labor statistics.* Washington, DC: Government printing office.

U.S. Department of Labor, Women's Bureau. (2007). *Employment status of women and men in 2006.* Washington DC. http://www.dol.gov/. Retrieved June 3, 2009.

U.S. News Bulletins (1997). Kevorklian assists in suicide of Canadian from British Columbia. http://www.rights.org/deathnet/usnews_9709. Retrieved June 2, 2002.

U.S. News Bulletins. (1997). Kevorkian assists in suicide of Canadian from British Columbia. http://www.rights.org/deathnet. 21 September.

U.S. News Bulletins. (1998). Kevorkian says: "Charge me—or leave me alone!" http://www.rights.org/deathnet. 1 March.

Uhlenberg, P. (1996). Mutual attraction: Demography and life-course analysis. *The gerontologist, 36* (2), 226–229.

Umberson, D., Williams, K., Powers, D. A., Chen, M. D., & Campbell, A. M. (2005). As good as it gets? A life course perspective on marital quality. *Social forces, 84*, 493–511.

Utchitelle, L. (2001). Economic view: Rumbles of warning as housing weakens. *New York Times, p4.* November 4.

van Blaricom, D. (2009). Payroll taxes: Basic information for employers. http://taxes.about.com. Retrieved September 18, 2009.

van den Hoonaard, D. (1997). Identity forclosure: Women's experiences of widowhood as expressed in autobiographical accounts. *Aging and society, 17*, 533–551.

Vanderbeck, R. M. (2007). Intergenerational geographies: Age relations, segregation and re-engagements. *Geography Compass, 1_2*, 200–221.

Vardi, N. (2005). Rx for fraud. *Forbes.com.* www.forbes.com/forbes. Retrieved September 4, 2009.

Vasavada, T., Masand P. S., & Nasra G. (1997). Evaluations of competency of patients with organic mental disorder. *Psychological Reports, 80*(2), 107–113.

Vasil, L., & Wass, H. (1993). Portrayal of the elderly in the media: A literature review and implications for educational gerontologists. *Educational Gerontology, 19*, 71–85.

Vaux, G. (2004). Social Security. *Research Matters, 17.*

Vertinsky, P. (1991). Old age, gender and physical activity: the biomedicalization of aging. *Journal of Sport History, 18*(1), 64–80.

Verwoerdt, A., Pfeiffer, E., & Wang, H. S. (1969). Sexual behaviorism in senescence. *Geriatrics, 24*, 137–153.

Vincent, J. A. (2008). The cultural construction old age as a biological phenomenon: Science and anti-ageing technologies. *Journal of Aging Studies, 22*, 331–339.

Vitaliano, P. P., Young, H. M., & Zhang, J. (2004). Is caregiving a risk factor for illness? *Current Directions in Psychological Science, 13*, 13–16.

Vitt, L. A. (1998). Home equity conversion financing: A recipe for well-being. *Innovations in Aging, 27*(1), 10–12.

Volger, G. P. (2006). *In* J. E. Birren & K. W. Schaie (Eds.), *Handbook of the psychology of aging, 6e,* pp. 41–50. San Diego: Elsevier.

Wade, N. (2007). Gene links longevity and diet, scientists say. *New York Times,* May 2007, 20.

Wailoo, K. (2006). Stigma, race, and disease in 20th century America. *Lancet, 367,* 531–533.

Waldinger, R. J., Vaillant, G. E., & Orav, E. J. (2007). Childhood sibling relationships as a predictor of major depression in adulthood: a 30-year prospective study. *American Journal of Psychiatry, 164*(6), 949–954.

Walford, Roy L. (1983). *Maximum life span.* New York: Norton & Co. 34. Walter, Paul. (1997). Effects of vegetarian diets on aging and longevity. *Nutrition Reviews, 55* (1), S61–S65.

Walker, B. L., & Harrington, D. (2002). Effects of staff training on staff knowledge and attitudes about sexuality. *Educational Gerontology, 28*(8), 639–654.

Walker, R. B., & Luszcz, M. A. (2009). The health and relationship dynamics of late-life couples: A systematic review of the literature. *Ageing & Society, 29*(3), 455–480.

Walls, C., & Zarit, S. (1992). Informal support from black churches and well-being of elderly blacks. *The Gerontologist, 31,* 490–495.

Walz, T. (2002). Crones, dirty old men, sexy seniors: Representations of the sexuality of older persons. *Journal of Aging and Identity, 72),* 99–112.

Wang, M. (2007). Profiling retirees in the retirement transition and adjustment process: Examining the longitudinal change patterns of retirees' psychological well-being. *Journal of Applied Psychology, 92,* 445–474.

Warner, H., Miller, R. A., & Carrington, J. (2002). Meeting report: National Institute on Aging Workshop on the Comparative Biology of Aging. *Science of Aging Knowledge Environments, 2002* (17), 5.

Watson, K. (1997). Current trauma in the lives of older adults: Surviving and healing the wounds of a changing self-concept and self-neglect. Unpublished Manuscript, Gerontology of Aging. Rohnert Park, CA: Sonoma State University.

Weaver, C. N. (1997). Has the work ethic in the USA declined? Evidence from nationwide surveys. *Psychological report, 81,* 491–495.

Weber, M. (1958). *The Protestant ethic and spirit of capitalism.* T. Parsons (transl.) New York: Scribner. [original work published 1904–1905].

Welch, J. (1997) Intel faces fight over "termination quotas." *People Management,* 26 June, 9.

Weller, E., Emslie, G., Kratochyil, C., Vitiello, B., Silva, S., Mayes, T., McNulty, S., Waslick, B., Casat, C. Walkup, J. Pathak, S., Rohde, P. Posner, K. & March, J. (2006). Treatment for adolescents with depression study (TADS): Safety results. *Journal of the American Academy of Child and Adolescent Psychiatry, 45*(12), 1440–1455.

Wells, R. M. (1997). Subsidies for Section 8 program are on the chopping block. *Congressional Quarterly,* 539–541.

Westinghouse Electric, Northrop settle suits on age discrimination. (1997). *Wall Street Journal,* 3 November, A2.

White, N., & Cunningham, W. R. (1988). Is terminal drop pervasive or specific? *Journals of Gerontology, 42,* P141–P144.

Whitehead, B. D., & Popenoe, D. (2006). *The state of our unions: The social health of marriage in America. The national marriage project.* New Brunswick NH: Rutgers University Press.

Whitley, B. E., & Kite, M. E. (2006). *The psychology of prejudice and discrimination.* Belmont CA: Thompson/Wadsworth Higher Education.

Whittstein, I. S., Thiemann, D. R., Lima, J. A. C., & Baughman, K. L. (2005). Neurohumoral features of myocardial stunning due to sudden emotional stress. *New England Journal of Medicine, 352,* 539–639.

Widiger, T. A., & Seidlitz, L. (2002). Personality psychopathology, and aging. *Journal of Research in Personality, 36,* 335–362.

Wilcox, S. (1997). Age and gender in relation to body attitudes: Is there a double standard of aging? *Psychology of Women Quarterly, 21* (4), 549–565.

Williams, L. S., Ghose, S. S., & Swindle, R. W. (2004). Depression and other mental health diagnoses increase mortality risk after ischemic stroke. *American Journal of Psychiatry, 161,* 1090–1905.

Williams, A., Coupland, J., Folwell, A., & Sparks, L. (1997). Talking about Generation X: Defining them as they define themselves. *Journal of Language and Social Psychology, 16* (3), 251–277.

Willis, S. L., Tennestedt, S. L., Marsiske, M., Ball, K., Elias, J., Koepke, K. M., Morris, J. N., Rebok, G. W., Unverzage, F. W., Stoddard, A. M., & Wright, E. (2006). Long-term effects of cognitive training on everyday functional outcomes in older adults. *JAMA: Journal of the American Medical Association, 296*(23), 2805–2814.

Wilson, R. S., Beck, T. L., Bienias, J. L., & Bennett, D. A. (2007). Terminal cognitive decline: Accelerated loss of cognitive function in the last years of life. *Psychomatic Medicine, 69,* 131–137.

Wiscott, R., & Kopera-Frye, K. (2000). Sharing of culture: Adult grandchildren's perceptions of intergenerational relations. *International Journal of Aging and Human Development, 51*(3), 199–215.

Witkin, G. (1994). Ten myths about sex. *Santa Rosa Press Democrat, Parade Magazine,* January 17.

Wolf, D. A., Freedman, V., & Soldo, B. J. (1997). The division of family labor: Care for elderly parents. *Journals of Gerontology, 52B* (special issue), 102–109.

Wolfson, C., Handfield-Jones, R., Glass, K. C., McClaran, J., & Keyserlingk, E. (1993). Adult children's perceptions of their responsibility to provide care for dependent elderly parents. *Gerontologist, 33*(3), 315–323.

Wolters, G., & Prinsen, A. (1997). Full versus divided attention and implicit memory performance. *Memory and Cognition, 25* (6), 764–771.

Wong, P. K., Christie, J. J., & Wark, J. D. (2007). The effects of smoking on bone health. *Clinical Science, 113*(5), 233–241.

Wood, S., & Liossis, P. (2007). Potentially stressful life events and emotional closeness between grandparents and adult grandchildren. *Journal of family issues, 28,* 380–398.

World Health Organization [WHO]. (1991). *World health statistics annual, 1990.* Geneva: Author.

World Health Organization [WHO] (2007). *The world health report, 2007.* World Health Organization. Retrieved March 2, 2009 from http://www.who.int/why/2007/en/index.html.

Worsnop, R. I. (1997). Caring for the dying. *Congressional Quarterly Researcher, 7* (33), 769–792.

Yali, A. M., & Revenson, T. A. (2004). How changes in population demographics will impact health psychology: Incorporating a broader notion of cultural competence into the field. *Health Psychology, 23 (2),* 0278–6133.

Young, Old Differ on Using Surplus to Fix Social Security (1998). http://people-press.org. Retrieved July 22, 2009.

Zacks, R. T., Hasher, L., & Li, K. Z. H. (2000). Human memory. *In* F. I. M. Craik & T. A. Salthouse (Eds.), *Handbook of Aging and Cognition* (2nd ed., 293–357). NJ: Erlbaum.

Zakzanis, K. K., Graham, S. J., & Campbell, Z. (2003). A meta-analysis of structural and functional brain imaging in dementia of the Alzheimer's type: A neuroimaging profile. *Neuropsychological Review, 13,* 1–18.

Zandi, T., Mirle, J., & Jarvis, P. (1990). Children's attitudes toward elderly individuals: A comparison of two ethnic groups. *International Journal of Aging and Human Development, 30* (3), 161–174.

Zeiss, A. M., & Breckenridge, J. S. (1997). Treatment of late life depression: A response to the NIH consensus conference. *Behavior Therapy, 28,* 3–21.

Zeiss, A. M., & Kasl-Godley, J. (2001). Sexuality in older adults' relationships. *Generations, 25* (2), 18–25.

Zezima, K. (2008). More women than ever are childless, census finds. *New York Times,* August 19, 12.

Ziegler, R., & Mitchell, D. B. (2003). Aging and fear of crime: an experimental approach to an apparent paradox. *Experimental aging research, 29,* 173–187.

Zweig, R. A. (2008). Personality disorder in older adults: Assessment challenges and strategies. *Professional Psychology: Research and Practice, 39* (3), 298–305.

PHOTO CREDITS

1: © AP Photo/Gerald Herbert; 5: © Hill Street Studios/ Blend Images/Jupiter images; 9: © Joe Raedle/Getty Images News/Getty Images; 11: © Radius Images/Jupiter Images; 15: © DaZo Vintage Stock Photos/Images.com/ Alamy; 20: © Manfred Bail/imagebroker.net/PhotoLibrary; 24: © Yuri Arcurs, 2009/Used under license from Shutterstock.com; 26: © AP Photo/Francois Mori; 28 © ERproductions Ltd/Blend Images/Jupiter Images; 30: © Ken Wramton/Digital Vision/Jupiter Images; 33: © Photodisc/Jupiter Images; 36: © Eamonn McCormack/ WireImage/Getty Images; 38: © Pattie Steib/istockphoto. com; 41: © Barbara Stitzer/PhotoEdit; 48: © Medioimages/ Photodisc/Jupiterimages; 52: © Rolf Bruderer/Blend Images/Jupiter Images; 56: © Donna Day/Digital Vision/ Jupiter Images; 60: © Mary Evans/SIGMUND FREUD COPYRIGHTS/ Alamy; 65: © Image Source/Jupiter Images; 67: © Inspirestock/Jupiter Images; 70: © Ariel Skelley/Blend Images/Jupiter Images; 78: © Ian Shaw/ Alamy; 87: © Ariel Skelley/Blend Images/Jupiter Images; 90: © Ellen Isaacs/Alamy; 95: © Peter Weber, 2009/Used under license from Shutterstock.com; 100: © Marcin Mory, 2009/Used under license from Shutterstock.com; 101: © Carl Glassman/The Image Works; 102:© Claude Dagenais/Photodisc/Jupiter Images; 107: Luc Ubaghs, 2009/Used under license from Shutterstock.com; 117: © Rob Lacey/ vividstock.net/Alamy; 127: © Christy Thompson, 2009/ Used under license from Shutterstock. com; 129: © Andersen Ross/Blend Images/Jupiter Images; 132: © Peter Teller / Photodisc/Jupiter Images; 137: © Yuri Arcurs, 2009/Used under license from Shutterstock.com; 138: © Corbis/Jupiter Images; 142: © AP Photo; 149: © Design Pics/Jupiter Images; 152: © Michel de Nijs/ istockphoto.com; 153: © Clarissa Leahy/Digital Vision/ Jupiter Images; 157:© BananaStock/Jupiter Images; 160: © istockphoto.com; 161: © Jon Feingersh/Blend Images/ Jupiter Images; 165: © Digital Vision/Jupiter Images; 166: © Hill Street Studios/Blend Images/Jupiter Images; 170: © Kris Timken/Blend Images/Alamy; 175: © Sergei Bachlakov, 2009/Used under license from Shutterstock. com; 179: © Sheryl Griffin/istockphoto.com; 181: © Ellen Isaacs/Alamy; 185: © BananaStock/Alamy; 186: © Getty Images/Comstock/Jupiterimages; 190: © GoGo Images/ Jupiter images; 195: © Purestock/Jupiter Images; 197: © Image Source/Jupiter Images; 199: © Uview, 2009/Used under license from Shutterstock.com; 203: © Liz Gregg/ Photodisc; 207: © Andersen Ross/Digital Vision/Jupiter Images; 211: © Jim West/Alamy; 217: © Michael Blann/ Digital Vision/Jupiter Images; 222: © Michael DeYoung/ Alaskastock/PhotoLibrary; 226: © Steve Mason/ Photodisc/Jupiter Images; 230: © Hans Georg Roth/ Documentary Value/Corbis; 235: © Photos.com/Jupiter Images; 238: © K. Preuss/The Image Works; 241: © Thinkstock/Jupiter Images; 248: © Radius Images/Jupiter Images; 257: © Stockbyte/Jupiter Images; 260: © Digital Vision/Jupiter Images; 263: © SW Productions/Photodisc/ Jupiter Images; 266: © Juan Collado/istockphoto.com; 271: © Jupiterimages/Thinkstock/Jupiterimages; 273: © ALAN ODDIE/PhotoEdit; 282: © Keith Brofsky/ Photodisc/Jupiter Images; 289: © Ryan McVay/Photodisc/ Jupiter Images; 290: © Dundanim, 2009/Used under license from Shutterstock.com; 298: © David Young-Wolff/PhotoEdit ;303: © Purestock/Jupiterimages; 308: © Andrew Fox/Encyclopedia/Corbis; 313: © BlueMoon Stock/SuperStock; 320: © Photodisc/Jupiter Images; 323: © Bob Daemmrich/The Image Works; 325: © Mikhail Tchkheidze, 2009/Used under license from Shutterstock. com; 328: © AP Photo/Oscar Sosa; 333: © George Gardner/The Image Works; 337: © David Taylor/Alamy; 343: © Corbis Super RF/Alamy; 344: © AP Photo/Damian Dovarganes; 348: © Corbis/Jupiter Images; 350: © Studio Zanello/Ken Chernus Photography/Blend Images/Jupiter Images; 355: © Mario Tama/Getty Images; 357: © Ariel Skelley/Blend Images/Jupiter Images; 358: © mistydawnphoto, 2008/Used under license from Shutterstock.com; 360: © Terry Vine/Blend Images/Jupiter Images; 363: © Purestock/Jupiter Images; 366: © Chris Hellier/Alamy; 367: © Robert Fried/ Alamy; 369: © David Strickler/The Image Works; 373: © Design Pics/Design Pics/Jupiter Images; 380: © E. Dygas/Jupiterimages; 385: © Erwin Wodicka, 2009/Used under license from Shutterstock.Com; 294: © Corbis/zefa RF/Jupiter Images; 396: © Arthur Tilley/Creatas/Jupiter Images; 397: © AP Photo/Teh Eng Koon; 399: © Sheryl Griffin/istockphoto. com; 402: © Lee Snider/The Image Works; 407: © Jupiterimages; 412: © Dennis MacDonald/PhotoEdit; 415: © AP Photo/Al Golub; 417: © Design Pics/Jupiter Images; 420: © Dimitri Vervitsiotis/Digital Vision/Jupiter Images

NAME INDEX

Abel, Steven A., 210
Achenbaum, W. A., 11, 40
Aday, R. H., 40
Adelman R., 285
Adler, J. M., 80
Agee, James, 173
Aiken, L., 398
Akincigil, A., 122
Akiyama, H., 163, 164
Albert M. S., 124
Albright, Madeleine, 51
Aldrin, Buzz, 51
Allaire, J. C., 124
Alzheimer, Alois, 142
Amenta, E., 405
Amir, Y., 40
Amri, M., 104
Anderson, S. W., 133
Andrus, Ethel Percy, 410
Antonucci, T. C., 163
Aouani, E., 104
Apple, Dorothy Dorian, 159
Aquino, J. A., 297
Arcelus J., 186
Aries, Philippe, 377
Aristotle, 44, 191
Arndt, S., 123
Arnold, S. E., 140
Arye, L., 351
Atchley, R., 202
Attar-Schwartz, S., 158, 159

Badgett, Bonita, 315
Baines, Gertrude, 344
Baker, Russell, 409

Baltes, B., 4, 34, 38, 51, 127
Banks, Marion, 290
Banks, William, 290
Barbato, C. A., 44
Barg, C., 40
Bargh, J. A., 32
Barnes, P. M., 23
Barrow, G. M., 41
Bastien, S., 214
Bauer, J. J., 39, 153
Beard, B. B., 114
Bechara, A., 133
Beck, P., 251
Beck, T. L., 131
Bellandi, D., 336
Bengtson, V. L., 40, 155
Bennett, D. A., 131
Bennett, K. M., 184, 328
Bensoussan, A., 100
Berzlanovich, M., 105
Beyrer, Aloysius Joseph, 314
Bianco, David, 416
Bickell N. A., 176
Bienias, J. L., 131
Binstock, Robert, 34
Blackwell, C., 40
Blagov, P. S., 80
Blanchard-Fields, F., 61
Blando, J. A., 128
Blazer, D., 339
Blieszner, R., 175
Blouin, Ann Scott, 209
Boich, L. H., 48
Bolland, M. J., 100, 101
Booth, A., 165

Botkin, D. R., 187
Bouman W. P., 186
Bower, Bruce, 31
Bowman, P. J., 59
Boxer, A. M., 190
Bragger, Lydia, 46
Brant L. J., 180
Brecht S. B., 267
Breckenridge, J. S., 113
Brent, Nancy J., 209
Bretschneider, J. G., 186, 194
Brody, E., 308
Brown, Bob, 119
Brown, Brownie, 119
Brown, David N., 121
Brown, Helen Gurley, 353
Brown, M. M., 41
Buchanan, A., 159
Buckwalter, J. A., 145
Buettner, Dan, 97
Buhrich, N., 264
Burger, J. M., 80
Burkhauser, R. V., 204, 205
Burtless G., 202
Butler, Robert, 12, 186, 381
Butrica B. A., 152
Byock, Ira, 395

Cacioppo J., 175
Cahill, S., 191
Caldwell, E. C., 39
Calment, Jeanne, 25, 103, 106
Calvin, John, 218
Camp, C., 325
Camp, G., 325

SUBJECT INDEX

intellectual development, adult, Primary Mental Abilities (PMA) for, 129
intelligence, 128
  crystallized, 129, 130
  dimensions of, 128
  fluid, 129, 130
  of older persons, 51
intelligence quotient (IQ), 128–129
intelligence tests, 130–131
intergenerational relationships, 158, 302–303, 311
  adult children/parent obligations, 299
  building of, 416
intergroup perspective, 68
intimacy
  friendship relationships and, 172
  kin relationships and, 172
  need for, 172–173
intimacy and affiliation, development, 68–69
"Intimate Relationships", 171
intrinsic causality, 82
introverted personality, 139
investment fraud, 334
invisible consumers, 220
IRAs, 245
ischemic heart disease, 98
"It's a Wonderful Life", 227

Japan
  IBM, 205
  life expectancy in, 93
  response to grief, 399
Japanese Alien Land Law, 364
Japanese Americans, 364–365
Japanese diet, 108
Japanese-American, 103
Jean's Way (Humphry), 320
job discrimination, against women, 406
job search
  Internet and, 212
  older workers and, 221
job search, age discrimination and, 210–212
job sharing, 223

Job Training Partnership Act (JTPA), 224
Johns Hopkins Medical Letter, 194
Journal of Health Psychology, 121
Journal of Humanistic Psychology, 120
The Journal of Sex Research, 171
Judeo-Christian culture, 218
Judging Amy, 47
Jung's midlife development, 66

Keeping Up Appearances, 47
Kelly Services: Encore, 212
kin relationships, 151
  intimacy in, 172
  types of, 156
kin structure
  blue-collar families, 156
  changing family structure and, 167
  divorce and, 159–160, 161, 184
  family types and, 156
  gender differences in, 156
  grandparents and, 156–161, 168
  individualism and, 160
  never married singles, 162, 184–185
  sibling relationships and, 168
  step-families and, 161
  white-collar families, 156
kin structure see also culture
kin structure, 151–152
  elder siblings, 152–155
  variations in, 156
Kinsey studies, 191, 193
Kraft General Foods, 212

labeling see stereotypes
labor force participation, 202
language, 43
  ageist, 44
  problems with, 141
language structures consciousness., 44
late life, 75
late life, transitions, 76

late midlife crisis, 71
late retirement incentive programs (LRIPs), 205
later adulthood, crisis of, 63
later life, importance of family and, 151
layoffs, age discrimination and, 212
layoffs, impact on older workers, 213
le troisième âge, 75
learning, 131
  anxiety and, 132
  The meaningfulness of the material, 132
  mental functioning and, 131
  motivation, 132
  pacing and, 132
  physical health, 132
legal services to seniors, 415
leisure, 219–220, 224
  the aged and, 220–221
Letter from Birmingham Jail (King), 201
Levinson's concept of life structure, 59
life, attitudes towards, 52
life altering events, 84
Life Care at Home (LCH), 275
life course, 66
life cycle, 60
  Levinson's concept of life structure, 64
life expectancy
  contributing factors, 22–23
  and ethnicity, 20
  gender and, 22
  increased, 20, 22–24
  of minority groups, 371
  survival rates, USA, 23
  in U.S., 104
  Levinson's concept of life structure, 64
life review process of, 382
life satisfaction, 77, 296–297, 381–383
  nursing home residence and, 383
life span, 92–93
life stories, 59

Senior Power, 407
Senior Wimbledon West, 52
seniors
    housing needs of, 263
    living environments and, 259
    membership in voluntary
        organizations, 162–163
    political powers of, 408, 421
    public office and, 408–409
    social life of, 162
    sporting events for, 51
seniors and crime, fighting back,
        324–325
sensation, 124
senses, the, 124–125
sensory loss, 125, 126
sensory memory, 125, 132–133
sensory threshold, 126
Seventy-two-year-old longitudinal
        study, 112
sex role stereotypes, 82
sexual attractiveness, aging
        and, 45
*Sexual Behavior in the Human
    Female* (Kinsey), 191
*Sexual Behavior in the Human
    Male* (Kinsey), 191
"Sexual Desire in Later Life"
        (DeLamater), 171
sexual function for procreation,
        189
sexual functioning, 47
sexual identity, 187, 188, 198
sexual invisibility, 188
sexual orientation, 155
sexual performance, in old
        age, 46
sexuality
    impact of care facilities on,
        196–197
    in later life, 186, 194
    older adults and, 198
    physical disability and, 195
    procreation and, 189
    research on
        Duke longitudinal study, 193
        Kinsey studies, 191
        Masters and Johnson,
            191–192

romantic love and, 189
special circumstances,
        196–197
stereotypes of aging and, 171
terminally ill and, 195
understanding of, 187
youth and, 188–189
Shadowlands, 49
shared housing, 262, 276
sibling relationship, 152–155
sibling relationships in old
        age, 168
    sexual orientation on, 155
"silent generation", 187
Silver Hair on the Silver Screen,
        50
Silver Pages, 255
"Silver Threads Among the
        Gold", 45
single room occupancy (SRO)
        hotels, 251, 252
SIR2, 105
sitting time, 79
"Sixty is the New Thirty!", 104
skid rows, 250
Skilled Nursing Facilities
        (SNFs) or convalescent
        hospitals., 262
skin care market, 337
small strokes, 144
social and environmental factors,
        impact on longevity,
        113–114
social bonds, strengthening of,
        166–167
social comparisons, 189
social construction of self, 38
social gerontology, 75
social group identification, 68
social isolation, 126, 180
    rural elders and, 252–253
social networks, 163–164
    women and, 168
social problems approach, 28
social relationships, health
        benefits of, 175
Social Security, 10, 27, 45, 203,
        205, 225
    affluence tests and, 237

baby boomer generation
        and, 234
benefits, 234, 413
benefits, statistics and, 234
COLA, 234, 235, 240
cost of living adjustments
        (COLA) and, 234, 240
disability benefits and, 419
history of, 233–234
inequities in, 237–238
as old age security, 237
penalties and older workers,
        237
pressure on, 236–237
privatization of, 237
retirement age and, 207, 220
solutions to problems of,
        236–237
Social Security Act (U.S.), 37,
        204, 233, 405
Social Security beneficiaries, 230
    increase in, 236
social security system, in
        Germany, 37
Social Security taxes, 27, 235
social support, 163
social world, disengagement
        from, 382
Sonoma State University, 403,
        404
Southeast Asian Americans,
        365–366
Southern Baptist Convention
        (SBC)., 200
    position on women, 200
    slavery and, 200
spiritual development, 166
spirituality, 164
spirituality and longevity, 121
sporting events for seniors, 51, 54
spousal caregiving, 303–304
SSI, 255
    Medicaid and, 243
step-grandchildren, 161
step-grandparents, 161
stereotypes, 32
    based on age, 32
    compassionate, 41
    definition, 32

University of Colorado, 120

University of Colorado, Boulder, 121

University of Madison Wisconsin, 171

University of Michigan, 171

University of Missouri, 121

University of Southern California, Los Angeles, 112

University of Wisconsin, 108

University Retirement Community (URC) in Davis, California,, 279

urban blight, 265–266

U.S. Army Special Forces, 120

U.S. Census Bureau, 252

U.S. Department of Housing & Urban Development, 232

U.S. Navy SEALs, 120

U.S. Supreme Court, 208

USA, life expectancy in, 93

vascular dementia, 144

very old, the, 75

Viagra, 46, 47, 171

Viagra Generation, the, 171

victims of crime, 341–342

Vietnam generation, 187, 188

View from Eighty, the, (Cowley), 297

vigilance, age-related differences in, 125–126

vitamin D metabolite calcitrol, 101

voter turnout, 408

voters, 408

Wage and Hour Division of the Department of Labor, 211

WAIS-R, 129, 130

waist-to-hip ratio (WHR), 103

War on Poverty, 406

wealth span, 230

Wechsler Adult Intelligence Scale (WAIS), 129

well-being, subjective, 53

wellness movement, 114, 115

Wernicke-Karsakoff's syndrome, 144

Western Illinois University, 119

Westinghouse Electric and Northrop Grumman Corporation, 212

"When I'm 64,", 45

"When It's Smart to Lie About Your Age" (Redbook magazine), 45

"Where've You Been?", 45

white Americans
divorce and, 183
poverty rates and, 230

Widowed Persons Service (WPS), 182

widowers, outreach services and, 182

widowhood
adapting to loss and, 179–182
adjustment to, 181
childless widows, 178
divorce and, 174
family dynamics and, 178–179
financial status of, 350
gender differences and, 182–183
"identity foreclosure" and, 181
old old and, 177
older adults, 177–178
oldest-old and, 296
outreach services and, 182
pension and, 167
remarriage and, 167, 181
Widowed Persons Service (WPS), 182
women, 349

widow-to-widow programs, 167

Wii Sports, 150

"Will You Love Me When My Face Is Worn and Old?", 45

Winter 2002 Olympics, 121

withdrawal, aging and, 77–78

women
African Americans and, 357

caregiving for elderly, 349

death of a spouse and, 180

discrimination against, 200

divorce and, 174
financial status of, 350–351

The Elder-Liker Study and, 85

employment and, 204

fear of aging on, 69

financial status of, 348

friendships and elders, 162

homelessness and, 250

human rights violations against, 200

impact of aging on, 45–47

job discrimination, 406

labor force and, 212

labor force participation and, 202

masturbation in later life, 192–193

as minorities, 371

percentage of oldest-old, 311

physical beauty and, 352

poverty rates of, 229

and retirement, 202

retirement and, 220

role in maintaining social networks, 163–164

sexual issues and aging, 194

sexuality and, 192

single, 349

single women, 349

social and economic impact on, 229

Social Security, 238

stress and, 118

stress as caregiver, 308

suicide, physician-assisted, 392

suicide and, 317–318

widowhood and, 182–183

women, older, financial status of, 351

women and pension plans, 229

women as minority, 347

work, impact on stress, 119–120